Lecture Notes in Computer S

T0238497

Commenced Publication in 1973
Founding and Former Series Editors:
Gerhard Goos, Juris Hartmanis, and Jan van Leeuwen

Shipeng Li Abdulmotaleb El Saddik
Meng Wang Tao Mei
Nicu Sebe Shuicheng Yan
Richang Hong Cathal Gurrin (Eds.)

Advances in Multimedia Modeling

19th International Conference, MMM 2013
Huangshan, China, January 7-9, 2013
Proceedings, Part I

Springer

Volume Editors

Shipeng Li, Microsoft Research Asia, Beijing, China
E-mail: spli@microsoft.com

Abdulmotaleb El Saddik, University of Ottawa, ON, Canada
E-mail: elsaddik@uottawa.ca

Meng Wang, Hefei University of Technology, China
E-mail: eric.mengwang@gmail.com

Tao Mei, Microsoft Research Asia, Beijing, China
E-mail: tmei@microsoft.com

Nicu Sebe, University of Trento, Italy
E-mail: sebe@disi.unitn.it

Shuicheng Yan, National University of Singapore
E-mail: eleyans@nus.edu.sg

Richang Hong, Hefei University of Technology, China
E-mail: hongrc@hfut.edu.cn

Cathal Gurrin, Dublin City University, Ireland
E-mail: cgurrin@computing.dcu.ie

ISSN 0302-9743   e-ISSN 1611-3349
ISBN 978-3-642-35724-4   e-ISBN 978-3-642-35725-1
DOI 10.1007/978-3-642-35725-1
Springer Heidelberg Dordrecht London New York

Library of Congress Control Number: 2012954016

CR Subject Classification (1998): H.5.1, H.3.1, H.3.3-5, H.4.1, I.4, I.5, H.2.8, H.5.2,
H.5.5

LNCS Sublibrary: SL 3 – Information Systems and Application, incl. Internet/Web
and HCI

Typesetting: Camera-ready by author, data conversion by Scientific Publishing Services, Chennai, India

Printed on acid-free paper

Springer is part of Springer Science+Business Media (www.springer.com)

Shipeng Li   Abdulmotaleb El Saddik
Meng Wang   Tao Mei
Nicu Sebe   Shuicheng Yan
Richang Hong   Cathal Gurrin (Eds.)

Advances in Multimedia Modeling

19th International Conference, MMM 2013
Huangshan, China, January 7-9, 2013
Proceedings, Part I

Springer

Volume Editors

Shipeng Li, Microsoft Research Asia, Beijing, China
E-mail: spli@microsoft.com

Abdulmotaleb El Saddik, University of Ottawa, ON, Canada
E-mail: elsaddik@uottawa.ca

Meng Wang, Hefei University of Technology, China
E-mail: eric.mengwang@gmail.com

Tao Mei, Microsoft Research Asia, Beijing, China
E-mail: tmei@microsoft.com

Nicu Sebe, University of Trento, Italy
E-mail: sebe@disi.unitn.it

Shuicheng Yan, National University of Singapore
E-mail: eleyans@nus.edu.sg

Richang Hong, Hefei University of Technology, China
E-mail: hongrc@hfut.edu.cn

Cathal Gurrin, Dublin City University, Ireland
E-mail: cgurrin@computing.dcu.ie

ISSN 0302-9743   e-ISSN 1611-3349
ISBN 978-3-642-35724-4   e-ISBN 978-3-642-35725-1
DOI 10.1007/978-3-642-35725-1
Springer Heidelberg Dordrecht London New York

Library of Congress Control Number: 2012954016

CR Subject Classification (1998): H.5.1, H.3.1, H.3.3-5, H.4.1, I.4, I.5, H.2.8, H.5.2, H.5.5

LNCS Sublibrary: SL 3 – Information Systems and Application, incl. Internet/Web and HCI

Typesetting: Camera-ready by author, data conversion by Scientific Publishing Services, Chennai, India

Printed on acid-free paper

Springer is part of Springer Science+Business Media (www.springer.com)

Preface

The 19th International Conference on Multimedia Modeling (MMM 2013) was held in Huangshan, China, during January 7–9, 2013, and hosted by the Hefei University of Technology (HFUT) at Hefei, China. MMM is a leading international conference for researchers and industry practitioners to share their new ideas, original research results, and practical development experiences from all multimedia-related areas.

It was a great honor for HFUT to host MMM 2013, one of the most longstanding multimedia conferences, in Huangshan, China. HFUT, located in the capital of Anhui province, is one of the key universities administrated by the Ministry of Education, China. Recently its multimedia-related research has been attracting increasing attention from the local and international multimedia community. The conference venue was the Huangshan International Hotel, located very close to Huangshan, which is well known for its scenery, sunsets, spectacularly shaped granite peaks, Huangshan pine trees, and unique views of the clouds from above. Furthermore, Huangshan is a UNESCO World Heritage Site and one of China's major tourist destinations. We hope that the choice of venue for MMM 2013 resulted in a memorable experience for all participants.

MMM 2013 featured a comprehensive program including three keynote talks, eight oral presentation sessions, one poster session, one demo session, seven special sessions, and the Video Browser Showdown. The 184 submissions from authors of 20 countries included a large number of high-quality papers in multimedia content analysis, multimedia signal processing and communications, and multimedia applications and services. We thank our 140-member Technical Program Committee who put in considerable effort in both reviewing papers and in providing valuable feedback to the authors. For the main conference, there were 111 submissions, each receiving at least three reviews. After the extensive reviewing process, the Program Chairs decided to accept 30 regular papers (27%) and 20 poster papers (18%). In total, 46 papers were accepted for seven special sessions after 53 were submitted, and 15 submissions were accepted for the demo session from a total of 22 submissions. Six teams competed in the Video Browser Showdown. The authors of accepted papers come from 16 countries. This volume of the conference proceedings contains the abstracts of three invited talks and all the regular, poster, special session, and demo papers, as well as special demo papers of the Video Browser Showdown (VBS). MMM 2013 included the following awards: the Best Paper Award, the Best Student Paper Award, two Best Demo Awards, and the VBS Competition Award, which were all sponsored by KAI Square.

The technical program is an important aspect but only has full impact if complemented by challenging keynotes. We were extremely pleased and grateful to have had three exceptional keynote speakers, Xian-Sheng Hua, Kiyoharu

Aizawa, and Ralf Steinmetz, accept our invitation and present interesting ideas and insights at MMM 2013.

We are also heavily indebted to many individuals for their significant contributions. We thank the MMM Steering Committee for their invaluable input and guidance on crucial decisions. We wish to acknowledge and express our deepest appreciation to the Honorary Chairs, Tat-Seng Chua, Phoebe Chen, and Wen Gao, the Local Organizing Chairs, Benoit Huet, Fei Wu, and Huaming Feng, the Special Session Chairs, Richang Hong and Changsheng Xu, the Demo Chairs, Ke Lu and Yi Yang, the VBS Chairs, Klaus Schöffmann and Werner Bailer, the Publicity Chairs, Kiyoharu Aizawa, Houqiang Li, Qingming Huang, Winston Hsu, and Xuelong Li, the Publication Chairs, Shuqiang Jiang and Cathal Gurrin, the Sponsorship Chairs, Yan-Tao Zheng, Shi-Yong Neo, Zhiwei Gu, and Qiong Liu, and last but not least the Webmaster, Xiaobin Yang. Without their efforts and enthusiasm, MMM 2013 would not have become a reality. Moreover, we want to thank our sponsors: Hefei University of Technology, National Natural Foundation of China, Beijing Ricoh Research Center, Microsoft Research Asia, FX Palo Alto Laboratory, and KAI Square Pte Ltd. We wish to thank all committee members, reviewers, session chairs, student volunteers, and supporters. Their contributions are much appreciated. Finally, we would also like to thank our local support team, Han Zhao, Xiaoping Liu, Yuanfa Zhu, Xiang Sun, Jianguo Jiang, Xuezhi Yang, Na Zhao, Yuetong Chen, Changzhi Luo, for their support and contribution to the conference organization.

January 2013

Tao Mei Nicu Sebe
Shuicheng Yan
Shipeng Li
Abdulmotaleb Ei Saddik
Meng Wang

Organization

MMM 2013 was organized by Hefei University of Technology, China.

MMM 2013 Organizing Committee

Honorary Co-chairs

Tat-Seng Chua	National University of Singapore, Singapore
Phoebe Chen	La Trobe University, Australia
Wen Gao	Peking University, China

General Co-chairs

Shipeng Li	Microsoft Research Asia, China
Abdulmotaleb Ei Saddik	University of Ottawa, Canada
Meng Wang	Hefei University of Technology, China

Program Co-chairs

Tao Mei	Microsoft Research Asia, China
Nicu Sebe	University of Trento, Italy
Shuicheng Yan	National University of Singapore, Singapore

Organizing Co-chairs

Benoit Huet	EURECOM, France
Fei Wu	Zhejiang University, China
Huaming Feng	Beijing Electronic Science and Technology Institute, China

Publicity Co-chairs

Kiyoharu Aizawa	University of Tokyo, Japan
Houqiang Li	University of Science and Technology of China, China
Qingming Huang	China Academy of Science, China
Winston Hsu	National Tai Wan University, Taiwan
Xuelong Li	XIOPM of Chinese Academy of Science, China

Sponsorship Co-chairs

Yantao Zheng	Google Corporation, USA
Shi-Yong Neo	Kai Square Co. Ltd., Singapore
Zhiwei Gu	Yahoo Corporation, USA
Qiong Liu	FX Palo Alto Laboratory, Inc., USA

Publication Co-chairs

Shuqiang Jiang	Institute of Computing, CAS, China
Cathal Gurrin	Dublin City University, Ireland

Special Session Co-chairs

Richang Hong	Hefei University of Technology, China
Changsheng Xu	Institute of Automation, China

Demo Co-chairs

Ke Lu	Chinese Academy of Science, China
Yi Yang	Carnegie Mellon University, USA

VBS Co-chairs

Klaus Schoeffmann	Klagenfurt University, Austria
Werner Bailer	Joanneum Research, Austria

Web Chair

Xiaobin Yang	Hefei University of Technology, China

US Liaisons

Qi Tian	University of Texas, San Antonio, USA
Alexander Hauptmann	Carnegie Mellon University, USA

Asian Liaisons

Chong-Wah Ngo	City University of Hong Kong, SAR China
Jialie Shen	Singapore Management University, Singapore

European Liaisons

Susanne Boll	University of Oldenburg, Germany
Alan Hanjalic	Delft University of Technology, The Netherlands

Technical Program Committee

Laurent Amsaleg	CNRS-IRISA, France
Xavier Anguera	Telefonica R&D, Spain
Yannis Avrithis	National Technical University of Athens, Greece
Bing-Kun Bao	CAS, China
Jenny Benois-Pineau	University of Bordeaux 1, France
Susanne Boll	University of Oldenburg, Germany
Laszlo Boszormenyi	Klagenfurt University, Austria
Liangliang Cao	IBM T.J. Watson Research, USA
Andrea Cavallaro	Queen Mary University of London, UK
Vincent Charvillat	University of Toulouse, France
Xiangyu Chen	National University of Singapore, Singapore
Gene Cheung	National Institute of Informatics, Japan
Liang-Tien Chia	Nanyang Technological University, Singapore
Wei-Ta Chu	National Chung Cheng University, Taiwan
Tat-Seng Chua	National University of Singapore, Singapore
Matthew Cooper	FX Palo Alto Laboratory, USA
Ajay Divakaran	Sarnoff Corporation, USA
Lingyu Duan	Peking University, China
Jianping Fan	University of North Carolina, USA
Yue Gao	National University of Singapore, Singapore
William Grosky	University of Michigan, USA
Cathal Gurrin	Dublin City University, Ireland
Martin Halvey	University of Glasgow, UK
Allan Hanbury	Technical University of Vienna, Austria
Andreas Henrich	University of Bamberg, Germany
Steven Hoi	Nanyang Technological University, Singapore
Richang Hong	Hefei University of Technology, China
Jun-Wei Hsieh	National Taiwan Ocean University, Taiwan
Winston Hsu	National Taiwan University, Taiwan
Benoit Huet	EURECOM, France
Wolfgang Hurst	Utrecht University, The Netherlands
Ichiro Ide	Nagoya University, Japan
Alejandro Jaimes	Yahoo!, USA
Rongrong Ji	Columbia University, USA
Yu-Gang Jiang	Columbia University, USA
Shuqiang Jiang	Chinese Academy of Sciences, China
Alexis Joly	INRIA, France
Mohan Kankanhalli	National University of Singapore, Singapore
Yoshihiko Kawai	NHK, Japan
Lyndon Kennedy	Yahoo! Research, USA
Yiannis Kompatsiaris	Informatics and Telematics Institute, Greece
Martha Larson	Delft University of Technology, The Netherlands
Duy-Dinh Le	National Institute of Informatics, Japan

Houqiang Li   University of Science and Technology of China,
                                China
Chia-Wen Lin   National Tsing Hua University, Taiwan
Xiaobai Liu   University of California, Los Angeles, USA
Dong Liu   Columbia University, USA
Yan Liu   Hong Kong Polytechnic University, Hong Kong,
                                SAR China
Zhu Liu   AT&T Laboratories, USA
Yuan Liu   Ricoh Software Research Center, China
Alexander Loui   Kodak Research Laboratories, USA
Guojun Lu   Monash University, Australia
Nadia Magnenat-Thalmann   University of Geneva, Switzerland
Jose Martinez   Universidad Autonoma de Madrid, Spain
Henning Mueller   HES-SO Valais, Switzerland
Francesco Natale   University of Trento, Italy
Chong Wah Ngo   City University of Hong Kong, Hong Kong,
                                SAR China
Naoko Nitta   Osaka University, Japan
Noel O'Connor   Dublin City University, Ireland
Wei-Tsang Ooi   National University of Singapore, Singapore
Vincent Oria   New Jersey Institute of Technology, USA
Marco Paleari   EURECOM, France
Fernando Pereira   Instituto Superior Tecnico, Portugal
Guo-Jun Qi   University of Illinois at Urbana-Champaign, USA
Shin'ichi Satoh   National Institute of Informatics, Japan
Klaus Schoffmann   Klagenfurt University, Austria
Heng Tao Shen   University of Queensland, Australia
Jialie Shen   Singapore Management University, Singapore
Koichi Shinoda   Tokyo Institute of Technology, Japan
Mei-Ling Shyu   University of Miami, USA
Alan Smeaton   Dublin City University, Ireland
Cees Snoek   University of Amsterdam, The Netherlands
Yongqing Sun   NTT Cyber Space Laboratories, Japan
Jinhui Tang   Nanjing University of Science and Technology,
                                China
Qi Tian   University of Texas at San Antonio, USA
Dian Tjondronegoro   Queensland University of Technology, Australia
Shingo Uchihashi   Carnegie Mellon University, USA
Xin-Jing Wang   Microsoft Research Asia, China
Zhiyong Wang   University of Sydney, Australia
Jingdong Wang   Microsoft Research Asia, China
Marcel Worring   University of Amsterdam, The Netherlands
Peng Wu   Hewlett-Packard, USA
Qiang Wu   University of Technology, Sydney, Australia
Xiao Wu   Southwest Jiaotong University, China

Feng Wu	Microsoft Research Asia, China
Changsheng Xu	Institute of Automation, Chinese Academy of Sciences, China
Keiji Yanai	University of Electro-Communications, Japan
Zheng-Jun Zha	National University of Singapore, Singapore
Zhongfei Zhang	State University of New York at Binghamton, USA
Yongdong Zhang	Institute of Computing Technology, CAS, China
Cha Zhang	Microsoft Research, USA
Roger Zimmermann	National University of Singapore, Singapore
Haojie Li	Dalian University of Technology, China

Additional Reviewers

Werner Bailer	Joanneum Research, Austria
Manfred del Fabro	Klagenfurt University, Austria
Frank Hopfgartner	DIA Laboratory, Technical University of Berlin, Germany
Mario Taschwer	Klagenfurt University, Austria
Wolfgang Weiss	Joanneum Research, Austria
Zhen Li	University of Illinois at Urbana-Champaign, USA
Ansgar Scherp	University of Koblenz-Landau, Germany
Makoto Okabe	University of Electro-Communications, Japan
Masaki Takahashi	NHK Science and Technology Research lab, Japan
Wen-Huang Cheng	Academia Sinica, Taiwan
Jiashi Feng	National University of Singapore, Singapore
Jitao Sang	Institute of Automation, Chinese Academy of Sciences, China
Congyan Lang	Beijing Jiaotong University, China
Si Liu	National University of Singapore, Singapore
Jian Cheng	Institute of Automation, Chinese Academy of Sciences, China
Jinqiao Wang	Institute of Automation, Chinese Academy of Sciences, China
Min-Hsuan Tsai	University of Illinois at Urbana, USA
Ming Yang	Northwestern University, USA
Peng Yang	Rutgers University, USA
Quan Fang	Institute of Automation, Chinese Academy of Sciences, China
Shiyang Lu	University of Sydney, Australia
Zhaowen Wang	University of Illinois at Urbana, USA
Weiqing Min	Institute of Automation, Chinese Academy of Sciences, China

Darui Li   University of Science and Technology of China,
                            China
Yang Yang   The University of Queensland, Australia
Ming Yan   Institute of Automation, Chinese Academy
                            of Sciences, China
Zhen Li   Dolby Laboratories, Inc., USA
Zhaoquan Yuan   Institute of Automation, Chinese Academy
                            of Sciences, China
Yan Wang   Columbia University, USA
Xian-Ming Liu   University of Illinois at Urbana, USA
Pengfei Xu   Harbin Institute of Technology, China
Xiaoshuai Sun   Harbin Institute of Technology, China
Liujuan Cao   Harbin Engineering University, China

Special Session Co-chairs

Richang Hong   Hefei University of Technology, China
Changsheng Xu   Institute of Automation, China

Special Session Committee

Haojie Li   Dalian University of Technology, China
Shiguo Lian   Huawei Technologies, Co. Ltd., China
Yongqing Sun   NTT Cyber Space Laboratories, Japan
Jialie Shen   Singapore Management University, Singapore
Haiyan Miao   IHPC, A*STAR, Singapore
Liangliang Cao   IBM Watson Research Center, USA
Chang Wen Chen   SUNY at Buffalo, USA
Zhen Wen   IBM T.J. Watson Research Center, USA
Lu Fang   University of Science and Technology of China,
                            China
Ngai-Man Cheung   Singapore University of Technology and Design,
                            Singapore
Jingjing Fu   Microsoft Research Asia, China
Rongrong Ji   Columbia University, USA
Yue Gao   National University of Singapore, Singapore
Qingshan Liu   Nanjing University of Information Science
                            and Technology, China
Wei-Ta Chu   National Chung Cheng University, Taiwan
Keiji Yanai   University of Electro-Communications, Japan
Bingkun Bao   Institute of Automation, Chinese Academy
                            of Sciences, China
Jitao Sang   Institute of Automation, Chinese Academy
                            of Sciences, China
Jinjun Wang   Epson Research and Development, USA

Best Paper Award Committee

Sponsors List

Hefei University of Technology

National Natural Science Foundation of China

KAI Square Pte. Ltd

Microsoft Research Asia

Ricoh Beijing Software
Research Center

FX Palo Alto Laboratory

Google Inc.

Springer Publishing

Keynote 1:
Perspective on Adaptive Video-Streaming

Prof. Dr.-Ing. Ralf Steinmetz

Department of Electrical Engineering and Information Technology
and Department of Computer Science in Technische Universität Darmstadt, Germany

Abstract. This talk covers perspectives on adaptive video streaming and how such techniques are essential for systems with heterogeneous devices. Adaptation is possible using flexible video coding techniques, such as the H.264 Scalable VIdeo Coding (SVC). In this context, it is important to consider various aspects of the video coding system (interdependencies, quality layers, QoE, etc) as well of the delivery architectures (client server, P2P, connectivity, etc). The first part relates to quality adaptation algorithms that match the video quality with available local and system resources without any a-priori knowledge about those resources. Subsequently in the second part, mechanisms that use Quality of Experience (QoE) metrics to enhance its performance for the users will be shown. The decision of which SVC quality to choose is usually driven by QoS metrics, such as throughput. Instead, it will be presented how objective QoE of the different SVC qualities can be used in the decision process. The talk concludes by presenting the major further research activities in this research area.

Keynote 2:
Multimedia FoodLog: Easiest Way to Capture and Archive What We Eat

Prof. Kiyoharu Aizawa

Department of Information and Communication Engineering
and Interfaculty Initiative of Information Studies of the University of Tokyo

Abstract. Eating is one of the most fundamental aspects of one's daily life, but at the same time, it is one of the most difficult aspects to manage by oneself. Recording what we eat is vital for our health care. We have been investigating the "FoodLog" multimedia food-recording tool, whereby users upload photos of their meals and a food diary is constructed by using image processing functions such as food image detection, dietary balance estimation, calory estimation etc. Foodlog is available in http://www.foodlog.jp, and to the best of our knowledge, it is currently the only publicly available multimedia food-recording application that makes use of image processing for dietary assessment. In addition to the PC-based interface, we have developed a few smartphone applications which makes easier to make detiled recording with the assist of image processing. In the talk, I would like to outline the current status of our FoodLog, and present various subjects on multimedia processing of foodlog.

Keynote 3:
Towards Web-Scale Content-Aware Image Search

Dr. Xian-Sheng Hua

Microsoft Corporation

Abstract. In recent years, remarkable progress has been made towards large-scale content-aware image search. However, there is still a long way to go to bridge the two "gaps": semantic gap and intention gap. Large-scale data brings us both challenges and opportunities to tackle these difficulties. In this presentation, we will review existing image search schemes and then focus on large-scale content-aware image search. We discuss the connections and differences among different large-scale CBIR techniques such as trees, clustering, hashing, graph and BoW, and then introduce a few exemplary scalable approaches including graph based search, color map based search, line sketch based search, and concept map based search. For each exemplary approach, we will discuss how to make it work for billions of images. The limitations will then be analyzed for these techniques, followed by introducing indexing and search schemes based on web-scale image content understanding. Connections between search and content understanding will be also discussed. And last we will talk about challenges and opportunities along this direction.

Table of Contents – Part I

Classification, Recognition and Tracking I

Classification, Recognition and Tracking II

Ranking in Search

Multimedia Representation

Multimedia Systems

Posters Papers

Table of Contents – Part II

Special Session Papers

Mobile-Based Multimedia Analysis

Multimedia Retrieval and Management with Human Factors

Location Based Social Media

3D Video Depth and Texture Analysis and Compression

Large-Scale Rich Media Search and Management in the Social Web

Multimedia Content Analysis Using Social Media Data

Cross-Media Computing for Content Understanding and Summarization

Demo Session Papers

Video Browser Showdown

Semi-supervised Concept Detection by Learning the Structure of Similarity Graphs

Symeon Papadopoulos[1], Christos Sagonas[1], Ioannis Kompatsiaris[1], and Athena Vakali[2]

[1] Information Technologies Institute, CERTH, Greece
{papadop,sagonas,ikom}@iti.gr
[2] Informatics Department, Aristotle University of Thessaloniki, Greece
avakali@csd.auth.gr

Abstract. We present an approach for detecting concepts in images by a graph-based semi-supervised learning scheme. The proposed approach builds a similarity graph between both the labeled and unlabeled images of the collection and uses the Laplacian Eigemaps of the graph as features for training concept detectors. Therefore, it offers multiple options for fusing different image features. In addition, we present an incremental learning scheme that, given a set of new unlabeled images, efficiently performs the computation of the Laplacian Eigenmaps. We evaluate the performance of our approach both on synthetic datasets and on MIR Flickr, comparing it with high-performance state-of-the-art learning schemes with competitive and in some cases superior results.

1 Introduction

Concept detection in images is typically conducted by learning a mapping between a set of features extracted from the input images and a set of concepts. In conventional settings, features extracted from individual labeled images are provided to classifiers (e.g. SVM) in order to learn the features-to-concept mapping. Once a new image appears, concept detection is conducted based on its feature vector, thus considering the image in isolation. In this work, we tackle concept detection by leveraging the similarity of the unknown image with labeled images of the collection. We call our proposal the Graph Structure Features (GSF) approach. GSF is based on building a similarity graph between the images of the collection and mapping them to low-dimensional features based on the eigenvectors of the graph Laplacian. These features correspond to semantically coherent groups of images, and are thus used to train concept classifiers.

The GSF approach is expressed as a combination of a semi-supervised with a supervised learning algorithm: at the first step the similarity graph and the graph structure features are computed for both the labeled and the unlabeled images, while at the second step the graph structure features are used to train a supervised learning algorithm. The approach leads to high concept detection performance, since it utilizes information on the similarities of unlabeled images with the labeled ones. It can also accommodate a variety of fusion techniques

S. Li et al. (Eds.): MMM 2013, Part I, LNCS 7732, pp. 1–12, 2013.

for combining different sets of features to further improve performance. To make the approach applicable in online settings, we also devise an incremental scheme for computing the graph structure features of unknown images in online mode.

Contributions: The paper proposes a novel semi-supervised concept detection approach that does not rely on features extracted from isolated content items, but from features that capture the similarity structure between the unknown and the labeled items. Thanks to the transformation of original content features into a similarity graph and the extraction of low-dimensional graph structure features, GSF is well-suited to high-dimensional features, and offers several options for performing fusion between multiple feature sets. GSF is tested in comprehensive experiments, in which it is found to outperform or to compete closely with two high performing state-of-the-art approaches. In addition, we propose an incremental implementation of the proposed scheme that achieves similar performance as the batch learning scheme, and enables application of the framework in online learning settings.

2   Related Work

Utilizing the implicit relational structure by computing similarities between images of a collection has been proposed before. In [1], an extended similarity measure is proposed that leverages local neighborhood structures of images, computed from the content and label information of similar images. The extended similarity was used on top of well-known semi-supervised learning methods [2] and shown to improve performance. However, this approach is not amenable to online learning settings, as it relies on global graph computations [2,3].

A different approach is presented in [4], where a sparse similarity graph is constructed based on a convex optimization method. Then, an additional λ_1-norm minimization problem is solved to perform semi-supervised inference on the noisy image tags, and to derive a compact concept space. Online concept detection is possible in the derived concept space. The approach of [4] is shown to yield superior concept detection accuracy in a standard benchmark. However, it suffers from a computationally intensive training step. Another semi-supervised concept detection approach is grounded on the notions of hashing-based λ_1-graph construction and KL-based multi-label propagation [5]. Despite being applicable to large-scale datasets and yielding high performance, this approach is not practical in online learning settings since it relies on a computational scheme that requires 50 iterations (as stated in [5]) for convergence during inference.

Our work is mostly related to [6] that introduces the concept of "social dimensions", i.e. the top-k Laplacian eigenvectors, as a means to tackle the relational classification problem [7]. We adopt a similar feature representation, but apply it in a different problem, i.e. concept detection in multimedia. Moreover, we consider an extension that renders our approach practical in online settings in contrast to [6] that is only applicable in transductive learning settings.

3   Proposed Approach

The basic formulation of the GSF approach is provided in a transductive learning setting, where both the labeled and the unlabeled samples are available at training time (subsection 3.1). To render our proposal more practical in real-world learning settings, we describe an incremental extension that can predict the concepts of unknown items that were not available during training (subsection 3.2). Finally, we describe a set of possibilities for fusing multiple features with the goal of improving the concept detection performance (subsection 3.3).

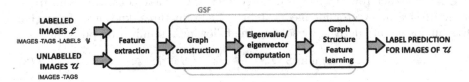

Fig. 1. Overview of proposed approach

3.1   Transductive Learning Setting

The GSF approach is illustrated in Figure 1. Given a set of K target concepts $\mathcal{Y} = \{Y_1, ..., Y_K\}$ and an annotated set $\mathcal{L} = \{(\boldsymbol{x_i}, \boldsymbol{y_i})\}_{i=1}^{l}$ of training samples, where $\boldsymbol{x_i} \in \mathbb{R}^D$ stands for the feature vector extracted from content item i and $\boldsymbol{y_i} \in \{0,1\}^K$ for the corresponding concept indicator vector, a transductive learning algorithm attempts to predict the concepts associated with a set of unknown items $\mathcal{U} = \{\boldsymbol{x_j}\}_{j=l+1}^{l+u}$, by processing together the sets $\mathcal{L}$ and $\mathcal{U}$.

Based on the features of the input items, a graph $G = (V, E)$ is constructed that represents the similarities between all pairs of items. The nodes of the graph include the items of both sets ($\mathcal{L}$ and $\mathcal{U}$), i.e. $V = V_L \cup V_U$ with $|V| = n$. There are different options for constructing such a graph:

- **Full graph:** All possible edges between the items of V are inserted as edges using weights to determine the degree of similarity for each pair. For two feature vectors x_i, x_j, a popular weighting scheme is the heat kernel:

$$w_{ij} = \exp\left(-\frac{||x_i - x_j||^2}{t}\right) \tag{1}$$

  Full graph is not considered for use with the GSF approach, since GSF requires a sparse adjacency matrix to compute the graph structure features.
- **kNN graph:** An edge is inserted between items i and j as long as one of them belongs to the set of top-k most similar items of the other. Two basic variants of this scheme are possible, symmetric and asymmetric, depending on whether both items i, j belong to the similar set of each other or not.
- **ϵNN graph:** A global distance threshold ϵ is defined and then an edge is inserted between items i and j if $||x_i - x_j||^2 < \epsilon$.

Having constructed the similarity graph between the input media items, GSF proceeds with mapping the graph nodes to feature vectors that represent the associations of nodes with latent groups of nodes that form densely connected clusters. In order to extract such features, we first construct the normalized graph Laplacian from the degree (D) and adjacency (A) matrices of the graph:

$$\tilde{L} = D^{-1/2}LD^{-1/2} = I - D^{-1/2}AD^{-1/2}, \tag{2}$$

where $L = D - A$ is the graph Laplacian. Computing the eigenvectors of $\tilde{L}$ corresponding to the C_D smallest non-zero eigenvalues of the matrix results in n vectors of C_D dimensions, which when concatenated form the input matrix $S \in \mathbb{R}^{n \times C_D}$, each row of which is denoted as $S_i \in \mathbb{R}^{C_D}$ and constitutes the graph structure feature vector for media item i. These features, known as Laplacian Eigenmaps [8], are derived by solving the following minimization problem:

$$\underset{S}{\text{argmin}}\ S^T \tilde{L} S, \qquad \text{s.t.}\ S^T S = I. \tag{3}$$

In the final step, a classifier is trained using the structure feature vectors of the labeled items as input. In our implementation, we opted for the use of SVM. Apart from classification performance considerations, it is important for retrieval applications that the classifier produces real-valued prediction scores for unlabeled items, so that they can be ranked per concept. Producing real-valued prediction scores is also important for performing result fusion when multiple sets of features are available (see subsection 3.3).

3.2   Incremental Learning

The scheme of subsection 3.1 requires both labeled and unlabeled items to be available at train time for constructing a similarity graph from both sets and computing the respective structure features. In case new unlabeled samples were provided as input, it would be necessary to reconstruct the similarity graph for the extended set of items and recompute the eigenvectors of $\tilde{L}$ (Equation 2). This is clearly impractical in settings where new media items are regularly arriving to the indexing system. Thus, it is necessary to devise incremental means for computing the graph structure features of unknown items.

Linear Projection (LP): A simple approach for deriving the graph structure features of a new item $n+1$ is to determine the set N_{n+1} of k most similar items and then to compute the weighted mean of their graph structure features:

$$\hat{S_{n+1}} = \frac{\sum_{j \in N_{n+1}} w_{n+1,j} S_j}{\sum_{j \in N_{n+1}} w_{n+1,j}} \tag{4}$$

where $w_{n+1,j}$ denotes the similarity score between the new item and neighbouring item j, which may be computed by use of the heat kernel (Equation 1).

Submanifold Analysis (SA): A more accurate technique for estimating the graph structure feature vector of item $n+1$ relies on the analysis of the graph submanifold around it [9]. Initially, the $(k+1) \times (k+1)$ similarity matrix W_S is constructed between the new item and the k most similar items. Then, the sub-diagonal and sub-Laplacian matrices are derived as follows:

$$D_S(i,i) = \sum_j W_S(j,i), \qquad L_S = D_S - W_S$$

We compute the eigenvalues $0 = \lambda_S^0 \leq \lambda_S^1 \leq ... \leq \lambda_S^d$ and d eigenvectors $v_1, ..., v_d$ of the non-zero eigenvalues. This computation is lightweight since k is selected to be small (cf. experiments in Section 4). Finally, a weight vector $C = [c_1...c_k] \in \mathbb{R}^k$ is determined by minimizing the following reconstruction error:

$$\underset{c_i}{\mathrm{argmin}} \left| v_{k+1} - \sum_{i=1}^{k} c_i v_i \right|^2 \quad \text{s.t.} \quad \sum_i c_i = 1$$

by using a non-linear constaint optimization method [10]. Once the weight vector C is computed, the new feature vector is computed as $\hat{S}_{n+1} = \sum_{i=0}^{k} c_i S_i$.

3.3 Fusion of Multiple Features

GSF offers several options for fusing different sets of input features:

- **Feature fusion (F-FEAT):** This constitutes a common early fusion technique. In its simplest form, this is implemented by simple vector concatenation and by an optional feature normalization step.
- **Similarity graph fusion (F-SIM):** Combining two similarity graphs G_1 and G_2 constructed from two feature sets, this technique produces a fused graph G_F. The combination can be implemented by means of an elementwise additive or multiplicative operation between G_1 and G_2.
- **Graph structure feature fusion (F-GSF):** According to this, graph structure feature vectors are computed separately from each similarity graph and are combined by concatenating the corresponding vectors.
- **Result fusion (F-RES):** This widely used late fusion technique is implemented by training a second-level classifier with the prediction scores of individual feature concept detectors as inputs.

4   Evaluation

The proposed approach was evaluated on both synthetic and real data.

4.1   Synthetic Data

GSF was thoroughly tested using synthetic 2D distributions with limited number of samples to enable visualization of input data and classification results, and to make possible the exploration of a large number of experimental settings and the repetition of each experiment multiple times for deriving reliable performance estimates. Four kinds of distributions were used: (a) Two moons, (b) Lines, (c) Circles, (d) Gaussians. The following performance aspects were investigated:

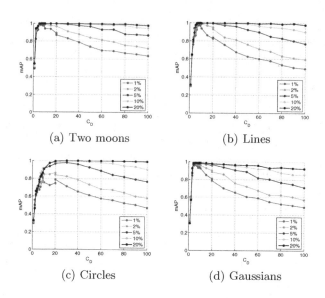

Fig. 2. Role of number of graph structure features (C_D) and training samples (α)

- **Parameters of GSF:** Number of graph structure features C_D, graph construction technique (kNN or ϵNN) and associated parameters (k, ϵ).
- **Inductive learning:** GSF was evaluated in an inductive learning setting with the use of the baseline scheme of subsection 3.2.
- **Fusion method:** Four fusion methods of subsection 3.3.

Parameters of GSF: To test the behaviour of the proposed approach for different number of graph structure features, for each distribution, the number of features C_D was varied from 1 to 100 and three settings were tested: low, medium and high noise. In the case of Two moons and Gaussians distributions, we found that the learning performance is almost insensitive to C_D even when large amounts of noise are added to the input data. In the case of the Lines distribution, a similar behaviour is observed with the exception of the high noise setting, where we found that increasing the number of graph structure features harms performance. Finally, in the Circles distribution, a larger number of graph structure features was found to be beneficial, especially in high noise settings. Furthermore, performance was measured for different training set sizes (Figure 2). When 10% or more of the input data are available for training, performance appears to be insensitive to the number of features C_D. However, for smaller training sizes, GSF appears to be increasingly sensitive to the selection of C_D.

The role of the graph construction technique was examined by graphing the performance of GSF using both kNN and ϵNN for different values of k and ϵ. Out of the two competing options, kNN appears as less sensitive to the selection of k, than ϵNN is to ϵ. Only in the case of the Circles distribution, increasing the value of k seems to have an adversary effect on performance. In case of ϵNN,

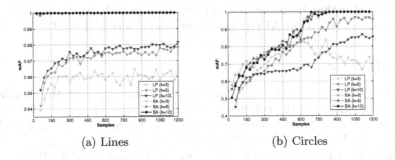

(a) Lines   (b) Circles

Fig. 3. Inductive learning performance

increasing the value of ϵ seems to harm performance, whereas in the case of the Circles distribution the relation between performance and ϵ is non-monotonic.

Additionally, we investigated the computational requirements of GSF when an increasing number of input features are provided as input. As the number of input features increases, so does the execution time of SVM-RBF and for large number of features it is considerably slower than GSF (for instance for 500 features SVM-RBF was 4x slower than GSF). Given the multitude of features (> 1000) typically used in multimedia concept detection, GSF may be considered a much more efficient option for practical problem settings.

Inductive Learning: We generated 3000 samples from the four distributions and $\alpha\%$ were used for training. The rest were not used in the similarity graph construction. Once they were provided as input, the two incremental schemes of subsection 3.2 (LP and SA) were used to estimate their graph structure features. The estimation of graph structure features was conducted by setting k equal to 3, 5 and 10. The results are presented in Figure 3 (for Lines and Circles). Sub-manifold Analysis (SA) consistently outperforms Linear Projection (LP). In the case of Two moons and Gaussians, their performance is comparable. However, for Lines and Circles, there is a clear improvement when using SA over LP. An additional benefit of SA is that its performance seems to be only marginally affected by k in contrast to LP that is sensitive to it. The above observation coupled with the fact that the overall performance rates for SA are exceptional demonstrate that the devised incremental scheme can effectively be used in online learning settings. We recognize that online learning with synthetic data is an easier problem compared to real-world settings due to the fact that the training and test samples are drawn from the same distribution.

Fusion Method: To evaluate the fusion techniques of subsection 3.3, we generated feature vectors from two 2D independent distributions. In the first experiment, both distributions were generated according to the Lines pattern, and the free parameter was the size of the training set. In the second experiment, Circle distributions were used and increasing amounts of noise were added to one of the two feature sets. In both experiments, the two late fusion techniques, namely GSF-level fusion (F-GSF) and result fusion (F-RES) led to the best performance, higher than both feature-level performance and early fusion methods, namely feature-level (F-FEAT) and similarity-based fusion (F-SIM).

4.2 MIR-Flickr

MIR-Flickr [13] was used as a real-world benchmark. Ground truth is available for all 25,000 images of the collection in three variants, strict relevance of image to concept (REL), loose relevance (POT) and aggregate annotation (ALL) [13]. REL and POT contain annotations for 14 concepts, while ALL includes annotations for 24 more concepts (38 in total). We compared GSF with two state-of-the-art approaches, the Semantic Spaces (SESPA) approach [14] and Multiple Kernel Learning (MKL) [15], which were selected for two reasons:

- High performance: SESPA reports better than average performance for the ImageCLEF 2009 photo annotation task. MKL reports higher performance compared to the best method in the PASCAL VOC'07 dataset.
- Reproducibility: The authors of both approaches have disclosed all necessary information and data (e.g. visual features of the dataset images) making possible the replication of their experimental settings.

In all experiments that follow, the similarity graphs were built with the kNN technique, setting $k = 2000$ (empirically found to produce slightly higher mAP scores compared to $k = 500, 1000$). We used different k values during the testing of incremental methods (LP, SA). In addition, we experimentally selected $C_D = 500$ due to its higher performance compared to $C_D = 100, 200, 1000$.

GSF vs SESPA. SESPA [14] is based on the concept of creating a high-dimensional space, in which closely positioned images are expected to be semantically related. The authors use three well known visual descriptors based on SIFT. Apart from the SESPA approach, the authors of [14] specify a complete MIR-Flickr-specific evaluation protocol. Following that protocol, we computed the performance of GSF for three different train-test splits and over all three annotation sets (REL, POT, ALL). The performance was quantified in terms

Table 1. GSF vs SESPA

Method	$|\mathcal{L}| = 5000$			$|\mathcal{L}| = 10000$			$|\mathcal{L}| = 15000$		
	REL	POT	ALL	REL	POT	ALL	REL	POT	ALL
SESPA	0.216	0.313	0.276	0.237	0.324	0.287	0.240	0.327	0.292
GSF-F$_1$	0.202	0.313	0.272	0.225	0.339	0.297	0.237	0.350	0.308
GSF-F$_2$	0.146	0.257	0.216	0.160	0.276	0.233	0.171	0.286	0.244
GSF-F$_3$	0.142	0.265	0.142	0.162	0.285	0.240	0.174	0.295	0.251
GSF-C	0.222	0.332	0.291	0.247	0.361	0.319	0.269	0.380	0.339
GSF-D$_1$	0.227	0.340	0.292	0.256	0.371	0.325	0.275	**0.388**	0.345
GSF-D$_2$	**0.237**	**0.357**	**0.314**	**0.263**	**0.379**	**0.337**	**0.278**	0.384	**0.350**

Table 2. GSF vs MKL ($|\mathcal{L}| = 12500$, ALL)

	MKL	GSF	GSF*
Visual	**0.530**	0.511	0.511
Tag	0.424	**0.433**	**0.474**
Visual+Tag	0.623	**0.624**	**0.641**

of mean Average Precision (mAP) and Area Under the Curve (AUC). Table 1 presents the obtained results (only mAP). Several variants of GSF were evaluated: (a) single feature GSF for each of the three features (GSF-F_1, GSF-F_2 and GSF-F_3), (b) multi-modal fusion (GSF-C) relying on the concatenation of the Laplacian eigenvectors of the three similarity graphs , (c) two result fusion variants (GSF-D_1 and GSF-D_2). Result fusion in GSF-D_1 relied on linear SVM [11], while in GSF-D_2 it relied on SVM-RBF [12].

The results of Table 1 indicate that GSF clearly outperforms SESPA. Across all train-test splits and all annotation sets, one or more of the feature fusion variants of GSF yield higher mAP scores. In terms of AUC, the difference in performance is less pronounced and in one case ($|\mathcal{L}| = 5000$, REL), SESPA outperforms GSF by a small margin. Out of the GSF variants, the highest performing ones are those relying on result fusion. There is a tendency for bigger improvement in performance (compared to SESPA) as the training set size increases, and the performance increase is larger on the POT set than on REL.

GSF vs MKL. The MKL approach by [15] leverages Multiple Kernel Learning for combining a kernel based on image content with a second kernel that encodes tag information. To derive the visual kernel, the authors of [15] make use of 15 features, while the tag-based kernel is computed by selecting the 457 most frequent tags as features. Table 2 presents the comparison of the mAP-based performance between MKL and GSF on a 50-50 train-test split (ALL annotation set). According to it, MKL outperforms GSF by 3.7% when only the visual features are used as input, while GSF outperfoms MKL by 2.1% when tag features are used. When all features are provided as input, GSF outperforms MKL by a very small margin. The third column of Table 2 reports the results obtained by GSF with the use of tag features computed by use of inverse document frequency. Using the MKL approach with this feature would be impractical due to its very high dimensionality (68,894). Instead, GSF could make use of the feature, since its complexity is not significantly affected by its dimensionality.

Figure 4 illustrates the top eight results from GSF using visual and tag features for the first five REL concepts of MIR-Flickr. All returned results (apart from the first in the Baby concept) are highly relevant to the query concepts. Manual inspection of the top 50 results for other concepts of the dataset further confirms their high relevance.

baby

car

clouds

dog

flower

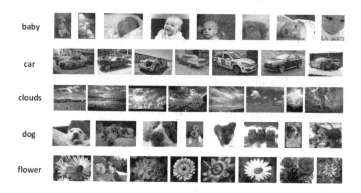

Fig. 4. Examples of top retrieved images per concept using the batch mode GSF

Contribution of Unlabeled Samples. In this experiment, we evaluate the gains in learning performance when more unlabeled samples are available. Out of the 25,000 images, 5,000 were reserved for training and 10,000 for testing. The remaining 10,000 images were used as unlabeled items. Figure 5 presents the results for three different features (GIST, DenseSiftV3H1, TagRaw50). The features are representative of three feature types: global visual, local visual, and text-based. The experiment is repeated for different values of k, i.e. the number of top-k most similar neighbours used by SA.

Adding more unlabeled samples together with the labeled ones leads to significant performance gains. In the case of GIST, the largest performance benefits are observed when $k = 10$: compared to using only the labeled examples (mAP $= 0.21$), the performance of the concept detector rises to 0.235, i.e. a relative increase of 11.9%, when 10,000 unlabeled samples are added during the computation of the graph structure features. For larger values of k, the performance benefits appear to be much less pronounced. In the case of DenseSift3VH1, significant performance gains are observed across all values of k. For instance, for $k = 10$ a 16.3% increase in performance is recorded, while for $k = 50$ the performance benefit still amounts to 11.8%. In case of the tag-based features, the situation is more similar to the one described for GIST. The performance improvements are clearer for smaller values of k and appear to level off after 3000-4000 unlabeled samples are added together with the labeled samples.

In the same experiment, we could also conclude that the performance of the SA online learning scheme is comparable to the one of the transductive learning scheme. For instance, when using $k = 100$ (which leads to the best performance for SA among the tested values), the map score of SA using the GIST features and no unlabeled images is 0.243, while the score achieved by the transductive version of GSF is 0.2456 (1.1% higher). When 5000 unlabeled images are provided together with the labeled ones, then the SA performance (again for GIST features) rises to 0.2552, slightly higher than 0.2522, which was achieved by the transductive implementation of GSF. Similar observations were made when using other features as well.

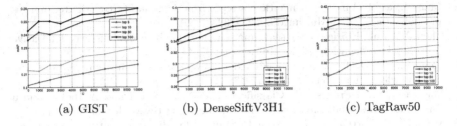

(a) GIST   (b) DenseSiftV3H1   (c) TagRaw50

Fig. 5. Performance gains by adding unlabeled samples

5 Conclusions

We presented GSF, a multimedia annotation approach leveraging the structure of image similarity graphs. GSF relies on the assumption that images with similar positions on the graph tend to carry the same concepts. Concept detection is then conducted by training classifiers using the graph Laplacian eigenvectors as features. Apart from the transductive formulation of the problem, we proposed two incremental versions, one based on Linear Projection and the other on Submanifold Analysis, and described four fusion techniques applicable to GSF. The transductive version of our approach was evaluated on a wide range of synthetic distributions, and was also compared against two state-of-the-art learning approaches on the MIR-Flickr dataset, giving superior or comparable results. The two incremental implementations were compared on synthetic data, with SA method yielding superior performance. SA was also evaluated on MIR-Flickr in a semi-supervised learning setting, resulting in mAP rates very close to the ones achieved with the transductive version. In addition, SA was found to be quite fast; the average time for predicting the concepts of an image using SA with $k = 5$ was measured to be 38.4ms (not including feature extraction). In the future, we plan to further study the computational characteristics of the proposed approach by applying it to larger scale problems.

Acknowledgements. This work is supported by the SocialSensor FP7 project, partially funded by the EC under contract number 287975.

References

1. Wang, M., Hua, X.-S., Tang, J., Hong, R.: Beyond distance measurement: constructing neighborhood similarity for video annotation. TMM 11(3), 465–476 (2009)
2. Zhu, X.: Semi-supervised learning with graphs. PhD Thesis, Carnegie Mellon University (2005) 0-542-19059-1
3. Zhou, D., Bousquet, O., Navin Lal, T., Weston, J., Schölkopf, B.: Learning with Local and Global Consistency. In: Advances in NIPS, vol. 16, pp. 321–328. MIT Press (2004)

4. Tang, J., et al.: Inferring semantic concepts from community contributed images and noisy tags. ACM Multimedia, 223–232 (2009)
5. Chen, X., et al.: Efficient large scale image annotation by probabilistic collaborative multi-label propagation. ACM Multimedia, 35–44 (2010)
6. Tang, L., Liu, H.: Leveraging social media networks for classification. Data Mining and Knowledge Discovery 23(3), 447–478 (2011)
7. Macskassy, S.A., Provost, F.: Classification in Networked Data: A Toolkit and a Univariate Case Study. Journal of Machine Learning Research 8, 935–983 (2007)
8. Mikhail, B., Partha, N.: Laplacian Eigenmaps for dimensionality reduction and data representation. Neural Computing 15(6), 1373–1396 (2003)
9. Jia, P., Yin, J., Huang, X., Hu, D.: Incremental Laplacian eigenmaps by preserving adjacent information between data points. PR Letters 30(16), 1457–1463 (2009)
10. Leyffer, S., Mahajan, A.: Nonlinear Constrained Optimization: Methods and Software. Preprint ANL/MCS-P1729-0310 (2010)
11. Fan, R., Chang, K., Hsieh, C., Wang, X., Lin, C.: LIBLINEAR: A Library for Large Linear Classification. Journal of ML Research 9, 1871–1874 (2008)
12. Chang, C.-C., Lin, C.-J.: LIBSVM: A library for support vector machines. ACM Transactions on Intelligent Systems and Technology 2(3), 27:1–27:27 (2011)
13. Huiskes, M.J., Michael, S., Lew, M.S.: The MIR Flickr Retrieval Evaluation. In: Proceedings of ACM Intern. Conference on Multimedia Information Retrieval (2008)
14. Hare, J.S., Lewis, P.H.: Automatically annotating the MIR Flickr dataset. In: ACM ICMR, pp. 547–556 (2010)
15. Guillaumin, M., Verbeek, J., Schmid, C.: Multimodal semi supervised learning for image classification. In: Proceedings of IEEE CVPR Conference, pp. 902–909 (2010)

Refining Image Annotation by Integrating PLSA with Random Walk Model

Dongping Tian[1,2], Xiaofei Zhao[1], and Zhongzhi Shi[1]

[1] Key Laboratory of Intelligent Information Processing, Institute of Computing Technology, Chinese Academy of Sciences, Beijing, 100190, China
[2] Graduate University of the Chinese Academy of Sciences, Beijing, 100049, China
{tiandp,zhaoxf,shizz}@ics.ict.ac.cn

Abstract. In this paper, we present a new method for refining image annotation by integrating probabilistic latent semantic analysis (PLSA) with random walk (RW) model. First, we construct a PLSA model with asymmetric modalities to estimate the posterior probabilities of each annotating keywords for an image, and then a label similarity graph is constructed by a weighted linear combination of label similarity and visual similarity. Followed by a random walk process over the label graph is employed to further mine the correlation of the keywords so as to capture the refining annotation, which plays a crucial role in semantic based image retrieval. The novelty of our method mainly lies in two aspects: exploiting PLSA to accomplish the initial semantic annotation task and implementing random walk process over the constructed label similarity graph to refine the candidate annotations generated by the PLSA. Compared with several state-of-the-art approaches on Corel5k and Mirflickr25k datasets, the experimental results show that our approach performs more efficiently and accurately.

Keywords: Refining Image Annotation, PLSA, EM, Random Walk, Image Retrieval.

1 Introduction

With the rapid development of multimedia information technology, image retrieval has become more and more important in Internet and other multimedia platforms. As we known, image annotation is a previous and vital step when it comes to the semantic based image retrieval. Traditional method for image annotation is to let people manually annotate the images by some keywords. However, this method is onerous and time-consuming. Furthermore, the annotating result is subjective to different people. To address these limitations, automatic image annotation (AIA) has become a focus and received extensive investigation, whose purpose is to automatically assign some keywords to an image that can well describe the content of it. Subsequently many methods have been developed for AIA, and most of them can be roughly classified into two categories, i.e. classification-based methods and probabilistic modeling methods.

S. Li et al. (Eds.): MMM 2013, Part I, LNCS 7732, pp. 13–23, 2013.

The representative works of the former are automatic linguistic index for pictures [1] and content-based annotation method with SVM [2] etc. The probabilistic modeling methods include the translation model (TM) [3], the cross-media relevance model (CMRM) [4], the continuous-space relevance model (CRM) [5], the multiple-Bernoulli relevance model (MBRM) [6] and the latent aspect model PLSA [7], etc. Unfortunately, all the mentioned annotation methods, to some extent, can achieve relative success compared to the manual annotation, but they are still far from satisfaction due to the well-known semantic gap problem.

In recent years, some researchers propose to refine the image annotation by taking the word correlation into account. As a pioneer work, Jin et al. [8] implement image annotation refinement based on WordNet by pruning the irrelevant annotations. In their work, however, only global textual information is employed and the refinement process is independent of the target image, which means that different images with the same candidate annotations would obtain the same refinement results. Subsequently, Wang et al. [9] apply random walk with restarts model to refine candidate annotations by integrating word correlations with the original candidate annotation confidence together. Followed by they propose a content based approach by formulating the annotation refinement as a Markov process [10]. Recently Liu et al. [11] rank the image tags according to their relevance with respect to the associated images by tag similarity and image similarity in a random walk model. Xu et al. [12] come up with a new graphical model termed as regularized latent Dirichlet allocation (rLDA) for tag refinement. In addition, Zhu et al. [13] put forward an efficient iterative approach for image tag refinement by pursuing the low-rank, content consistency, tag correlation and error sparsity, which constitute a constrained yet convex optimization problem and an efficient accelerated proximal gradient method is utilized to resolve it. More recently, Zhuang et al. [14] propose a two-view learning approach for image tag ranking by effectively exploiting both textual and visual contents of social images to discover the complicated relationship between tags and images.

Most of these approaches can achieve state-of-the-art performance and motivate us to explore image annotation with the help of their excellent experiences and knowledge. So in this paper, we present a new method for refining image annotation by means of combining PLSA and random walk model (PLSA-RW). To begin with, a PLSA model with asymmetric modalities is constructed to estimate the scores (i.e. posterior probabilities. For simplicity, we use the terminologies score and posterior probability interchangeably in the rest of this paper) of all the annotating keywords, and this can be seen as the initial annotation for the image. And then a label [1] similarity graph is constructed by a weighted linear combination of label similarity and visual similarity. Followed by a random walk process over the label similarity graph is implemented to further mine the words correlation. Once the random walk reaches the steady-state probability distribution, the top several candidates with the highest probabilities can be seen as the refining annotation. Our method can boost the annotating performance by introducing a two-stage annotation refinement process. We evaluate our method

---

[1] Here label means the initial annotation generate by the PLSA.

on Corel5k and Mirflickr25k datasets and their experimental results compare favorably with several state-of-the-art approaches. To the best of our knowledge, this is the first study to try to integrate PLSA with random walk in the task of refining image auto-annotation.

The rest of the paper is organized as follows. Section 2 presents how to apply PLSA to model annotated images. In section 3, the construction of label similarity graph is first introduced, and then a random walk over the graph is elaborated. Experimental results on Corel5k and Mirflickr25k datasets are reported and analyzed in section 4 respectively. Finally, we end this paper with some important conclusions and future work in section 5.

2 PLSA Model

PLSA [15] is a statistical latent class model which introduces a hidden variable (latent aspect) z_k in the generative process of each element x_j in a document d_i. Given this unobservable variable z_k, each occurrence x_j is independent of the document it belongs to, which corresponds to the following joint probability:

$$P(d_i, x_j) = P(d_i) \sum_{k=1}^{K} P(z_k|d_i)P(x_j|z_k) \qquad (1)$$

The model parameters of PLSA are the two conditional distributions: $P(x_j|z_k)$ and $P(z_k|d_i)$. $P(x_j|z_k)$ characterizes each aspect and remains valid for documents out of the training set. On the other hand, $P(z_k|d_i)$ is only relative to the specific documents and cannot carry any prior information to an unseen document. An EM algorithm is used to estimate the parameters through maximizing the log-likelihood of the observed data.

$$L = \sum_{i=1}^{N} \sum_{j=1}^{M} n(d_i, x_j) \log P(d_i, x_j) \qquad (2)$$

where $n(d_i, x_j)$ is the count of element x_j in document d_i. The steps of the EM algorithm can be succinctly described as follows.

E-step. The conditional distribution $P(z_k|d_i, x_j)$ is computed from the previous estimate of the parameters:

$$P(z_k|d_i, x_j) = \frac{P(z_k|d_i)P(x_j|z_k)}{\sum_{l=1}^{K} P(z_l|d_i)P(x_j|z_l)} \qquad (3)$$

M-step. The parameters $P(x_j|z_k)$ and $P(z_k|d_i)$ are updated with the new expected values $P(z_k|d_i, x_j)$:

$$P(x_j|z_k) = \frac{\sum_{i=1}^{N} n(d_i, x_j)P(z_k|d_i, x_j)}{\sum_{m=1}^{M} \sum_{i=1}^{N} n(d_i, x_m)P(z_k|d_i, x_m)} \qquad (4)$$

$$P(z_k|d_i) = \frac{\sum_{j=1}^{M} n(d_i, x_j)P(z_k|d_i, x_j)}{\sum_{j=1}^{M} n(d_i, x_j)} \quad (5)$$

If one of the parameters ($P(x_j|z_k)$ or $P(z_k|d_i)$) is known, the other one can be inferred by using fold-in method, which updates the unknown parameters with the known parameters kept fixed, so that it can maximize the likelihood with respect to the previously trained parameters. In this paper, we construct a PLSA model with asymmetric modalities since the textual modality is more appropriate to learn a semantically meaningful latent space [7], and the joint probability between an image and the semantic concepts is calculated from two linked PLSA models sharing the same distribution over aspects. Given an unseen image visual features $v(d_{new})$, the conditional probability distribution $P(z_k|d_{new})$ can be inferred with the previously estimated model parameters $P(v|z_k)$, then the posterior probability of words can be computed by the following equation.

$$P(w|d_{new}) = \sum_{k=1}^{K} P(w|z_k)P(z_k|d_{new}) \quad (6)$$

3   Random Walk-Based Refining Annotation

As a latent aspect model, PLSA has been successfully applied in automatic image annotation, such as the representative PLSA-WORDS and PLSA-FEATURES [7] as well as the PLSA-FUSION proposed by Li et al. [16], which uses two linked PLSA models to learn the mixture of aspects from both visual and textual modalities. However, since all the annotations are calculated independently in PLSA model and the relations among them are not exploited, which inevitably results in some ambiguity and inconsistency in the process of image annotation. In order to combine the prior confidence of candidate annotations and word correlations together, we present a two-stage image annotation refinement framework displayed in Figure 1. More details of it will be described in the following subsections.

3.1   Label Graph Construction

To construct the label graph, i.e. the initial annotation graph, each candidate is transformed to a vertex, and the pair-wise label similarity is used as the weight of the corresponding edge. For now we focus on how to reasonably estimate the similarities between pair-wise concepts related to an image, which is still a tough problem in multimedia information processing. The mostly used methods include WordNet [17] and normalized Google distance (NGD) [18]. From their definitions, we can easily see that NGD is actually a measure of the contextual relation while WordNet focuses on the semantic meaning of keyword itself. What is more, both of them build word correlations only based on textual descriptions, and the visual information of images in the dataset is not utilized for refinement, which also plays a key role in precise image annotation. So in this paper, the

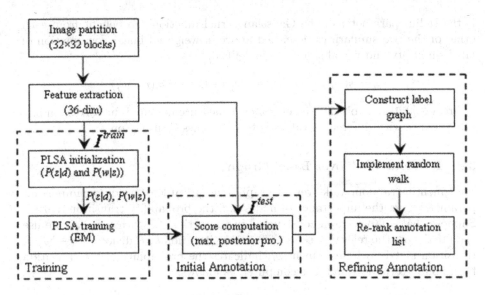

Fig. 1. The proposed refining annotation framework in this paper

pair-wise annotation similarity is calculated by a weighted linear combination of label similarity and visual similarity, which can effectively avoid the phenomenon that different images with the same candidate annotations would obtain the same refinement results. The label similarity between w_i and w_j is defined as follows:

$$s_l(w_i, w_j) = \exp(-d(w_i, w_j)) \tag{7}$$

where $d(w_i, w_j)$ represents the distance between two labels w_i and w_j and it is defined similarly to NGD as:

$$d(w_i, w_j) = \frac{\max(\log f(w_i), \log f(w_j)) - \log f(w_i, w_j)}{\log G - \min(\log f(w_i), \log f(w_j))} \tag{8}$$

where $f(w_i)$ and $f(w_j)$ are the numbers of images containing labels w_i and w_j respectively, and $f(w_i, w_j)$ is the number of images containing both w_i and w_j, G is the total number of images in the dataset. Similar to [11], for a label w associated with an image x, we collect the K nearest neighbors from the images containing w, and these images can be regarded as the exemplars of the label w with respect to x. Thus from the point view of labels associated with an image, the visual similarity between labels w_i and w_j is given as follows:

$$s_v(w_i, w_j) = \exp(-\frac{1}{K \times K} \sum_{x \in \Gamma_{w_i}, y \in \Gamma_{w_j}} \frac{\|x - y\|^2}{\sigma^2}) \tag{9}$$

where Γ_w is the representative image collection of label w, x and y denote image features corresponding to the respective image collections of label w_i and w_j, σ

is the radius parameter of the Gaussian kernel function. To benefit from each other of the two similarities described above, a weighted linear combination of label similarity and visual similarity is defined:

$$s_{ij} = s(w_i, w_j) = \lambda s_l(w_i, w_j) + (1 - \lambda)s_v(w_i, w_j) \tag{10}$$

where $\lambda \in [0, 1]$ controls the weights for each measurement and the corresponding performance with different λ values is to be discussed in section 4.

3.2   Random Walk over Label Graph

Implementing random walk over the graph structure at least needs two important parameters, i.e. the importance of nodes and the probability transition matrix. Suppose that a label graph constructed in subsection 3.1 with n nodes, we use $r_k(i)$ to denote the relevance score of node i at iteration k, P denotes a $n - by - n$ transition matrix, whose element p_{ij} indicates the probability of the transition from node i to node j and it is computed as follows:

$$p_{ij} = s_{ij} \Big/ \sum_k s_{ik} \tag{11}$$

where s_{ij} is the pair-wise label similarity (defined in Eq.10) between node i and node j. Then the random walk process is formulated as:

$$r_k(j) = \alpha \sum_i r_{k-1}(i)p_{ij} + (1 - \alpha)v_j \tag{12}$$

where $\alpha \in (0, 1)$ is a weight parameter to be determined, v_j denotes the initial annotation probabilistic scores calculated by the PLSA. In the process of refining annotation, random walk proceeds until it reaches the steady-state probability distribution and then the top several candidates with the highest probabilities can be seen as the final refining image annotation results.

4   Experimental Results and Analysis

For the purpose of comparison, we first conduct our experiments on the Corel5k dataset, which consists of 5000 images from 50 Corel Stock Photo CD's provided by [3]. Each CD contains 100 images with a certain theme, of which 90 are designated to be in the training set and 10 in the testing set, resulting in 4500 training images and a balanced 500-image test collection. Since the focus of this paper is not on image feature selection, we use similar features extracted by [6] to make a fair comparison with the state-of-the-art approaches. First of all, we simply decompose images into a set of 32×32-sized blocks, then compute a 36 dimensional feature vector for each block, consisting of 24 color features (auto-correlogram) computed over 8 quantized colors and 3 Manhattan Distances,12 texture features (Gabor filter) computed over 3 scales and 4 orientations. As a result, each block is represented as a 36-dim feature vector. Then each image is represented as a bag of features, i.e., a set of 36 dimensional vectors.

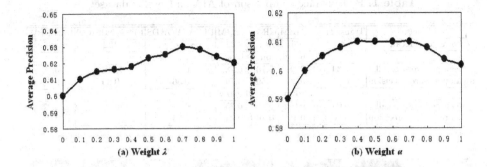

Fig. 2. Evaluation for weight parameters λ and α

4.1   Evaluation for the Weights

Since there are two variable weights λ and α to be determined, we should first fix one of them so as to observe the other's varied trend and vice versa. Suppose that α is set to 0.5, then we range λ from 0 to 1. As shown in Figure 2(a), we can clearly see that the performance is better when $\lambda \in (0, 1)$ than $\lambda = 0$ or $\lambda = 1$ individually. Particularly, the best result is achieved when $\lambda = 0.7$, which demonstrates the complementary nature of label similarity and visual similarity. On the other hand, we set $\lambda = 0.7$ and range α from 0 to 1. From the curve in Figure 2(b), we note that the performance improves consistently before 0.5, followed by it almost keeps in a smooth state. The performance begins to reduce when α exceeds 0.7. Thus we choose $\alpha = 0.5$ as the optimal parameter in our experiment.

4.2   Refining Image Annotation on Corel5k

To show the effectiveness of our model (PLSA-RW) proposed in this paper, we make a direct comparison with several previous approaches [3,4,5,6,7,16]. Similar to [6], we compute the recall and precision of every word in the test set and use the mean of these values to summarize its performance. The experimental results listed in Table 1 are based on two sets of words: the subset of 49 best words and the complete set of all 260 words that occur in the training set. From table 1, it is easy to see that our model PLSA-RW outperforms all the others, especially the first three approaches. Meanwhile, it is also superior to MBRM, PLSA-WORDS and PLSA-FUSION.

Figure 3 shows some annotating results (only four cases are listed here due to the limited space) using PLSA-FUSION and PLSA-RW. It is worth noting that the annotations with the highest probabilities obtained in the last itera-tion of the random walk process are considered as the final annotation of the corresponding image. It is also important to note that the annotation order of the keywords for each image, which is very significant for semantic based image retrieval. Especially those with different annotating orders and enriched

Table 1. Performance comparison of AIA on Corel5k dataset

Models	Translation	CMRM	CRM	MBRM	PLSA-WORDS	PLSA-FUSION	PLSA-RW
#words with recall > 0	49	66	107	122	105	122	126
Results on 49 best words							
Mean per-word recall	0.34	0.48	0.70	0.78	0.71	0.76	0.78
Mean per-word precision	0.20	0.40	0.59	0.74	0.56	0.65	0.75
Results on all 260 words							
Mean per-word recall	0.04	0.09	0.19	0.25	0.20	0.22	0.27
Mean per-word precision	0.06	0.10	0.16	0.24	0.14	0.19	0.25

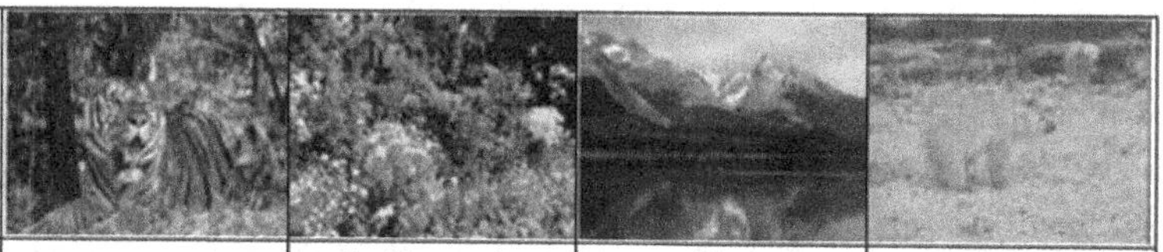

Images				
Ground Truth Annotation	tiger, forest, cat, trees	garden, flowers, trees, farm, plants	mountain, water, sky, clouds	polar, bear, snow, tundra
PLSA-FUSION Annotation	cat, tiger, trees, forest, leaves	flowers, trees, garden, farm, plants	sky, mountain, water, clouds, trees	snow, polar, bear, tundra, ice
PLSA-RW Annotation	_tiger_, trees, _leaves_, _forest_, cat	flowers, trees, garden, _plants_, _farm_	sky, mountain, water, clouds, _trees_	snow, _bear_, _polar_, tundra, _ice_

Fig. 3. Annotation comparison between PLSA-FUSION and PLSA-RW (Re-ranked and enriched annotations are underlined and italic)

annotating keywords compared to the PLSA-FUSION and the ground truth annotation are underlined and italic, respectively.

4.3   Refining Image Annotation on Mirflickr25k

To further demonstrate the effectiveness of PLSA-RW proposed in this paper, we also conduct experiment on Mirflickr25k dataset [2], which contains 25000 images with 1386 labels. For the sake of fair comparison with the state-of-the-art approaches in [10] and [13], we use similar features to reference [13], that is, a 428-dimension feature vector is extracted from each image, including 225-dim block-wise color moment features generated from 5×5 fixed partition, 128-dim wavelet texture features and 75-dim edge distribution histogram features. At the same time, we evaluate the performance on 18 tags in Mirflickr25k where the ground-truth annotation of these tags has been provided. In addition, we remove those tags whose occurrence numbers are less than 50, thus 205 unique tags are obtained in total for Mirflickr25k in our experiment.

Table 2 summarizes the average performances measured by F-value for different refinement methods. As can be seen from Table 2, the F-value of our method is 0.475 which gives significant better result than the value obtained by the original user-provided tags (UT) [19]. Furthermore, it compares favorably with the

---

[2] Download from http://press.liacs.nl/mirflickr/dlform.php

Fig. 4. Four exemplars of image annotation refinement on Mirflickr25k

state-of-the-art approaches proposed by Wang et al. (RWR) [10] and Zhu et al. (LR-ES-CC-TC) [13], which further proves that the PLSA-RW is efficient in refining image annotation.

Table 2. Performance comparison of different methods on Mirflickr25k

Methods	UT	RWR	LR-ES-CC-TC	PLSA-RW
F-value	0.221	0.338	0.477	0.475

Alternatively, some exemplars of image annotation refinement are depicted in Figure 4 (only four cases are listed here due to the limited space). It can be observed that our method PLSA-RW can generate more accurate annotation results compared with the original annotations as well as the ones provided in [13]. Taking the second image of the first row for example, there exists only one tag 'girl' in the original annotation. However, after refinement by PLSA-RW, its annotation is enriched by other three keywords 'face', 'child' and 'portrait', which are very appropriate and reasonable to describe the visual content of the image. Overall, the experiment on Mirflickr25k indicates that PLSA-RW is fairly stable and efficient with respect to its parameters setting.

5   Conclusions

In this paper, we have proposed a novel refining image annotation method by combining PLSA with random walk to enhance the annotating performance. We first construct a PLSA model with asymmetric modalities to estimate posterior probabilities of each annotating keyword for one image, and then employ a random walk process to mine the correlations of the keywords so as to capture the final refining annotation results. A weighted linear combination of label similarity and visual similarity is employed to calculate the pair-wise similarities between two candidate annotating keywords. Experimental results on Corel5k and Mirflickr25k datasets show that our model outperforms several state-of-the-art approaches. In the future, we intend to introduce semi-supervised learning into our approach and employ different image datasets to detect its performance comprehensively.

Acknowledgements. This work is supported by the National Natural Science Foundation of China (No.61035003, No.61072085, No.60933004, No.60903141), the National Program on Key Basic Research Project (973 Program) (No.2013CB329502), the National High-tech R&D Program of China (863 Program) (No.2012AA011003) and the National Science and Technology Support Program of China (2012BA107B02).

References

1. Li, J., Wang, J.: Automatic linguistic indexing of pictures by a statistical modeling approach. IEEE Transactions on Pattern Analysis and Machine Intelligence 25(9), 1075–1088 (2003)
2. Cusano, C., Ciocca, G., Schettini, R.: Image annotation using svm. In: Proceedings of Internet imaging IV. SPIE, vol. 5304, pp. 330–338 (2004)
3. Duygulu, P., Barnard, K., de Freitas, J.F.G., Forsyth, D.: Object Recognition as Machine Translation: Learning a Lexicon for a Fixed Image Vocabulary. In: Heyden, A., Sparr, G., Nielsen, M., Johansen, P. (eds.) ECCV 2002, Part IV. LNCS, vol. 2353, pp. 97–112. Springer, Heidelberg (2002)
4. Jeon, J., Lavrenko, V., Manmatha, R.: Automatic image annotation and retrieval using cross-media relevance models. In: Proceedings of the 26th Annual International ACM SIGIR Conference on Research and Development in Informaion Retrieval, pp. 119–126. ACM (2003)
5. Lavrenko, V., Manmatha, R., Jeon, J.: A model for learning the semantics of pictures. In: NIPS (2003)
6. Feng, S., Manmatha, R., Lavrenko, V.: Multiple bernoulli relevance models for image and video annotation. In: Proceedings of the 2004 IEEE Computer Society Conference on Computer Vision and Pattern Recognition, CVPR 2004, vol. 2, pp. 1002–1009. IEEE (2004)
7. Monay, F., Gatica-Perez, D.: Modeling semantic aspects for cross-media image indexing. IEEE Transactions on Pattern Analysis and Machine Intelligence 29(10), 1802–1817 (2007)
8. Jin, Y., Khan, L., Wang, L., Awad, M.: Image annotations by combining multiple evidence & wordnet. In: Proceedings of the 13th Annual ACM International Conference on Multimedia, pp. 706–715. ACM (2005)
9. Wang, C., Jing, F., Zhang, L., Zhang, H.: Image annotation refinement using random walk with restarts. In: Proceedings of the 14th Annual ACM International Conference on Multimedia, pp. 647–650. ACM (2006)

10. Wang, C., Jing, F., Zhang, L., Zhang, H.: Content-based image annotation refinement. In: IEEE Conference on Computer Vision and Pattern Recognition, CVPR 2007, pp. 1–8. IEEE (2007)
11. Liu, D., Hua, X., Yang, L., Wang, M., Zhang, H.: Tag ranking. In: Proceedings of the 18th International Conference on World Wide Web, pp. 351–360. ACM (2009)
12. Xu, H., Wang, J., Hua, X., Li, S.: Tag refinement by regularized lda. In: Proceedings of the 17th ACM International Conference on Multimedia, pp. 573–576. ACM (2009)
13. Zhu, G., Yan, S., Ma, Y.: Image tag refinement towards low-rank, content-tag prior and error sparsity. In: Proceedings of the 18th ACM International Conference on Multimedia, pp. 461–470. ACM (2010)
14. Zhuang, J., Hoi, S.: A two-view learning approach for image tag ranking. In: Proceedings of the 4th ACM International Conference on Web Search and Data Mining, pp. 625–634. ACM (2011)
15. Hofmann, T.: Unsupervised learning by probabilistic latent semantic analysis. Machine Learning 42(1), 177–196 (2001)
16. Li, Z., Liu, X., Shi, Z., Shi, Z.: Learning image semantics with latent aspect model. In: IEEE International Conference on Multimedia and Expo, ICME 2009, pp. 366–369. IEEE (2009)
17. Fellbaum, C.: Wordnet. Theory and Applications of Ontology: Computer Applications, 231–243 (2010)
18. Cilibrasi, R., Vitanyi, P.: The google similarity distance. IEEE Transactions on Knowledge and Data Engineering 19(3), 370–383 (2007)
19. Huiskes, M., Lew, M.: The mir flickr retrieval evaluation. In: Proceedings of the 1st ACM International Conference on Multimedia Information Retrieval, pp. 39–43. ACM (2008)

Social Media Annotation and Tagging
Based on Folksonomy Link Prediction
in a Tripartite Graph

Majdi Rawashdeh, Heung-Nam Kim, and Abdulmotaleb El Saddik

School of Electrical Engineering and Computer Science, University of Ottawa,
800 King Edward, Ottawa, Ontario, Canada, K1N 6N5
{majdi,hnkim,abed}@mcrlab.uottawa.ca

Abstract. Social tagging has become a popular way for users to annotate, search, navigate and discover online social media, resulting in the sheer amount of metadata collectively generated by people. This paper focuses on two tagging problems—(1) recommending the most suitable tags during a user's tagging process and (2) labeling latent tags relevant to a social media item—so that social media can be more browsable, searchable, and shareable by users. The proposed approach employs the Katz measure, a path-ensemble based proximity measure, to predict links in a weighted tripartite graph which represents folksonomy. From a graph-based proximity perspective, our method recommends appropriate tags for a given user-item pair, as well as uncovers hidden tags potentially relevant to a given item. We evaluate our method on real-world folksonomy collected from Last.fm. From our experiments, we show that not only does our algorithm outperform existing algorithms, but it can also obtain significant gains in cold start situations where relatively little information is known about a user or an item.

Keywords: Folksonomy, Graph-based ranking, Link prediction, Social tagging, Tag recommendation and annotation.

1 Introduction

With the recent proliferation of social media sites (e.g., Last.fm, Flicker, and You-Tube), users are overwhelmed by the huge amount of social media content available. Therefore, organizing and retrieving appropriate multimedia content is becoming an increasingly important and challenging task. This challenging task led a number of research communities to concentrate on social tagging systems that allow users to feely annotate their media items (e.g., music, images, or video) with any sort of arbitrary words, referred to as *tags* [8]. Tags assist users to organize their own content, as well as to find relevant content shared by other users. In social tagging systems, the suggestion of tags when a user is annotating a certain item can help users build more coherent folksonomies [3]. On the other hand, automatic tagging or annotation aims at making items more visible to users by uncovering relevant hidden tags that have not previously annotated to the item in the past [11]. The challenges in designing such tag recommenders and automatic tagging is due to the size and complexity of data as well

S. Li et al. (Eds.): MMM 2013, Part I, LNCS 7732, pp. 24–35, 2013.

as some noise in the tag vocabulary. In particular, it is often the cases in folksonomy that little information is known about a user or an item, posing the challenge problem of recommending/uncovering tags relevant to such a user/item.

Recent studies tackled the issue of tag recommendation from a graph-based perspective. Some of these studies viewed a folksonomy as a tripartite hypergraph in which each hyperedge connects a user, a tag and an item [3, 8, 13]. Since folksonomies grow and change rapidly, the need for an approach to predicting accurately the links in the folksonomy graph-based representation becomes an important issue to be addressed. As such, this paper presents a graph-based ranking algorithm that is designed for identifying suitable tags for an individual user (in the tag recommendation scenario) or an item (in the annotation scenario). As folksonomy can be represented by a tripartite graph, we formalize the tag recommendation and annotation problems as the link-prediction problems. On this graph, we exploit the Katz measure—a path-ensemble based proximity measure [10]—to quantify the proximity of two nodes based on weighted sums over collections of possible paths connecting those nodes. The underlying assumption behind this approach is that appropriate nodes for a given node would be in close proximity to that node from a graph viewpoint. In the context, for recommending suitable tags for a given user-item pair, our method speculates as to how a certain node is close not only to that user, but also to that item. Based on the proximity of nodes, we uncover triangle graphs that are likely to appear in the tripartite folksonomy graph. In the annotation scenario, we only consider the proximity from a given item to tags.

The remainder of this paper is organized as follows: Section 2 briefly reviews studies related to tag annotation and recommendation. In Section 3, we provide basic notations and formalizations. We then describe our approaches for finding disable tags in Section 4. In Section 5, we present the effectiveness of our method through experimental evaluations. Finally, we give our conclusions in Section 6.

2   Related Work

Research on social tagging has received considerable interest in the recent years. Different approaches have been applied to recommendations based on social tagging.

Some research has focused on tensor models. Rendle et al. [14] presented a new factorization model for tag recommendation as special case of the Tucker decomposition. The new factorization model extends the Bayesian personalized optimization criterion to the task of tag recommendation and explicitly reflects the pairwise interactions between user, items and tags. Wetzker et al. [16] tackled the same problem by introducing the User-centric Tag Model (UCTM). The UCTM uses a 3-order tensor to model the association between users, items, and tags, and maps individual tag vocabularies, called personomies, on the corresponding folksonomies with the tagged items.

A personalized tag recommendation based on collaborative filtering is proposed by Hamouda et al. [7]. The proposed system recommends tags based on similar users and similar bookmarks, also addresses the two limitations of collaborative filtering, first-time seen bookmark that have not been tagged before and cold start users with no sufficient history for recommendation. The work by Jäschke et al. [9] also applied collaborative filtering in social tagging, where he uses 2-dimensional projections to model relationships among users, items, and tags.

Krestel et. al. [11] investigated a method based on the latent topic model. This method applied Latent Dirichlet Allocation (LDA) to tag recommendation. The proposed method extracts latent topics from a dense folksonomy to be used to recommend additional tags for new resources. A similar work is addressed by Diaz-Aviles et al. [4]. For a new item, the proposed approach creates an ad hoc corpus of similar items, and then applies LDA to extract the latent topics for the item and the associated corpus. The automatic tagging for the item is based on the most likely tags derived from the latent topic identified.

Similar to our approach, graph-based approaches have been studied. Hotho et al. [8] proposed a variant of PageRank, called FolkRank, that performs on a tripartite graph induced by a folksonomy. FolkRank used a folksonomy structure for tag recommendations and searches requested within tagging systems. In recent work, Ramezani [13] modeled folksonomy as a weighted directed graph. They then applied the PageRank algorithm to this graph so as to enhance graph-based tag recommendation techniques. Budru et al. [1] recommended tags based on the neighborhood of a document-tag graph. The tag rank relies on the occurrence of the tag in the neighbors of active resource, co-occurrence of the tag to a set of tags inferred from active resource, distance between resources and between tags.

3   Preliminaries

We begin with the formulation of our study, presenting the underlying notations and definitions. We henceforth use the term "items" to refer to social media content.

3.1   Folksonomy Tripartite Graph

In a social tagging system, users annotate items with tags, creating ternary associations between users, tags, and items. For a set of users $U = \{u_1, u_2,...,u_{|U|}\}$, a set of tags $T = \{t_1,t_2,...,t_{|T|}\}$, and a set of items $I = \{i_1, i_2,...,i_{|I|}\}$, a folksonomy can be defined as a tuple $F := (U, T, I, Y)$ where $Y \subseteq U \times T \times I$ is a set of ternary relations, called tag assignments [8]. From the tag assignments, three matrices can be obtained by aggregating over items, tags, and users, respectively: a $|U| \times |T|$ user-tag matrix $\mathbf{M}_{UT}$, a $|U| \times |I|$ user-item matrix $\mathbf{M}_{UI}$, and a $|T| \times |I|$ tag-item matrix $\mathbf{M}_{TI}$. Each entry of the matrices represents the number of times that the corresponding row and column co-occurred in Y [16]. We apply a BM25 weighting scheme to $\mathbf{M}_{UT}$ and $\mathbf{M}_{TI}$ so as to increase/decrease the importance of tags within/among users/items [15]. In addition, we binarize $\mathbf{M}_{UI}$—if a certain user has tagged an item, we set the corresponding entry to 1 and 0 otherwise—since frequency information in this matrix would not be fairly fruitful. By using these three matrices, folksonomy can be converted into a weighted tripartite graph $G = (U \cup T \cup I, E)$ whose nodes can be partitioned into three disjoint sets, U, T, and I, such that every node of each set is adjacent to at least one node in each of the two other sets. This graph G can be represented by an adjacency matrix $\mathbf{A}$:

$$\mathbf{A} = \begin{pmatrix} 0 & \mathbf{M}_{UT} & \mathbf{M}_{UI} \\ \mathbf{M}_{TU} & 0 & \mathbf{M}_{TI} \\ \mathbf{M}_{IU} & \mathbf{M}_{IT} & 0 \end{pmatrix} \tag{1}$$

Fig. 1 shows an example of a weighted undirected tripartite graph that represents folksonomy.

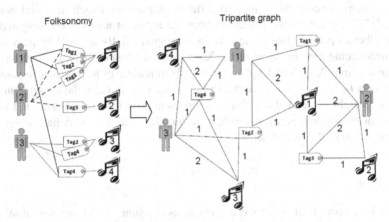

Fig. 1. Transforming folksonomy to a weighted undirected tripartite graph

3.2   Problem Formulation

Our study considers two types of tagging scenarios within social tagging systems: personalized tag recommendations and automatic tag annotations. Given the undirected tripartite representation described earlier, our problems can be viewed as the link-prediction problems.

In general, the main goal of the tag recommendation problem is to identify a set of tags suited to a given user-item pair (u, i). From a tripartite graph point of view, if user u is interested in tagging item i, then new link (u, i) between user u and item i appear in the graph. To recommend tags for a given link (u, i), our method attempts to predict the link (u, t) between user u and tag t, as well as to predict the link (i, t) between item i and tag t. Thereafter, our method calculates a ranking score of tag t based on such predicted links and thus generates a list of top N ranked tags suited to user u for item i. Formally, this problem is defined as follows: Given a folksonomy graph G and a pair of a user node $u \in U$ and an item node $i \in I$, identify an ordered set $T(u, i)$ of tag nodes that are most likely to appear with the pair (u, i) in the form of a triangle graph such that $|T(u, i)| \leq N$ and $T(u, i) \subseteq T$.

In the annotation problem, the objective is to uncover *latent tags* relevant to a given item, even if those tags have not been previously labeled to that item. To identify such tags from a tripartite graph viewpoint, our method attempts to predict the link (i, t) between item i and tag t that is likely to emerge from many user nodes.

4   Folksonomy Link Prediction

4.1   Proximity Estimation

Our study employs the Katz measure to estimate proximity between two nodes in the tripartite graph. The Katz measure is one of the most effective path-based measures that have been successfully applied to different applications such as social network analysis [12]. It measures the proximity between a pair of nodes via a weighted sum of the number of paths between the two nodes, exponentially damped by path length. Before introducing how to calculate the Katz score, we first define a *path* and its *path weight* in a graph. A path of length n, $n \geq 1$, from node x to node y is a sequence of directed links $l_1, l_2, \ldots, l_n$ such that the initial   node of l_1 is x and the terminal node of l_n is y. The weight associated to a link l_k is represented as W_k where $1 \leq k \leq n$. Let $P^n_{x,y}$ be the set of all possible paths of length n from node x and node y. A path weight of a particular path $p \in P^n_{x,y}$ is defined as:

$$PW^n_p(x, y) = \prod_{k=1}^{n} W_k$$

(2)

Then, the Katz score from x and y is calculated as the sum of path weights of all paths connecting the two with varying lengths:

$$K_{x,y} = \sum_{n=1}^{\infty} \alpha^n \times \left(\sum_{p \in P^n_{x,y}} PW^n_p(x, y) \right)$$

(3)

where $\alpha \in (0, 1)$ is an attenuation parameter (usually a small value, e.g. 0.05 or 0.005). The Katz score of all pairs of nodes in the graph can be expressed in matrix form as follows [5]:

$$\mathbf{K} = \alpha\mathbf{A} + \alpha^2\mathbf{A}^2 + \alpha^3\mathbf{A}^3 + \cdots = (\mathbf{I} - \alpha\mathbf{A})^{-1} - \mathbf{I}$$

(4)

where $\mathbf{I}$ is the identity matrix and $\mathbf{A}$ is an adjacency matrix for the graph. The series expansion converges if $\alpha < 1/\lambda_{max}(\mathbf{A})$, where $\lambda_{max}(\mathbf{A})$ is the largest absolute value of any eigenvalue of $\mathbf{A}$; thus this condition determines how large α can be [10]. Each entry in the Katz matrix $\mathbf{K}$ represents the sum of the path weights of all lengths from the corresponding row node to the corresponding column node in the graph. We exploit the resultant matrix $\mathbf{K}$ to compute the tag ranking score.

4.2   Tag Recommendation and Annotation

In this section we present our tag ranking method based on the Katz matrix $\mathbf{K}$. Since users may use different tags to describe the same item according to their own interpretation of the content or their tagging purposes, we should take account not only of individual tagging behavior, but also of social choices of tags [6]. To identify top-N suitable tags for a given user-item pair, we assess how a certain tag is in close proximity not only to the user, but also to the item. To that end, we compute a ranking score as a sum of Katz scores. Given a user u and an item i, a ranking score for a particular tag t is computed by:

$$score_{u,i}(t) = K_{u,t} + K_{i,t}$$
(5)

where $K_{u,t}$ and $K_{i,t}$ are the Katz scores from user u to tag t and from item i to tag t, respectively. According to the sum of Katz scores, the set of N ordered tags with the highest values are recommended to user u in regard to item i.

In the tag annotation task, we are interested in determining the relevant tags for an item. These relevant tags are not labeled to the item in the past, but can be of great benefit to users in retrieving more appropriate items. To discover such tags for a given item, we estimate how a specific tag is in close to the item. Given an item, therefore, the ranking score for a tag is equal to the Katz score from the item to a tag, simply defined as:

$$score_i(t) = K_{i,t}$$
(6)

where $K_{i,t}$ is the Katz scores from item i to tag t. Tags that obtain the highest scores, but have not previously annotated in the item, are regarded as tags that are most likely to be relevant to that item. We can succinctly express all ranking scores of tags in matrix form as:

$$s = v^T K$$
(7)

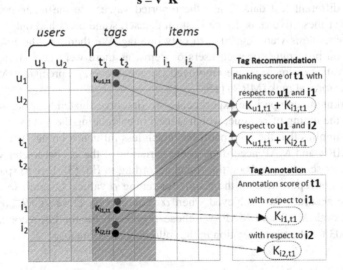

Fig. 2. The process of computing tag ranking scores based on the Katz measure

where $\mathbf{v}^T$ is a vector that has value ones at target entries—for example, user u and item i in the tag recommendation task while only item i in the annotation task—and zeros everywhere else. Fig. 2 illustrates an intuitive view of computing tag ranking scores.

5   Experimental Evaluation

The data used in our experiments was taken from Last.fm, a social music service that assists users to discover, tag, and share music. The Last.fm dataset[1] used in this study was collected by the Informational Retrieval Groups at Autónoma University of Madrid [2]. This dataset contains 186,479 tag assignments on 12,523 items (i.e., music artists) from 1892 users with 9749 tags. We also projected the tag assignments onto three two-dimensional matrices: 35,816 non-zero entries of the user-tag matrix (0.19% density), 71,064 non-zero entries of the user-item matrix (0.3% density), and 109,750 non-zero entries of the tag-item matrix (0.09% density).

5.1   Experimental Setup

To evaluate the performance of our proposed approach, we adapted the evaluation procedure described in [17]. In the tag recommendation scenario, for every user, we randomly eliminated one item and his/her tags assigned to that item from the training set. This eliminated set was used as the test set. In the tag annotation scenario, we withheld 20% of tags per item and subsequently used those as a test set. The remaining 80% of tags per item was used as a training set. We repeated this procedure five times with different test data. Thus, the reported values are the mean performance averaged over these five runs. In the Last.fm dataset, some users had only one tagged item and some items were tagged with only one tag. We therefore did not carry out the evaluation procedure for such users or items, as there were no training data for them. As evaluation metrics, we employed the mean average precision (MAP) [11] and mean reciprocal rank (MRR) [15].

Prior to running comparison experiments, we first investigated how our accuracy is sensitive to the tuning of α value, which is used in calculating the Katz score. A small value for α implies that the long paths has much less influence on the final score. We measured MRR and MAP according to the variation of the α parameter values because the Katz score depends on this value according to Eq. (3). Due to space limitation, we did not report here all the results for tuning α values. Overall, we identified that the best accuracy was achieved when α value is very close to $1/\lambda_{max}(\mathbf{A})$. Upon considering both MRR and MAP, we chose $\alpha = 0.006$ for the tag recommendation and $\alpha = 0.003$ for the tag annotation in the following experiments.

---

[1] http://www.grouplens.org/node/462

5.2   Evaluation of Recommendation

To evaluate the task of recommending personalized tags, we compare our method (denoted as *KaztTag*) with three other approaches: (i) User-Centric Tag Model (*UCTM*) [16], (ii) the *FolkRank* algorithm [9], and (iii) the most *Popular Tag* approach [9].

As stated earlier, the objective of this task is to suggest potentially useful tags to a user when he or she is asked to tag a particular item. In social tagging systems, it is often that some users are very active in utilizing tags while some other users use few tags. Accordingly, the recommendation performance for individual users could be affected by how many tags each user has used. To understand this impact on the performance, we divided users into five groups according to their tagging activities: (1) *Very Low* taggers who used less than 5 different tags; (2) *Low* taggers who used greater than or equal to 5 tags and less than 10 tags; (3) *Medium* taggers who used greater than or equal to 10 tags and less than 20 tags; (4) *Heavy* taggers who used greater than or equal to 20 tags and less than 40 tags; and (5) *Very Heavy* taggers who used greater than or equal to 40 tags. We then measured the MRR and MAP values with respect to each group of users. These evaluation measures help us see whether desirable tags are appearing at the very top of the ranked list. We note that on average users had approximately 3.5 tags in test data. Consequently, we considered on average the top-3 or top-4 ranked tags when calculating MAP, because the number of recommended tags for each test user depends on how many tags they assigned their test item.

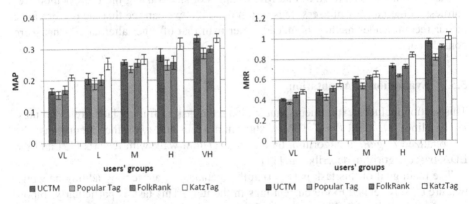

Fig. 3. The MAP and MRR result for tag recommendation at different groups of users

Fig. 3 shows the MAP and MRR results for different groups of users. Looking at the MAP and MRR results obtained with each method, we noticed that all algorithms were profoundly sensitive to the number of tags used by test users. It was determined that the more tags users used, the better recommendation quality they received. For instance, KatzTag achieved an MAP of 0.33 for the very heavy taggers while achieving an MAP of 0.21 for the very low taggers (an improvement of 12% on MAP). Similar improvements were also observed for the other methods.

We continued with comparisons to results obtained using each method within the same groups. In the case of UCTM, it tended to perform well particularly for users who had many tags, but provided users having few tags with poor recommendations. Similar to our method, FolkRank almost made good recommendations to the very heavy taggers. As for the popular tag method, it did evidence the worst performance on all occasions, thereby indicating that merely the popularity of tags annotated in a given item is not enough to fully reflect individual tagging behaviors. For the two evaluation metrics, KatzTag was consistently and clearly superior to the other methods in all cases of groups in terms of both MAP and MRR. That is, not only does our method perform well for the active users, but it is also capable of recommending more appropriate tags to the cold start users compared to the baselines. Since the recommendation problem for the cold start users is one of notable challenges in the field of recommender systems, the proposed approach can be beneficial to this problem.

Table 1. MRR and MAP results shown with standard deviations

Method	MAP±STD	MRR±STD
Popular Tag	0.212 ± 0.004	0.550 ± 0.003
UCTM	0.234 ± 0.006	0.619 ± 0.007
FolkRank	0.226 ± 0.004	0.654 ± 0.005
KatzTag	0.257 ± 0.004	0.695 ± 0.009

Table 1 shows overall MAP and MRR results obtained using the four methods regardless of the number of users' tags. From the table, we can see that the performance of our recommender method is indeed better than that of other alternatives that were considered.

5.3   Evaluation of Annotation

This section includes experiments that evaluate the task of finding latent relevant tags. As baselines, we used the *FolkRank* algorithm again and two other algorithms: the personalized *PageRank* algorithm based on a random walk with restarts [12] and the LDA-based algorithm described in [11].

The main goal of this task is to enrich the metadata of an item by adding new tags that are closely related to the current tags in the item. This task is particularly important to items being labeling with insufficient tags, so that such items can be more searchable, browsable, and shareable by users. We therefore focused on the performance regarding the number of tags labeled to items. To that end, we classified test items into three groups as follows: (1) *low* annotated items tagged with less than 4 different tags; (2) *medium* annotated items tagged with greater than or equal to 4 tags and less than 10 tags; and (3) *heavy* annotated items tagged with greater than or equal to 10 tags. One point that is worth noting is that our aim in this experiment was not to compare the results for one group with those for another group, due to the different sample size of the three groups and the different number of hidden test tags per group. We instead aimed to compare the performance of algorithms within each partitioned group.

Fig. 4 plots the MAP and MRR results for the low, medium, and heavy items. As shown by the results, it turns out that our approach is highly effective in finding relevant tags for the items labeled with sufficient tags, as well as in doing for the items with insufficient tags, as compared to the baseline methods. For example, we observed for the low annotated items that KatzTag improved MAP by 16.2% over the LDA, by 7.4% over the PageRank, and by 3.7% over the FolkRank, respectively. The results for the other groups validate the superiority of our method as well. Interestingly, the LDA method that models latent topics merely from the tag-item count matrix M_{TI} produced remarkably unsatisfactory performance, particularly in situations where items were being tagged with a few tags. This result may be explained by the fact that the LDA cannot observe explicit relations between users and tags, and between users and items from the tag-item matrix. On the other hand, the graph-based approaches that attempt to analyze the tripartite link structure A—PageRank, FolkRank, and KatzTag—appear to produce rather robust performance under "cold start" cases.

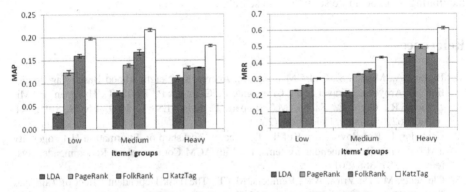

Fig. 4. The MAP and MRR result for tag annotation at different groups of items

We present a summary of the experimental results in Table 2. As shown, our method KatzTag outperformed the baseline methods for both MAP and MRR; overall, it obtained 12.7%, 6.9%, and 4.5% improvement on MAP compared to LDA, Page-Rank, and FolkRank, respectively. These experimental results support again that the proposed approach can be of great benefit to social media search area.

Table 2. MRR and MAP results shown with standard deviations

Method	MAP ± STD	MRR ± STD
LDA	0.075 ± 0.003	0.242 ± 0.007
PageRank	0.133 ± 0.002	0.343 ± 0.004
FolkRank	0.157 ± 0.003	0.351 ± 0.004
KatzTag	0.202 ± 0.002	0.440 ± 0.003

6   Conclusions and Future Work

In this paper we presented an approach to identifying relevant tags in social tagging systems. We model folksonomy as a weighted tripartite graph to capture the 3-dimensional data, i.e., users, tags, and items. We then apply the Katz measure to this tripartite graph for predicting link proximity between nodes via weighted sums over collections of possible paths connecting the nodes. Our experiments on Last.fm data demonstrate that not only can the proposed method accurately recommends/labels suitable tags for individual users or items, but it is also able to successfully position such tags at higher ranks. Additionally, our method is found to be fruitful in improving the performance for both the active taggers/heavy annotated items and the cold-start taggers/items, especially as compared to existing alternatives.

Though the Katz measure provides accurate proximity estimation, it is computationally expensive to calculate the ensemble of all paths between two nodes in large-scale graphs. We therefore need to study more efficient, scalable techniques that can approximate this proximity estimation for future work. Further experiments are also required in order to analyze how hidden tags identified via our method are useful for facilitating more accurate social media search.

References

1. Budura, A., Michel, S., Cudré-Mauroux, P., Aberer, K.: Neighborhood-Based Tag Prediction. In: Aroyo, L., Traverso, P., Ciravegna, F., Cimiano, P., Heath, T., Hyvönen, E., Mizoguchi, R., Oren, E., Sabou, M., Simperl, E. (eds.) ESWC 2009. LNCS, vol. 5554, pp. 608–622. Springer, Heidelberg (2009)
2. Cantador, I., Brusilovsky, P., Kuflik, T.: Second Workshop on Information Heterogeneity and Fusion in Recommender Systems. In: 15th ACM Conference on Recommender Systems, pp. 387–388 (2011)
3. Clements, M., De Vries, A.P., Reinders, M.J.T.: The Task-Dependent Effect of Tags and Ratings on Social Media Access. ACM Transactions on Information Systems 28(4), 21 (2010)
4. Diaz-Aviles, E., Georgescu, M., Stewart, A., Nejdl, W.: LDA for On-the-Fly Auto Tagging. In: 4th ACM Conference on Recommender Systems, pp. 309–312. ACM, New York (2010)
5. Foster, K., Muth, S., Potterat, J., Rothenberg, R.: A Faster Katz Status Score Algorithm. Computational & Mathematical Organization Theory 7(4), 275–285 (2001)
6. Fu, W.-T., Kannampallil, T., Kang, R., He, J.: Semantic Imitation in Social Tagging. ACM Transactions on Computer-Human Interaction 17(3), 12 (2010)
7. Hamouda, S., Wanas, N.: PUT-Tag: Personalized User-Centric Tag Recommendation for Social Bookmarking Systems. Social Network Analysis and Mining 1(4), 377–385 (2011)
8. Hotho, A., Jäschke, R., Schmitz, C., Stumme, G.: Information Retrieval in Folksonomies: Search and Ranking. In: Sure, Y., Domingue, J. (eds.) ESWC 2006. LNCS, vol. 4011, pp. 411–426. Springer, Heidelberg (2006)
9. Jäschke, R., Marinho, L., Hotho, A., Schmidt-Thieme, L., Stumme, G.: Tag Recommendations in Social Bookmarking Systems. AI Communications 21(4), 231–247 (2008)

10. Katz, L.: A New Status Index Derived from Sociometric Analysis. Psychometrika 18(1), 39–43 (1953)
11. Krestel, R., Fankhauser, P., Nejdl, W.: Latent Dirichlet Allocation for Tag Recommendation. In: 3rd ACM Conference on Recommender Systems, pp. 61–68 (2009)
12. Liben-Nowell, D., Kleinberg, J.: The link-prediction problem for social networks. Journal of the American Society for Information Science and Technology 58(7), 1019–1031 (2007)
13. Ramezani, M.: Improving Graph-based Approaches for Personalized Tag Recommendation. Journal of Emerging Technologies in Web Intelligence 3(2), 168–176 (2011)
14. Rendle, S., Schmidt-Thieme, L.: Pairwise Interaction Tensor Factorization for Personalized Tag Recommendation. In: 3rd International Conference on Web Search and Web Data Mining, pp. 81–90 (2010)
15. Vallet, D., Cantador, I., Jose, J.M.: Personalizing Web Search with Folksonomy-Based User and Document Profiles. In: Gurrin, C., He, Y., Kazai, G., Kruschwitz, U., Little, S., Roelleke, T., Rüger, S., van Rijsbergen, K. (eds.) ECIR 2010. LNCS, vol. 5993, pp. 420–431. Springer, Heidelberg (2010)
16. Wetzker, R., Zimmermann, C., Bauckhage, C., Albayrak, S.: I Tag, you Tag: Translating Tags for Advanced User Models. In: 3rd ACM International Conference on Web Search and Data Mining, pp. 71–80 (2010)
17. Zanardi, V., Capra, L.: Social Ranking: Uncovering Relevant Content using Tag-based Recommender Systems. In: 2nd ACM Conference on Recommender Systems, pp. 51–58 (2008)

Can You See It? Two Novel Eye-Tracking-Based Measures for Assigning Tags to Image Regions

Tina Walber[1], Ansgar Scherp[1,2], and Steffen Staab[1]

[1] Institute for Web Science and Technology, University of Koblenz-Landau, Germany
http://west.uni-koblenz.de
[2] Research Group on Data and Web Science, University of Mannheim, Germany
{walber,scherp,staab}@uni-koblenz.de
http://dws.informatik.uni-mannheim.de

Abstract. Eye tracking information can be used to assign given tags to image regions in order to describe the depicted scene in more details. We introduce and compare two novel eye-tracking-based measures for conducting such assignments: The segmentation measure uses automatically computed image segments and selects the one segment the user fixates for the longest time. The heat map measure is based on traditional gaze heat maps and sums up the users' fixation durations per pixel. Both measures are applied on gaze data obtained for a set of social media images, which have manually labeled objects as ground truth. We have determined a maximum average precision of 65% at which the segmentation measure points to the correct region in the image. The best coverage of the segments is obtained for the segmentation measure with a F-measure of 35%. Overall, both newly introduced gaze-based measures deliver better results than baseline measures that selects a segment based on the golden ratio of photography or the center position in the image. The eye-tracking-based segmentation measure significantly outperforms the baselines for precision and F-measure.

Keywords: Fixation measures, automatic segmentation, heat maps.

1 Introduction

The understanding of image content is still a challenge in automatic image processing. Often, tags are used to manually describe images. Another approach is to analyze the text surrounding an image, e.g., on web pages, to draw conclusions about the depicted scene. A better understanding of the objects depicted in an image can improve the handling of images in many ways, e.g., by allowing similarity search based on regions [8] or by serving as ground truth for computer vision algorithms [11]. It is intuitive for humans to identify objects depicted in an image. The human perception system can compensate perspective distortions, occlusions and can also identify objects with an unusual appearance. These adaptions of the perception system are hard tasks for algorithms and have not yet been solved.

S. Li et al. (Eds.): MMM 2013, Part I, LNCS 7732, pp. 36–46, 2013.

The idea of our work is to benefit from human abilities to perceive visual information in order to obtain a better understanding of depicted scenes. We notice a rapid development of sensor hardware (cameras) in devices like laptops and a decreasing of cost for hardware. Extrapolating this development into the future, eye tracking will be more widely available and can be performed using standard sensors like web cameras [12]. In this work, we investigate two new eye-tracking-based measures with regard to their capability of assigning a given tag to a region in an image such that a depicted object is correctly labeled. For this purpose, we have investigated how efficient measures applied on eye fixations may serve the region labeling task. Fixations are the phases in the gaze trajectories when the eyes are fixating a single location. The first measure is the *eye-tracking-based segmentation measure*. It is based on a standard image segmentation algorithm [2] and selects the image segment as most relevant for the given tag which the user fixates on for the longest time interval. The second measure is the *eye-tracking-based heat map measure*. It is based on a traditional heat map and sums up the duration of the fixations.

We compare the two new eye-tracking-based measures with two baseline measures. The baselines also make use of automatically computed segments, but not of additional information. The eye tracking data for our investigations is taken from a controlled experiment conducted with 30 subjects each viewing 51 social media images with given tags. The experiment is presented in [14]. First, the subjects where shown a specific tag. Subsequently, we have recorded their gaze path while they viewed the image and while they had to decide whether an object referring to that tag was depicted or not. The social media images have as ground truth manually labeled objects. We have used this experimental data to tackle the following core research questions:

- To which extent may the two new eye-tracking-based measures identify the correct position in the image for a given tag (maximum precision)?
- To which extent does the area determined by the two new measures cover the actual object depicted in the image (maximum F-measure)?

We show that the segmentation measure performs better for both questions, although the difference to the heat map measure is not significant. The segmentation measure delivers significantly better results for precision and F-measure than the baseline approaches.

In the subsequent section, we discuss the related work. In Section 3, we describe our two novel eye-tracking-based measures and the baselines. In Section 4, the experiment is described from which we have obtained the eye tracking data. The examination of the best parameters determined on a subset of the images is presented in Section 5 followed by the results obtained from our experiments in Section 6.

2   Related Work

Yarbus [15] has already shown in 1967 that image content strongly influences eye movements. The tendency of humans to fixate faces in images is well known and also the identification of parts of the faces from gaze paths can be performed [4]. Klami [9] investigates which parts of images are relevant for a user in a given task. In his work, relevance is calculated only from the gaze information and it is represented in a Gaussian mixture model, which resembles heat maps. The work reveals that the visual attention depends on the task given to the subject before viewing an image. The work of Ramanathan et al. [10] aims at localizing affective objects and actions in images by using gaze information. Areas that are affecting the users' attention are identified and correlated with given concepts from an affection model. The affective image regions are identified using segmentation and recursive clustering of the gaze fixations. General identification of image regions showing specific objects like it is aimed at in this work is not conducted. In a previous work [14], we have investigated the possibilities to assign tags to image regions, where these regions were manually labeled with hand-drawn polygons. Gaze paths of users looking at the images were analyzed by 13 different fixation measures to calculate the assignment. A tag was assigned to a correct image region for 63% of the image-tag-pairs.

Essig [6] takes user-relevance feedback, gained from gaze information, into account to improve the content-based image search. The feedback is calculated on the basis of image regions. He showed that the retrieval results of his approach received significantly higher similarity values than those of the standard approach, which is based only on automatically derived image features. Bartelma [3] investigated the combination of gaze control and image segmentation. He has implemented a system that is controlled by gaze to manually segment images. The gaze is exclusively used as a mouse replacement. The subjects were instructed to outline a given object with their gaze. Santella et al. [13] present a method for semi-automatic image cropping using gaze information in combination with image segmentation. Goal is to find the most important image region, independent of the objects in the image. Their work shows that the image cropping approach based on gaze information is preferred by the users to fully automatic cropping in 58.4% of the cases.

The related work shows that eye tracking information is exact enough to be used on the level of image regions and that this information can be of value in several use cases. To the best of our knowledge, no work is done on assigning given tags to image regions by using gaze information without a given ground truth segmentation.

3   Identifying Objects in Images

We suggest two methods for assigning tags to image regions, thus identifying objects that correspond to a predefined tag. Both methods, as well as the baseline methods, proceed using the following input:

- An image I is a set of pixels $P(x, y)$, $0 \leq x < width$, $0 \leq y < height$
- A tag t, describing an object depicted in I
- A set of users U that have viewed the images during the experiment

- Set of gaze paths provided by users $u \in U$, to which the tag t was shown and who had to decide whether an object described by t can be seen in the image or not

Gaze paths consist of fixations and saccades. Fixations F are short stops that constitute the phases of the highest visual perception, while saccades are quick movements between the fixations. Every gaze path G_t consists of a set of fixations F, provided by user $u \in U$. Every fixation $f = (x_f, y_f, d)$ is described by a fixated point in the image (x_f, y_f) and a duration d. To measure the human visual attention, the fixations are analyzed by so called fixation measures. From these measures, a value ν is calculated for given regions R of an image I. Example eye tracking measures are the fixationCount, a standard measure which counts the number of fixations on a region and the lastFixationDuration, which sums up the duration of the last fixation on an image region. We have compared 13 fixation measures with respect to their ability to identify a concrete image region for a tag t given to the users [14]. Derived from the results of this work, we use the measure lastFixationDuration, which has delivered the best results.

Subsequently, we present the two novel eye-tracking-based measures and the baseline measures; we also describe the method for evaluating the proposed eye-tracking-based measures.

Eyetracking-Based Segmentation Measure: The idea of this approach is to calculate ν for the fixation measure lastFixationDuration for all regions $r \in R$ gained from an automatically segmented algorithm. $\nu(r, u)$ is calculated for every user $u \in U$ viewing the image. The values ν are summed up for every region over all users and the favorite region r_{fav} is determined by the highest value:

$$r_{fav} = \arg\max_{r \in R} \sum_{u \in U} \nu(r, u) \tag{1}$$

Eyetracking-Based Heat Map Approach: Heat maps are two-dimensional graphical representations of a number of gaze information. They visualize the frequency of fixations for every pixel $P = (x, y)$ in an image. Different colors symbolize how many times or how long a pixel was fixated. The advantage of heat maps is that they can summarize a large quantity of data and are easy to comprehend by humans. Thus, they are often used in usability experiments to visualize users' attention. Different kinds of heat maps can be created based on different measures, e.g., a fixationCount or an absoluteDuration heat map [5]. As the lastFixationDuration was the best measurement for the region identification in our previous work [14], we use this measure as basis for our approach. A radius rd has to be defined for the creation of a heat map. We use a default value of 50 pixels, taken from Tobii Studio [1]. A maximum value of $h_{max} = 100$ is assigned to the pixel fixated by a fixation $f = (x_f, y_f, d)$. Starting from this point, values are added to the pixel in the surrounding of the fixation, based on a linear interpolation between h_{max} and 0. The result is multiplied by the fixation duration d. An example is visualized in Figure 1. For a single fixation, we calculate the heat map values h of all pixels $P = (x, y)$ in the surrounding of the fixation:

$$h(P, f) = \begin{cases} d * (h_{max} - (dist(P, f) * \frac{h_{max}}{rd})) & \text{, if } dist(P, f) \leq s \\ 0 & \text{, otherwise} \end{cases} \quad (2)$$

All last fixations f_{last} of all gaze paths provided by the users $u \in U$ are summed up in the final heat map H:

$$H(P) = \sum_{u \in U} h(P, f_{last}) \quad (3)$$

From all heat map values H, the highest value $max(H)$ is determined. To obtain the favorite region from the heat map, we set a threshold $0 < t \leq 100\%$. For example a $t = 5\%$ means that only heat map values are considered that belong to the highest 5% of all values. This procedure can be described by an analogy of a flooded region with valleys and elevations. The threshold t symbolizes the water level. With a level of $t = 5\%$, only the highest 5% of the landscape are visible above the water level or in our case all pixels with $H(P) > 0.95 * max(H)$ are determined as possible favorite regions. The biggest area of connected pixels is selected as favorite region r_{fav}. An illustration of this thresholding is presented in Figure 7.

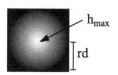

Fig. 1. Heat Map Values

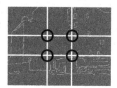

Fig. 2. Golden Sections

Baseline Approach: Initially, we had investigated a random baseline approach as used in our previous work [14], which is randomly selecting one segment of an automatically segmented image as favorite region. As the results of this baseline were very weak, we decided to improve the baseline approach by taking into account the position of the segments in the image in two different ways. As the pictures used in our analysis are taken by humans, we can suppose an inherent photographic bias. The golden ratio rule is a very basic rule in photography [7]. Taking images based on this rule can improve the aesthetics of a photograph and it is often met instinctively to achieve aesthetically appealing pictures. According to the golden ratio, width and height of an image are divided into two parts in the ratio 1 to 1.618. This results into four intersections, at which important objects in the images are often placed. In Figure 2, the golden sections are highlighted by black circles. Another typical bias is to position the important object in the center of the image. For each picture, the golden ratio and the center baselines are calculated. The segment placed at the golden section respectively the center point is selected as favorite region r_{fav}.

Evaluation Method: After obtaining favorite regions with one of the two new measures or the baseline measures, the results have to be evaluated by means of comparing them with ground truth object labels. In information retrieval, precision, recall, and F-measure are standard approaches to measure the relevance of search results. We use these measures to evaluate the covering of the ground truth object region r_{gt} by the favorite region r_{fav} at pixel level. The algorithm runs through the image and classifies every pixel as tp (true positive), fp (false positive), fn (false negative), and tn (true negative) as described in Figure 3.

		r_{gt} from the ground truth image	
		Pixel belongs to r_{gt}	Pixel does not belong to r_{gt}
r_{fav} calculated from heat map, segmentation or baseline measure	Pixel belongs to r_{fav}	**tp**	**fp**
	Pixel does not belong to r_{fav}	**fn**	**tn**

Fig. 3. Definition of tp, fp, fn, and tn

4   Experimental Data

This work is based on the data gained in an experiment described in [14]. More details about the experiment setup, the subjects, and the used data set, can be found there. The experiment data was gained in a controlled experiment, performed with 30 subjects organized in three groups. The experiment was performed on a screen with a resolution of 1680x1050 pixels while the subjects' eye movements were recorded with a Tobii X60 eye-tracker at a data rate of 60Hz and an accuracy of 0.5 degree.

The gaze data of the first two groups are used for parameter fitting, while the data from third group is used to verify the results of our measures. The experiment sequence consisted of three steps conducted for each image: First, a tag was presented to the subjects with the experiment task "Can you see the following thing in the image?". After pressing a button, users had to fixate a small blinking dot in the upper middle for one second. In a third step, the image was shown to the subjects. Viewing the image, the subjects had to judge whether the tag shown in the first screen would have an object counterpart in the image or not by pressing the "y" (yes) or "n" (no) key. We used images from LabelMe[1] with 182.657 user contributed images (download August 2010) to create three sets of images I, one for each group of subjects. The LabelMe community has manually created image regions by drawing polygons over the images and tagging them. The labels were used as tags t and the polygons as ground truth image segmentation. For every image selected, we randomly chose

---

[1] http://labelme.csail.mit.edu/

a "true" (describing an object in the image) or "false" tag. About 50% of the given tags corresponded to an object displayed in the image ("true" tag), while the other half did not. In our analysis of the gaze data, we consider only data belonging to images with a given "true" tag and a correct answer by the user.

5   Determining Best Parameter Settings

The data set is split into two subsets: a training set for the parameter fitting (56 images-tag-pairs each viewed by 10 users) and a test set for the evaluation of the approaches (29 images-tag-pairs each viewed by 10 users). In this section, we investigate different parameters for our approaches and identify the parameters leading to the best results. The outcome is applied to the test data set and used to compare the different measures from Section 6.

Eye-Tracking-Based Segmentation Measure: The segmentation is performed by using the bPb-owt-ucm algorithm [2]. Different hierarchy levels for $k = 0 \ldots 1$ are calculated, each representing a different level of detail. An example is presented in Figures 4, showing the segmentation results for different k-values. The first segmentation level $k = 0$ delivers 1831 segments, the segmentation with $k = 0.4$ the least number of segments, namely six.

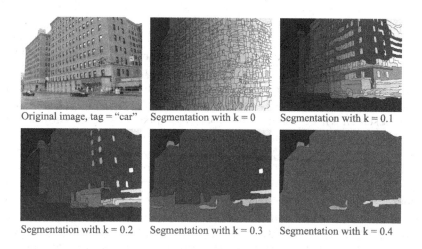

Original image, tag = "car"   Segmentation with k = 0   Segmentation with k = 0.1

Segmentation with k = 0.2   Segmentation with k = 0.3   Segmentation with k = 0.4

Fig. 4. Segmentations with Different Parameters k

Applying eye-tracking-based segmentation measures to those segmentations provides the favorite region r_{fav} from all segments, as described in Section 3. In Figure 5(a), an example for a gaze path of a single user is shown. The fixations are displayed as circles, the fixation duration is presented by the diameter of the circles. The saccades are depicted as lines between the fixations. The brightness of the image segments encodes the fixation measure values $\nu(R)$. The order of the viewed regions is encoded from the favorite region in white to the segments with few fixations in dark gray. The black segments have not been fixated at all. Figure 5(b) shows the results for one image aggregating the gaze paths of all users. To determine the best hierarchy level k, we have compared the results for different levels $k = 0 \ldots 1$ by calculating precision, recall, and F-measure. For $k > 0.4$, the number of segments is too low to obtain a reasonable favorite region r_{fav}. Basically the result is a very large segment, covering almost the entire image plus a few very small segments.

(a) Gaze path with inter-sected regions for $k = 0.2$

(b) Favorite regions over all users for $k = 0.2$

Fig. 5. Identification of r_{fav} for one user (a) and aggregated for 10 users (b)

The results for all investigated k values are depicted in Figure 6(a). The best precision with 50% is obtained for the smallest sizes of segments for $k = 0$. The best recall with 54% for $k = 0.4$. The maximum F-measure of 25% is reached with $k = 0.1$. It is calculated from a precision of 4% and a recall of 34%. One can see that the F-measure is relatively stable between the $k = 0.1$ and $k = 0.4$, because of the rising recall and the falling precision values.

Eye-Tracking-Based Heat Map Measure: For the heat map measure, described in Section 3, we have investigated different thresholds $t = 1 \ldots 100\%$. Some examples are depicted in Figure 7. It shows the original image, next to a

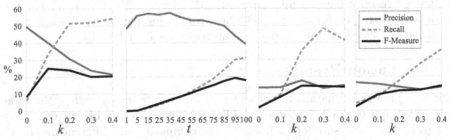

(a)   Segmentation measure with k

(b) Heat map measure with parameter t

(c) BL   Golden Section with k

(d) BL   Center with k

Fig. 6. Precision, recall, and F-measure for the two gaze-based and the two baseline measures (BL)

classical heat map visualization of gaze information from all 10 users. The next four images show different potential favorite areas after applying the threshold t to the heat map. If several areas appear, the biggest one (i.e., the one with the most pixels) is supposed to be the favorite region r_{fav}.

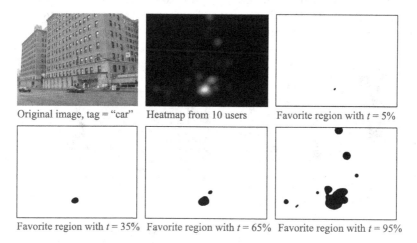

Original image, tag = "car"   Heatmap from 10 users   Favorite region with $t = 5\%$

Favorite region with $t = 35\%$ Favorite region with $t = 65\%$ Favorite region with $t = 95\%$

Fig. 7. Visualization of the Heat Map Measure

Precision and F-measure are calculated, comparing the computed favorite region r_{fav} with the ground truth object region r_{gt}. An overview of the results is presented in Figure 6(d). The highest precision value is obtained for $t = 35\%$ with 57%. Even with constantly high precision values of more than 44% the F-measure values cannot get very high because of the poor recall results (maximum: 31%). The best F-measure result is 19% with $t = 95\%$.

Baseline Measures: For the baseline measures, we also compute the segmentation using the bPb-owt-ucm algorithm [2]. For both baselines, we investigate the best parameters $k = 0 \ldots 0.4$. For the golden section baseline, we obtain the highest precision value over all images with 18% for $k = 0.2$ and the highest the F-Measure with 14% for $k = 0.2$. The best results for the center baseline are a precision of 16% for $k = 0.1$ and a F-Measure of 13 % for $k = 0.4$.

6   Evaluation Results

The best performing parameters from the training data set for each of the measures are applied to the test data set. For each measure, we obtain values for precision and F-measure for each image. For comparing the different measures, we have conducted a Kolmogorov-Smirnov test to determine if the precision values and F-measure values exhibit a normal distribution. As most of our computed values do not exhibit a normal distribution (details of the test results omitted for brevity), we have conducted a Friedman test to investigate for a statistical significance in the difference of the obtained precision values and F-measure values. We found that the differences between the four assignment measures (segmentation, heat map, and two baselines) are significant ($\alpha < .05$) for precision ($\chi^2(3) = 32.668$, $p = .000$) and F-measure ($\chi^2(3) = 15.891$, $p = .001$). Thus, post-hoc analyses with pairwise Wilcoxon Tests are conducted with a Bonferroni correction for the significance level (now: $\alpha < .017$). The values used in the pairwise Wilcoxon Tests are presented in Figure 8. We obtain the best precision with 65% for the segmentation measure and the second best with 48% for the heat map measure. These results significantly outperform the two baselines with $Z = -4,059, p = .000$ for the segmentation measure compared to the golden section baseline, respectively $Z = -4,090, p = .000$ for the center baseline. The results for the heatmap measure are $Z = -3,438, p = .001$ and $Z = -3,286, p = .001$, respectively. There is a weakly significant difference between the two eye-tracking-based measures ($Z = -1.905, p = .057$). For 12 of 29 images, r_{fav} lies completely inside r_{gt}. For 20 images at least 1% of r_{fav} intersects the ground truth object region r_{gt}. The highest F-measure is obtained again by the segmentation measure with 35%. The result for the heat map measure is 22% and for the baselines 11% (golden section) and 14% (center). A significant difference is recognized between the segmentation measure and the baselines with $Z = -2,943, p = .003$ for both baseline. The other results do not differ significantly (segmentation - heat map: $Z = -.934, p = .350$, heat map - golden section: $Z = -2,345, p = .019$, heat map - center: $Z = -2,186, p = .029$).

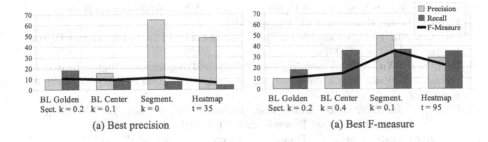

(a) Best precision   (a) Best F-measure

Fig. 8. Comparison of the two gaze-based measures and the baseline

7   Conclusion

For 63% of the images, we were able to identify the correct image region, described by a given tag. The assignment of tags to regions becomes much harder without the given, manually created regions like the automatically computed segments considered in this work. The reason lies in the inaccuracies involved with the automatic image segmentation. However the results obtained from the eyetracking-based measures are still very good, with an average precision of 65% over all images for the segmentation-based measure and 48% for the heatmap-based measure. The eyetracking data in this work was gained in a controlled experiment where users had to identify regions for predefined, given tags (see Section 4). To relax this constrain, we will conduct in a next step an experiment with users tagging images using an application in the style of a real online image annotation tool like Flickr. We have introduced a segmentation measure and heat map measure that both use gaze information as source of information. The results show that the new measures perform better in the assignment of tags to image regions than a baseline approach without gaze information. The segmentation measure performs best for both evaluations: precision and F-measure. The segmentation measure significantly outperforms the baseline. The segmentation measure can easily be adapted to different needs by modifying the parameter k to maximize the precision or F-measure.

References

1. Tobii studio 2.x - user manual (2010), http://www.tobii.com
2. Arbeláez, P., Maire, M., Fowlkes, C., Malik, J.: Contour detection and hierarchical image segmentation. IEEE TPAMI 33(5), 898–916 (2011)
3. Bartelma, J.M.: Flycatcher: Fusion of gaze with hierarchical image segmentation for robust object detection. PhD thesis, Massachusetts Institute of Technology (2004)
4. Belle, W.V., Laeng, B., Brennen, T., et al.: Anchoring gaze when categorizing faces sex: Evidence from eye-tracking data. Vision Research 49(23), 2870–2880 (2009)
5. Bojko, A.: Informative or misleading? heatmaps deconstructed. In: Human-Computer Interaction. New Trends, pp. 30–39 (2009)
6. Essig, K.: Vision-Based Image Retrieval (VBIR)-A New Approach for Natural and Intuitive Image Retrieval. PhD thesis (2008)
7. Freeman, M.: The Photographer's Eye: Composition and Design for Better Digital Photos. Focal Press (2007)
8. Kim, D.H., Yu, S.H.: A new region filtering and region weighting approach to relevance feedback in content-based image retrieval. Journal of Systems and Software 81(9), 1525–1538 (2008)
9. Klami, A.: Inferring task-relevant image regions from gaze data. In: Workshop on Machine Learning for Signal Processing. IEEE (2010)
10. Ramanathan, S., Katti, H., Huang, R., Chua, T., Kankanhalli, M.: Automated localization of affective objects and actions in images via caption text-cum-eye gaze analysis. In: Multimedia. ACM, New York (2009)
11. Russell, B.C., Torralba, A., Murphy, K.P., Freeman, W.T.: LabelMe: a database and web-based tool for image annotation. J. of Comp. Vision 77(1), 157–173 (2008)
12. San Agustin, J., Skovsgaard, H., Hansen, J.P., Hansen, D.W.: Low-cost gaze interaction: ready to deliver the promises. In: CHI, pp. 4453–4458. ACM (2009)
13. Santella, A., Agrawala, M., DeCarlo, D., Salesin, D., Cohen, M.: Gaze-based interaction for semi-automatic photo cropping. In: CHI, p. 780. ACM (2006)
14. Walber, T., Scherp, A., Staab, S.: Identifying Objects in Images from Analyzing the Users' Gaze Movements for Provided Tags. In: Schoeffmann, K., Merialdo, B., Hauptmann, A.G., Ngo, C.-W., Andreopoulos, Y., Breiteneder, C. (eds.) MMM 2012. LNCS, vol. 7131, pp. 138–148. Springer, Heidelberg (2012)
15. Yarbus, A.L.: Eye movements and vision. Plenum (1967)

Visual Analysis of Tag Co-occurrence
on Nouns and Adjectives

Yuya Kohara and Keiji Yanai

Department of Informatics, The University of Electro-Communications, Tokyo
1-5-1 Chofugaoka, Chofu-shi, Tokyo, 182-8585 Japan
{kohara-y,yanai}@mm.cs.uec.ac.jp

Abstract. In recent years, due to the wide spread of photo sharing Web sites such as Flickr and Picasa, we can put our own photos on the Web and show them to the public easily. To make the photos searched for easily, it is common to add several keywords which are called as "tags" when we upload photos. However, most of the tags are added one by one independently without much consideration of association between the tags. Then, in this paper, as a preparation for realizing simultaneous recognition of nouns and adjectives, we examine visual relationship between tags, particularly noun tags and adjective tags, by analyzing image features of a large number of tagged photos in social media sites on the Web with mutual information. As a result, it was turned out that mutual information between some nouns such as "car" and "sea" and adjectives related to color such as "red" and "blue" was relatively high, which showed that their relations were stronger.

1   Introduction

In recent years, due to the wide spread of digital cameras and mobile phones with camera, the amount of images on the Web has increased explosively. At the same time, because photo sharing sites such as Flickr have become common where users post their images with tags, there are so many tagged images on the Web. These tag information are used as a keyword on image search. However, most of the tags are added one by one independently without much consideration of association between the tags. This sometimes causes irrelevant results when we do AND-search with multiple keywords. For example, we obtain a photo showing blue sky and a red car for the query with "blue AND car". To remove such a photo and to obtain only the photos including blue cars, simultaneous image recognition of multiple tags such as "blue cars" is needed. If we can automatically eliminate irrelevant images, search results for multiple keywords become more correct. In addition, we can create dataset easily with less noise. In order to perform more accurate image acquisition and image search, it is necessary to take into account the relationship between the tags and to focus more on contents of images. Then, in this paper, we analyze visual relationship between nouns and adjectives using a large number of tagged images in social media sites on the Web such as Flickr. To do that, we use entropy and mutual information based on

S. Li et al. (Eds.): MMM 2013, Part I, LNCS 7732, pp. 47–57, 2013.
© Springer-Verlag Berlin Heidelberg 2013

visual features extracted from image regions. Moreover, the result of analysis in this paper can be used in simultaneous recognition where we recognize a certain object by a noun and the state of the object by an adjective further like "there is a car and the color of the car is red".

In the rest of this paper, we describe related work in Section 2. We explain the overview in Section 3 and the detail of the proposed system in Section 4. We show experimental results in Section 5. In Section 6, we conclude this paper.

2   Related Work

In the community of object recognition, recently, recognition of attributes of objects such as adjectives are paid attention to. In this paper, we are inspired by the current trends of the object recognition research, and focus on "attributes" in the tags of social media photos.

T. L. Berg et al.[1] focused on attributes on color, shape, and texture. They extracted words associated with the attributes from the texts which were listed in the shopping site, and labeled local regions represented by attributes corresponding text description.

D. Parikh et al.[2] focused on the attributes from the perspective of "nameable". "Nameable" means whether human can understand and represent attributes automatically extracted from images by language. They discovered attributes of "nameable" by an interactive approach by using Amazon Mechanical Turk.

A. Farhadi et al.[3] described images by a set of attributes. They recognized not only "dog" but "spotty dog". By using attributes, we are able to describe "dog" which has "spots" as "spotty dog", when we have no knowledge about the subordinate categories of "dog". In addition, it has become possible to mention also distinctive attributes by using this description. That is, if attributes which "dog" has but "sheep "does not have exists, they will be the attributes which discriminate a dog from a sheep. Moreover, discovering distinctive attributes enables us to mention the attributes which should exist or shouldn't exist in each of the given classes. Therefore, a car whose doors are not visible can be recognized as a car with doors. In addition, they also estimated bounding box regions to which the given attributes correspond.

S. Dhar et al.[4] focused on the particular attributes of aesthetics and interestingness. They dealt with two types of attributes about composition and contents of objects in a given image as a guideline of aesthetics and interestingness. The attribute of composition contains saliency, location, and color of objects. The attributes of contents contain type, place, and scene where objects are shown. In this research, they mainly focused on subjective attributes.

The paper [1,2] recognized a single attribute, and the paper [3] dealt with attributes as parts which should exist in the corresponding objects like "a door is a part of a car", while we limit attributes to only adjectives, and also focus on the relationship between nouns and adjectives. The paper [4] analyzed the specific adjectives such as aesthetics and interestingness, while we dealt with more general adjective.

Next, we introduce related works on visual concept analysis. Here, we cite the papers of Yanai et al. [5], Akima et al. [6], and Kawakubo et al. [7]. Yanai et al. proposed the entropy as a way to quantify the relation of visual concept, and referred to the visual relation about 150 adjectives [5]. We use the method of quantifying the visualness of the word by entropy and calculation of entropy. Akima et al. built a database with hierarchical structure of between concepts from distance relationship and hierarchical relationship [6]. They used entropy and calculate the distribution of the images to determine the hierarchical relationship. In addition, tag information which is given to the images was also used. Kawakubo et al. analyzed the visual and geographical distribution in word concepts [7]. In this research, they calculated the image distribution of the class of concepts such as a noun or an adjective to evaluate visualness by using calculation of entropy and region segmentation. The difference between this research and the above-mentioned works is that we define concept classes with the combinations of two words, and pursuit the visual relation of the combinations of nouns and adjectives.

3   Overview

In this paper, the visual relationship between a nouns and an adjective is evaluated by the distribution of the image features of the corresponding image regions. We judge that there is a high relation if visual distribution is narrow enough. The extent of the distribution is quantified using the concept of entropy. Entropy is used to calculate the local features obtained from a set of image regions. Mutual information is the difference between entropy of a noun and entropy of combination with an adjective and the noun. Mutual information becomes higher, when visual relation between a noun and an adjective becomes higher.

Processing procedure of the experiment in this paper is shown below.

Procedure

1. Image acquisition from the Flickr by tag-based search
2. Image segmentation
3. Feature extraction and creating BoF for each region
4. Positive region detection
5. Calculation of feature distribution in each positive region by PLSA
6. Calculation of entropy and mutual information

In this paper, we also calculated similarity by co-occurrence of tags by the Normalized Google Distance (NGD) for comparison with the visual relation by entropy.

4   Proposed Method

In this section, we describe the methods used in the experiment. In this experiment, we calculated the entropy to refer to visual relation between an adjective and a noun. In addition, we calculated the similarity by co-occurrence of tags for comparison.

4.1   Image Acquisition

We collect 200 positive tagged images for each class by using the Flickr API. At the time, we use AND-search with a noun word and an adjective word. In addition, we prepare the 800 negative images which have neither of the tags. Note that we collected only one image from the same Flickr contributor for one query, since the same user tends to upload many near-duplicated photos, which sometimes causes irrelevant bias on feature distribution.

4.2   Region Segmentation

To select regions directly related to the given words and remove background regions, we perform region segmentation with JSEG [8]. In the experiment, we set the number of the maximum regions as 10, and carry out post-processing to unify relatively smaller regions into larger regions.

4.3   Feature Extraction

As feature representation of each region, we use Bag-of-Features (BoF) [9] with Color-SIFT [10].

At first, we extract the Color-SIFT feature as local features. To extract Color-SIFT, we extract the SIFT [10] features from each of the color channels such as R, G and B, regarding each keypoint, and create new feature vectors by concatenating SIFT vectors of the three channels. Therefore, the dimension size of this feature vector is 128×3. When extracting Color-SIFT features, we use dense sampling where all the local features are extracted from multi-scale grids.

Bag-of-Features (BoF) [9] is a standard feature representation to convert a set of local features into one feature vector. To convert a set of local feature vectors into a BoF vector, we vector-quantize them against the pre-specified codebook. After that, all the BoF vectors are L1-normalized. In the experiments, we built a 1000-dim codebook by k-means clustering with local features sampled from all the images. Note that we construct a BoF vector for each region by using the local features extracted inside the corresponding region.

4.4   Positive Region Selection

To select positive regions with the region-based BoF vectors, we use mi-SVM [11] which is a method of multiple instance learning. The mi-SVM is a support vector machine modified for multiple instance setting, and it is carried out by iterating a training step and a classification step using a standard SVM.

Under the multiple instance setting, training class labels are associated with a set of instances instead of individual instances. A positive set, which is called as a "positive bag", has one positive instance at least, while a negative set, which is called as a "negative bag", has only negative instances. This multiple instance setting fits well with the situation where an image consists of several foreground regions and background regions. Since we can regard foreground and background regions as positive and negative instances, respectively, by using multiple-instance learning methods we can classify regions into either foregrounds or backgrounds without explicit knowledge on foregrounds.

The process of positive region selection is shown below.

The process of positive region selection

1. Initially, regard all the regions in the positive images as positive instances, and regard all the regions in the negative images as negative instances.
2. Train a standard SVM using the positive and negative instances.
3. Classify all the instances in the positive images with the trained SVM.
4. Regard the instances assigned the higher scores by the SVM as positive instances in the next step, and regard the instances having the lower scores as negative instances in the next step.
5. Repeat from 2 through 4 several times.

4.5 Calculation of Feature Distribution

To calculate the feature distribution, we perform probabilistic clustering of feature vectors using the Probabilistic Latent Semantic Analysis (PLSA) [12].

pLSA. The calculation of pLSA is performed as follows: First, the joint probability of an image and an element of the BoF vector, which corresponds to "visual words", are represented as

$$P(d_i, w_j) = \sum_{k=1}^{K} P(d_i|z_k)P(w_j|z_k)P(z_k) \tag{1}$$

where $d_i(i = 1, 2, \ldots, I)$ is an image, $w_j(j = 1, 2, \ldots, J)$ is an element of BoF feature vectors (visual word frequency), and $z_k(k = 1, 2, \ldots, K)$ is a latent topic variable. Then, the probability that the word will be generated within the document is given by

$$P(w_j|d_i) = \sum_{k=1}^{K} P(w_j|z_k)P(z_k|d_i) \tag{2}$$

using latent topic variable z_k. In addition, if the number of the word w_j within the document d_i is defined as $n(d_i, w_j)$, the log-likelihood of the data is represented as the following expression:

$$L = \sum_{i=1}^{I} \sum_{j=1}^{J} n(d_i, w_j) \log P(d_i, w_j) \tag{3}$$

We determine $P(z_k)P(d_i|z_k)$ and $P(w_j|d_i)$ such as to maximize this log-likelihood by the iterative EM algorithm.

4.6 Calculation of Entropy and Mutual Information

The entropy was calculated using the probability obtained by the pLSA. The value of the entropy increases, when the distribution of the BoF feature vectors corresponding to the positive regions becomes wider. On the other hand, the entropy value decreases, when the distribution becomes narrower. Therefore, the value of the entropy represents the size of the image distribution belonging to a given class concept which is the combination of a noun and an adjective. That is, calculating the entropy leads to searching for visual relation on the combinations of nouns and adjectives.

We calculate the entropy based on $P(z_k|d_i)$ which is estimated by pLSA. First, we calculate the probability of z_k over a given concept X by:

$$P(z_k|X) = \frac{\sum_{d_i \in X} P(z_k|d_i)}{|X|}, \tag{4}$$

where X represents a set of the images corresponding to a given concept, for each latent topic variable. Then, we calculate the entropy of a given concept X by

$$H(X) = -\sum_{k=1}^{K} P(z_k|X) \log P(z_k|X). \tag{5}$$

The mutual information is a value represented by the difference between the entropy and the conditional entropy, which indicates the relevance between the tags. We calculate the mutual information as

$$MI(X;Y) = H(X) - H(X|Y), \tag{6}$$

where $H(X)$ is the entropy of one class, and $H(X|Y)$ is the entropy of the class combined two classes. If image distribution becomes narrow by combining the tag X with the tag Y, we judge the visual relevance become higher from the increase of mutual information.

4.7 Calculation of Similarity by Co-occurrence of Tags

For comparison, we calculate similarity by co-occurrence of tags using the Normalized Google Distance (NGD) [13] as well. The formula is

$$NGD = \frac{\max\{\log f(x), \log f(y)\} - \log f(x, y)}{\log N - \min\{\log f(x), \log f(y)\}}, \tag{7}$$

Table 1. The 20 nouns used in experiment

beach	bird	boat	bridge	car
cat	cloud	cup	dog	flower
fruit	house	people	sea	sky
snow	sun	tower	train	tree

Table 2. The 15 adjectives used in experiment

red	blue	green	black	white
circle	square	morning	night	winter
summer	new	old	beautiful	cool

where x is a noun, y is an adjective, $f(x)$ and $f(y)$ are the image number of tag search by a noun and an adjective in Flickr, and $f(x, y)$ is the image number of AND-search by combination of a noun and an adjective. Moreover, N is the number of all images in Flickr. However, we assume N is 50 billion since it is unable to get to know the exact number.

5 Experiments

5.1 Dataset

Images were collected using the API from Flickr. We collected 800 negative images and 200 positive images under the restriction that we obtained only one image from the same uploader. In addition, we retrieved positive images in order from the top in the search ranking of Flickr. Negative images were selected from among the images obtained at random from Flickr, which does not have the tags of nouns and adjectives of a particular class. In this experiment, we selected 20 nouns as shown in Table 1, and 15 adjectives as shown in Table 2. Thus, we calculate the entropy about 20×15 classes which are the combination of each noun and each adjective, as well as 20 classes which are only noun.

5.2 Experimental Results

According to the procedure explained in the previous section, we calculated mutual information for each class. Figure 1 shows calculation result of the entropy values in the second columns and the mutual information values after the third columns. On the other hand, Figure 2 shows the calculated results on the similarity of NGD using tag co-occurrence. We summarized the combinations of nouns and adjectives which are judged to have high relation by mutual information values in Table 3 and by NGD in Table 4, respectively.

With these experimental results, we compare mutual information of each class. Mutual information decreases when the distribution of images in each class spreads, and increases when the distribution of images in each class is narrow. Then, we can judge that the classes which have amount of mutual information have high visual relation between nouns and adjectives. Moreover, we determine the classes which have small NGD have high visual relation between nouns and adjectives.

noun/adjective	-	red	blue	green	black	white	circle	square	morning	night	winter	summer	new	old	beautiful	cool
beach	5.383	0.198	0.099	-0.009	0.027	-0.059	0.018	0.181	0.338	0.305	0.101	-0.058	0.037	-0.045	0.075	0.011
bird	5.478	0.147	0.193	0.182	0.029	-0.045	-0.009	0.115	0.321	0.034	0.103	0.212	-0.023	-0.012	0.063	0.082
boat	5.398	0.193	0.123	-0.065	0.110	-0.034	-0.045	0.122	0.440	0.297	0.095	0.020	0.065	-0.050	0.197	-0.053
bridge	5.466	0.071	0.354	0.161	0.232	0.078	-0.018	0.151	0.336	0.143	0.003	0.042	0.085	-0.028	0.016	-0.022
car	5.486	0.139	0.105	0.003	0.130	0.118	0.131	0.035	0.101	0.129	0.049	0.044	0.150	-0.003	0.018	0.039
cat	5.521	0.003	0.061	0.046	0.145	0.117	0.061	0.092	0.032	0.063	0.069	0.064	0.046	0.044	0.070	0.048
cloud	5.334	0.078	0.066	-0.020	0.154	-0.024	0.030	0.217	0.220	0.135	0.063	-0.064	0.069	-0.005	0.086	0.014
cup	5.431	0.105	0.137	0.100	0.121	0.150	0.073	0.096	0.169	0.103	0.132	0.013	-0.027	-0.060	-0.015	-0.005
dog	5.522	0.027	0.024	0.069	0.120	0.124	0.144	0.086	0.137	0.211	0.069	0.048	0.050	0.038	0.066	0.008
flower	5.357	0.096	0.185	0.145	0.082	0.055	-0.040	0.175	0.153	0.088	0.011	0.077	0.106	-0.128	0.018	0.030
fruit	5.474	0.112	0.113	0.157	0.242	0.085	0.042	0.113	0.006	0.050	0.117	0.149	0.007	-0.048	0.061	0.018
house	5.536	0.114	0.170	0.163	0.161	0.060	0.040	0.091	0.224	0.078	0.129	0.033	-0.011	0.093	-0.003	-0.001
people	5.519	0.084	0.047	0.024	0.114	0.078	0.035	0.013	0.164	0.153	0.093	0.020	0.134	0.058	0.090	0.040
sea	5.368	0.211	-0.022	-0.038	0.198	-0.030	-0.032	0.108	0.439	0.237	0.198	-0.066	0.056	-0.021	0.077	-0.006
sky	5.387	0.188	0.108	0.016	0.146	0.030	-0.026	0.036	0.287	0.237	0.011	0.048	0.053	-0.022	0.006	-0.002
snow	5.490	0.036	0.261	0.038	0.084	0.044	-0.014	0.167	0.279	0.159	0.054	-0.009	-0.013	0.047	0.084	0.077
sun	5.380	0.278	0.044	0.027	0.069	-0.008	0.176	0.042	0.237	0.248	0.069	0.007	-0.016	-0.067	0.179	0.015
tower	5.473	0.113	0.234	0.051	0.151	0.046	0.022	0.063	0.443	0.101	0.044	0.012	0.043	0.037	0.015	0.015
train	5.535	0.056	0.133	0.054	0.242	0.128	0.149	0.071	0.036	0.145	0.026	0.016	0.040	0.045	0.050	0.023
tree	5.437	0.014	0.137	0.072	0.183	0.058	-0.022	0.173	0.376	0.186	0.164	0.056	0.046	0.056	-0.003	0.011

Fig. 1. Calculation result of mutual information (red: high relevance class, blue: low relevance class)

noun/adjective	red	blue	green	black	white	circle	square	morning	night	winter	summer	new	old	beautiful	cool
beach	0.678	0.492	0.650	0.646	0.630	0.802	0.983	0.620	0.639	0.669	0.445	0.669	0.715	0.539	0.717
bird	0.587	0.518	0.566	0.550	0.563	0.798	0.960	0.606	0.629	0.776	0.802	0.708	0.757	0.616	0.726
boat	0.616	0.516	1.556	0.683	0.647	0.676	0.954	0.606	0.629	0.725	0.575	0.690	0.593	0.618	0.710
bridge	0.646	0.579	0.616	0.623	0.619	0.665	0.798	0.600	0.487	0.613	0.680	0.584	0.567	0.640	0.728
car	0.508	0.556	0.613	0.523	0.573	0.766	0.918	0.747	0.573	0.704	0.683	0.623	0.425	0.666	0.542
cat	0.666	0.624	0.630	0.462	0.518	0.884	0.934	0.761	0.735	0.754	0.774	0.794	0.759	0.660	0.714
cloud	0.579	0.422	0.552	0.588	0.532	0.666	0.859	0.462	0.616	0.640	0.630	0.731	0.651	0.548	0.617
cup	0.659	0.721	0.720	0.711	0.679	0.671	0.943	0.628	0.853	0.853	0.858	0.700	0.734	0.831	0.770
dog	0.638	0.621	0.646	0.477	0.528	0.828	0.925	0.744	0.765	0.817	0.684	0.746	0.720	0.697	0.708
flower	0.405	0.480	0.379	0.579	0.408	0.724	0.878	0.666	0.730	0.709	0.523	0.739	0.765	0.517	0.707
fruit	0.508	0.687	0.534	0.694	0.663	0.667	0.890	0.699	0.809	0.779	0.647	0.812	0.735	0.707	0.671
house	0.594	0.583	0.555	0.604	0.543	0.722	0.895	0.689	0.597	0.618	0.649	0.521	0.434	0.623	0.657
people	0.597	0.589	0.600	0.525	0.524	0.763	0.788	0.700	0.506	0.640	0.527	0.625	0.576	0.474	0.604
sea	0.541	0.394	0.571	0.579	0.560	0.788	0.917	0.588	0.600	0.614	0.472	0.700	0.615	0.525	0.699
sky	0.463	0.226	0.415	0.535	0.444	0.696	0.306	0.489	0.450	0.504	0.480	0.645	0.599	0.498	0.635
snow	0.633	0.560	0.669	0.644	0.435	0.732	0.922	0.603	0.567	0.157	0.763	0.667	0.717	0.653	0.717
sun	0.495	0.408	0.468	0.530	0.496	0.673	0.871	0.420	0.606	0.515	0.416	0.664	0.582	0.405	0.585
tower	0.679	0.573	0.663	0.666	0.629	0.722	0.728	0.649	0.522	0.683	0.725	0.688	0.557	0.670	0.713
train	0.697	0.694	0.731	0.664	0.681	0.748	0.910	0.690	0.649	0.884	0.759	0.692	0.571	0.742	0.711
tree	0.483	0.447	0.376	0.536	0.483	0.709	0.826	0.541	0.565	0.447	0.601	0.691	0.574	0.558	0.654

Fig. 2. Calculation result of co-occurrence of tags by NGD (red: high relevance class, blue: low relevance class)

5.3   Discussion on Visual Relations

First, we discovered the tendency that the mutual information becomes large in the class where a lot of images have been greatly affected by color (see Figure 3). For example, "morning sea", "morning sky", and "blue bridge" class would be cited. Looking at the positive region of the image to be included in the "morning sky" class, there are the regions which have a lot of red region of the morning glow and blue region of the sunny sky. We consider that image distribution in these classes becomes narrow, and their mutual information increases for that reason. However, visual relevance in "blue bridge" class becomes high, although the class has few images of a blue bridge. We consider that visual relevance increases, because the positive regions include blue of sea, river, and sky around bridge in order to take the entire bridge in the photos.

Next, when we pay attention about the class in combination with the adjective about a color, it turns out that mutual information becomes relatively larger in the class where the adjective about color qualifies directly to the object being

Table 3. Main classes of high similarity by mutual information

sea+morning	sky+morning
sun+red	sky+night
sun+beautiful	car+red
flower+blue	sun+circle

Table 4. Main classes of high similarity by co-occurrence of tag

snow+winter	sky+blue
tree+green	flower+green
sea+blue	sun+beautiful
flower+red	sun+blue

Fig. 3. Positive regions in the class where the color influenced greatly

Fig. 4. Positive regions in the class combined with an adjective about color

indicated by the noun (see Figure 4). We would mention "red sun" and "red car" class as examples of large mutual information, and "red cat" and "red dog" class as examples of small mutual information in the class which combined with the adjective about a color. In such classes where mutual information is greater, and positive regions of that class contains the particular color and object. Whereas, in the class where mutual information is smaller, positive regions of that class do not contain the particular color and object. Therefore, it can be thought that visual relation has been correctly calculated, which is consistent with our intuition.

5.4 Comparison with Tag Co-occurrence

Some classes have high visual relevance while their co-occurrence relevance by tag is low (see Figure 5). As an example, we cite "old people" class. Both visual relevance and relevance by tag are low in the class combined with the "old". However, there is a tendency that visual relevance of "old" becomes higher than other classes, when it combined with the noun in connection with artificial things and living things such as "house" and "people".

On the other hand, there are classes which have low visual relation, although their relation by tag is high (see Figure 6). As an example, we cite "summer beach" and "green sky" class. It is thought that visual relation became low in "summer beach" class, because there are not only the image of a beach but many images of the people who are doing sea bathing. Meanwhile, it is thought that the relation by tag became high in "green sky" class, because the "green sky" class contains many images of grass, and the co-occurrence of "sky" and "grass"

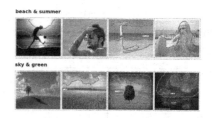

Fig. 5. Positive regions of the class which the relevance of co-occurrence is low, and visual relevance is high

Fig. 6. Positive regions of the class which the relevance of co-occurrence is high, and visual relevance is low

the color of which is green is higher. However, their visual relation is shown as being low.

6   Conclusion and Future Work

In this paper, first, we collected images tagged with both particular nouns and adjectives from Flickr. Then, we extracted local features from images, and calculated the distribution of image as the numeric value by the entropy. Finally, we performed comparison and consideration about the visual relation between a noun and an adjective from the change in entropy for each class which combined a noun and an adjective.

As a result, we obtained the results that on mutual information represents intuitive visual similarity. Therefore, it turned out that there was a tendency that the pairs of nouns and adjectives related to color have the stronger visual relation. Regarding tag-based similarity, the degree of similarity by the co-occurrence of tag using NGD showed the results which fitted our intuition as well.

For future work, we plan to use other kinds of visual features than Color-SIFT BoF. In addition, we would like to utilize the results obtained in this paper to improve performance on simultaneous recognition of a noun and an adjective.

References

1. Berg, T.L., Berg, A.C., Shih, J.: Automatic Attribute Discovery and Characterization from Noisy Web Data. In: Daniilidis, K., Maragos, P., Paragios, N. (eds.) ECCV 2010, Part I. LNCS, vol. 6311, pp. 663–676. Springer, Heidelberg (2010)
2. Parikh, D., Grauman, K.: Interactively building a discriminative vocabulary of nameable attributes. In: Proc. of IEEE Computer Vision and Pattern Recognition (2011)
3. Farhadi, A., Endres, I., Hoiem, D., Forsyth, D.: Describing objects by their attributes. In: Proc. of IEEE Computer Vision and Pattern Recognition, pp. 1778–1785 (2009)
4. Dhar, S., Ordonez, V., Berg, T.L.: High level describable attributes for predicting aesthetics and interestingness. In: Proc. of IEEE Computer Vision and Pattern Recognition, pp. 1657–1664 (2011)

5. Yanai, K., Barnard, K.: Image region entropy: A measure of "visualness" of web images associated with one concept. In: Proc. of ACM International Conference Multimedia (2005)

6. Kawakubo, H., Akima, Y., Yanai, K.: Automatic construction of a folksonomy-based visual ontology. In: Proc. of International Symposium on Multimedia, pp. 330–335 (2010)

7. Yanai, K., Kawakubo, H., Qiu, B.: A visual analysis of the relationship between word concepts and geographical locations. In: Proceedings of the ACM International Conference on Image and Video Retrieval (2009)

8. Deng, Y., Manjunath, B.S.: Unsupervised segmentation of color-texture regions in images and video. IEEE Transactions on Pattern Analysis and Machine Intelligence 23(8), 800–810 (2001)

9. Csurka, G., Bray, C., Dance, C., Fan, L.: Visual categorization with bags of keypoints. In: Proc. of ECCV Workshop on Statistical Learning in Computer Vision, pp. 59–74 (2004)

10. Lowe, D.G.: Distinctive image features from scale-invariant keypoints. International Journal of Computer Vision 60(2), 91–110 (2004)

11. Andrews, S., Tsochantaridis, I., Hofmann, T.: Support Vector Machines for Multiple-Instance Learning. In: Advances in Neural Information Processing Systems, pp. 577–584 (2003)

12. Hofmann, T.: Unsupervised learning by probabilistic latent semantic analysis. Machine Learning 43, 177–196 (2001)

13. Cilibrasi, R.L., Vitanyi, P.M.B.: The google similarity distance. IEEE Transactions on Knowledge and Data Engineering 19(3), 370–383 (2007)

Verb-Object Concepts Image Classification via Hierarchical Nonnegative Graph Embedding

Chao Sun[1,2], Bing-Kun Bao[1,2], and Changsheng Xu[1,2]

[1] National Laboratory of Pattern Recognition, Institute of Automation,
Chinese Academy of Sciences, Beijing, China
[2] China-Singapore Institute of Digital Media, Singapore
{csun,bkbao,csxu}@nlpr.ia.ac.cn

Abstract. Most existing image classification methods focus on handling images with only "object" concepts. At the same time, in real-world cases, there exists a great variety of images which contain "verb-object" concepts, rather than only "object" ones. The hierarchical structure embedded in these "verb-object" concepts can help to enhance classification. However, traditional feature representing methods cannot utilize it. To tackle this defect, we present in this paper a novel approach, called Hierarchical Nonnegative Graph Embedding (HNGE). By assuming that those "verb-object" concept images which share the same "object" part but different "verb" part have a specific hierarchical structure, we make use of this hierarchical structure and employ an effective technique, named nonnegative graph embedding, to perform feature extraction as well as image classification. Extensive experiments compared with the state-of-the-art algorithms on nonnegative data factorization demonstrate the feasibility, convergency and classification power of proposed approach on "verb-object" concept images classification.

Keywords: hierarchical nonnegative graph embedding, verb-object.

1 Introduction

Image understanding and classifying applications have been wildly researched for decades [14] [1]. Most existing image classification methods focus on handling images with only "object" concepts [4] [3], such as "horse", "car" and "tree" etc. However, in real-world, a huge number of images contain "verb-object" concepts, such as "ride horse", "repair boat" , and "cut tree", rather than only "object" concepts. Moreover, some concepts, like "ride horse" and "feed horse", "repair boat" and "row boat", "cut tree" and "plant tree", are sharing the same "object" part. Figure 1 illustrates several images under a set of "verb-object" concepts sharing same "object" but different "verb" part. Intuitively, this kind of set of concepts very likely share a common latent information or pattern in images, which can be helpful for image classification. By using traditional image representation techniques, each concept in the set will be treated individually and the shared common "object" part will be ignored. In such a way, the classification performance would be discounted.

S. Li et al. (Eds.): MMM 2013, Part I, LNCS 7732, pp. 58–69, 2013.

Motivated by these observations, in this paper, we present a approach, named Hierarchical Nonnegative Graph Embedding (HNGE), to better interpret "verb-object" concept images. We assume that those "verb-object" concept images which share the same "object" but different "verb" part have a specific hierarchical structure, which can be utilized for image classification. By applying the hierarchical structure, one "verb-object" concept not only can be used to separate a set of concepts from other concepts which have different "object" part, but also can be used to discriminated itself from concepts in the same set by the different "verb" part. For example, images of "row boat" can help to discriminate images of "repair boat" from images of "cut tree", while itself should be classified from images of "repair boat". We regard this structure as hierarchical structure and utilize it in classification.

Repair boat   Build boat   Row boat   Carry boat

Fig. 1. An illustration of a set of "verb-object" concept images

A similar work was proposed by Zhang *et* al. [19], which tried to simultaneously learn a set of classifiers for "verb-object" concepts in the same group. However, in this paper, we do not focus on designing classifiers, but focus on feature extraction and representation, which refer to nonnegative and sparse representation techniques. Nonnegative and sparse representation techniques have been well researched in recent decades to find nonnegative basis of data features with few nonzero elements [6]. A pioneer work for such a purpose is Nonnegative Matrix Factorization (NMF) [8]. It decomposed the data features matrix as the arithmetical product of two matrices which possessed only nonnegative elements. Generally, NMF belongs to the techniques of feature extraction and dimensionality reduction, as it results in a dimension-reduced representation of the primal data features [9]. In recent years, there are some works to extend and apply NMF in many different fields, such as face and object recognition [11] [18], biomedical applications [5] [7], color science [10], and so on. Recently, beyond the original nonnegative data factorization, Yang *et* al. [16] proposed a approach, named Nonnegative Graph Embedding (NGE), which obtained customized nonnegative data factorization by simultaneously realizing the specific purpose characterized by the intrinsic and penalty graphs. This work was further refined by Wang *et* al. [12] with the efficient multiplicative updating rule. As possessing the algorithmic properties in convergency, sparsity, and classification power, nonnegative matrix factorization (NMF) and its variants, nonnegative graph embedding (NGE) and its refined version multiplicative nonnegative graph embedding (MNGE) [12] have been proved effective in applications of image classification. In our work, we combine hierarchical structure with nonnegative graph embedding to develop a new formulation for nonnegative data factorization. We conduct experiments on a web image corpus composed of 9000 images on 45 "verb-object" concepts. The experimental results demonstrate the effectiveness of our approach.

Our work seems to be similar to human action recognition, like [17], as "verb-object" concepts look like actions. However, the main difference between our work and human action recognition is, human action recognition treats every action as unrelated to each others while our HNGE considers "verb-object" concepts having hierarchical structure to explore a new layer of linkage among actions. Besides, in human action recognition, an action does not always have an object, while HNGE focuses on handling images containing both "verb" and "object".

The rest of the paper is structured as follows: We elaborate HNGE and its formulation in Section 2. Experimental results are reported in Section 3. Finally, we give conclusions in Section 4.

2   Hierarchical Nonnegative Graph Embedding

In this section, we elaborate hierarchical nonnegative graph embedding and formulate the problem within the framework of nonnegative data decomposition.

2.1   Motivation

Practically, we believe that the "verb-object" concept images contain hierarchical structure. Figure 2 illustrates the hierarchical structure in "verb-object" concept images. As shown, we divide whole "verb-object" concepts into two levels. On the first level, those "verb-object" concepts containing the same "object" are treated as a class, here "class" is as the same meaning as "group" or "set" mentioned above. On the second level, those "verb-object" concepts in the same class on first level are divided into sub-classes, according to the different "verb" parts they have. Although the final aim of our classification is to discriminate all the sub-classes on the second level, we do not directly perform classification on only second level. Instead, on one hand, we enlarge interclass distance for classes on the first level. On the other hand ,we reduce intraclass distance and enlarge interclass distance for sub-classes on the second level. These two step are

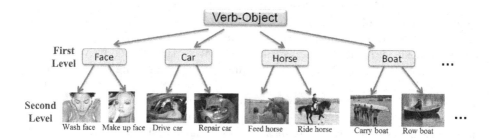

Fig. 2. An illustration of hierarchical structure

performed simultaneously while the one on the first level will compensate the one on the seconde level and improve the performance on the final "verb-object" concepts classification.

2.2   Objective for Nonnegative Graph Embedding

Before introduce the objective for nonnegative graph embedding, we first list the notations used in this paper here. Let $\mathbf{X} = [x_1, x_2, \cdots, x_N]$ denote the data sample set, in which $x_i \in \mathbb{R}^k$ denotes the feature descriptor of the ith sample and N is number of total samples. Here we assume that the matrix X is nonnegative. Letting m be the dimension of the desired dimension-reduced feature space,the task of our data factorization is to derive a nonnegative basis matrix $W \in \mathbb{R}^{k \times m}$ and a nonnegative encoding coefficient matrix $H \in \mathbb{R}^{m \times N}$, while the data matrix X can be approximated as the product of matrices W and H. In this paper, we utilize the following rule to facilitate presentation: for any matrix A, A_i denotes the ith row vector of A, its corresponding lowercase version a_i denotes the ith column vector of A. A_{ij} denotes the element of A at the ith row and jth column, and $A_{p \times q}$ means that A has p rows and q columns.

Non-Negative Matrix Factorization(NMF) factorizes the data matrix X into one lower-rank nonnegative basis matrix W and one nonnegative coefficient matrix H. Its usual objective function is:

$$\min_{W,H} \|X - WH\|_F^2, \quad s.t. \quad W, H \geq 0, \tag{1}$$

Meanwhile, as proposed by Yan *et* al. in [15], most dimensionality reduction algorithms can be explained within a unified framework, called graph embedding. Let $G = \{X, S\}$ be an undirected weighted graph with vertex set X and similarity matrix $S \in \mathbb{R}^{N \times N}$. Each element of the real symmetric matrix S measures similarity between a pair of vertices. The Laplacian matrix L and diagonal matrix D of a graph G are defined as:

$$L = D - S, \quad D_{ii} = \sum_{j \neq i} S_{ij}, \quad \forall i \tag{2}$$

Graph embedding generally involves an intrinsic graph G, which characterizes the favorite relationship among the data samples, and a penalty graph $G^u = \{X, S^u\}$, which characterizes the unfavorable relationship among the data samples. Correspondingly, penalty graph G^u also has its Laplacian matrix L^u and diagonal matrix D^u. In our work, we assume that the sample data set has two levels. Therefore, there are an intrinsic graph and a penalty graph on the first level, and an intrinsic graph and a penalty graph on the second level, respectively.

Let $G = \{X, S\}$ be the intrinsic graph on the first level, $G^u = \{X, S^u\}$ be the penalty graph on the first level, $\tilde{G} = \{X, \tilde{S}\}$ be the intrinsic graph on the second level, and $\tilde{G}^u = \{X, \tilde{S}^u\}$ be the penalty graph on the second level. Their Laplacian matrices and diagonal matrices are L, D, L^u, D^u, $\tilde{L}$, $\tilde{D}$, $\tilde{L}^u$, and $\tilde{D}^u$, respectively.

To serve for graph embedding, we first divide the coefficient matrix H into two parts, namely,

$$H = \begin{bmatrix} H^1 \\ H^2 \end{bmatrix} \tag{3}$$

where $H^1 = [h_1^1, h_2^1, \cdots, h_N^1] \in \mathbb{R}^{d \times N}$, $d < N$, denotes the desired low-dimensional representations for the training data on the first level, and $H^2 = [h_1^2, h_2^2, \cdots, h_N^2] \in \mathbb{R}^{(m-d) \times N}$.

As existing hierarchical structure, we then divide the matrix H^1 into two parts, namely,

$$H^1 = \begin{bmatrix} H^{11} \\ H^{12} \end{bmatrix} \tag{4}$$

where $H^{11} = [h_1^{11}, h_2^{11}, \cdots, h_N^{11}] \in \mathbb{R}^{r \times N}$, $r < d$, denotes the desired low-dimensional representations for the training data on the second level, and $H^{12} = [h_1^{12}, h_2^{12}, \cdots, h_N^{12}] \in \mathbb{R}^{(d-r) \times N}$.

Therefore, to clearly reveal the structure, we rewrite matrix H as:

$$H = \begin{bmatrix} \begin{bmatrix} H^{11} \\ H^{12} \end{bmatrix} \\ H^2 \end{bmatrix} \tag{5}$$

Meanwhile, to facilitate presentation, we define matrix $\bar{H}^2$ as the combination of H^{12} and H^2:

$$\bar{H}^2 = \begin{bmatrix} H^{12} \\ H^2 \end{bmatrix} \tag{6}$$

where $\bar{H}^2 = [\bar{h}_1^2, \bar{h}_2^2, \cdots, \bar{h}_N^2] \in \mathbb{R}^{(m-r) \times N}$.

Correspondingly, the basis matrix W is also divided as:

$$W = \begin{bmatrix} W^1 & W^2 \end{bmatrix} \tag{7}$$

and:

$$W^1 = \begin{bmatrix} W^{11} & W^{12} \end{bmatrix} \tag{8}$$

where $W^1 \in \mathbb{R}^{k \times d}$, $W^2 \in \mathbb{R}^{k \times (m-d)}$, $W^{11} \in \mathbb{R}^{k \times r}$ and $W^{12} \in \mathbb{R}^{k \times (d-r)}$.

According to graph embedding, the target of graph-preserving on the first level is:

$$\max_{H^1} \sum_{i \neq j} \|h_i^1 - h_j^1\|^2 S_{ij}^u \tag{9}$$

As (W^2, H^2) are considered as the complementary space of (W^1, H^1). From the complementary property between H^1 and H^2, the objective is transformed into:

$$\min_{H^2} \sum_{i \neq j} \|h_i^2 - h_j^2\|^2 S_{ij}^u \tag{10}$$

On the second level, our two targets of graph-preserving are given as:

$$\begin{cases} \min_{H^{11}} \sum_{i \neq j} \|h_i^{11} - h_j^{11}\|^2 \tilde{S}_{ij} \\ \max_{H^{11}} \sum_{i \neq j} \|h_i^{11} - h_j^{11}\|^2 \tilde{S}_{ij}^u \end{cases} \tag{11}$$

As $(\bar{W}^2, \bar{H}^2)$ are considered as the complementary space of (W^{11}, H^{11}). From the complementary property between H^{11} and $\bar{H}^2$, the second objective above is transformed into:

$$\min_{\bar{H}^2} \sum_{i \neq j} \|\bar{h}_i^2 - \bar{h}_j^2\|^2 \tilde{S}_{ij}^u \tag{12}$$

2.3 Unified Formulation

To achieve the above three objectives required for hierarchical non-negative graph embedding, we can have the unified objective function as:

$$\min_{W,H} \sum_{i \neq j} \|h_i^2 - h_j^2\|^2 S_{ij}^u + \sum_{i \neq j} \|h_i^{11} - h_j^{11}\|^2 \tilde{S}_{ij} + \sum_{i \neq j} \|\bar{h}_i^2 - \bar{h}_j^2\|^2 \tilde{S}_{ij}^u$$

$$+ \lambda \|X - WH\|_F^2, \quad s.t. \quad W, H \geq 0 \tag{13}$$

where λ is a positive parameter to balance the two parts for graph embedding and data reconstruction.

From the definitions of Laplacian matrix and diagonal matrix, we have:

$$\sum_{i \neq j} \|h_i^2 - h_j^2\|^2 S_{ij}^u = Tr(H^2 L^u H^{2^T}) \tag{14}$$

$$\sum_{i \neq j} \|h_i^{11} - h_j^{11}\|^2 \tilde{S}_{ij} = Tr(H^{11} \tilde{L} H^{11^T}) \tag{15}$$

$$\sum_{i \neq j} \|\bar{h}_i^2 - \bar{h}_j^2\|^2 \tilde{S}_{ij}^u = Tr(\bar{H}^2 \tilde{L}^u \bar{H}^{2^T}) \tag{16}$$

Furthermore, as W is the basis matrix, it is natural to require that each column vector of W is normalized, that is $\|W_i\| = 1, \forall i$. But this constraint makes the optimization problem much more complicated. Hence, we compensate the norms of the bases into the coefficient matrix and rewrite objectives as:

$$Tr(H^2 L^u H^{2^T}) \Rightarrow Tr(E_2 H^2 L^u H^{2^T} E_2^T) \tag{17}$$

$$Tr(H^{11} \tilde{L} H^{11^T}) \Rightarrow Tr(E_{11} H^{11} \tilde{L} H^{11^T} E_{11}^T) \tag{18}$$

$$Tr(\bar{H}^2 \tilde{L}^u \bar{H}^{2^T}) \Rightarrow Tr(\bar{E}_2 \bar{H}^2 \tilde{L}^u \bar{H}^{2^T} \bar{E}_2^T) \tag{19}$$

where the matrix $E_2 = diag\{\|w_1^2\|, \|w_2^2\|, \cdots, \|w_{m-d}^2\|\}$, $E_{11} = diag\{\|w_1^{11}\|, \|w_2^{11}\|, \cdots, \|w_r^{11}\|\}$, and $\bar{E}_2 = diag\{\|\bar{w}_1^2\|, \|\bar{w}_2^2\|, \cdots, \|\bar{w}_{m-d}^2\|\}$. w_i^k denotes the ith column vector of matrix W^k, $k = 1, 2, 11, 22$.

By combining formulations (14) to (19), the final objective function is then reformulated as:

$$\min_{W,H} Tr(E_2 H^2 L^u H^{2^T} E_2^T) + Tr(E_{11} H^{11} \tilde{L} H^{11^T} E_{11}^T)$$

$$+ Tr(\bar{E}_2 \bar{H}^2 \tilde{L}^u \bar{H}^{2^T} \bar{E}_2^T) + \lambda \|X - WH\|_F^2, \quad s.t. \quad W, H \geq 0 \tag{20}$$

2.4   Convergent Iterative Procedure

As the final objective function is biquadratic, and generally there does not exist closed-form solution. We use an iterative procedure to get the non-negative solution.

Most iterative procedures for solving high-order optimization problems transform the original intractable problem into a set of tractable sub-problems, and finally obtain the convergence to a local optimum. Our proposed iterative procedure also follows this philosophy and optimizes W and H alternately.

Optimize W for Given H

For a fixed matrix H, the final objective function with respect to basis matrix W can be rewritten as:

$$F(W) = Tr(WQ^1W^T) + Tr(WQ^2W^T) + Tr(WQ^3W^T) + \lambda\|X - WH\|^2 \quad (21)$$

Where:

$$Q^1 = \begin{bmatrix} 0_{d\times d} & 0_{d\times(m-d)} \\ 0_{(m-d)\times d} & H^2 L^u H^{2T} \end{bmatrix} \circ I, \quad Q^2 = \begin{bmatrix} H^{11}\tilde{L}H^{11T} & 0_{r\times(m-r)} \\ 0_{(m-r)\times r} & 0_{(m-r)\times(m-r)} \end{bmatrix} \circ I,$$

$$Q^3 = \begin{bmatrix} 0_{r\times r} & 0_{r\times(m-r)} \\ 0_{(m-r)\times r} & \bar{H}^2 \tilde{L}^u \bar{H}^{2T} \end{bmatrix} \circ I$$

and $Q^1, Q^2, Q^3 \in \mathbb{R}^{m\times m}$, operator $\circ$ indicates the element-wise matrix multiplication, I indicates the identity matrix.

After deduction, we have the update rule of W as:

$$W_{ij} \leftarrow W_{ij} \frac{[\lambda XH_T + WQ_-^1 + WQ_-^2 + WQ_-^3]_{ij}}{[\lambda WHH^T + WQ_+^1 + WQ_+^2 + WQ_+^3]_{ij}} \quad (22)$$

where:

$$Q_+^1 = \begin{bmatrix} 0_{d\times d} & 0_{d\times(m-d)} \\ 0_{(m-d)\times d} & H^2 D^u H^{2T} \end{bmatrix} \circ I, \quad Q_+^2 = \begin{bmatrix} H^{11}\tilde{D}H^{11T} & 0_{r\times(m-r)} \\ 0_{(m-r)\times r} & 0_{(m-r)\times(m-r)} \end{bmatrix} \circ I,$$

$$Q_+^3 = \begin{bmatrix} 0_{r\times r} & 0_{r\times(m-r)} \\ 0_{(m-r)\times r} & \bar{H}^2 \tilde{D}^u \bar{H}^{2T} \end{bmatrix} \circ I$$

and:

$$Q_-^1 = \begin{bmatrix} 0_{d\times d} & 0_{d\times(m-d)} \\ 0_{(m-d)\times d} & H^2 S^u H^{2T} \end{bmatrix} \circ I, \quad Q_-^2 = \begin{bmatrix} H^{11}\tilde{S}H^{11T} & 0_{r\times(m-r)} \\ 0_{(m-r)\times r} & 0_{(m-r)\times(m-r)} \end{bmatrix} \circ I,$$

$$Q_-^3 = \begin{bmatrix} 0_{r\times r} & 0_{r\times(m-r)} \\ 0_{(m-r)\times r} & \bar{H}^2 \tilde{S}^u \bar{H}^{2T} \end{bmatrix} \circ I$$

and

$$Q^1 = Q_+^1 - Q_-^1, \quad Q^2 = Q_+^2 - Q_-^2, \quad Q^3 = Q_+^3 - Q_-^3$$

Optimize H for Given W

After updating the matrix W, we normalize the column vectors of it and consequently convey the norm to the coefficient matrix H, namely,

$$H_{ij} \leftarrow H_{ij} \times \|w_i\| \quad \forall i, j \tag{23}$$

$$w_i \leftarrow w_i / \|w_i\| \quad \forall i \tag{24}$$

Note that the updating of H and W here will not change the value of the objective function in (20).

Then based on the normalized W, the objective function in (20) with respect to H is then rewritten as:

$$F(H) = Tr(R^1 H L^u H^T R^{1^T}) + Tr(R^2 H \tilde{L} H^T R^{2^T}) + Tr(R^3 H \tilde{L}^u H^T R^{3^T}) \\ + \lambda \|X - WH\|_F^2 \tag{25}$$

where

$$R^1 = [0_{(m-d) \times d} \quad I_{(m-d) \times (m-d)}], \quad R^2 = [I_{r \times r} \quad 0_{r \times (m-r)}], \quad R^3 = [0_{(m-r) \times r} \\ I_{(m-r) \times (m-r)}]$$

After deduction, we have the update rule of H as:

$$H_{ij} \leftarrow H_{ij} \frac{[\lambda W^T X + R^{1^T} R^1 H S^u + R^{2^T} R^2 H \tilde{S} + R^{3^T} R^3 H \tilde{S}^u]_{ij}}{[\lambda W^T W H + R^{1^T} R^1 H D^u + R^{2^T} R^2 H \tilde{D} + R^{3^T} R^3 H \tilde{D}^u]_{ij}} \quad \forall i, j \tag{26}$$

Algorithm 1.

1: Input: Image representation matrix X, graphs $G, G^u, \tilde{G}, \tilde{G}^u$
2: Initialization: Randomly choose W^0, H^0 as non-negative matrices.
3: For $t = 0, 1, 2, \cdots, T_{max}$, Do
    1) For given $H = H^t$, update matrix W as:

$$W_{ij} \leftarrow W_{ij} \frac{[\lambda X H_T + W Q_-^1 + W Q_-^2 + W Q_-^3]_{ij}}{[\lambda W H H^T + W Q_+^1 + W Q_+^2 + W Q_+^3]_{ij}} \quad \forall i, j$$

    2) Normalize the column vectors of W^{t+1}

$$H_{ij} \leftarrow H_{ij} \times \|w_i^{t+1}\| \quad \forall i, j$$

$$w_i^{t+1} \leftarrow w_i^{t+1} / \|w_i^{t+1}\| \quad \forall i$$

    3) For given $W = W^{t+1}$, update the matrix H as:

$$H_{ij} \leftarrow H_{ij} \frac{[\lambda W^T X + R^{1^T} R^1 H S^u + R^{2^T} R^2 H \tilde{S} + R^{3^T} R^3 H \tilde{S}^u]_{ij}}{[\lambda W^T W H + R^{1^T} R^1 H D^u + R^{2^T} R^2 H \tilde{D} + R^{3^T} R^3 H \tilde{D}^u]_{ij}} \quad \forall i, j$$

    4) If $\|W^{t+1} - W^t\| < \epsilon$ and $\|H^{t+1} - H^t\| < \epsilon$ (ϵ is a small positive number), then break.
4: Output: $W = W^t$ and $H = H^t$

Based on these update rules, the whole procedure can be concluded as algorithm 1.

3   Experiments

In this section, we evaluate the effectiveness of our proposed hierarchical non-negative graph embedding compared with other subspace algorithms including Linear Discriminant Analysis (LDA) [2], Marginal Fisher Analysis (MFA) [15], and other nonnegative basis pursuit algorithm Nonnegative Matrix Factorization (NMF) [8] and Multiplicative Nonnegative Graph Embedding (MNGE) [12].

3.1   Database

The most existing image databases do not satisfy the requirement of containing "verb-object" concepts with hierarchical structure. Hence we setup a "verb-object" concept images database and conduct experiments on it. We predefined 45 "verb-object" concepts which involves 19 different "object", while each "object" containing 2-3 "verb-object" concepts. Table 1 lists all the "verb-object" concepts in our database. Then a total number of 9000 images were collected from Google Image[1] and Flickr[2], with 200 images on each "verb-object" concept, while every image was labeled manually. Each image should have two labels, one indicates class category on the first level, another indicates class category on the first level. Figure 3 illustrates some "verb-object" concepts and their corresponding image samples.

Table 1. All "verb-object" Concepts

		Concept names		
answer phone	play phone	feed horse	ride horse	build boat
repair boat	row boat	buy car	drive car	repair car
buy vegetable	cook vegetable	cut vegetable	carry bike	ride bike
repair bike	carry water	drink water	pour water	clap hands
wave hand	comb hair	cut hair	wash hair	cut tree
plant tree	prune tree	iron clothes	wash clothes	use computer
fix computer	put on shoes	fix shoe	feed dog	walk dog
sit on chair	repair chair	water flowers	arrange flower	make up face
wash face	lie on bed	sit on bed	read book	write on book

3.2   Settings

To describe the image content, each image was resized to 256×256 pixels, then a set of 21504-dimensional $((1^2 + 2^2 + 4^2) \times 1024)$ Locality-constrained Linear

---

[1] http://images.google.com
[2] http://www.flickr.com/

Play phone   Answer phone   Row boat   Repair boat   Build boat

Fix computer   Use computer   Prune tree   Plant tree   Cut tree

Fig. 3. Sampling Images in Our Database

Coding (LLC) features [13] were extracted from each image. We randomly choose 60% images from each "verb-object" concept and combine them as training data. The rest 40% images are then used as testing data.

As LDA ,MFA, NMF, and MNGE algorithms can not handle hierarchical structure, to serve these algorithms, we just treat each "verb-object" concept in the second level as a class by ignoring the first level when performing experiments using these algorithms. For MFA, NMF and MNGE algorithms, the intrinsic graph and penalty graph are set as the same, and each dimension of the desired dimension-reduced feature space, m, is set as 1000. In our HNGE, parameter m is also set as 1000, parameter d is set as $0.8 \times m$, parameter r is set as $0.8 \times d$, and parameter λ is set as 1.

3.3 Results

We conduct all experiments on our self-built database. Table 2 shows the comparison experimental results of different algorithms. The fist line illustrates precisions over all the "verb-object" concepts on our proposed HNGE as well as four baselines. The next ten lines are the precisions on randomly selected 10 "verb-object" concepts from 4 different "objects". It is shown that HNGE achieves

Table 2. Precisions (%) of Different Algorithms

	LDA	NMF	MFA	MNGE	**HNGE**
Precision	26.51	31.96	37.15	41.58	**45.28**
Play phone	11.58	10.35	8.37	12.45	**17.20**
Answer phone	14.75	21.81	34.95	**35.84**	35.12
Row boat	67.81	69.98	**89.25**	75.38	88.34
Repair boat	27.19	26.64	27.53	29.44	**35.48**
Build boat	23.74	26.89	32.57	30.83	**37.75**
Fix computer	32.57	37.78	39.19	44.72	**48.56**
Use computer	15.11	29.42	21.39	25.92	**35.67**
Prune tree	24.28	31.78	45.67	41.47	**47.82**
Plant tree	21.37	39.11	**57.49**	50.02	55.28
Cut tree	34.28	41.38	67.41	60.83	**72.45**

better than those classic feature dimension reduction algorithms. From these results, we can also obtain following observations: 1) The accuracy of LDA is lower than other NMF-based algorithms, which demonstrates the discriminative power of nonnegative data factorization. 2) By taking hierarchical structure into consideration and jointly optimizing on both two levels, our HNGE algorithm outperforms all other algorithms.

4   Conclusions

In this paper, we proposed the hierarchical nonnegative graph embedding algorithm for "verb-object" concept images classification. Our HNGE takes hierarchical structure involved in "verb-object" concepts into consideration, and develop the method of feature extraction and dimensionality reduction based on nonnegative graph embedding. The entire "verb-object" concept images classification problem is formulated within the nonnegative data factorization framework, and an efficient iterative procedure is proposed for optimizing the objective function with theoretically and practically convergency. Experiments on the self-collected "verb-object" concept image database demonstrate the effectiveness of our algorithm in "verb-object" concept images classification.

Acknowledgement. This work is supported by National Program on Key Basic Research Project (973 Program, Project No. 2012CB316304), the National Natural Science Foundation of China (Grant No. 61201374, 90920303, 61003161), and China Postdoctoral Science Foundation (Grant No. 2011M500430).

References

1. Bao, B., Ni, B., Mu, Y., Yan, S.: Efficient region-aware large graph construction towards scalable multi-label propagation. Pattern Recognition 44(3), 598–606 (2011)
2. Belhumeur, P., Hespanha, J., Kriegman, D.: Eigenfaces vs. fisherfaces: Recognition using class specific linear projection. IEEE Transactions on Pattern Analysis and Machine Intelligence 19(7), 711–720 (1997)
3. Carneiro, G., Chan, A., Moreno, P., Vasconcelos, N.: Supervised learning of semantic classes for image annotation and retrieval. IEEE Transactions on Pattern Analysis and Machine Intelligence 29(3), 394–410 (2007)
4. Gao, Y., Fan, J., Xue, X., Jain, R.: Automatic image annotation by incorporating feature hierarchy and boosting to scale up svm classifiers. In: Proceedings of the 14th Annual ACM International Conference on Multimedia, pp. 901–910 (2006)
5. Heger, A., Holm, L.: Sensitive pattern discovery with fuzzy alignments of distantly related proteins. Bioinformatics 19, 130–137 (2003)
6. Hu, C., Zhang, B., Yan, S., Yang, Q., Yan, J., Chen, Z., Ma, W.: Mining ratio rules via principal sparse non-negative matrix factorization. In: Fourth IEEE International Conference on Data Mining, pp. 407–410 (2004)
7. Kim, P., Tidor, B.: Subsystem identification through dimensionality reduction of large-scale gene expression data. Genome research 13(7), 1706–1718 (2003)

8. Lee, D., Seung, H., et al.: Learning the parts of objects by non-negative matrix factorization. Nature 401(6755), 788–791 (1999)
9. Liu, X., Yan, S., Jin, H.: Projective nonnegative graph embedding. In: IEEE Conference on Image Processing, vol. 19(5), pp. 1126–1137 (2010)
10. Ramanath, R., Kuehni, R., Snyder, W., Hinks, D.: Spectral spaces and color spaces. Color Research and Application 29(1), 29–37 (2004)
11. Ramanath, R., Snyder, W., Qi, H.: Eigenviews for object recognition in multispectral imaging systems. In: Proceedings on Applied Imagery Pattern Recognition Workshop, pp. 33–38 (2003)
12. Wang, C., Song, Z., Yan, S., Zhang, L., Zhang, H.: Multiplicative nonnegative graph embedding. In: IEEE Conference on Computer Vision and Pattern Recognition, pp. 389–396 (2009)
13. Wang, J., Yang, J., Yu, K., Lv, F., Huang, T., Gong, Y.: Locality-constrained linear coding for image classification. In: IEEE Conference on Computer Vision and Pattern Recognition, pp. 3360–3367 (2010)
14. Wang, M., Ni, B., Hua, X., Chua, T.: Assistive tagging: A survey of multimedia tagging with human-computer joint exploration. ACM Computing Surveys (CSUR) 44(4), 25 (2012)
15. Yan, S., Xu, D., Zhang, B., Zhang, H., Yang, Q., Lin, S.: Graph embedding and extensions: A general framework for dimensionality reduction. IEEE Transactions on Pattern Analysis and Machine Intelligence 29(1), 40–51 (2007)
16. Yang, J., Yang, S., Fu, Y., Li, X., Huang, T.: Non-negative graph embedding. In: IEEE Conference on Computer Vision and Pattern Recognition, pp. 1–8 (2008)
17. Yao, B., Jiang, X., Khosla, A., Lin, A., Guibas, L., Fei-Fei, L.: Human action recognition by learning bases of action attributes and parts. In: IEEE International Conference on Computer Vision, pp. 1331–1338 (2011)
18. Yun, X.: Non-negative matrix factorization for face recognition. PhD thesis, Hong Kong Baptist University (2007)
19. Zhang, X., Zha, Z., Xu, C.: Learning verb-object concepts for semantic image annotation. In: Proceedings of the 19th ACM International Conference on Multimedia, pp. 1077–1080 (2011)

Robust Semantic Video Indexing
by Harvesting Web Images

Yang Yang[1,2], Zheng-Jun Zha[1], Heng Tao Shen[2], and Tat-Seng Chua[1]

[1] National University of Singapore, Singapore
{zhazj,chuats}@comp.nus.edu.sg
[2] The University of Queensland, Brisbane QLD 4072, Australia
{yang.yang,shenht}@itee.uq.edu.au

Abstract. Semantic video indexing, also known as video annotation, video concept detection in literatures, has attracted significant attentions recently. Due to the scarcity of training videos, most existing approaches can scarcely achieve satisfactory performances. This paper proposes a robust semantic video indexing framework, which exploits user-tagged web images to assist learning robust semantic video indexing classifiers. The following two challenges are well studied: (a) domain difference between images and videos; and (b) noisy web images with incorrect tags. Specifically, we first estimate the probabilities of images being correctly tagged as confidence scores and filter out the images with low confidence scores. We then develop a robust image-to-video indexing approach to learn reliable classifiers from a limited number of training videos together with abundant user-tagged images. A robust loss function weighted by the confidence scores of images is used to further alleviate the influence of noisy samples. An optimal kernel space, in which the domain difference between images and videos is minimal, is automatically discovered by the approach to tackle the domain difference problem. Experiments on NUS-WIDE web image dataset and Kodak consumer video corpus demonstrate the effectiveness of the proposed robust semantic video indexing framework.

1 Introduction

Recent years have witnessed an explosive growth of video data driven by the wide availability of massive storage devices, video cameras, fast networks, and media sharing sites (e.g., Youtube[1]). There is a compelling need for effective and efficient retrieval of video content. Many commercial search engines, such as Google[2] and Microsoft Bing[3], offer video search based on indexing textual metadata associated with videos. However, the problem with text-based video search is that the textual metadata are usually noisy, incomplete, and inconsistent with video content.

---

[1] http://www.youtube.com/
[2] http://www.google.com/
[3] http://www.bing.com/

S. Li et al. (Eds.): MMM 2013, Part I, LNCS 7732, pp. 70–80, 2013.

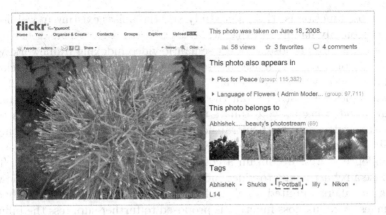

Fig. 1. A "flower" image tagged with *"football"* incorrectly

Recent research moves beyond text-based search and utilizes a set of semantic concepts as intermediate semantic descriptors to facilitate video search. The task of predicting the presence of semantic concepts in video clips is known as semantic video indexing, video annotation, or video concept detection in literatures [1,2,3,4,5,6,7]. Most existing semantic video indexing approaches are developed based on machine learning techniques, where large number of training samples are often required in order to achieve reasonable performance due to the complexity of visual content [8]. However, manual labelling of video data is extremely time-consuming and labor-intensive.

On the other hand, it is noticed that abundant user-tagged images are available on the Web and easy to collect. These images are valuable resources for assisting learning reliable semantic video indexing classifiers. However, it is nontrivial to involve these images in the learning process and two challenges are imposed: (a) **Domain difference**: images and videos are usually of different qualities and/or come from different domains, leading to great difference between their distributions; and (b) **Noisy samples**: many user-contributed tags of Flickr images are imprecise [9,10,11,12] and there are only around 50% tags actually related to the images [13]. For example, as illustrated in Fig. 1, a "flower" image is tagged with "football" incorrectly. If we involve such noisy image in learning a "football" classifier, the performance will be severely deviated. Recently, Jiang *et al.* [14] employed Flickr images with tags to calculate a new semantic metric to facilitate semantic video indexing. Yang *et al.* [15,16] integrated semi-supervised learning and transfer learning techniques to exploit Flickr images for video indexing. Although these methods studied the domain difference, none of them takes the noisy sample problem into account. A preliminary work that attempts to tackle both the domain difference and noisy sample problem was proposed in [17], where a cross-media transfer learning approach was proposed for improving video event recognition with noisily-tagged web images. An $\ell_{2,1}$-norm loss function was used instead of the traditional Frobenius loss to alleviate the influence of noisy samples. However, the denoising feature of

$\ell_{2,1}$-norm loss function is still under study and the noises existing in web images are not thoroughly handled in that work.

In this paper, we propose a robust semantic video indexing framework, which is able to not only tolerate noises in user-tagged web images, but also tackle the domain difference between images and videos. In order to alleviate the influence of noisy samples, we first filter out the images that are likely incorrectly-tagged. For each target concept, we estimate the visual relevance of the images to the concept. The relevance are then transformed to confidence scores indicating the probabilities that the images are correctly tagged. By filtering the images with low confidence scores, we obtain a collection of relatively clean images. Then, we propose a robust image-to-video indexing method which learns reliable video indexing classifiers a limited number of videos together with the relatively clean Web images. A robust loss function is proposed to further suppress the influence of noisy images. The loss function explicitly penalizes the ℓ_1 loss weighted by the confidence scores of images. The underlying intuition is that images with larger confidence scores should contribute more in the learning process. Moreover, the proposed approach discovers an optimal kernel space in which the domain difference between image and video data is minimal. By performing semantic indexing in this space, the domain difference problem is well addressed.

The rest of this paper is organized as follows. We elaborate the proposed robust semantic video indexing framework in Section 2. Section 3 reports experimental results and analysis, followed by conclusions in Section 4.

2   Robust Semantic Video Indexing Framework

In this section, we elaborate the proposed robust semantic video indexing framework. We first introduce the problem formulation and framework overview. Then, we depict the main components of the framework, including confidence score estimation, the robust image-to-video indexing algorithm and the solution.

2.1   Problem Formulation and Framework Overview

Given a semantic concept, let $\mathcal{Z} = \{(z_i, y_i)\}|_{i=1}^{N_V}$ denote a limited number of training samples, i.e., video keyframes. $z_i \in \mathbb{R}^d$ represents the visual feature of the ith video keyframe and $y_i \in \{-1, 1\}$ is the corresponding label. As aforementioned, it is difficult to learn a reliable classifier from the limited training videos, while the abundant user-tagged web images can be used as auxiliary training samples. Let $\mathcal{X} = \{(x_j, t_j)\}|_{j=1}^{N_I}$ denote a set of web images, in which some images are tagged with the given concept while the rest are not. $x_j \in \mathbb{R}^d$ represents the visual feature of the jth image and $t_j \in \{-1, 1\}$ indicates whether the image is tagged with the concept. The objective is to develop an effective semantic video indexing approach, which can learn a robust classifier from the limited training videos $\mathcal{Z}$ together with abundant user-tagged web images $\mathcal{X}$.

Fig. 2 illustrates the flowchart of the proposed framework. Given a set of concepts, we collect Flickr images tagged with each concept. Then, we employ a probabilistic approach to estimate the relevance of all the images to each concept. The relevance are further transformed to confidence scores which indicate the probabilities of the images being correctly tagged. We remove the images with low confidence scores to obtain a set of relatively clean images. These images together with the limited number of training videos are then fed into the proposed robust image-to-video indexing model, come out with reliable indexing classifiers. In online prediction, a test video sample is annotated with corresponding concepts by the offline-learned classifier.

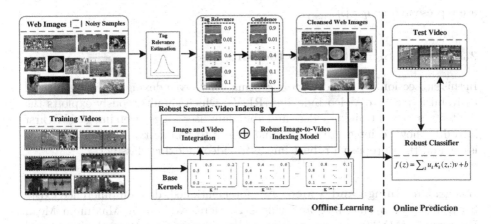

Fig. 2. Flowchart of the proposed robust semantic video indexing framework

2.2 Confidence Score Estimation

As aforementioned, a certain number of images in $\mathcal{X}$ are incorrectly tagged. In order to facilitate semantic video indexing with $\mathcal{X}$, it is required to alleviate influence from the noisy samples in $\mathcal{X}$. A key challenge is that we are not aware of which images are or not correctly tagged. Here, we propose to estimate the confidence scores of the images being correctly tagged.

Denote (x, t) as a tagged image, where $t \in \{-1, 1\}$ indicates if x is tagged with a particular concept c. Let $\eta(x, t|c) \in [0, 1]$ be a random variable indicating the probability that t is a correct tag for x. To obtain the estimation of $\eta(x, t|c)$, we first compute the visual relevance of x to c. We use the classic Kernel Density Estimation (KDE) to estimate the relevance $p(x|c)$ [13] as:

$$p(x|c) = \frac{1}{|\mathcal{X}_c|} \sum_{x_j \in \mathbf{X}_c} \kappa(x, x_j) \tag{1}$$

where $\mathcal{X}_c$ is the set of images that are tagged with c and $|\cdot|$ is the cardinality of a set. $\kappa(\cdot, \cdot)$ is a kernel function that measures the similarity between two images. Then, we estimate the confidence $\eta(x, t|c)$ from the relevance $p(x|c)$ as:

$$\eta(x, t|c) = \begin{cases} p(x|c), & \text{if } t = +1. \\ 1 - p(x|c), & \text{if } t = -1. \end{cases} \tag{2}$$

The underlying intuition of the above formula is as follows: (a) if x is tagged with c, i.e., $t = 1$, and it has high relevance to c, the probability of x being correctly tagged, i.e., $\eta(x, t|c)$ should be large; and (b) if x is not tagged with c, i.e., $t = -1$, and it has low relevance to c, the probability $\eta(x, t|c)$ should also be large. For brevity, hereafter we consistently use η_i short for $\eta(x_i, t_i|c)$. We filter out the images with low confidence scores. Given a threshold τ, we eliminate images with confidence scores less than τ from $\mathcal{X}$. As a result, we obtain a set of relatively clean images, denoted as $\tilde{\mathcal{X}} = \{(x_i, t_i)\}|_{i=1}^{\tilde{N}_I}$, where $\tilde{N}_I$ is the number of images in $\tilde{\mathcal{X}}$ and $\tilde{N}_I < N_I$. The corresponding confidence scores of the images are represented as $\eta = \{\eta_i\}|_{i=1}^{\tilde{N}_I}$.

2.3   Robust Semantic Video Indexing

In this subsection, we propose a robust semantic video indexing approach, termed as Robust Image-to-Video Indexing (RI2V), which simultaneously exploits the limited number of training videos and abundant user-tagged web images. We first introduce how to integrate images and videos to handle the domain difference issue, and then propose a robust image-to-video indexing classifier.

Integrating Image and Video. As aforementioned, domain difference exists between the web images $\tilde{\mathcal{X}}$ and videos $\mathcal{Z}$. Here, we employ Maximum Mean Discrepancy (MMD) to measure the domain difference in a Reproducing Kernel Hilbert Space (RKHS) $\mathcal{H}$ as:

$$mmd(\tilde{\mathcal{X}}, \mathcal{Z}) = \left\| \frac{1}{\tilde{N}_I} \sum_{i=1}^{\tilde{N}_I} \phi(x_i) - \frac{1}{N_V} \sum_{j=1}^{N_V} \phi(z_j) \right\|_{\mathcal{H}} \tag{3}$$

where $\phi : \mathbb{R}^d \to \mathcal{H}$ is a kernel mapping function and $\| \cdot \|_{\mathcal{H}}$ is the ℓ_2-norm in $\mathcal{H}$. By defining a column vector $\mathbf{s} = [\underbrace{1/\tilde{N}_I, \ldots, 1/\tilde{N}_I}_{\tilde{N}_I}, \underbrace{-1/N_V, \ldots, -1/N_V}_{N_V}]^T$, we can rewrite the above MMD criterion as below:

$$mmd^2(\tilde{\mathcal{X}}, \mathcal{Z}) = \text{Tr}(\mathbf{KS}) \tag{4}$$

where $\mathbf{S} = \mathbf{ss}^T$ and $\mathbf{K} = \begin{bmatrix} \mathbf{K}_{xx} & \mathbf{K}_{xz} \\ \mathbf{K}_{zx} & \mathbf{K}_{zz} \end{bmatrix}$ is a compound kernel matrix with $\mathbf{K}_{xx}$ and $\mathbf{K}_{zz}$ corresponding to the kernel matrices over images and videos, respectively. $\mathbf{K}_{xz} = \mathbf{K}_{zx}^T$ is the kernel matrix between images and videos. $\text{Tr}(\cdot)$ is the trace of a matrix.

In order to handle the domain difference issue, we intend to discover an optimal kernel space in which the distribution difference of $\tilde{\mathcal{X}}$ and $\mathcal{Z}$ is minimal. We here define the optimal kernel space as an optimized linear combination of a set of base kernels, denoted as $\{\mathbf{K}^{(k)}\}|_{k=1}^m$. The multiple kernel Maximum Mean

Discrepancy (MK_MMD) criterion is written as follows:

$$mkmmd^2(\tilde{\mathcal{X}}, \mathcal{Z}) = \mathrm{Tr}\left(\sum_{k=1}^{m} u_k \mathbf{K}^{(k)})\mathbf{S}\right) = \mathbf{u}^T \mathbf{p} \tag{5}$$

where $\mathbf{u} = [u_1,\ldots,u_m]^T$ is the to-be-optimized combination coefficients. $u_k \geq 0, k = 1, 2, \ldots, m$ and $\sum_k u_k = 1$. $\mathbf{p} = [\mathrm{Tr}(\mathbf{K}^{(1)}\mathbf{S}),\ldots,\mathrm{Tr}(\mathbf{K}^{(m)}\mathbf{S})]^T$.

Robust Image-to-Video Indexing Classifier. We aim to learn a robust classifier $f : \mathbb{R}^d \to \{-1, 1\}$ for semantic video indexing:

$$f(z) = \kappa_{\mathbf{u}}(z, :)v + b \tag{6}$$

where $\kappa_{\mathbf{u}}(z, :) = [\kappa_{\mathbf{u}}(z, x_1),\ldots, \kappa_{\mathbf{u}}(z, x_{\tilde{N}_I}), \kappa_{\mathbf{u}}(z, z_1),\ldots, \kappa_{\mathbf{u}}(z, z_{N_V})]$ and $\kappa_{\mathbf{u}}(\cdot, \cdot) = \sum_k u_k \phi_k^T(\cdot)\phi_k(\cdot), k = 1, 2, \ldots, m$. $\{\phi_k\}|_{k=1}^{m}$ are the kernel mapping functions corresponding the m base kernels. $v \in \mathbb{R}^{\tilde{N}_I + N_V}$ is the classification parameter and $b \in \mathbb{R}$ is the bias.

In order to learn v, b and $\mathbf{u}$, we propose a robust loss function as below:

$$\mathcal{L}(\tilde{\mathcal{X}}, \eta, \mathcal{Z}|\mathbf{u}, v, b) = \sum_{i=1}^{\tilde{N}_I} \eta_i |t_i - f(x_i)| + \sum_{j=1}^{N_V} |y_j - f(z_j)| + \lambda \Omega(\mathbf{u}, v) \tag{7}$$

where the first two terms penalize the errors of $\tilde{\mathcal{X}}$ and $\mathcal{Z}$, respectively. $\Omega(\mathbf{u}, v)$ is a regularization term and λ is a tradeoff parameter. In this work we choose the ridge regularization, i.e., $\Omega(\mathbf{u}, v) = v^T(\sum_k u_k \mathbf{K}^{(k)})v$.

Note that we use ℓ_1 loss rather than the squared loss. Thus, the errors from noisy samples will not be squared. Moreover, we weight the ℓ_1 loss by the confidence scores of samples to further suppress the influence of noisy samples. For the clean training videos, their confidence scores are all set to 1. Then, by combining Eq.(5) and Eq.(7) we arrive at:

$$\min_{v,b,\mathbf{u},\mathbf{u} \geq 0 \wedge \mathbf{u}^T \mathbf{1}=1.} \alpha(\mathbf{u}^T \mathbf{p}\mathbf{p}^T \mathbf{u}) + \sum_{i=1}^{\tilde{N}_I} \eta_i |t_i - \kappa_{\mathbf{u}}(x_i, \cdot)v - b|$$
$$+ \sum_{j=1}^{N_V} |y_j - \kappa_{\mathbf{u}}(z_j, \cdot)v - b| + \lambda(v^T(\sum_k u_k \mathbf{K}^{(k)})v) \tag{8}$$

where $\alpha \geq 0$ is a balance parameter.

Solution. We propose an alternating algorithm to solve the above problem. For brevity, we rewrite Eq.(8) as:

$$\min_{v,b,\mathbf{u},\mathbf{u} \geq 0 \wedge \mathbf{u}^T \mathbf{1}=1.} \alpha(\mathbf{u}^T \mathbf{p}\mathbf{p}^T \mathbf{u}) + \sum_{i=1}^{N} \tilde{\eta}_i |\tilde{y}_i - \kappa_{\mathbf{u}}(x_i, \cdot)v - b| + \lambda(v^T(\sum_k u_k \mathbf{K}^{(k)})v) \tag{9}$$

where $N = \tilde{N}_I + N_V$, $\tilde{\eta} = [\tilde{\eta}_1, \ldots, \tilde{\eta}_N]^T = [\eta_1, \ldots, \eta_{\tilde{N}_I}, 1, \ldots, 1]^T$, $\tilde{y} = [\tilde{y}_1, \ldots, \tilde{y}_N]^T = [t_1, \ldots, t_{\tilde{N}_I}, y_1, \ldots, y_{N_V}]^T$ and $x_{\tilde{N}_I + j} = z_j, j = 1, 2, \ldots, N_V$.

First, by fixing $\mathbf{u}$ we obtain the following sub-problem:

$$\min_{v,b} \sum_{i=1}^{N} \tilde{\eta}_i |\tilde{y}_i - \kappa_{\mathbf{u}}(x_i, \cdot)v - b| + \lambda(v^T \mathbf{K}_{\mathbf{u}} v) \tag{10}$$

where $\mathbf{K}_{\mathbf{u}} = \sum_k u_k \mathbf{K}^{(k)}$. By setting the derivative w.r.t. b to 0, we arrive at:

$$b = (\tilde{\eta}^T D(\tilde{y} \circ \tilde{\eta}) - \tilde{\eta}^T D \tilde{\mathbf{K}}_{\mathbf{u}} v)/(\tilde{\eta}^T D \tilde{\eta}) \tag{11}$$

where $D \in \mathbb{R}^{N \times N}$ is a diagonal matrix with $D_{ii} = 1/(2\tilde{\eta}_i |\tilde{y}_i - \kappa_{\mathbf{u}}(x_i, \cdot)v - b|)$. $\tilde{\mathbf{K}}_{\mathbf{u}} = \mathbf{K}_{\mathbf{u}} \circ (\tilde{\eta} \mathbf{1}_N^T)$ and $\circ$ denotes Hadamard product. $\mathbf{1}_N$ is an all-one column vector. Then we substitute b into Eq.(10) and get:

$$\min_{v,b} (H \tilde{\mathbf{K}}_{\mathbf{u}} v - H(\tilde{y} \circ \tilde{\eta}))^T D(H \tilde{\mathbf{K}}_{\mathbf{u}} v - H(\tilde{y} \circ \tilde{\eta})) + \lambda(v^T \mathbf{K}_{\mathbf{u}} v) \tag{12}$$

where $H = I_n - \tilde{\eta} \tilde{\eta}^T D/(\tilde{\eta}^T D \tilde{\eta})$. By setting the derivative w.r.t v to 0, we get:

$$v = (\tilde{\mathbf{K}}_{\mathbf{u}} H D H \tilde{\mathbf{K}}_{\mathbf{u}} + \lambda \mathbf{K}_{\mathbf{u}})^{-1} \tilde{\mathbf{K}}_{\mathbf{u}} H D H(\tilde{y} \circ \tilde{\eta}) \tag{13}$$

Then, by fixing v and b we get the sub-problem as below:

$$\min_{\mathbf{u}, \mathbf{u} \geq 0 \wedge \mathbf{u}^T \mathbf{1} = 1.} \alpha(\mathbf{u}^T \mathbf{p} \mathbf{p}^T \mathbf{u}) + \sum_{i=1}^{N} \tilde{\eta}_i |\tilde{y}_i - \kappa_{\mathbf{u}}(x_i, \cdot)v - b| + \mathbf{u}^T \mathbf{h} \tag{14}$$

where $\mathbf{h} = [\lambda v^T \mathbf{K}^{(1)} v, \cdots, \lambda v^T \mathbf{K}^{(m)} v]^T$. We may rewrite the above problem as the following QP problem:

$$\min_{\mathbf{u}, \mathbf{u} \geq 0 \wedge \mathbf{u}^T \mathbf{1} = 1.} \mathbf{u}^T A \mathbf{u} + \mathbf{u}^T \mathbf{d} \tag{15}$$

where $A = \alpha \mathbf{p} \mathbf{p}^T + M$. $M = [q_j^{(1)}, \ldots, q_j^{(m)}]^T D[q_j^{(1)}, \ldots, q_j^{(m)}]$ and $q^{(k)} = v^T \tilde{\mathbf{K}}^{(k)}$, for $k = 1, 2, \ldots, m$. $\mathbf{d} = \mathbf{h} - 2\mathbf{g}$. $\mathbf{g} = [r^T D \tilde{\mathbf{K}}^{(1)} v, \ldots, r^T D \tilde{\mathbf{K}}^{(m)} v]^T$. $r = \tilde{y} \circ \tilde{\eta} - \tilde{\eta} b$. $\tilde{\mathbf{K}}^{(k)} = \mathbf{K}^{(k)} \circ (\tilde{\eta} \mathbf{1}_N^T), k = 1, 2, \ldots, m$.

Finally we summarize the optimization of the problem in Algorithm 1.

3   Experiments

We conduct experiments to evaluate the proposed semantic video indexing framework on two multimedia datasets, i.e., NUS-WIDE web image dataset and Kodak consumer video collection.

Algorithm 1. An alternating algorithm for optimizing our model

Input: Cleansed Images $\tilde{\mathcal{X}}$, Video Dataset $\mathcal{Z}$ and confidence scores $\tilde{\eta}$;
Output: $v, b, \mathbf{u}$;
1: Generate m basic kernels $\{\mathbf{K}^{(1)}, \mathbf{K}^{(2)}, ..., \mathbf{K}^{(m)}\}$ based on both $\tilde{\mathcal{X}}$ and $\mathcal{Z}$;
2: Randomly initialize $\mathbf{u}$;
3: **repeat**
4:    Compute the diagonal matrix D with $D_{ii} = 1/(2\tilde{\eta}_i|\tilde{y}_i - \kappa_{\mathbf{u}}(x_i, \cdot)v - b|)$;
5:    Compute v according to Eq.(13);
6:    Compute b according to Eq.(11);
7:    Update D with $D_{ii} = 1/(2\tilde{\eta}_i|\tilde{y}_i - \kappa_{\mathbf{u}}(x_i, \cdot)v - b|)$;
8:    Update $\mathbf{u}$ by solving the QP problem in Eq.(15);
9: **until** convergence
10: **return** $v, b, \mathbf{u}$;

3.1 Data and Experimental Setting

In our experiments, the web images are from the NUS-WIDE [18] which consists of $269, 648$ Flickr images associated with $5, 018$ user-provided tags. The experimental videos are from the Kodak consumer video corpus [19], which contains 25 concepts and $5, 166$ keyframes extracted from $1, 358$ videos. We selected 18 concepts[4] that are covered by both NUS-WIDE and Kodak datasets as our experimental concepts. For each concept, we randomly sampled $1, 000$ images from NUS-WIDE as experimental images. We divided the video keyframes into two subsets: 30% as training samples and 70% for testing. We used the samples annotated with at least one of the 18 concepts. This gives rise to $51, 150$ images and 746 videos ($2, 996$ video keyframes). For each image or video keyframe, we extracted four types of visual features, i.e., $1, 000$-D bag-of-visual-words based on SIFT descriptors, 150-D color moments, 100-D edge histogram and 108-D haar wavelet.

For each type of the visual features, we constructed 20 base kernel matrices based on the four types of kernels, i.e., Gaussian kernel (i.e., $k(\cdot, \cdot) = \exp(-\gamma d^2(\cdot, \cdot))$), Laplacian kernel (i.e., $k(\cdot, \cdot) = \exp(-\sqrt{\gamma}d(\cdot, \cdot))$), Inverse Square Distance kernel (i.e., $k(\cdot, \cdot) = 1/(\gamma d^2(\cdot, \cdot) + 1)$) and Inverse Distance kernel (i.e., $k(\cdot, \cdot) = 1/(\sqrt{\gamma}d(\cdot, \cdot) + 1))$, with γ varying as $\frac{1}{\bar{d}}\{2^{-4}, 2^{-2}, 2^0, 2^2, 2^4\}$, where $\bar{d}$ is the mean distance over all the sample pairs. In particular, we computed $d(\cdot, \cdot)$ by χ^2 distance for the histogram-based features, i.e., $1, 000$-D bag-of-visual words and 100-D edge histogram, and Euclidean distance for the other features. Moreover, we constructed two histogram intersection kernel matrices for $1, 000$-D bag-of-visual-words and 100-D edge histogram, respectively. This gives rise to 82 ($4 \times 20 + 2$) base kernel matrices in total.

To estimate tag relevance, we used Gaussian kernel to compute four similarities based on four types of visual features, respectively. The overall similarity

---

[4] There are 21 concepts co-occurring in NUS-WIDE and Kodak datasets. Three of them are audio-oriented tags, i.e., "cheer", "music" and "singing", and were not used in the experiments.

between two images was calculated as the mean value of the four similarities between them. The kernel bandwidth parameter was set as mean distance over all the sample pairs. The threshold τ was empirically set to 0.25. We compared the proposed robust semantic video indexing approach, i.e., RI2V, with the following methods: (a) Robust cross-media transfer (RCMT) [17] which attempts to handle the noisy sample problem simply relying on a ℓ_1 loss function; (b) Cross-media transfer (CMT) that adopts an ℓ_2 loss instead of the ℓ_1 loss in RCMT; and (c) I2V: a variant of the proposed RI2V method. I2V is different from RI2V in that it uses ℓ_2 loss instead of ℓ_1 loss. The tradeoff parameters in all the evaluated methods were tuned in the range of $\{10^{-4}, 10^{-2}, 10^0, 10^2, 10^4\}$, and the best results were reported. We adopted the Average Precision (AP) as the performance metric, and the mean Average Precision (MAP) over all the concepts was used as the overall performance metric.

3.2   Results

Fig. 3 shows the MAP over all the 18 concepts and the detailed AP on each individual concept. From these results, we can obtain the following observations:

1. The proposed RI2V method outperforms the other evaluated semantic video indexing approaches in terms of MAP. The relative improvements over RCMT, I2V, and CMT are 9.8%, 35.2%, and 98.7%, respectively. This implies that RI2V possesses better robustness towards noisy samples and domain difference.
2. RI2V and I2V consistently achieve better performances than RCMT and CMT, respectively. Although RI2V and I2V have less training images, they perform much more robustly. This is because (a) we eliminate the noisy samples with low confidence scores to obtain a set of relatively clean images, and (b) The learning errors on the training images in RI2V and I2V

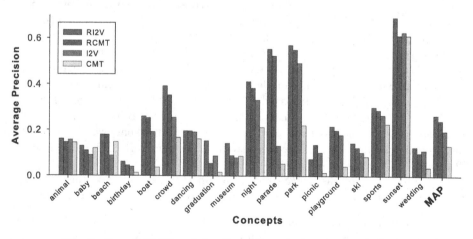

Fig. 3. Overall MAP and detailed AP on each individual concept

are weighted by the confidence scores of the images. This leads to better robustness towards noisy samples.
3. Moreover, the methods that employ ℓ_1 loss (i.e., RI2V and RCMT) perform better than their counterparts that use the squared ℓ_2 loss (i.e., I2V and CMT), respectively. This gives us a hint that ℓ_1 loss can help to prevent the unexpected errors caused by noisy samples.
4. RI2V achieves the best AP on most individual concepts. The improvements on some concepts are significant.

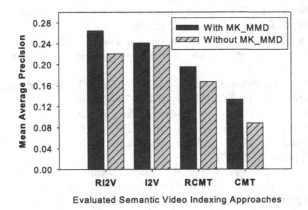

Fig. 4. MAPs of all the evaluated approaches vs. their corresponding counterparts without MK_MMD

Moreover, we investigated the effects of tackling the domain difference problem between images and videos by minimizing MK_MMD (see Eq.(5)). As illustrated in Fig.4, all the evaluated approaches with MK_MMD achieve better performances than their corresponding counterparts that do not use MK_MMD. The main reason is that minimizing MK_MMD leads to an optimal kernel space for learning reliable classifiers. The domain difference between images and videos is minimal in the space.

4   Conclusion

In this paper, we proposed a new robust semantic video indexing approach which leverages user-tagged web images to enhance semantic video indexing. In order to overcome the challenges imposed by noisy samples and domain difference between images and videos, we first estimated the confidence scores indicating the probabilities of images being correctly tagged. Images with low confidence scores were filtered out. We then proposed a robust image-to-video indexing method which can find an optimal kernel space in which domain difference between images and videos is minimal and learn a robust classifier from a limited number of training videos together with abundant user-tagged images. Experiments on two real-world multimedia datasets demonstrated the superiority of the proposed approach.

References

1. Zha, Z.J., Wang, M., Zheng, Y.T., Yang, Y., Hong, R., Chua, T.S.: Interactive video indexing with statistical active learning. TMM 14(1), 17–27 (2012)
2. Wang, M., Hua, X., Tang, J., Hong, R.: Beyond distance measurement: Constructing neighborhood similarity for video annotation. TMM 11(3), 465–476 (2009)
3. Geng, B., Li, Y., Tao, D., Wang, M., Zha, Z.J., Xu, C.: Parallel lasso for large-scale video concept detection. TMM 14(1), 55–65 (2012)
4. Zha, Z., Mei, T., Wang, J., Wang, Z., Hua, X.: Graph-based semi-supervised learning with multiple labels. JCVIR 20(2), 97–103 (2009)
5. Shen, J., Tao, D., Li, X.: Modality mixture projections for semantic video event detection. TCSVT 18(11), 1587–1596 (2008)
6. Mei, T., Zha, Z., Liu, Y., Wang, M., Qi, G., Tian, X., Wang, J., Yang, L., Hua, X.: Msra at trecvid 2008: High-level feature extraction and automatic search. In: TRECVID (2008)
7. Wang, M., Hua, X., Hong, R., Tang, J., Qi, G., Song, Y.: Unified video annotation via multigraph learning. TCSVT 19(5), 733–746 (2009)
8. Yang, Y., Yang, Y., Huang, Z., Shen, H., Nie, F.: Tag localization with spatial correlations and joint group sparsity. In: CVPR, pp. 881–888 (2011)
9. Zha, Z.J., Yang, L., Mei, T., Wang, M., Wang, Z.: Visual query suggestion. In: ACM MM, pp. 15–24 (2009)
10. Zha, Z., Yang, L., Mei, T., Wang, M., Wang, Z., Chua, T., Hua, X.: Visual query suggestion: Towards capturing user intent in internet image search. TOMCCAP 6, 13:1–13:19 (2010)
11. Yang, Y., Huang, Z., Shen, H.T., Zhou, X.: Mining multi-tag association for image tagging. WWW 14(2), 133–156 (2011)
12. Wang, M., Yang, K., Hua, X., Zhang, H.: Towards a relevant and diverse search of social images. TMM 12(8), 829–842 (2010)
13. Liu, D., Hua, X., Yang, L., Wang, M., Zhang, H.: Tag ranking. In: WWW, pp. 351–360 (2009)
14. Jiang, Y., Ngo, C., Chang, S.: Semantic context transfer across heterogeneous sources for domain adaptive video search. In: ACM MM, pp. 155–164 (2009)
15. Yang, Y., Yang, Y., Huang, Z., Shen, H.T.: Transfer tagging from image to video. In: ACM MM, pp. 1137–1140 (2011)
16. Yang, Y., Yang, Y., Shen, H.: Effective transfer tagging from image to video. TOMCCAP (2012)
17. Yang, Y., Yang, Y., Huang, Z., Ma, Z.: Robust cross-media transfer for visual event detection. In: ACM MM. (2012)
18. Chua, T.S., Tang, J., Hong, R., Li, H., Luo, Z., Zheng, Y.T.: Nus-wide: A real-world web image database from national university of singapore. In: CIVR (2009)
19. Yanagawa, A., Loui, A.C., Luo, J., Chang, S.F.: Kodak consumer video benchmark data set: concept definition and annotation (2008)

Interactive Evaluation of Video Browsing Tools

Werner Bailer[2], Klaus Schoeffmann[1], David Ahlström[1], Wolfgang Weiss[2],
and Manfred Del Fabro[1]

[1] Alpen-Adria-Universität Klagenfurt, Austria
{ks,manfred}@itec.aau.at, david.ahlstroem@aau.at
[2] DIGITAL – Institute for Information and Communication Technologies
JOANNEUM RESEARCH Forschungsgesellschaft mbH, Graz, Austria
{werner.bailer,wolfgang.weiss}@joanneum.at

Abstract. The Video Browser Showdown (VBS) is a live competition
for evaluating video browsing tools regarding their efficiency at known-
item search (KIS) tasks. The first VBS was held at MMM 2012 with eight
teams working on 14 tasks, of which eight were completed by expert users
and six by novices. We describe the details of the competition, analyze
results regarding the performance of tools, the differences between the
tasks and the nature of the false submissions.

1 Introduction

The Video Browser Showdown (VBS) is a live video browsing competition where
international researchers, working in the field of interactive video search, evaluate and demonstrate the efficiency of their tools in presence of the audience. The
aim of the VBS is to evaluate video browsing tools for efficiency at known-item
search (KIS) tasks with a well-defined data set in direct comparison to other
tools. For each task the moderator presents a target clip on a shared screen that
is visible to all participants. The participants use their own computer to perform an interactive search in the specified video file taken from a common data
set. The performance of participating tools is evaluated in terms of successful
submissions and search time. The first VBS was held at the 18th International
Conference on MultiMedia Modeling (MMM 2012) in Klagenfurt, Austria, where
eight international teams participated (see [1] for descriptions of their systems).
The setup of the room is shown in Figure 1. In this paper, we describe the details of the competition, analyze results and evaluate data collected during the
competition for further insights.

2 Competition Details

The VBS consisted of 14 tasks. The first eight tasks were performed by experts
(the developers of the tools) and the last six tasks were performed by eight
volunteers randomly assigned to one of the tools. For each task the corresponding
video containing a randomly selected target clip was mentioned first, in order to

S. Li et al. (Eds.): MMM 2013, Part I, LNCS 7732, pp. 81–91, 2013.
© Springer-Verlag Berlin Heidelberg 2013

Fig. 1. Photo of the VBS competition at MMM 2012 (credit: Rene Kaiser)

allow the participants to set up their systems for that video. After that, the 20 seconds long target clip was projected on a wall and the sound was played on loudspeakers. After the presentation, participants were given a maximum of two minutes to find the target clip. Participants were allowed to start their search already while the target clip was presented but could miss important information as the target clip did not necessarily contain redundant content. The collection from which videos for the tasks were randomly selected (see Table 1) consisted of 30 videos with an average length of 77 minutes (min: 31, max: 139).

Participants submitted found segments to an HTTP server with an URI that contained the following information: team, video, start and end frame number of the submitted segment. The server was responsible for three tasks: (1) checking whether the submitted segment was correct (i.e., overlapped with the target clip), (2) measuring the task solve time and (3) computing scores for all teams and tasks. In each run (expert and novice), the team with the highest sum of scores was the winner. A submitted segment was considered as correct if $(S_i - 125) \leq s_i \leq e_i \leq (E_i + 125)$, where s_i and e_i are the submitted start and end frame numbers and S_i and E_i are the begin and end frame numbers of the target clip for task i. For each task a maximum score of 100 could be obtained. The score was dependent on the task solve time (t_i) and a penalty based on the number of wrong submissions (w_i) for a task i before the correct submission was received: $score = (1/\max(1; w_i - 1))(100 - 50\frac{t_i}{T_{max}})$. T_{max} is the maximum time allowed for each task, i.e. 120 seconds. During the search, the current score, overall score, number of correct and wrong submission for all teams were projected on the wall.

Table 1. Videos used in VBS (expert tasks at top, novice tasks at bottom)

Task	Video	Language	Genre	Duration
1	8	JP+EN	Documentation	01:15:01
2	11	NL	News	01:07:30
3	1	IT	Talkshow	01:50:02
4	26	IT	Talkshow	01:27:56
5	14	NL	News	01:05:06
6	9	IT	Talkshow	01:32:33
7	20	DE	News	00:55:32
8	13	NL	News	00:30:36
9	5	JP+EN	Documentation	01:15:01
10	20	DE	News	00:55:32
11	15	EN	News	00:55:59
12	29	JP+EN	Documentation	01:15:01
13	13	NL	News	00:30:36
14	28	JP+EN	Documentation	01:15:00

3   Evaluation

In the following, we describe the use of the data gathered during the VBS to compare the performance of tools, the different tasks and analyze the false submissions.

3.1   Comparison of Tools

Given the low number of completed tasks used in the VBS – eight for the experts and six for the novices – we limit our analyses to descriptive statistics. Figure 2 shows the combined team scores and scores from the expert and novice runs separately. Overall, the median score was 85.5 points (mean 65.7, s.d. 38.9). Team 1 had the best total score (expert + novice) of 1130 points and was closely followed by Team 2 and Team 8 with 1061 and 1048 points respectively. With scores between 853 and 933 points, there were close calls in the middle field between Team 3, 4, 5 and 6. Team 7 had the lowest score which was 55% lower than the winning Team 1 and 40% lower than the second last team, Team 6. A similar order among the teams is also visible in the data from the expert run. In the novice run the teams were positioned somewhat closer together with Team 5 and 8 having the highest scorers.

The box plot in Figure 3 provides a more detailed overview of how the different browsers performed. Team 1 was impressively consistent and scored 92 or above in 11 of the 14 tasks, resulting in a median score of 95.5. Task scores for the other teams were more varied, as indicated by the rather large interquartile ranges shown in Figure 3. Five of the expert users managed to score 100 points in at least one of the tasks and all of them failed in finding the correct segment in one or more tasks, and thus ended up with no points in these tasks. Generally, the novices were more consistent in their scoring than the experts were (however,

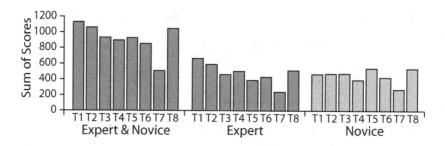

Fig. 2. Total team scores (T1 to T8) and scores in the expert run and novice run separately

note that experts performed two more tasks). All but one of the novices (Team 2) had a higher median score than their expert team colleague. Except from the novice user in Team 2, all novices scored full points in at least one task and three of them (Team 2, 5 and 8) were successful in all of their six tasks, achieving scores of 20 points or above.

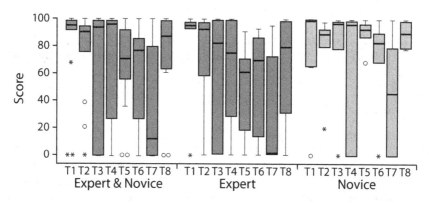

Fig. 3. Box plot of combined team scores (expert+novice), and separated by expert and novice (o: >1.5 IQR, ⋆: >3 IQR)

In summary, with the limited data collected, no inferential statistical procedure is applicable that let us draw any conclusions about performance differences between the browsers. However, the above analysis provides indications that some browser designs might have been more suitable than others for the video browsing tasks we used. In particular, we see a tendency of consistent and good performance of Team 1's browser. Similarly, the low scores for Team 7 might also indicate a sub-optimal design. A detailed description of these two browsers is provided in Section 4.

Figure 4 shows the relation between submission time and score. It is apparent that the submission time is the most important component in the score, resulting in a linear dependency along the maximum score possible at a certain time into

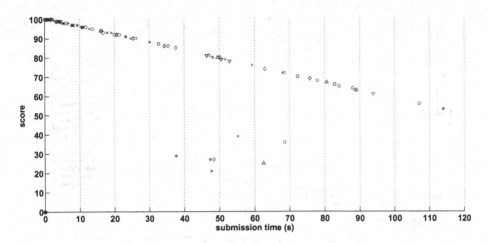

Fig. 4. Score vs. submission time of correct results (red = experts, blue = novices, the symbols denote the different teams)

the task. Only 7 of the 88 correct submissions have reduced score due to prior false submission (5 from the expert run, and of those 2 from the same team, and 2 from the novice run). It is interesting that none of the false submissions after which a correct result was submitted, were submitted before 70s into the task time (i.e., mostly in the first half of the working time, while after that users seem to have a clear idea about the distinctive features of the target clip).

3.2   Comparison of Tasks

Figure 5 shows the correct and false submissions for expert and novice runs. It is evident that novice users submitted a lower number of false results than expert users at a comparable number of correct ones. Furthermore, novice users made their correct submissions no later than 85s after trial start, whereas expert users often submitted both correct and false results until just before the allowed 120 seconds had elapsed. One interesting observation from these results is that the lines for correct and false submission in the figure rarely cross after the first 10-20 seconds, i.e., for the rest of the task duration, the number of false submissions always either stays below or above the number of correct submissions. Only for the first two tasks in the expert run several crossings occur, while in the other cases one can already predict the average success rate quite well after 20 seconds. As these were the first two tasks performed in the competition, we probably cannot draw any conclusions from this fact.

Can we draw conclusions regarding the difficulty of each task from the measured results? Looking at the scores per task (Figure 6), we can see that tasks 3, 4, 6 and 12 have lower mean scores than all others, and that at least the lower quartile of scores is 0. We see a similar pattern in the submission times of the correct results (setting it to 120 seconds if no correct result has been

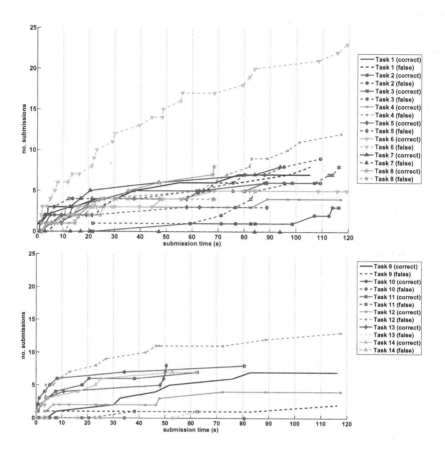

Fig. 5. Correct (solid) and false (dashed) submissions over time per task for expert (top) and novice runs (bottom)

submitted): Tasks 3, 4, and 12 have clearly higher mean duration (for task 6 it is only slightly higher than the others), but for all of them at least the upper quartile is at the maximum duration of the task. The false submissions show a slightly different picture: Tasks 4, 6 and 12 have the highest number of false submissions. However, task 3 has only one false submission, probably due to the fact that the video contains no other segment sufficiently similar to the target clip than the correct one. The contrary is true for tasks 2 and 8: both have a rather high number of false submissions, but the mean submission time is only 30 seconds, and more than 75% of the teams completed the task in time (some after an initial false submissions). It seems that the submission time and the score (which is linearly dependent on submission time for tasks with few false positives) are good indicators of tasks for which it is hard to locate a relevant segment. However, these measures do not capture the cases where a number of similar candidate segments increases the risk of making a false submission. Looking at the submissions over time in Figure 5, we see that the tasks that

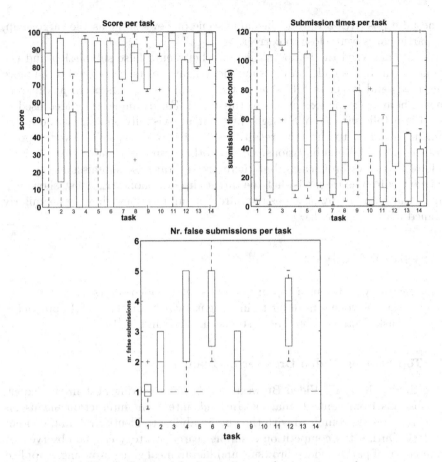

Fig. 6. Scores, submission time and misses per task

are salient in the plots in Figure 6 (task 3, 4, 6, 12) have consistently a higher number of false than correct submissions from about 20s into the task time to the end.

3.3 Analysis of False Submissions

In total, 172 results were submitted for all the tasks, 88 were correct and 84 were false. In order to answer the question "how false are the false submissions" we manually examined each of the false submissions. It turns out that 82% of them are globally visually similar to the target clip (i.e., have a similar layout and colors), 38% have a similar background as the target clip, and 49% show a similar object as the target clip. This means that depending on the type of material request, many of these results could already be useful results for a material search task in a media production scenario. Apart from visually similarity, a request could be to find a video clip about a certain topic. Here we see that 46%

of the false positives are from the same scene or news story as the target clip, i.e., participants were close to correct.

We also analyzed the temporal distance between a false submission and the ground truth. It is smaller in news and documentaries (as they tend to be more structured) and larger in talk shows, where similar segments occur at different times. There are some exceptions to this trend. For example, the video used in task 6 is a talk show, but all segments that are visually similar to the query segment are close to the true result, resulting in an atypically low temporal distance of the false submissions. On the other hand, the video in task 11 is made of two news broadcasts, but there are several false submissions far from the true result. The reason is that the target clip is a phone interview, showing a map of Haiti, and many false positives are from an interview showing a similarly designed map of Sri Lanka.

4 System Analysis

In this section we present a top and a low scoring video browsing system that attended the showdown in order to investigate which are successful approaches for a KIS task. Details about the systems can be found in [1].

4.1 Top Scoring Video Browser of Team 1

Interestingly, the AAU Video Browser (see Fig. 7), scoring best in the expert run, abstains from content analysis. Instead, intelligent interaction means for video browsing are combined with the human ability to identify relevant content very fast. During the competition a simple search strategy can be observed. A combination of parallel video browsing and hierarchical video browsing is applied to all tasks. First the parallel mode was used to divide each video into four or even nine equal parts to be shown in parallel. The decision to use four or nine video windows depends on type of the video. If a video contains a lot of scenes with similar content, only four parts were used, whereas for videos containing a lot of scenes with different content nine parallel windows were chosen. By dragging and pulling the timeline slider, all parts can be observed in parallel and thus it is possible to quickly get an overview of a video. If a candidate segment has been located, the hierarchical browsing mode can be used to investigate it in detail. This is done by recursively applying the parallel mode only to the candidate segment.

The VBS showed that this simple search strategy, combined with intelligent and easy to use interaction means, is very efficient for KIS tasks. In contrast to the human mind, an artificial concept classifier is only able to identify a limited amount of trained concepts. The dataset contained a broad range of different videos. It seems that the concept-based approaches are too limited regarding the range of concepts they are able to identify in these videos. Therefore, not the most sophisticated tool was successful, but the one that supported the interactive search task best.

Fig. 7. Screenshot of the best scoring tool (AAU Video Browser)

4.2 Low Scoring Video Browser of Team 7

Team 7 used a video browsing application targeted for content management in post-production of film and TV. It works based on automatic content analysis that performs camera motion estimation, visual activity estimation, extraction of global color features and estimation of object trajectories. The central component of the browsing tool's user interface is a light table that shows the current content set and cluster structure using a number of representative frames for each of the clusters (see also Fig. 8). The users apply an iterative content selection process, where they cluster content or search similar keyframes based on the extracted features and can then select relevant keyframes to reduce the content set.

Analysis on task level: The team ended at the bottom of the list with 510 points. The expert finished 3 out of 8 tasks successfully and needed on average 51 seconds for a successful task. The novice needed nearly the same amount of time of 49 seconds on average for a successful task and completed 4 out of 6 tasks successfully.

Analysis on submission level: At the expert run, the user made 16 false and 3 successful submissions. A wrong submission took in average 27 seconds where the user did about 10 interactions (e.g., clustering, similarity search, preview) on the user interface. In 94% of the cases of wrong submissions the user submitted a visually similar keyframe and 75% of the wrong submissions were in a scene of the target clip. This means the user was temporally and visually near the searched item but still failed. A successful submission took on average 42 seconds with 19 interactions. The novice made 6 wrong and 4 successful submissions. In contrast to the expert, the novice submitted only in 17% of the wrong submissions a visually similar keyframe and 50% of the wrong submissions were in a scene of the target clip. A wrong submission took 32 seconds on average with 11.5 interactions and a successful submission took 38 seconds on average with 14.5 interactions.

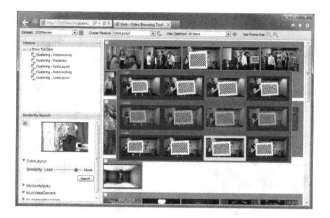

Fig. 8. Screenshot of the video browsing tool from team 7

5   Conclusion

We have presented an analysis of the results of the Video Browser Showdown, an interactive evaluation of video browsing tools by solving known-item search (KIS) tasks, where eight teams worked on 14 tasks. The intent of the Video Browser Showdown is to demonstrate the efficiency of interactive video search tools for the real world scenario, where users want to quickly find content in videos, which is known to the user but cannot be found or effectively described by a query with a typical video retrieval tool. Because the competition addresses both expert and novice users, it should promote research on interactive video search tools that allow efficient search for the majority of users. It is an encouraging result that a correct keyframe was submitted before the allowed 120 seconds in 79% of the trials. Also, many of the false results were visually very similar and/or belonged to the same semantic unit of the content.

We have analyzed the tools of the best and the worst performing teams in order to get additional insigths about what approaches work and do not work for KIS tasks. It seems that for KIS tasks working with a small dataset, simplicity is the key. This observation holds for both expert and novice users, which showed surprisingly little difference in their performance. The analysis of the different tasks shows that four of the fourteen tasks were particularly difficult, an observation that is supported by several of the measured parameters. Users tend to solve tasks with a first correct submission or not at all. The majority of all false submissions (72 out of 84) were registered in trials where no correct result has been submitted. Out of 88 correct submissions, 19 (22%) were made after a false submission, but only 7 (8%) received a reduced score due to penalties for multiple prior false submissions. As the scores were mainly related to the submission time, it could make sense to penalize also the first false submission in future competitions. It would also be interesting to compare different types of tasks, e.g., KIS on multi-item collections, as well as to consider related task settings motivated by problems in media production workflows. Furthermore, in order to allow for a deeper analysis and better understanding of how users interact with the system for particular tasks, systems participating in future events of this competition should implement an appropriate logging feature and provide log data for post-hoc analysis.

In summary, besides providing useful insights regarding the performance of various video browsing tools, the VBS competition did also allow participating teams to demonstrate their tools in a fun and exiting setting that was highly appreciated by other conference participants as a suitable round-up of the conference days.

Acknowledgments. The research leading to these results has received funding from the European Union's Seventh Framework Programme (FP7/2007-2013) under grant agreement n° 287532, "TOSCA-MP".

Reference

1. Schoeffmann, K., Merialdo, B., Hauptmann, A.G., Ngo, C.-W., Andreopoulos, Y., Breiteneder, C. (eds.): MMM 2012. LNCS, vol. 7131. Springer, Heidelberg (2012)

Paint the City Colorfully: Location Visualization from Multiple Themes

Quan Fang[1,2], Jitao Sang[1,2], Changsheng Xu[1,2], and Ke Lu[3]

[1] National Lab of Pattern Recognition,
Institute of Automation, CAS, Beijing 100190, China
[2] China-Singapore Institute of Digital Media, Singapore, 139951, Singapore
[3] College of Computing & Communication Engineering,
Graduate University of CAS, Beijing 100049, China
{qfang,jtsang,csxu}@nlpr.ia.ac.cn, luk@gucas.ac.cn

Abstract. The prevalence of digital photo capturing devices has generated large-scale photos with geographical information, leading to interesting tasks like geographically organizing photos and location visualization. In this work, we propose to organize photos both geographically and thematically, and investigate the problem of location visualization from multiple themes. The novel visualization scheme provides a rich display landscape for location exploration from all-round views. A two-level solution is presented, where we first identify the highly photographed places (POI) and discover their distributed themes, and then aggregate the lower-level themes to generate the higher-level themes for location visualization. We have conducted experiments on a Flickr dataset and exhibited the visualization for the Singapore city. The experimental results have validated the proposed method and demonstrated the potentials of location visualization from multiple themes.

Keywords: Geotagged photo, Point of Interest, Location visualization, Multiple theme discovery.

1 Introduction

Recent years have witnessed the prosperity of photo sharing websites. Vast amount of photos are being uploaded and shared online, associated with valuable metadata such as capturing geographical position and semantic tags. The rich user-generated content has opened up great possibilities for novel multimedia research and applications, among which georeferenced multimedia data mining is a most attractive one. With early work focusing on geographically organizing photos by directly utilizing their geographical positions [11,6], recent work of georeferenced multimedia has been devoted to exploiting semantic tags for semantic knowledge mining, e.g., place tag extraction [8], landmark recognition [13] and travel route summarization [5]. Along this research line, we move one step further in this work, to study the problem of *Location Visualization from Multiple Themes* (LVMT).

S. Li et al. (Eds.): MMM 2013, Part I, LNCS 7732, pp. 92–105, 2013.
© Springer-Verlag Berlin Heidelberg 2013

The concept "theme" is introduced to denote certain interesting topic or representative pattern in a location. It is natural that each location will have multiple themes for visualization. For example, Singapore is expected to be visualized from its finance, shopping, natural scene, landmark, intersecting culture, etc. In addition to "theme", another important concept in our work is Point of Interest (POI), which refers to a highly photographed place and is marked by its centroid. POI is the basic unit in the proposed problem, where each POI contributes to a dominant theme and location visualization is enriched by aggregating POIs inside this location to generate an all-round display from multiple themes. Therefore, by location visualization from multiple themes, the outcomes we manage to obtain are two-level: (1) POI visualization - the identified POIs inside a location, the POIs' theme distribution and exemplary photos; (2) Location visualization - the summarized location theme distribution, the representative POIs and associated exemplary photos for each theme. In particular, we use Singapore city as the running example. Fig. 1 shows the expected multiple-theme visualization map for Singapore city. Fig. 1(b) illustrates a representative POI with its theme distribution and exemplary photos [1]. In Fig. 1(c), popular themes of Singapore are visualized by aggregating the identified POIs, where one color corresponds to one theme. For each theme, several representative POIs and their photos are shown, with each POI pointing to its geographical position on the map. We can see that location visualization from multiple themes can facilitate an intuitive thematical as well as geographical exploration, which enables various applications like semantic location navigation and POI search, location-based POI or theme suggestion and the tour assistant.

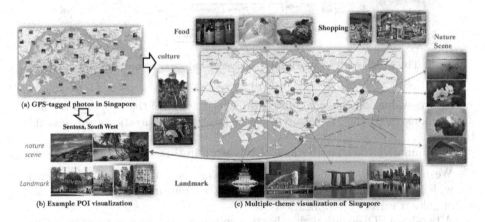

Fig. 1. Problem illustration (best viewed in color)

Mining multiple themes for location visualization is not trivial. The challenges derive from three-fold: (1) The identification of POI from user-generated content is difficult. Users tend to take photos of POI from arbitrary angles and views they like, which leads to large visual variance among photos of the same POI. (2) POI theme list is not available beforehand . The uploaded photos and associated textual metadata are rather noisy and sparse for extracting the underlying themes for each POI. (3) Lower POI-level theme needs to be fused to yield location theme distribution.The derived theme spaces vary between POIs, making the aggregation of POI themes a necessity. To address these issues, we formulate the problem of location visualization from multiple themes as follows (see Fig. 2). The input is a set of photos within the target location, associated with their capturing positions, textual metadata and annotator user IDs. Our solution for LVMT contains three components: (1) POI identification, (2) POI theme discovery (3) location theme aggregation and visualization. Since the geographical position is only available for part of the photos, we conduct POI identification by first constructing POI vocabulary from the geo-tagged photos and then estimating the belonging POI for the rest photos. POI theme discovery is the core component. We propose an incremental learning scheme for automatically discovering the underlying themes in a POI. For location theme aggregation and visualization, we extract location themes by aggregating similar POI themes via clustering based method. With the discovered POI and location theme, we can easily visualize a POI and location from multiple themes. We have conducted experiments on a Flickr dataset of the Singapore photos. Objective as well as

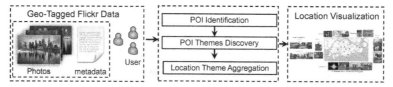

Fig. 2. The proposed framework for location visualization from multiple themes

subjective evaluations have validated the effectiveness of the proposed method, and demonstrated the great potentials of location visualization from multiple themes.

We summarize the contributions of this work in three-fold:

- We introduce the problem of location visualization from multiple themes, which offers a novel vision for georeferenced multimedia data mining and enables a variety of applications.
- A two-level POI-location visualization solution framework is presented. Considering POI level not only enriches the visualization landscape but guarantees a compact theme representation.
- We propose an incremental learning scheme to discover the POI themes, which can automatically mine the underlying themes by exploiting and combining textual and visual content of photos in a principled way.

2   Location Visualization from Multiple Themes

Formally, given a set of photos within a location defined as $\mathcal{P} = \{p\}$, where p is a tuple $(\theta_p, \ell_p, x_p, t_p, u_p)$ containing: (1) a unique photo ID θ_p, (2) the photo's capture location, represented by its latitude and longitude ℓ_p, (3) the photos's visual content feature vector x_p; (4) tagged text t_p including the title, tags and description, and (5) the ID of the user that contributed the photo u_p. The problem of location visualization from multiple themes is formulated as follows: Given a collection of geotagged photos of an interesting location, we aim to mine the POIs with multiple underlying themes to visualize the location. Our approach to address this problem contains three components: POI identification, POI theme discovery, and location theme aggregation and visualization. We first detect POIs using the geo-tagged photos. Then for each POI, we build up a POI vocabulary to estimate the POIs of the photo without GPS-coordinates. After that, we exploit both textual and visual content of photos to discover multiple themes for each POI in an incremental learning manner. Finally, we aggregate the POI themes to obtain the location themes for visualization of the location. We discuss each of these key components in the following.

2.1   POI Identification

In this component, we first present how to detect POIs by aggregating geo-tagged photos and then describe the method to estimate the POI distribution of photos which are not geo-tagged.

POI Detection. Given a large collection of geotagged photos we want to automatically find popular places at which people take photos. In measuring how popular a place is we consider both the number of distinct photographers and their corresponding number of photos taken there. We apply mean shift clustering to the GPS coordinates of the images which are geotagged. Each cluster represents a POI l_k. Then, we compute a popularity score $F(l_k)$ for POI l_k as:

$$F(l_k) = \sum_{i=1}^{N_{user}} \log(N_{photo}(u_i) + 1) \tag{1}$$

where $N_{photo}(u_i)$ is the number of photos provided by the owner u_i. The popular POIs are obtained by ranking the popular scores of POIs and selecting the top-k candidates.

POI Estimation for Images: By exploring the semantic metadata of geo-tagged photos, we can extract representative tags for each identified POI, which we refer as POI vocabulary. Denote the POI vocabulary as $L = \{l_1, \ldots, l_H\}$, where l_h is the hth POI and we have H POIs. Each POI is represented as $l_h = \{t_1, \ldots, t_N\}$, where t_n is the nth place tag. Referred to the TagMaps method [8], we use a TF-IDF like method to detect the place tags and construct the POI vocabulary. We compute a score for tag t with POI l as

$$Score(l, t) = \mathrm{tf}(l, t) \cdot \mathrm{idf}(t) \cdot \mathrm{uf}(l, t) \tag{2}$$

where $\text{tf}(l,t) = |N_{l,t}|$ is tag frequency of t and $\text{idf}(t) = |N|/|N_t|$ is the inverse document frequency for a tag t in POI l. $\text{uf}(l,t) = |U_{l,t}|/|U_l|$ is a factor for users to guard against the tags frequently used only by a small number of users. N and U are the counts of corresponding terms. We set a threshold s_t and recognize those tags $Score(l,t) > s_t$ as the place tags.

With the derived POI vocabulary, we can estimate the most probable POI l^* for a given photo without GPS coordinates by majority voting:

$$l^* = \arg\max_l \sum_{t \in l} E(t|T_I)w(t|l) \tag{3}$$

where $E(t|T_I)$ is 1 if the associated text T_I of the input photo I contains the place tag t otherwise 0. $w(t|l)$ is the weight of tag t in the POI l and modeled by using the place tag score obtained in the POI tag construction stage, i.e., $w(t|l) = \frac{Score(t,l)}{\sum_t Score(t,l)}$. The scheme takes into the account the importance of each place tag in a POI. With the geo-referenced photos, we move our effort to discover multiple themes in each POI of the interesting location in the next section.

2.2   POI Theme Discovery

Now we consider the problem of discovering multiple themes in a POI. As aforementioned, it suffers from large variance of visual content and high noise of photos. To deal with these difficulties, we develop an incremental learning scheme to automatically discover the potential themes in a POI reflected by salient tags and images. This learning scheme consists of two parts: salient tags extraction and image theme assignment. The procedure for multiple theme discovery in a POI is summarized in Algorithm 1.

Algorithm 1. INCREMENTAL LEARNING FOR MULTIPLE THEME DISCOVERY IN A POI

Input: A data set of images $\mathcal{I}$ with visual content features X and associated textual features T in a POI.

Output: Multiple themes in a POI $\mathcal{A} = \{a_k\}_{k=1}^K$.

1: **repeat**
2:     **Extract** salient tags $\mathcal{S}_T$ for updated theme a
3:     **Determine** the theme of the input image I using the existing themes with corresponding salient tags and images.
4:     **Augment the dataset** with the newly input image
5: **until** images exhausted

Salient Tags Extraction: Given all images $\mathcal{I}$ in a POI l and a couple of images $\mathcal{I}_a = \{I_i\}_{i=1}^N \in \mathcal{I}$ of one theme a, we aim to extract salient tags $\mathcal{S}_T$ for this theme from the whole associated tags T_a with images $\mathcal{I}_a$. These salient tags should be able to well describe the visual content in this theme. We propose our technique to extract salient tags by making two assumptions [9] as follows:

- **Separation:** If a tag t well describes a visual concept c, then the probability of observing the visual concept c among images $\mathcal{I}_a$ of theme a is larger than the probability of observing it among all images $\mathcal{I}$ in the POI.
- **Cohesion:** A tag t is visually representative if its annotated images $\mathcal{I}_a$ are visually similar to each other, containing a common visual concept c such as an object or a scene.

Based on these two assumptions, we present a formulation that simultaneously integrates the above two assumptions in a single framework. Considering the first assumption, we measure the observation probability between $t_i \in \mathcal{S}_T$ and images $\mathcal{I}_a$ of theme a with tag co-occurrence. We calculate the co-occurrence of t_i in images $\mathcal{I}_a$ as the observation probability for tag t_i.

$$p(t_i|\mathcal{I}_a) = \frac{Q(t_i \cap \mathcal{I}_a)}{Q(\mathcal{I}_a)} \tag{4}$$

where $Q(\mathcal{I}_a)$ denotes the total number of images in theme a, while $Q(t_i \cap \mathcal{I}_a)$ is the number of images associated with tag t_i in theme a. Likewise, we compute the observation provability for tag t_i in all images $\mathcal{I}$ in a POI. $p(t_i|\mathcal{I}) = \frac{Q(t_i \cap \mathcal{I})}{Q(\mathcal{I})}$, where $Q(\mathcal{I})$ denotes the total number of all images and $Q(t_i \cap \mathcal{I})$ is the number of images containing tag t_i in all images. Then, we have the constraint between $p(t_i|\mathcal{I}_a)$ and $p(t_i|\mathcal{I})$ as

$$\mathcal{R}(t_i, \mathcal{I}_a) = f(\tau), \tau = p(t_i|\mathcal{I}_a) - p(t_i|\mathcal{I}) > 0 \tag{5}$$

where $f(\cdot)$ is certain monotonically increasing function and defined $f(\cdot)$ as the standard sigmoid function, i.e., $f(x) = \frac{1}{1+e^{-x}}$. Accordingly, the separation of a salient tag set $\mathcal{S}_T$ in images $\mathcal{I}_a$ of theme a is given by $\mathcal{R}(\mathcal{S}_T, \mathcal{I}_a) = \sum_{t_i \in \mathcal{S}_T} \mathcal{R}(t_i, \mathcal{I}_a)$.

Considering the second assumption, we adopt pairwise similarity constraints to measure the cohesion for images of theme a. Denote $\mathbf{x}_u \in X_{t_i}$ as the visual content feature vectors of image I_u associated with tag t_i in images $\mathcal{I}_a$. $\mathcal{I}_{t_i}$ is the set of images containing tag t_i in images $\mathcal{I}_a$ and X_{t_i} are corresponding feature vectors of $\mathcal{I}_{t_i}$. We adopt the classical Kernel Density Estimation (KDE) method [2] to measure visual similarity between image pairs in images $\mathcal{I}_{t_i}$ as

$$Sim(\mathcal{I}_t) = \frac{1}{|X_t|^2} \sum_{u,v=1}^{|X_t|} K(x_u - x_v) \tag{6}$$

where X_{t_i} is the cardinality of X_{t_i} and K is the Gaussian kernel function with radius parameter σ, i.e., $K(x_u - x_v) = \exp\left(\frac{-\|x_u - x_v\|^2}{2\sigma^2}\right)$ where σ is a scaling parameter and adaptively assigned as the median value of all pair-wise Euclidean distances between images. Accordingly, we define the cohesion of tag t_i with $\mathcal{I}_a$ as

$$C(t_i, \mathcal{I}_a) = 1 - g(Sim(\mathcal{I}_{t_i})) \tag{7}$$

where $g(\cdot)$ is defined as $g(x) = \frac{1}{1+e^{-x}}$. The cohesion of a salient tag set $\mathcal{S}_T$ in theme images $\mathcal{I}_a$ is computed as $\sum_{t_i \in \mathcal{S}_T} C(t_i, \mathcal{I}_a)$.

Based on the definitions of terms regarding separation and cohesion for salient tag set $\mathcal{S}_T$, we now present formulation for extracting salient tags $\mathcal{S}_T$ as follows:

$$\mathcal{S}_T^* = \arg\max_{\mathcal{S}_T}\left\{\frac{1}{N}\sum_{t_i \in \mathcal{S}_T}\phi(t_i)\right\}, \phi(t_i) = \lambda\mathcal{R}(t_i, \mathcal{I}_a) + (1-\lambda)\mathcal{C}(t_i, \mathcal{I}_a) \tag{8}$$

where $N = |\mathcal{S}_T|$ is the number of the selected tags. $\lambda \in [0,1]$ is a weighting parameter that is used to module two contributions; $\phi(t_i)$ is the saliency score of tag t_i in theme a. However, it is computationally intractable to solve Eqn. (8) directly since it is a non-linear integer programming (NIP) problem [3]. Alternatively, we resort to a greedy strategy which is simple but effective to solve the problem. In reality, for each theme in a POI, only a small set of tags are salient and valuable. Therefore, we perform a pre-filtering step to obtain the tags T with large values of $\mathcal{R}(t_i, \mathcal{I}_a)$. This can reduce the computational cost to favor the salient tags extraction. For salient tags extraction, we first select the tag $t \in T$ with largest value of $\mathcal{R}(t, \mathcal{I}_a)$ and then choose the next tag t_i from $T\setminus\mathcal{S}_T^*$ by solving $\arg\max_{t_i}\phi(t_i)$. The salient tags set is updated by $\mathcal{S}_T^* = \mathcal{S}_T^*\cup\{t_i\}$ and obtained until $T = \emptyset$.

Image Theme Assignment: Given salient tags and images for each existing theme in a POI, we aim to assign a theme label for a new input image I. We first use a salient tag correlation method to find the most potential correlated themes from the existing theme sets. Then, we employ sparse representation to predict the most probable theme for the new input image. We elaborate this process as follows.

Given a set of salient tags $\mathcal{S}_T^a$ and images $\mathcal{I}_a$ for a certain theme a in a POI l, we use a majority voting scheme to decide the correlated themes based on measuring the tag relatedness between the associated tags T_I of the input image I and the salient tags $\mathcal{S}_T^a$ of theme a. It can be represented as:

$$TR(T_I, \mathcal{S}_T^a) = \sum_{t_i \in \mathcal{S}_T^a} H_{saliency}(t_i|T_I)P(t_i|\mathcal{S}_T^a) \tag{9}$$

where $H_{saliency}(t_i|T_I)$ outputs 1 if there exist a salient tag $t_i \in \mathcal{S}_T^a$ in T_I otherwise 0. $P(t_i|\mathcal{S}_T^a)$ is the salient weight of tag t_i in theme a and computed by using the saliency score in salient tags extraction stage, i.e., $P(t_i|\mathcal{S}_T^a) = \frac{\phi(t_i)}{\sum\phi(t_j)}$. Then, we select the correlated themes $A_{correlated}$ for input image I as follows:

$$A_{correlated} = \{a|, TR(T_I, \mathcal{S}_T^a) > thre_A\} \tag{10}$$

where $thre_A$ is the threshold value set for all themes. The themes whose tag relatedness score is larger than $thre_A$ are selected as correlated themes and used to determine the theme of input image via sparse representation classification. If an input image is not correlated with all existing themes, we assign a new theme label a_{new} for the new input image.

For input image which has correlated themes, we model the reconstruction sparsity to assign the theme label. Inspired by the great performance of sparse representation in face recognition [12] and existence of similar or duplicate visual contents coming from dominant views in a POI, we model the theme assignment for an input image as a visual content reconstruction cost minimization problem:

The theme assignment of an input image I is treated as the cost of using the images $\mathcal{I}_a$ from existing themes in a POI l to reconstruct the visual content feature of I.

Given an input image I, $\mathbf{x} \in \mathbb{R}^m$ is its visual content feature vector. Denote a matrix of existing samples from theme images $B = [B_1, B_2, \ldots, B_K] \in \mathbb{R}^{m \times n}$ for K themes as the corresponding visual content feature vectors of existing theme images $\mathcal{I}$ in POI l. We reconstruct $\mathbf{x}$ by using B as the coding dictionary as follows:

$$\mathbf{x} = \sum_{k=1}^{K} B_k \alpha_k + \epsilon \tag{11}$$

where $\alpha = [\alpha_1, \alpha_2, \ldots, \alpha_K]$ are the reconstruction coefficients of $\mathbf{x}$ and $\epsilon \in \mathbb{R}^m$ is a noise vector. Following the principle of sparse coding for face recognition [12], we learn the reconstruction coefficients by solving the optimization problem of least square error and ℓ^1-norm regularization. We formulate this as:

$$\alpha^* = \arg\min_{\alpha} \|B\alpha - \mathbf{x}\|_F^2 + \gamma \|\alpha\|_1 \tag{12}$$

where γ is the regularization parameter. Here we use L1-Homotopy [1] to solve the optimization problem in Eqn.(12). Then the residuals $r_k(\mathbf{x})$ is computed as $r_k(\mathbf{x}) = \|\mathbf{x} - B_k \alpha_k\|_2$, where $k = 1, \ldots, K$. For theme assignment, the rule is decided in favor of the theme with the lowest total reconstruction error:

$$a^* = \arg\min_{a} \|\mathbf{x} - B_k \alpha_k\|_2 \tag{13}$$

However, if the minimum residual is large ($\min(r_k(\mathbf{x})) > e$, e is a threshold), it indicates that the new input image I cannot be well represented with low error as a sparse linear combination of dictionary B, which is a good indicator of novelty of the input image I. Therefore, if $\min(r_k(\mathbf{x})) > e$, we would also assign a new theme label a_{new} for the new input image I.

In summary, the theme of a new input image is determined as:

$$a_I = \begin{cases} a_{new} & \text{if } \mathcal{L}^* > thre_A \text{ or } \min(r_k(\mathbf{x})) > e, \\ a^* & \text{otherwise.} \end{cases} \tag{14}$$

Hence, a_I is the final theme label for the input image I.

Through the incremental learning scheme for POI theme discovery, we can obtain multiple POI themes $\mathcal{A}_l = \{a_k\}_{k=1}^K$, each of which is a POI theme containing salient tags $\mathcal{S}_T$ and images $\mathcal{I}$. For a POI l, we rank the popularity of its different themes by considering the number of users and their photos individually. The popularity score is computed using Eqn.(1). We select the top-k themes of a POI as its representative themes. The dominant theme $a_{dominant}$ is obtained by choosing the theme with the largest score. To visualize POI l, we extract the top-k salient tags and representative images via the affinity propagation method [4] in each theme.

2.3 Location Theme Aggregation and Visualization

Location Theme Aggregation: After obtaining POIs with multiple themes, we aim to discover representative themes at the location level by aggregating similar POI themes via a clustering algorithm. Each cluster represents a location theme. For POI l, after POI theme discovery based on the incremental learning scheme, we have K POI themes $\mathcal{A}_l = \{a_k\}_{k=1}^{K}$ and each theme is represented as $a_k = \{\mathcal{S}_T, \mathcal{I}\}$, where $\mathcal{S}_T = \{t_i\}_{i=1}^{M}$ is set of salient tags and $\mathcal{I} = \{I_j\}_{j=1}^{N}$ is the corresponding images. We first represent each theme as a feature vector in a feature space. Since the salient tags can well describe the semantics of POI themes, we exploit these tags to aggregate themes. A tag vocabulary is constructed by extracting the top-m tags from $\mathcal{S}_T$ of each theme considering the tag saliency. Let $V = \{t_d\}_{d=1}^{D}$ be the vocabulary with D distinct tags. Then each theme can be represented as a D dimensional vector $\theta_{l,k}$ based on the tag occurrence of $\mathcal{S}_T$ and the vocabulary. We perform the k-means clustering method on the D-dimensional points $\theta_{l,k}, l \in 1, \ldots, L; k \in 1, \ldots, K_l$. The number of clusters is empirically predefined. Consequently, we aggregate the POI themes into k clusters, each of which has a couple of images and tags. In addition, We try to annotate each cluster of POI themes with some semantic terms, which should well describe the visual content of images within each cluster. We take a simple method by aggregating the tag feature vectors and using the most frequent tags in each cluster to annotate a location theme.

Location Visualization: After clustering and annotation, each cluster is termed as a location theme containing a couple of POI themes. We compute a popularity score for each location theme using Eqn.(1) and select the top-k themes as its representative themes for the location. For each representative location theme, the POIs contributing a dominant POI theme for this location theme are selected as representative POIs. Hence, the location can be visualized by using representative location themes, which is reflected by annotated tags and exemplary photos from the dominant themes of the representative POIs.

3   Experiments

We have conducted extensive experiments to evaluate the effectiveness and usefulness of the proposed framework.

Our dataset was collected by crawling images and photo metadata using Flickr API. We use Singapore as a search word and all queried images are collected together with their associated information, including title, tags, description, and GPS coordinates. The initial dataset consists of 263,953 images. A pre-filtering process is performed on the dataset to remove the duplicate images and images with GPS coordinates outside Singapore, etc. We also restrict that the images should have complete associated textual contents of title, tags, description. In this way, 110,846 images in total are obtained. This dataset contains 26,623 geo-tagged photos and 9,044 users.

3.1   Evaluation of POI Identification

We first illustrate the distribution of mined POIs in Singapore. Fig. 3 presents the distribution map of geo-tagged images, which follows the perfect contour of

Singapore. The bandwidth for meanshift as 0.023 according to the scale of photo distribution. Each small circle denotes a photo, while each big circle denotes a detected POI. The photo density around a POI indicates the POI popularity. 40 clusters are obtained. By filtering out the POIs with less than 10 photos, we obtain 24 POIs in total, which are used in the subsequent experiments. Table 1 shows the top-8 highly photographed places with their popularity scores computed using Eqn. (1) in Singapore. The POI name is obtained by issuing the GPS -coordinates of each POI cluster center to the Flickr API. It can be seen that the detected POIs uncover the most popular places in Singapore.

We evaluate the performance of estimating POI for images using proposed method. Two methods are compared - simply voting with no weight ($w(t|l) = 1$ or 0) and weighted voting ($w(t|l) \in [0,1]$). The results are shown in Table 1. The weighted tag correlated multiple voting utilizing the difference of place tag importance achieves much better performance (improve AP by 18%) than voting with no weight, which benefits the multiple theme discovery. The reason may be that many images in our dataset are not geo-tagged while they contain important content for POIs.

Table 1. The popular Top-8 POIs in Singapore and their popularity scores

POI	popularity score	accuracy (no weight)	accuracy (weighted)
Singapore, Central Singapore	3861.84	0.4828	0.8079
Central Catchment Reserve	599.699	0.1069	0.1961
Sentosa, South West	562.57	0.3001	0.6908
Jurong Town, South West	330.998	0.2431	0.2339
Singapore Changi Airport	328.480	0.5487	0.8453
Jalan Kayu, Central Singapore	275	0.1336	0.1940
Queenstown, Central Singapore	263.69	0.14814	0.288
Central Catchment Area, North West	194.45	0.5358	0.6603

3.2 Evaluation of Multiple Theme Discovery

In the following we will evaluate the performance of multiple theme discovery at the POI and location level.

Experiment Setting: To represent the image content, we extract five types of visual features to form a 809-dimension vector for each image, including 81-dimension color moment, 37-dimension edge histogram, 120-dimension wavelet texture feature, 59-dimension LBP feature [7] and 512-dimension GIST feature [10]. The tradeoff parameters in Eqn. (8) is empirically set to 0.7. We build a vocabulary of irrelevant tags including stop words, meaningless words, time and number related words, camera related words, and some general frequent tags, which are all removed before further operations. The threshold parameters in Eqn. (14),i.e., $thre_A$ and e, are chosen as 0.005 and 0.8 respectively, through qualitative cross-validation measuring and are fixed in subsequent experiments.

Fig. 3. Distribution of geotagged photos in Singapore from Flickr and the detected POIs

We extract the top 50 salient tags of each representative POI theme to construct the vocabulary for location theme aggregation. The number of k in K-means is set as 20.

Qualitative Evaluation: At POI level, Fig. 4 shows the salient tags and exemplary images of the Top-3 POIs in Singapore detected by our approach. For each POI, we present 3 popular themes with the salient tags and exemplary images. Clearly, these themes are related to *nature scene, landmark, animals, food,* and *culture*. We can see that the extracted images and tags are consistent and the discovered themes are adequate to represent the POIs.

At location level, we exhibit the visualization of Singapore from multiple themes in Fig. 5. Seven representative themes are presented, which correspond to *scene, landmark, people, airplane, culture, animals birds,* and *flowers* respectively. Each aggregated theme is denoted by its salient tags and enriched by the belonging POIs and the associated images. From this visualized location graph with POIs and multiple themes, we can have easy access to an intuitive geographical exploration and a better thematic understanding of Singapore.

User Study: We conduct a small-scale user study to evaluate the effectiveness of the proposed method and the user experience of the novel visualization form. Three criteria are considered: (1) *consistency,* the level of consistency between the visual content and the salient tags (0: Not consistent, 10: Very consistent); (2) *relatedness,* the extent that the mined themes relates to the POI (0: Not related, 10: Very related); and (3) *satisfaction,* how satisfactory are the aggregated multiple themes for location visualization (0: Not satisfied, 10: Very satisfied)? We invited 20 participants for the user study experiment. The themes depicted in Fig. 5 are selected for evaluation. The results are averaged over all participants for each theme and shown in Fig. 6. It is obvious that the participants gave positive feedback to the novel location visualization scheme, which further validates the potential of LVMT in advanced location exploration and related applications.

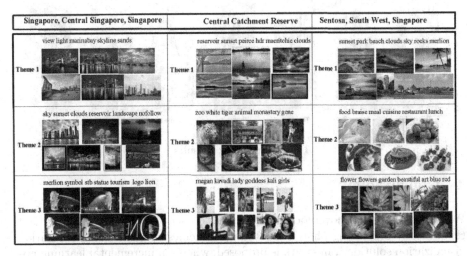

Fig. 4. Salient tags and representative images of the top-3 POIs in Singapore discovered by our approach

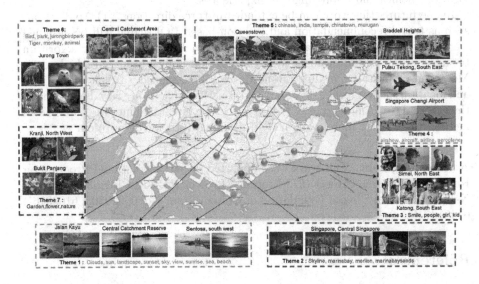

Fig. 5. Visualizing Singapore from multiple themes

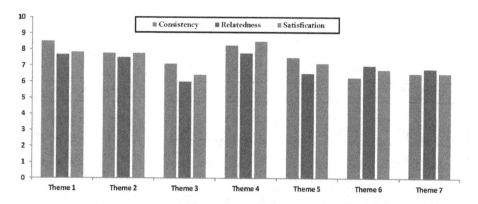

Fig. 6. User study results

4   Conclusions

We have presented a novel location visualization scheme by geographically and thematically organizing photos into multiple themes. A two-level POI-location visualization solution framework is proposed, where an incremental learning approach is introduced for POI theme discovery and location theme is then obtained by aggregating the involved POI themes. Experiments on a dataset of Singapore collected from Flickr have shown its advantage in deriving compact location themes and improving user experiences.

In the future, this work can be extended in three directions. (1) Applications such as travel recommendation and location-based POI search will be designed based on the proposed visualization scheme. (2) We will apply our algorithm to more locations and evaluate its generalization ability in large-scale dataset. Moreover, quantitative evaluation metrics will be designed. (3) The mined POI and location themes will be described in a concise and accurate way. For example, the external topic ontology, e.g., wikipedia can be integrated into the data-driven method to generate more readable and meaningful theme description.

Acknowledgment. This work was supported in part by National Program on Key Basic Research Project (973 Program, Project No. 2012CB316304) and the National Natural Science Foundation of China (Grant No. 90920303, 61003161, 61201374).

References

1. Asif, M.S., Romberg, J.K.: Dynamic updating for l1 minimization. CoRR, abs/0903.1443 (2009)
2. Bishop, C.: Pattern Recognition and Machine Learning. Information Science and Statistics. Springer (2006)
3. Boyd, S., Vandenberghe, L.: Convex optimization. Cambridge Univ. Pr. (2004)

4. Frey, B.J., Dueck, D.: Clustering by passing messages between data points. Science 315, 972–976 (2007)
5. Hao, Q., Cai, R., Wang, C., Xiao, R., Yang, J.-M., Pang, Y., Zhang, L.: Equip tourists with knowledge mined from travelogues. In: WWW, pp. 401–410 (2010)
6. Jaffe, A., Naaman, M., Tassa, T., Davis, M.: Generating summaries and visualization for large collections of geo-referenced photographs. In: Multimedia Information Retrieval, pp. 89–98 (2006)
7. Ojala, T., Pietikäinen, M., Harwood, D.: A comparative study of texture measures with classification based on featured distributions. Pattern Recognition 29(1), 51–59 (1996)
8. Rattenbury, T., Naaman, M.: Methods for extracting place semantics from flickr tags. TWEB 3(1), 1 (2009)
9. Sun, A., Bhowmick, S.S.: Quantifying tag representativeness of visual content of social images. In: MM 2010: Proceedings of the International Conference on Multimedia, pp. 471–480. ACM, New York (2010)
10. Torralba, A., Murphy, K.P., Freeman, W.T., Rubin, M.A.: Context-based vision system for place and object recognition. In: ICCV, pp. 273–280 (2003)
11. Toyama, K., Logan, R., Roseway, A.: Geographic location tags on digital images. In: ACM Multimedia, pp. 156–166 (2003)
12. Wright, J., Yang, A.Y., Ganesh, A., Sastry, S.S., Ma, Y.: Robust face recognition via sparse representation. IEEE Trans. Pattern Anal. Mach. Intell. 31(2), 210–227 (2009)
13. Zheng, Y.-T., Zhao, M., Song, Y., Adam, H., Buddemeier, U., Bissacco, A., Brucher, F., Chua, T.-S., Neven, H.: Tour the world: building a web-scale landmark recognition engine. In: Proceedings of International Conference on Computer Vision and Pattern Recognition, Miami, Florida, U.S.A (June 2009)

Interactive Video Advertising:
A Multimodal Affective Approach

Karthik Yadati, Harish Katti, and Mohan Kankanhalli

School of Computing, National University of Singapore
{nyadati,harishk,mohan}@comp.nus.edu.sg

Abstract. Online video advertising (video-in-video) strategies are typically agnostic to the video content (ex.: advertising on YouTube) and the human viewer's preferences. How to assess the emotional state and engagement of the viewer to place an advertisement? Where to insert an advertisement based on the content in an advertisement and a specific target video stream? Surely these are relevant questions that should be addressed by a good model for video advertisement placement. In this paper, we propose a novel framework to address two important aspects of (a) multi-modal affective analysis of video content and viewer behavior (b) a method for interactive personalized advertisement insertion for a single user. Our analysis and framework is backed by a systematic study of literature in marketing, consumer psychology and affective analysis of videos. Results from the user-study experiments demonstrate that the proposed method performs better than the state-of-the-art in video-in-video advertising.

1 Introduction

There has been an explosion of video content on the world wide web accompanied by an ever growing audience. YouTube, for example, is streaming over 4 billion online videos every day and this is 25 % increase over just eight months[1] Online video distribution in websites such as Youtube, Hulu and MySpace are fast becoming a potential alternative for video content generation and distribution. Video advertising usually has at least four key players involved (a) advertiser (b) media broadcast company (c) viewer and (d) program content provider. Conventional video broadcast (ex: Television) typically involves large spending from advertisers and professional editing for advertisement placement in the program content. Online video advertising, on the other hand, has content providers very often from amongst the viewer community and the sheer volume of uploaded video rules out any possibility of manual editing or ad selection and insertion. Furthermore, viewers and often program content uploaders pay nominal amounts to access the video distribution service. Our target in this paper is to cater to such online video content, where the user is logged into an online video distribution service (ex: YouTube) and the system observes the user's behavioral

---

[1] http://www.reuters.com/article/2012/01/23/
us-google-youtube-idUSTRE80M0TS20120123

S. Li et al. (Eds.): MMM 2013, Part I, LNCS 7732, pp. 106–117, 2013.
© Springer-Verlag Berlin Heidelberg 2013

responses. We propose a single user interactive human-in-loop model for real-time personalization of advertisement delivery. This is now made tangible by devices such as the new eye-tracking enabled laptop from Lenovo® and TOBII® that can monitor user's eye-gaze and expressions.

Budget spent on video advertising is up by 19% over the last year, according to a recent survey involving close to 600 advertisers, publishers, and video technology providers. In this paper, we bring forth and address two important complementary aspects of **affect**[2] and **interaction** for the online video-in-video advertising problem. We focus on the placement of video advertisements in a video program stream (**video-in-video** advertising). Unless otherwise stated, advertisements are discussed in reference to video-in-video context, in the rest of this paper.

1.1 Affective Advertising Is Effective

Marketing and consumer psychology literature suggests that advertising that can connect to people at an emotional level, is more compelling for the viewer and plays an important role in consumer decision making [2]. The primary objective of advertising is to be informative about the product and influence the user towards an action of buying the product being advertised. In this regard, emotive advertising has been shown to be more effective than it's rational counterpart.[3] Apart from the advertisements being emotive, studies [5] suggest the affective context in which an advertisement can be placed, in order to extract favorable reactions from the users. We identify some important results from experiments in consumer psychology and marketing:

1. In a high arousal program context, an assimilation effect is expected to occur wherein viewers shift their evaluations of the ads in the direction of their evaluation of the program [8].
2. In a low arousal program context, viewers shift their evaluations of ads in the direction opposite to their evaluation of the program [8].
3. To obtain optimal effect, an emotionally neutral commercial should be placed in a negative program context [14].
4. A positive commercial viewed in the context of a positive program is evaluated better when compared to the same commercial viewed in a negative program context [8].

We now realize these observations into thumb rules for identifying insertion points as well as selecting appropriate advertisements. The rules *(1)* and *(2)* are used to identify the advertisement insertion points and the rules *(3)* and *(4)* are used to identify the appropriate advertisements to be placed at given insertion points in the video. Figure 1 and 2 show valence(pleasantness / unpleasantness) and arousal (excitability) of the video program (*spiderman 3*) from our

---

[2] Affect is measured in terms of arousal (intensity of the emotion) and valence (pleasantness/unpleasantness)

[3] http://www.neurosciencemarketing.com/blog/
articles/emotional-ads-work-best.htm

evaluation dataset. Pre/post-roll, our method *MyAds* and VideoSense [13] are then used to insert 2 advertisements into this program. The right panel shows choices made by *MyAds*, Rule *(2)* (low arousal and low valence) is followed for the first advertisement insertion point (Ad 1) and rule *(1)* (high arousal and high valence) is followed for the second advertisement insertion point (labeled Ad 2, right panel; Figure 2). We also observe that for the first insertion point, a neutral advertisement is selected (rule *(3)*) and a moderately positive advertisement (rule *(4)*) is selected for the second insertion point. The left panel (Figure 1) shows the advertisement insertions made by the affect agnostic pre/post-roll strategies as well as VideoSense [13].

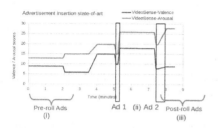 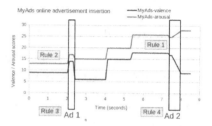

Fig. 1. Valence-Arousal plot for (i) pre-roll, (iii) post-roll and (ii) VideoSense [13]

Fig. 2. Affective, online advertisement insertion in *MyAds*

1.2 Importance of Interaction and Personalization

Successful advertising needs user engagement during the advertisement delivery and it is intuitive that successful personalization of advertising strategy can come about by involving the viewer in a human-in-loop manner with the advertisement delivery system. The effectiveness of interaction is also well supported by work in attentive user interfaces [4] that sense and reason on viewer's attention and regulate subsequent interaction.

1.3 Key Contributions and Organization

The important contributions of this paper are,

- A novel multimodal approach to model affective state of the viewer, our method fuses content-based arousal with pupillary dilation and content-based valence with facial expression and gives good agreement with human annotated ground truth.
- An interactive model for single-user personalization based video-in-video advertisement insertion and selection strategy, which effectively addresses the conflicting goals of minimizing disturbance caused to the user and maximizing impact of the advertisement on the viewer.

The paper is organized as follows, section 2 an overview of related work in online video advertising. Analysis of content based on behavioral signal based features is presented in section 3. The interactive online advertising is presented in section 4 followed by details of evaluation and user studies in section 5. Important results are discussed in section 6.

2 Related Work

Current online video advertisement still relies on *content-agnostic* strategies as seen on Youtube, where video advertisements are played before (*pre-roll*) or after (*post-roll*) the video playback. A notable work that explores contextual video advertising is *VideoSense*[13], where advertisement insertion points correspond to shot change points with high audio-visual discontinuity and low change in video attractiveness(attention). VideoSense also computes contextual similarity between a video scene and the available advertisements, using low-level audio-visual features like motion activity, color histogram, tempo of the audio signal, prior to advertisement insertion and selection. We address the relatively less explored domain of interaction and affect-based personalized advertising.

2.1 Affective Analysis and Behavioral Signals

Mehrabian and Russel [7] have proposed the representation of affect in terms of two continuous dimensions - *arousal* (intensity of the emotion) and *valence* (type of emotion). Traditionally, arousal and valence corresponding to any audio-visual stimuli are measured using subjective user-studies, where users are shown the stimuli and asked to give their subjective rating for valence and arousal using the self-assessment manikin (SAM) [3]. Computational models for arousal and valence, using relevant low-level features in an image, audio and video content [1] have also been proposed.

Affective state of a viewer is typically manifested in various behavioral signals viz., pupillary dilation, facial expression, gestures etc.. For example eye-tracking has been used in studying human emotions in [10], where pupillary dilation is used as a measure of physiological arousal. With recent technological developments, modalities such as eye-tracking equipment and face expression analysis have become nearly non-intrusive to the user. We propose a fusion strategy to combine the two channels of information - content-based features and behavioral signals (pupillary dilation and facial expression in the context of this paper), which provides a better assessment of the user's affective state.

3 Feature Selection and Fusion

We first discover affect related features from video content [1] and behavioral information such as pupillary dilation and facial expression and collect valence and arousal ground-truth for each scene in 9 video programs with ratings of

1(0) for high(low) arousal and 1(0) for high(low) valence, given by 5 annotators. Pupillary dilation and facial expression data are recorded for an independent group of subjects as they free-view the chosen videos. We then evaluate behavioral signal based features and content based features [1] with human annotated valence and arousal ground-truth.

Facial expressions result in characteristic deformations in the face and can convey the emotional state of an individual. We use six canonical emotions (neutral, happy, surprise, anger, sad, fear and disgust)[11] for our analysis. The out of the box *eMotion* emotion analyzer is employed to give continuous probability scores on the chosen canonical emotions. High(low) valence is defined as follows, If p_{pos} and p_{neg} are the sum of probability scores over positive emotions (happy, surprise, neutral) and negative emotions (angry, sad, disgust) respectively, over the time window of a video segment (scene)

$$valence(p_{pos}, p_{neg}) = \begin{cases} 1 \, , p_{pos} \geq p_{neg} \\ 0 \, , p_{pos} < p_{neg} \end{cases} \tag{1}$$

Pupillary dilation information needs additional preprocessing to discover high(low) arousal segments of the video, we extend the method from [6] for this purpose. Signal variance σ over the smoothed pupillary dilation signal is computed and video segments having significant proportion τ of points with deviation greater than $k \times \sigma$, $k \in [0,1], \tau \in [0,1]$, are identified as high arousal. We explore the space spanned by k and τ and find that $k = 0.5, \tau = 0.4$ are good choices for affective video content and yield an agreement of upto 76% with high arousal segments of the video. Given a fraction $\delta_{PD(s)} \in [0,1]$ of the video found to have corresponding pupillary dilation value greater than $k \times \sigma$, we define the indicator function for high(low) arousal in videos,

$$arousal(PD_s) = \begin{cases} 1 \, , \delta_{PD(s)} \geq \tau \\ 0 \, , \delta_{PD(s)} < \tau \end{cases} \tag{2}$$

We find an average agreement of 74%(71%) between content-based arousal (valence) [1] and the ground-truth arousal(valence) and define indicator functions for content based valence and arousal for a video scene s as follows,

$$content\_arousal(s) = \begin{cases} 1 \, , \delta_{CArousal(s)} \geq \tau_{CArousal} \\ 0 \, , \delta_{CArousal(s)} < \tau_{CArousal} \end{cases} \tag{3}$$

$$content\_valence(s) = \begin{cases} 1 \, , \delta_{CVal(s)} \geq \tau_{CVal} \\ 0 \, , \delta_{CVal(s)} < \tau_{CVal} \end{cases} \tag{4}$$

The values $\delta_{CArousal(s)}, \delta_{CVal(s)}, \tau_{CArousal}$ and τ_{CVal} are obtained from content-based affect model [1]. Similarly, we label the advertisement as positive or neutral by applying the equations 3 and 4 over advertisements.

3.1 Multimodal Fusion

An appropriate fusion strategy is an important step, which fuses information from multiple modalities viz., content, eye-tracking. In out method, we have facial expression based valence and pupillary dilation based arousal scores which need to be combined with content based valence and arousal scores respectively. We evaluate a variety of fusion strategies including $argmin(x, y)$, $argmax(x, y)$, $(x \oplus y)$, $product(x \otimes y)$ over pairs of $content\_valence(s)$, $valence(p_{pos}, p_{neg})$ and $content\_arousal(s)$, $arousal(PD_s)$ values, $\oplus$ and $\otimes$ denote point wise addition and multiplication between two signals. We compare the labels (arousal and valence) given to each video segment using each of the fusion strategies and select the strategy which compares best to the ground-truth labels for arousal and valence. In this case, we find the best fusion strategy to be point wise multiplication $\otimes$ and realize it as a logical AND between indicator functions defined earlier. Contrary to previous research [15], we find comparatively poor correlation between eye blink rate and eye movement related parameters such as fixation frequency and saccades and valence or arousal ground-truth. Hence we do not include these parameters in our model.

4 Interactive Online Advertising Framework

We propose an online advertisement insertion algorithm $MyAds$, to insert advertisements on the fly, as the user is watching the video. We propose an advertising framework taking into account, the affective impact of the video on the user and place affectively similar advertisement in videos, the overall schematic of $MyAds$ is shown in Figure 3. Given a video V containing a set of scenes

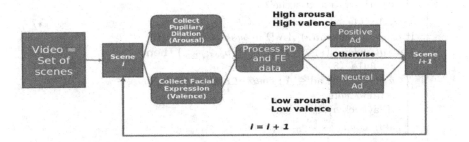

Fig. 3. Interactive online advertisement insertion

$S = \{s_1, s_2, ..., s_n\}$ obtained using automatic segmentation [16] and a set of advertisements $A = \{a_1, a_2, ..., a_m\}$, video scenes s_i and advertisements a_j have varying arousal and valence values computed from low level features. Let PD_i and FE_i be the pupillary dilation and facial expression values buffered over a time window T that ending at the last frame of scene $s_i \in S$ and assuming a recommended number of advertisements N to be placed in the video stream. We aim to place emotionally neutral advertisement in a negative program context [14] and a positive advertisement in a positive program context [8]. We select 9 videos and 50 advertisements for our evaluation, each video yielding an average of 12 scenes. The insertion of upto 3 advertisements per video stream gives a solution space of $C^3{}_{12} \times C^3{}_{50} \times 3!$, the proposed $MyAds$ algorithm has linear time complexity $\| S \|$) and is described in Alg. 1.

4.1   Experimental Setup

As we can see in Figure 4, our setup includes 3 different components (labeled in Fig. 4) with which the user interacts. The primary equipment is an eye-tracker which tracks the user's eye movements as well as the pupillary responses. The eye-tracker used in the setup is a binocular infra-red based remote eye-tracking device SMI RED 250 (http://www.smivision.com), which can record eye-movement data at a frequency of 250Hz. In addition to the eye-tracking device, the setup also consists of a camera which observes the user's face and tracks the different facial expressions shown by the user using an out-of-the-box software *eMotion*. The videos are shown to the user on a stimulus monitor labeled as (d) in Figure 4.

Algorithm 1. MyAds:

---

    input : Video V, Set of advertisements A

$S \leftarrow$ scenes from V [16];
$i \leftarrow 1$; /* $skip\_count$, `minimum inter-advertisement interval` */;
$skip\_count \leftarrow 0$; **while** $i \leq \parallel S \parallel$ **do**

    Playback s_i;

    if $skip\_count > 0$ **then**
        $skip\_count \leftarrow skip\_count - 1$;
        Continue to next scene s_{i+1} ;

    /* `Rule (1)` */
    if $content\_arousal(s)$ AND $arousal(PD_s)$ AND
    $content\_valence(s)$ AND $valence(p_{pos}, p_{neg})$ **then**

        /* `Rule (3)` */
        $a_j \leftarrow$ unused ad $\in A$: $content\_arousal(a_j) AND$
        $content\_valence(a_j)$;
        Playback a_j;
        $skip\_count \leftarrow \frac{\parallel S \parallel + 1}{N}$;

    /* $\tilde{}$ `is the complement operator` */;
    /* `Rule (2)` */
    if $\widetilde{content\_arousal}(s)$ AND $\widetilde{arousal}(PD_s)$ AND
    $\widetilde{content\_valence}(s)$ AND $\widetilde{valence}(p_{pos}, p_{neg})$ **then**

        /* `Rule (4)` */
        $a_k \leftarrow$ unused ad $\in A$: $\widetilde{content\_arousal}(a_k) AND$
        $\widetilde{content\_valence}(a_k)$;
        Playback a_k;
        $skip\_count \leftarrow \frac{\parallel S \parallel + 1}{N}$;

    $i \leftarrow 1 + 1$;

---

5   Evaluation and User Studies

We now evaluate our multimodal, interactive, online affective placement strategy *MyAds* using a thorough user-study. In order to check the effectiveness of the advertisement campaign, we need to quantify how well the user has noticed the advertisements. We find that conventional metrics such as Cost-Per-Impression (CPI) and Click-Through Rate (CTR) do not suit video-in-video advertising so well, instead the suitable measure is *recall* which assesses how well a viewer remembers advertisement content. The assumption here is that the user is more likely to purchase a product whose advertisement is recalled at the time of purchasing a commodity. We evaluate our interactive online affective advertisement insertion strategy against the standard pre-roll/post-roll advertising and relevance and attention based method in [13]. Participants are shown a video program with advertisement insertion in one of 3 modes (a) pre-roll and post-roll advertising (*Mode I*) (b) VideoSense [13] (*Mode II*) and (c) *MyAds* (*Mode III*). Three interlinked dimensions of recall are investigated across the three advertising strategies,

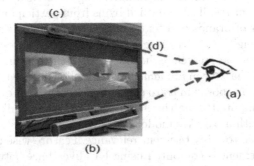

Fig. 4. Experimental setup. (a) The user; (b) Eye-tracker; (c) Camera; (d) Stimulus monitor

1. Advertisement/Brand recall: How well does the user recall advertisement content and product brands ?
2. Cued/Uncued recall: After the session, the subject is presented a list containing some of the brands(cued) or asked to recall purely from memory(uncued) ?
3. Short-term/Long-term memory: The user is asked to do cued / uncued brand / advertisement recall (a) immediately after viewing the video program (b) a day after.

In addition to the recall, the users also answer subjective questions based on the following parameters (i) whether the advertisements are distributed uniformly over the video program (ii) how disturbing are ad-insertions (iii) how contextual relevance of inserted ad (iv) overall viewing experience. This part of the evaluation is similar in spirit to [13].

A set of 9 (of total duration 130 minutes) videos covering different kinds - movie clips, tv shows in both english and chinese (foreign language, with english subtitles), documentaries, animated clips and 50 advertisements (of average duration 20 seconds) were chosen. The advertisements cover a wide range of products including popular brands such as Coke® and also relatively unfamiliar brands. The video data used in our experiments is similar to broadcast content, but is accessed online. We recruit our participant pool from university graduate, undergraduate and staff, there were 26 participants in total (14 male, 12 female) with mean age= 24.2 (stdev=4.2). All participants sign consent forms and receive a token payment on completion of the experiment.

The experiments follow a block-based design, where each participant sees 3 randomly chosen videos (program+inserted advertisements), all videos in a block are generated using one of (Mode I/II/III). The participants are ignorant about the strategy being used in placing advertisements. As the proposed advertising strategy *MyAds* is online and involves noticeable editing artifacts when compared to the other two strategies which are offline, users are shown videos stemming from the same advertising strategy. Each video lasts for about 12-15 minutes and the participants then answer systematically chosen subjective questions captured on a 5 point scale. Short/long term cued, uncued brand and advertisement content recall is quantified by us from participant responses using pictorial depictions of brands and text-based descriptions of the advertisements and brands. The long term advertisement/brand recall is assessed offline by the users a day after they have watched the video. All of the baseline methods including VideoSense [13], have been implemented and following are the details. We use the attention model proposed in [9] to compute *attractiveness* of video scenes and *discontinuity* is computed using the merge level in visual similarity based video segmentation algorithm [16]. We model tempo, motion and colour based similarity to compute scene-to-advertisement relevance. Feature wise distance between scene and advertisement is computed using KL-divergence between corresponding histograms and $arg_{min}(\cdot)$ over feature wise divergence scores is chosen as local relevance score. Please note that we do not model concept based relevance for fair comparison with our method *MyAds* and (a) pre-roll/post-roll advertising.

6   Results and Discussion

We now discuss some key results from the evaluation of the proposed *MyAds* algorithm. Normality of the underlying distribution for all scores obtained from the user-study and the recall experiments is ensured using KS(Kolmogorov-Smirnov) tests. We then analyze the performance of MyAds (Mode III) using t-tests with scores from pre/post-roll insertion (Mode I) and VideoSense [13] (Mode II).

Brand and Advertisement Recall. Does brand and advertisement recall improve with *MyAds* ? We derive cued brand recall (CBR), uncued brand recall (UBR), cued advertisement recall (CAR) and uncued advertisement recall

(UAR), for short-term over all videos and users; Table 1. *MyAds* in row 3, performs well (p=0.1) as compared to pre/post-roll insertion (CAR, column 1).*MyAds* performs better for uncued advertisement recall, than pre/post-roll (p=0.1) and better than VideoSense [13](p=0.05) (UAR, column 2). The recall scores vary from 0(very poor recall) to 1(complete recall), MyAds performs significantly better at content recall (CBR, column 3) than both VideoSense [13] and post/pre-roll (p=0.05). Our method also performs at par with VideoSense and better than pre/post-roll strategies (p=0.01), for uncued brand recall (UBR, column 4).

Table 1. Cued and uncued recall over brands and advertisement content

Mode	CAR	UAR	CBR	UBR
Pre/post-roll	0.7 (± 0.09)	0.6 (±0.096)	0.75 (±0.064)	0.6 (±0.1)
VideoSense	0.7 (±0.097)	0.67 (±0.075)	0.71 (±0.08)	0.71 (±0.09)
MyAds	0.78 (±0.075)	0.75 (±0.08)	0.72 (±0.08)	0.73 (±0.1)

Subjective Experience. Does perceived advertisement placement uniformity(Q1), program flow(Q2), advertisement relevance(Q3) and overall experience(Q4) improve with *MyAds*? We measure subjective viewer responses across all users and videos; Table 2. When compared to Mode I, *MyAds* in row 3 performs better on all parameters (p=0.1, 0.05, 0.01 for columns 2, 3, 4) except uniformity of advertisement placement (column 1), where its performance is similar to [13]. Since Mode I, by definition, places advertisements in a uniform manner, it outperforms the other two methods. When compared to VideoSense [13], *MyAds* performs consistently better on program flow (p=0.1, column 2) and advertisement relevance (p=0.05, column 3), where the average rating across users is equal. The scores vary from 0 (bad) to 5 (good).

Table 2. Subjective user-responses on a 5-point scale

Mode	Uniform distribution	Program flow	Ad relevance	Overall viewing experience
Pre/post-roll	4.6 (±0.5)	3.1 (±0.56)	2.7 (±0.6)	3 (±0.47)
VideoSense	3.63 (±0.6)	3.27 (±0.46)	3.45 (±0.5)	3.54 (±0.5)
MyAds	3.72 (±0.46)	3.45 (±0.68)	3.45 (±0.52)	3.65 (±0.6)

Impact on Long-Term Recall. Long-term recall is measured a day after the user has taken part in the experiment. We observe that the *MyAds* in row 3 performs better than the Mode I (p=0.01, 0.01: columns 2, 3) and II (p=0.01, 0.05: columns 2, 3), for both advertisement and brand recall respectively; Table 3. The scores vary from 0-very poor recall to 1-complete recall.

Table 3. Long-term (Day-after) Advertisement/Brand recall

Mode	Advertisement Recall	Brand Recall
Pre/post-roll	0.3 (±0.05)	0.29 (±0.05)
VideoSense	0.3 (±0.06)	0.31 (±0.09)
MyAds	0.36 (±0.08)	0.39 (±0.1)

7   Conclusion and Future Work

We propose an interactive, multimodal affective online video-in-video advertisement strategy and demonstrate good performance on recall statistics. The model is aimed primarily towards individual personalization. We are working towards enhancing scene-to-advertisement relevance in *MyAds* and bringing in saliency and eye-tracking based models of user engagement. Interactive advertising in commercial *billboard* like scenarios is another direction of ongoing work. Consumer generated videos like home videos provide a different set of challenges which would be addressed in future.

Acknowledgments. This work is funded by the A*STAR PSF Grant No. 102-101-0029 on "'Characterizing and Exploiting Human Visual Attention for Automated Image Understanding and Description"'.

References

1. Hanjalic, A., Xu, L.-Q.: Affective video content representation and modeling. IEEE Transactions on Multimedia 7(1), 143–154 (2005)
2. Mellers, B.A., McGraw, A.P.: Anticipated emotions as guides to choice. Current Directions in Psychological Science 10(6), 210–214 (2001)
3. Bradley, M.M., Lang, P.J.: Measuring emotion: The self-assessment manikin and the semantic differential. Journal of Behavior Therapy and Experimental Psychiatry 25, 49–59 (1994)
4. Roel Vertegaal, C.R., Shell, J.S., Mamuji, A.: Designing for augmented attention: Towards a framework for attentive user interfaces, vol. 22, pp. 771–789 (2006)
5. Forgas, J.P.: Toward Understanding the Role of Affect in Social Thinking and Behavior. Psychological Inquiry, 90–102 (2002)
6. Katti, H., Yadati, K., Kankanhalli, M., Tat-seng, C.: Affective VideoSummarization and Story Board Generation Using Pupillary Dilation and Eye-Gaze. In: 2011 IEEE International Symposium on Multimedia, pp. 319–326 (2011)
7. Russell, J.A.: A circumplex model of affect. Journal of Personality and Social Psychology 39(6), 1161–1178 (1980)
8. Broach, J., Carter, V., Page, J., Thomas, J., Wilson, R.D.: Television programming and its influence on viewers' perceptions of commercials: The role of program arousal and pleasantness. Journal of Advertising 24(4), 45–54 (1995)
9. Harel, J., Koch, C., Perona, P.: Graph-based visual saliency. In: Advances in Neural Information Processing Systems 19, pp. 545–552. MIT Press (2007)

10. De Lemos, J., Sadeghnia, G.R., Ólafsdóttir, Í., Jensen, O.: Measuring emotions using eye tracking. In: Spink, A. (ed.) 6th International Conference on Methods and Techniques in Behavioral Research (2008)
11. Ekman, P.: Basic Emotions. In: Handbook of Cognition and Emotion. John Wiley and Sons (1996)
12. Hazlett, R.L., Hazlett, S.Y.: Emotional response to television commercials: Facial emg vs. self-report. Journal of Advertising Research 39(2), 7–23 (1999)
13. Mei, T., Hua, X.-S., Yang, L., Li, S.: Videosense: towards effective online video advertising. In: Proceedings of the 15th International Conference on Multimedia, MULTIMEDIA 2007, pp. 1075–1084 (2007)
14. Bolhuis, W.: Commercial breaks and ongoing emotions: Effects of program arousal and valence on emotions, memory and evaluation of commercials. Masters Thesis (2006)
15. Peng, W.-T., Chang, C.-H., Chu, W.-T., Huang, W.-J., Chou, C.-N., Chang, W.-Y., Hung, Y.-P.: A real-time user interest meter and its applications in home video summarizing. In: 2010 IEEE International Conference on Multimedia and Expo (ICME), pp. 849–854 (2010)
16. Rasheed, Z., Shah, M.: Scene detection in hollywood movies and tv shows. In: Proceedings of IEEE Computer Society Conference on Computer Vision and Pattern Recognition, vol. 2, pp. II-343–II-348 (2003)

GPS Estimation from Users' Photos

Jing Li[1], Xueming Qian[1,*], Yuan Yan Tang[2], Linjun Yang[3], and Chaoteng Liu[1]

[1] Depart. Information and Communication Engineering, Xi'an Jiaotong University, China
lijing.1@stu.xjtu.edu.cn, qianxm@mail.xjtu.edu.cn
[2] FST of Macau University, Macau, China
yytang@umac.edu.cn
[3] Microsoft Research Asia, China
linjuny@microsoft.com

Abstract. Nowadays social media are very popular for people to share their photos with their friends. Many of the photos are geo-tagged (with GPS information) whether automatically or manually. Social media management websites such as Flickr allow users manually labeling their uploaded photos with GPS with the interface of dragging them into the map. However, manually dragging the photos to the map will bring more error and very boring for users to labeling their photos. Thus in this paper, a GPS location estimation approach is proposed. For an uploaded image, its GPS information is estimated by both hierarchical global feature classification and local feature refinement to guarantee the accuracy and computational cost. To guarantee the estimation performances, k-nearest neighbors are selected in global feature classification stage. Experiments show the effectiveness of our proposed approach.

Keywords: GPS Estimation, K-NN, Hierarchical Structure, BoW, Geo-tag.

1 Introduction

Considering you are in the situation of missing at some unknown places, little information is available outside, how will you try to find where you are. Sometimes you call you friends for help by telling them the appearances of landmarks around you. Moreover, we come across some beautiful photos in the internet and what is a pity is that there is little description of the locations that the images are taken. However, automatic GPS locations estimation for an image is possible with the help of the large scale geo-tagged photos shared by minions of worldwide users.

The main contributions of this paper are as follows: 1) Building a hierarchical structure for the dataset for the geo-tagged dataset using both their vision and their GPS information. 2) An unsupervised hierarchical global feature classification and local feature refinement based GPS estimation approach is proposed. 3) k-NN based approach is proposed to guarantee the estimation accuracy for an input image.

The rest of the paper is organized as follows: Related works on GPS estimations are overviewed in Section 2. Section 3 is the system overview and Section4 and

---

* Corresponding author.

S. Li et al. (Eds.): MMM 2013, Part I, LNCS 7732, pp. 118–129, 2013.
© Springer-Verlag Berlin Heidelberg 2013

Section 5 are details about the system. Experiment containing the comparison with the recently popular method and parameters discussion is shown in Section 6. In Section 7 the conclusions are drawn.

2   Related Work

Nowadays, many works are relative to the attached GPS of the image. For example Qian et al utilized the GPS information to tagging users' photos using their own vocabularies [11]. Mikolajczyk and Schmid proposed a direct feature matching based approach of GPS estimation for an input image [4]. The method computes the distances between the input image and all tagged images in the dataset. Then they compute the distances on different feature spaces and at last use K-NN (k-nearest-neighbor) technique to determine the detailed information of GPS of an input image. However images with high similarity in global feature do not guarantee their contents are similar. In contrast local feature based approach can be utilized to make sure that the input image and their visual similar images are with same content. However, its computational cost of local feature based approach is very high.

Both [1] and [6] provide methods for GPS estimation. Hays and Efros build a model for each of the test landmarks by training a classifier using photos taken at the landmark versus those taken elsewhere [1]. As they use the method of training, the selection of training dataset should be a crucial problem. Moreover the computational costs are very high both for model parameters' training and online GPS estimation. That is why we proposed to use an unsupervised method in this paper. Chum et al. also proposed an approach for estimating the location of the image by using local feature matching [6]. User interaction is required to confine the locations of the input images to really small ranges [6]. If the range of the image that the user assigned is wrong (or with large error) then the estimations results will be wrong too. Even after the rough geographic area is selected, the online GPS estimation for an image is still with high computational cost. The GPS estimation for an image can be converted to image classification and image retrieval tasks [7]-[10]. From this point of view, existing example based image retrieval approach can be utilized in GPS estimation for an input image. Zhou et al proposed a spatial coding based image retrieval approach by building the contextual visual vocabulary.

3   System Overview

To speed up the estimation, we proposed a fast GPS estimation algorithm by using a hierarchical structure for the dataset. The system of the proposed approach is shown in Fig.1 where the image dataset is represented as A. The preprocessed image dataset is denoted as S. Then, R categories of images obtained after clustering is $C_1 \sim C_R$. After that, each category C_i is divided into $c_{i,1} \sim c_{i,n}$ according to the images' GPS, where n represent the n-th location in the dataset. Accordingly, in the online system, $S_1, ..., S_M$ are the selected first layer candidates, and $S_{i,1}, ..., S_{i,n}$ is the selected second layer candidates for S_i. Our system consists of two parts namely the online and offline systems. The offline system aims at preprocessing the large scale geo-tagged image dataset according to the GPS information and removing the images with wrong GPS locations. The details steps of our system are provided in Section4 and Section 5.

4   The Offline System

The goal of the offline system is to build such a hierarchical structure. The two-layer structure partitions the image dataset into GPS location oriented clusters.

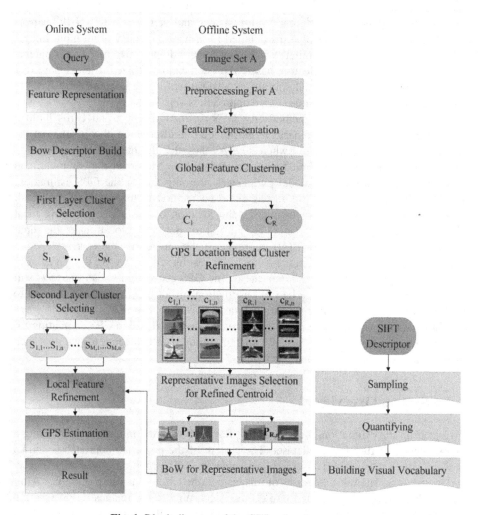

Fig. 1. Block diagram of the GPS estimation system

4.1   Preprocessing for the Dataset

Considering our method can be used for image dataset crawled from Internet randomly, the quality of the images is variant sharply. Some of the images might be too bright, too dark, blurred, or only containing persons etc. So preprocessing is needed before the clustering in order to delete the noise images with inappropriate lightness or too low quality.

4.2   Feature Representation

Different from [1], in this paper, global and local features are made full used of to improve the estimation performances and reduce computational costs. Here, color moment (CM) and hierarchical wavelet packet descriptor (HWVP) are used as the global features and SIFT as the local features.

45-D color Moment (CM)

Color feature has been proved the most GPS-informed feature [1]. It is used as global feature representation for the image in our method. An image is divided it into 4 equal sized blocks and a centralized image with equal-size. For each block, a 9-D color moment is computed, and thus the dimension of color comment for each image is 45. The 9-D color moment of an image segment is utilized, which contains values of mean, standard deviation and skewness of each channel in HSV color space.

170-D Hierarchical Wavelet Packet Descriptor (HWVP)

Texture feature has been shown to work well for texture description of image and for scene categorization and image recognition. The texture feature in our method is described by hierarchical wavelet packet descriptor (HWVP) [2]. A 170-D HWVP descriptor is utilized by setting the decomposition level to be 3 and the wavelet packet basis to be DB2.

Scale Invariant Feature Transform (SIFT)

The images are further represented via local interest point descriptors given by SIFT [3]. The SIFT match is used for assuring the images matched are from one certain place. In our paper, SIFT feature matching is utilized in the offline system and used for the best matched GPS selection for the input image in the online system.

4.3   Global Feature Clustering

In this paper, we propose to cluster the image dataset, and using the centroids in the online system instead of using low-level features of the images. Through image clustering, the whole dataset can be divided into small groups according to the appearance of the images. Our main purpose of clustering is to reduce the computation cost and improve GPS estimation performance.

As k-means clustering [7] has been approved a good method to divide dataset into small clusters, we choose to use k-means to cluster the global features. The global feature clustering is carried out on the 215d vector including 45d CM and 170d HWVP. To support fast online GPS estimation, the number of first layer clusters R in k-means should not be set to be too large. In this paper R is set according to the different appearances (in color and texture) of images in 4 seasons, daytime and night, landmark and landscape, modern and ancient. Thus we set R to 32. More detailed discussions are given in **Section 6**.

4.4   GPS Location Based Cluster Refinement

After obtaining the centroids $C_1,..,C_R$, then we classify the geo-tagged images into the R clusters. Assuming that the GPS location number of the images in $C_i(i=1,..,R)$ is $N_i(i=1,..,R)$, we further depart the images in each cluster $C_i(i=1,..,R)$ according to the corresponding GPS information. Then the corresponding refined second layer clusters $c_{i,j}(i=1,..,R, j=1,...,N_i)$ of its first layer centroid $C_1$are obtained by using the average centroid of the low-level features of the images in each GPS location as follows:

$$c_{i,j} = \frac{1}{n_{i,j}} \sum_{k=1}^{n_{i,j}} L_{i,j,k} \, , \quad i = 1, \cdots, R; j = 1, \cdots, N_i \qquad (1)$$

$$M_i = \sum_{j=1}^{N_i} n_{i,j} ; \qquad i = 1, \cdots, R; j = 1, \cdots, N_i \qquad (2)$$

$$M = \sum_{i=1}^{R} M_i ; \qquad i = 1, \cdots, R; j = 1, \cdots, N_i \qquad (3)$$

where $n_{i,j}$ is the image number of j-th GPS of the i-th centerC_i, M_i is the image number in the i-th cluster C_i, and M is the image number of the whole image set after preprocessing, $L_{i,j,k}$ is the 215 dimensional low-level global feature vector of k-th image with the j-th GPS from the i-th center $C_i(i=1,..,R)$.

4.5   Representative Images Selection for the Refined Centroids

The images in the GPS refined centroid have similar content but with some outliers with error geo-tagged images. Moreover some of the centroids many contain too many images, especially for some famous places such as Eiffel Tower and Leaning tower of Pisa etc. Thus selecting some representative images for each of the refined centroid $c_{i,j}$ is helpful for reducing the negative influences of erroneously geo-tagged images.

In this step, only when two images have sufficient matched SIFT point pairs, then the two images are considered match [5]. We first select the image that matches with most images in the cluster as the representative image for the cluster. Then image matches with most images (i.e. relevance) and has large visual differences (i.e. diversity) from the already selected representative images is selected iteratively. Then repeat the process until we select all the representative images out. The algorithm of representative images selection is as follows:

4.6   BoW for Representative Images

In this paper a BoW description is computed for the input image and all the refined clusters. Firstly, we randomly sampling the SIFT feature points from an images sets about 30 million images, and grouping the SIFT points into Q centroids (i.e. the BoW number is Q) using k-means. Then for the offline dataset, each SIFT point is quantized into one of the Q centroids by assigning it to the nearest centers.

Algorithm 1. Selecting Representative Images For the Cluster $c_{i,j}$

Input:
      All the images in $c_{i,j}$ denoted Un

Output:
      The selected representative image set $P_{i,j}$ for $c_{i,j}$

Initial:
      Delete noise images with matched image number less than the average;
      Denote the de-noised image cluster as Dn
      Select image A have most matched images in Dn;
      $P_{i,j} = \{A\}$ and $Dn=Dn-\{A\}$

The process:
      while Dn contains image
         Refresh A to the image with largest number of matched images in Dn;
         Denote the number of matched point pair between A and images from $P_{i,j}$ as np;
         Denote the number of SIFT for A as ns;
         if $np>ns/2$
            $Dn= Dn-\{A\}$
         else
            $P_{i,j} = P_{i,j} +\{A\}$;
            $Dn= Dn -\{A\}$;
         end
      end

Each centroid is represented by a set of representative images. Our approach is utilizing the normalized histogram of the BoW of the representative images in each refined centroids $c_{i,j}$ as follows:

$$HR_{i,j}(k) = \frac{1}{N_R}\sum_{m=1}^{N_R} H_{i,j}^m(k), k =1,\cdots,Q \tag{4}$$

where $H_{i,j}^m(k)$ is the BoW histogram of the m-th representative image of the refined centroid $c_{i,j}$, and NR is the representative images of the refined centroid $c_{i,j}$.

5   The Online System

The online system is concentrating on how to convert an image to its GPS location. The detailed GPS estimation for an input image is shown in Fig. 1. The details of how to estimate is shown as follows.

5.1   Fist Layer Cluster Selection

Let L_x denote the 215 dimensional global features of the input image. The selection of first layer cluster candidates' selection is according to the distance between L_x and the R centers $C_i (i=1,..,R)$. In this step, the distance D_i between the query image and i-th center C_i is computed as follow.

$$D_i = \|C_i - L_x\|, (i = 1, \cdots, R),\tag{5}$$

where $\|X\|$ denotes the norm of X.

In this paper the top ranked M (M≤R) centroids are selected. The reason that we choosing of M clusters rather than the most similar one is that images with same GPS may scattered into different clusters in the first layer, and the visual similar cannot guarantee the content are the same. Let $S=\{S_1,...,S_M\}$ denote the selected M candidate where $S_k \in \{C_1,...C_R\}, K \in \{1,..,M\}$, is the selected cluster candidates.

5.2   Second Layer Clusters Selection

Each $S_k \in \{C_1,...C_R\}$ has Z_k refined global centroids in the second layer. Thus there are totally N refined centroids for the second layer after selecting M coarse centroids in the first layer. Let $s=\{r_1,...,r_N\}$ (with $r_i \in \{c_{j,k}\}, j=1,...,R; k=1,...,n$)denote the set of candidate centroids in the second layer. More precise clusters can be determined by computing the distances of the input image with its global feature L_x with the centroids second layer r_i as follow.

$$d_i = \|r_i - L_x\|, i \in \{1, \cdots, N\}\tag{6}$$

In the second layer refined clusters selection, we firstly rank the distances d_i in ascending order, and then assume the number of select centroids as GPS estimation candidates is F. The selected candidates are denoted as $SG=\{g_1,...,g_F\}$with $g_k \in \{r_1,...,r_N\}$.

5.3   Local Feature Refinement

After the hierarchical global feature matching based coarse GPS location selection, several candidates $SG=\{g_1,...,g_F\}$ are selected. The above-mentioned cluster candidates' selection is mainly for the sake of speeding up the process which cannot ensure estimation accuracy. Local feature matching can improve GPS estimation performances.

In the offline system, the normalized BoW histograms of the representation images are built. Then the similarity of the input image with the refined centroids can be measured by utilizing the Cosine similarity (denoted as CS), mean abstract distance (denoted as F1), mean squared distance (denoted as F2) and histogram intersection (denoted as HI).

5.4   GPS Estimation

From Fig. 1, it is very likely that images taken from one certain place can be distributed into different clusters. For example, in the second layer, the images of Eiffel are divided into different clusters. So in the online system K-NN is necessary for improving the GPS estimation performance. When using K-NN based approach instead of using the one best match cluster's GPS as estimation (denoted 1-NN), we first choose K best match second layer clusters (each cluster corresponds a GPS). And then count the number of clusters for each occurred GPS. The majority GPS in the K clusters is assigned for the input image. In the 1-NN based GPS estimation, the GPS of the cluster with the highest similarity to the input image is assigned for it.

6   Experiments and Discussions

In order to test the performance of the proposed GPS estimation approach, comparisons are made with IM2GPS [1], and spatial coding based approach (denoted as SC) [7] and ours. Experiment is done on three datasets: COREL5000, OxBuild5000 and GOLD. All experiments are done on a server with 2.0 GHz CPU and 24 GB memory, and all the experiments are performed on the environment of C.

OxBuild5000 and COREL5000 are utilized to test the method. Each of the categories in OxBuild5000 and COREL5000 is severed as a GPS location. 100 images are selected randomly as the test set, while the rest is utilized in the offline system for the hierarchical structure building. GOLD contains more than 3.3 million images together with their Geo-tags and it covers 60k different cites in the world. 80 travel spots are randomly selected for testing. The test dataset for the 80 sites contains 52046 images.

6.1   Performance Evaluation

If the GPS of an input image is exact with its ground-truth GPS, it is correctly estimated, otherwise falsely estimated. Assuming that A_i is the correct, then average recognition rate (AR) is utilized to evaluate the GPS estimation performance which is given as follows:

$$AR = \frac{1}{G}\sum_{i=1}^{G} A_i \tag{7}$$

$$A_i = \frac{NC_i}{NA_i}\times 100\%, i \in \{1,\cdots,G\} \tag{8}$$

where NC_i is the correct estimated image number, NA_i is the test image number and A_i is the correct recognition rate of the i-th spot. G is the number of GPS location, 14, 50, and 80 for OxBuild5000, COREL5000, and GOLD respectively.

6.2   GPS Estimation Performance Comparisons

In SC, K-NN is utilized and K is set to be 120 under which best performance is achieved. As for IM2GPS, we chose the best parameters provided in [1], and use their method of K-NN in classification. We determine whether the classification is right by judge if the selected K images contain most images with the same GPS as the input. The performances of our approach under the similarity measurement method CS, F1, F2 and HI are evaluated.

It is clear from TABLE1-2 our method can outperform the other methods. The results of IM2GPS in the three test dataset are 45.98%, 39.67% and 53.06% while the results of SC in the three test dataset are 76.01%, 60.87% and 71.84%. Those of ours under CS for the three datasets are 97%, 91%, and 84.64% respectively. An average 102% improvement is achieved. Our approach of F2, HI also performs better.

The average computational costs of IM2GPS on the three test sets are 60.46ms, 33.74ms and 64927ms, while that of SC are respectively 7.3ms, 5.51ms and 39.60ms. And time costs of CS, F1, F2, HI are all lower than both SC and IM2GPS. Even though the SC is more efficient than IM2GPS, its computational cost is more than 10 times of CS for COREL5000 and OxBuild5000.

Table 1. Average Recognition Rates (%) of SC, IM2GPS, our approach under CS, F1, F2, and HI

Dataset	SC	IM2GPS	CS	F1	F2	HI
COREL5000	76.01	45.98	97.00	96.00	97.00	95.00
OxBuild5000	60.87	39.67	91.00	90.00	90.00	89.00
GOLD	71.84	53.06	84.64	84.05	85.02	84.21

Table 2. Average Computational Costs (in ms) of SC, IM2GPS, our approach under CS,F1, F2, and HI

Dataset	SC	IM2GPS	CS	F1	F2	HI
COREL5000	7.94	60.46	0.76	0.71	0.82	1.08
OxBuild5000	5.42	33.74	0.47	0.41	0.50	0.49
GOLD	47.00	64927	0.96	0.93	1.03	0.99

6.3   Impacts of Parameters

Here the impacts of parameters are discussed. The parameters in our baseline algorithm are set as R=32, M=10, K=50, V=100, and the size of BoW is 60K. In the following, discussion is done on GOLD.

Total Number of the First Layer Clusters-R

To study the impact of total number of the first layer, we experiment with different R on GOLD. The corresponding Average Recognition rates of CS, F1(mean absolute distance), F2 and HI with R={1,10,20,32,50,100,200,500} are shown in Fig.2(a), and their computational costs are shown in Fig.2(b). R= 1 means that no clustering is performed. With R increases, the performance first increases and then drops. When R is set to be in the range of [20,100], better performance can be achieved. However the time cost increases largely with the increase of R shown in Fig2(b).

Number of First Layer Candidate-M

Fig.3(a) shows the AR's change with the increasing of M (M≤R) and Fig.3(b) shows the time cost. It is clear that with the increase of M, the time increase sharply for our method of histogram based approaches. The results of CS with the change of M from 1 to 32 are 73.41%, 79.1472%, 84.64%, 85.26%, 85.63%, 86.11%, 86.91%, and 87.39%. It can be seen that with the increase of M the AR is first increasing sharply and then into small fluctuation for all the methods. The consuming time of CS are 0.12, 0.23, 0.96, 2.42, 3.92, 5.71, 8.42 and 9.41, all measured in ms. the other three histogram based methods' time costs are all range in the similar scale.

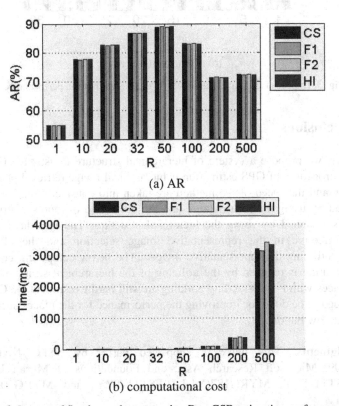

(a) AR

(b) computational cost

Fig. 2. Impact of first layer cluster number R to GSP estimation performances

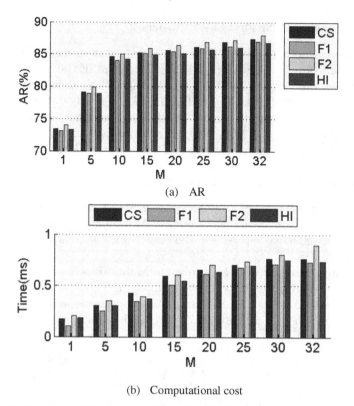

(a)   AR

(b)   Computational cost

Fig. 3. Impact of first layer candidate M to GPS estimation performances

7   Conclusions

In this paper, we propose a system of hierarchical structure to estimate GPS for an image. The procedure of GPS estimation is hierarchical image retrieval process. Both the accuracy and the speed of the method are taken into consideration. The accuracy is guaranteed by using the local feature and the utilization of representative images; the speed is obtained by using the hierarchal structure. The introduction of local feature is effective in the representative image selection and the effectiveness obtained by utilizing the representative images. The heavy computing cost of local feature computing is reduced by the utilizing of the hierarchical structure. However for some places with no outstanding building are still hard to estimate the GPS. Much more still need to be done for improving the performance for the places such as beach or street with few buildings.

Acknowledgments. This work is supported partly by NSFC No.60903121, No.61173109, Microsoft Research Asia, and Foundations of Macau University: SRG010-FST11-TYY, MYRG187(Y1-L3)-FST11-TYY, and MYRG205(Y1-L4)-FST11-TYY.

References

[1] Hays, J., Efros, A.A.: IM2GPS: estimating geographic information from a single image. In: CVPR (2008)

[2] Qian, X., Liu, G., Guo, D., Li, Z., Wang, Z., Wang, H.: Object Categorization using Hierarchical Wavelet Packet Texture Descriptors. In: Proc. ISM, pp. 44–51 (2009)

[3] Lowe, D.: Distinctive image features from scale-invariant key points. Int'l J. Computer Vision 2(60), 91–110 (2004)

[4] Mikolajczyk, K., Schmid, C.: A performance evaluation of local descriptors. IEEE Transactions on Pattern Analysis and Machine Intelligence 27(10), 1615–1630 (2005)

[5] Kennedy, L., Naaman, M.: Generating diverse and representative image search results for landmarks. In: WWW 2008 (2008)

[6] Chum, O., Philbin, J., Sivic, J., Isard, M., Zisserman, A.: Total recall: Automatic query expansion with a generative feature model for object retrieval. In: Proc. ICCV (2007)

[7] Zhou, W., Lu, Y., Li, H., Song, Y., Tian, Q.: Spatial Coding for Large Scale Partial-Duplicate Web Image Search. In: MM 2010, Firenze, Italy, October 25-29 (2010)

[8] Wang, M., Ni, B., Hua, X.-S., Chua, T.: Assistive Tagging: A Survey of Multimedia Tagging with Human-Computer Joint Exploration. ACM Computing Surveys 44(4) (2012)

[9] Wang, M., Yang, K., Hua, X.-S., Zhang, H.: Towards a Relevant and Diverse Search of Social Images. IEEE Transactions on Multimedia 12(8), 829–842 (2010)

[10] Xue, Y., Qian, X.: Visual Summarization of Landmarks via Viewpoint Modeling. In: Proc. ICIP 2012, pp. 2873–2876 (2012)

[11] Qian, X., Liu, X., Zheng, C., Du, Y.: Tagging Photos Using Users' Vocabularies. Neurocomputing

Knowing Who You Are and Who You Know: Harnessing Social Networks to Identify People via Mobile Devices

Mark Bloess, Heung-Nam Kim, Majdi Rawashdeh, and Abdulmotaleb El Saddik

School of Electrical Engineering and Computer Science, University of Ottawa,
800 King Edward, Ottawa, Ontario, Canada, K1N 6N5
{mbloe023,hkim,mrawa056,elsaddik}@uottawa.ca

Abstract. With more and more images being uploaded to social networks each day, the resources for identifying a large portion of the world are available. However the tools to harness and utilize this information are not sufficient. This paper presents a system, called PhacePhinder, which can build a face database from a social network and have it accessible from mobile devices. Through combining existing technologies, this is made possible. It also makes use of a fusion probabilistic latent semantic analysis to determine strong connections between users as well as social photos. We demonstrate a working prototype that can identify a face from a picture taken from a mobile phone using a database derived from images gathered directly from a social network and return a meaningful social connection to the recognized face.

Keywords: Personalized Mobile Services, Social Networks, Photo Tagging, Face Recognition.

1   Introduction

Social networking is a continually growing medium for online photo sharing. Facebook (http://facebook.com) alone claims 250 million new photos are uploaded each day. With the addition of "tagging," the act of annotating an image with a specific user's name and their location in the image, this makes these images rich with information. Combining this information with face recognition technologies, it is possible to build face recognition databases. With the majority of cell phones today coming equipped with cameras, the ability to capture your surroundings at any given time has never been more wide spread. Combining these two technologies would be a great step forward into the realm of eliminating personal anonymity. Using your phone to capture the people around you, you can then use face recognition to identify them.

Just identifying a person is intriguing; however, perhaps you would like to introduce yourself to them. Using additional information from social networks we can not only improve recognition, but find a social connection between two users. This could inform a user as to how they may know the person being recognizing. For example they could be a friend of a friend, or perhaps they have appeared in pictures with people you know. With this aspect in mind, we analyzes the potential of who is with

S. Li et al. (Eds.): MMM 2013, Part I, LNCS 7732, pp. 130–140, 2013.
© Springer-Verlag Berlin Heidelberg 2013

whom in photos for determining meaningful social connections between users, as well as for finding hidden new friends. To this end, we introduce in this paper a probabilistic fusion model that unifies social friend connections with people's co-occurrences in photos, based on Probabilistic Latent Semantic Analysis (PLSA) [4, 7].

This paper presents two specific contributions. First, we describe a methodology for collecting images from a social network and constructing a face recognition database. Second, we propose a method of modeling social connections between users. Additionally we present our working prototype of a mobile application which allows a user to take a picture of a person and recognize their face using a database derived from a social network using our described methodology. They can then see the social connection between themselves and the recognized face.

1.1 Related Work

Combining social networks with face recognition has been an area of high interest. Specifically improving the accuracy through use of context gathered from social networks. Often times the interest is in auto-annotation, which is the act of automatically tagging an image [14]. Using context such as which users are often tagged by a specific user, or which users often appear in photos together, you can make assumptions on who a specific user may be. However since this is for automatic annotation, they can also make the assumption that the user most likely resides in the friends list of the uploader. Similarly, Mavridis et al. [11] also demonstrates how using context gained from a social network can improves recognition accuracy. Their research is intended for use in the FaceBot system, which is a robot which can access social networks and perform real time face recognition. This is very similar to a mobile application, however again they use context such as friendships and co-occurrences to improve accuracy. Their methodology could be useful for recognizing multiple faces in the same image. They determine two users appearing in an image together have approximately 80% likelihood that they are friends. So if the confidence in the recognition of one face is high, we can use this as seed information to improve the recognition of the second face. However, if there is only one face to be recognized, we should not assume it is going to be a friend of the user taking the picture. The intended use for our research is to identify any person, which could mean any level of social separation between users. Another example of context helping improve recognition is shown in [5]. However, they build a dataset entirely from their mobile application. This ensures that the dataset will contain all the information they want. So although it does demonstrate that context can improve on pure computer vision face recognition, we must determine how to use the contextual data when there is no guarantee it even exists.

Finding connections between users is also a topic of interest for research, primarily in the field of friend recommendation. Kim et al. [8] suggest using image co-occurrence as a factor when trying to find strong connections between two users. Although our research targets this issue as well, we also deal with the scenario where the end user is known, and instead we are trying to find the best connection to this user. However we can still build a similar connection graph using both friendships and picture co-occurrences.

2   Building Social Face Databases

In our study, Facebook was used to build the database. In order to access a user's information we created a Facebook application. A user must first grant the application permission to access their Facebook information. In this case we require permission to access a user's photos, and their friend's photos. We will refer to the user that has granted the permissions as a primary user. Table 1 shows the size of the dataset after only gaining permission from two primary users.

Table 1. Statistics of collected data from Facebook

Total number of Friendships	65,160
Individual user entries	4774
Albums Searched	6027
Photos Gathered	11,089
Face Images Gathered	14,603

This illustrates that with only a few primary users granting permission we gain access to a large number of users and faces, so growth of the database will occur very quickly. In addition to building the face recognition database, we also collect information about the users. We attempt to gather information such as name, gender, birth date, hometown, current location, etc. The hope is that this contextual information can be used to improve accuracy [15]. However, the application only has access to the private information of a primary user and the users on the primary user's friends list. So we only access extra information on a small subset of the database, in this case 619 of the total 4774 users. However, name and gender are public information, so the only context we can assume we will have on most users is their gender. The only exception would be if the user never stated their gender on Facebook. In this data set 4648 of the total 4774 users had a gender stated. Although this is only one piece of information, it can effectively reduce the search space by half if the gender is known.

When collecting these face images, we only consider Facebook images in which there occur manual tags as well as detected faces. Any image that has tags but no face is detected, or that faces are detected but has no tags, is not considered at all. Since all the tags are done manually there are issues that arise. For example, if a tag is placed on the back of someone's head, there is now a tag that cannot be matched to the correct user. To make things worse, if there is a very clear face without a tag, it is possible that this face could be incorrectly matched with another tag. It is very difficult to decide what distance is too far to be considered anymore. As well, if a person were to incorrectly tag someone, there is no way to avoid adding the incorrect face to a specific user. We could attempt to perform face recognition to a face to ensure it matches the user we are adding the image to. However this could only be done once the database has grown to a reasonable size where it can adequately recognize users.

In order to solve the issue of mismatching tags and faces, we do three things. First, faces found on social networks are referred to as "in the wild" [14] or unconstrained. This means they can vary immensely such as in pose, lighting, and rotation. So any face which is detected must undergo a second face detection process in order to reduce the chance of a non-face being detected, and reduce the number of obscured faces. Second, we will then only consider a tag to belong to a face if it occurs within the *x*, *y* plane of the detected face, as is done in [2]. This is based on the assumption that most people will click on a face when adding a tag. Third, if more than one tag occurs within a detected face we calculate the distance from the center of the face to each of the tags, and assume the closest distance is the correct tag. Eq. (1) is used to calculate distance, where *tagX* and *tagY* are the *x* and *y* coordinates of the Facebook tag respectively, and *faceX* and *faceY* are the *x* and *y* coordinates of the center of the detected face respectively.

$$distance = \sqrt{(tagX - faceX)^2 + (tagY - faceY)^2} \tag{1}$$

This criteria is the reason why the number of photos and faces gathered is low. In reality, the number of images with tags and faces was more than double. However many were excluded because they did not pass the matching criteria.

3   PhacePhinder: Knowing Who You Are and Who You Know

We have designed and implemented a system to both build a database for face recognition and access it from a mobile device called PhacePhinder—pronounced "face finder." In this implementation we used Facebook as the social network, and developed a mobile application for the android operating system.

Fig. 1 illustrates the high level architecture for the PhacePhinder system. The client side resides on a mobile device. It communicates with Facebook only to grant

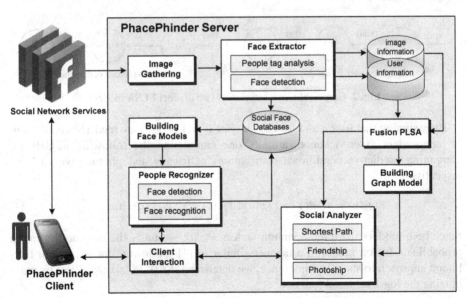

Fig. 1. High level architecture of the PhacePhinder system

permission to the PhacePhinder Facebook application, as well as verify the login of the user. All other communication is with the PhacePhinder server. The PhacePhinder server is broken down into four components and two databases. The face gathering component is responsible for going through photos and information on Facebook. The face detection and recognition component is responsible for all face detection and recognition, whether it be from the face gathering component or straight from the client interaction component. The client interaction component is a web server interface which deals with the http requests from a user and issues the appropriate responses. It passes images to the face recognition component to be recognized, as well as stores the user id and OAuth token received from the client. OAuth is an open standard for authorization [6] which is used by Facebook. The social graph constructor builds the weighted social graph which is used to find the connection between two users. The face recognition database stores all face images and hidden Markov models for users. The user information database stores all other information about a user, including personal information, which photos they occur in, and the social graph.

3.1   A Probabilistic Fusion Model for Connecting Users with Photos

The information we gather can help us gauge the connection between users and photos. In order to do so, we use the PLSA model [7] to determine a quantitative representation of a user's connection to other users and photos. The PLSA model attempts to explain a set of co-occurrence pairs in terms of a set of k latent variables (or factors), $Z = \{z_1, z_2, ..., z_k\}$. As shown in Fig. 2, in our study a latent variable $z \in Z$ is associated not only with every user–friend pair (u, f)—meaning that user u is connected to friend f—but also with every user–photo pair (u, i)—meaning that user u appears in photo i.

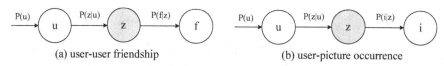

(a) user-user friendship   (b) user-picture occurrence

Fig. 2. Graphical representation of two distinct PLSA models

By assuming that user u and friend f or user u and photo i are rendered conditionally independent given a latent variable, one can define the following models for computing predictive conditional distributions of friends and photos given a user, respectively:

$$P(f \mid u) = \sum_z P(f \mid z)P(z \mid u) \, , \, P(i \mid u) = \sum_z P(i \mid z)P(z \mid u) \tag{2}$$

Since both models share the common factors $P(z|u)$, we unify the two models into a probabilistic fusion model in a manner similar to [4]. Following the maximum likelihood approach to statistical inference, we determine $P(f|z)$, $P(i|z)$, and $P(z|u)$ by maximizing the log-likelihood function:

$$L = \sum_{u} \left[\alpha \sum_{f} \log P(u)P(f \mid u) + (1 - \alpha) \sum_{i} \log P(u)P(i \mid u) \right] \tag{3}$$

where parameter α, $0 \le \alpha \le 1$, is a relative weight that grants more significance to either photo occurrence or friendship. The typical procedure for finding maximum likelihood parameters is to use the well-known Expectation Maximization (EM) algorithm. During the E-step, the posterior probabilities of the latent variables associated with each observation can be computed by:

$$P(z \mid f, u) = \frac{P(f \mid z)P(z \mid u)}{\sum_{z'} P(f \mid z')P(z' \mid u)} \ , \quad P(z \mid i, u) = \frac{P(i \mid z)P(z \mid u)}{\sum_{z'} P(i \mid z')P(z' \mid u)} \tag{4}$$

Then, the conditional distributions are recomputed in the M-step as follows:

$$P(f \mid z) = \frac{\sum_{u} P(z \mid f, u)}{\sum_{f', u} P(z \mid f', u)} \ , \quad P(i \mid z) = \frac{\sum_{u} P(z \mid i, u)}{\sum_{i', u} P(z \mid i', u)} \tag{5}$$

Along with the following fusion:

$$P(z \mid u) = \alpha \frac{\sum_{f} P(z \mid f, u)}{\mid F(u) \mid} + (1 - \alpha) \frac{\sum_{i} P(z \mid i, u)}{\mid I(u) \mid} \tag{6}$$

where $F(u)$ is the set of friends user u has and $I(u)$ is the set of photos user u occurs in. Once the model parameters are learned, we use probability distributions over potential friends for a given user u, i.e. $P(f \mid u)$, to recommend new friends as well as to estimate a relationship weight between user u and his/her current friends. It is worth noting that conditional distributions of photos given a particular user, i.e. $P(i \mid u)$, can also be used for recommending/ranking photos personally tailored to him/her.

3.2   Knowing Who You Are

When a user first uses the PhacePhinder Client application, they will be prompted to log into Facebook, using Facebook's provided android library. They will then be prompted to grant permissions to the application. Once accepted, the client will send the user's OAuth token to the server so that it can begin collecting their images. The permission prompt only occurs on first use for each primary user, unless the user revokes permission. A user can then take an image from their mobile device, which will be sent to the server to be analyzed. Before the image is sent, the user is prompted to state the gender of the face to be recognized. This is intended to reduce the search space and improve face recognition accuracy. Once the server receives the image, it will immediately try to detect a face. If no face is found, it will inform the client. Otherwise, if a face is detected, it will attempt to recognize it using Hidden Markov Model face recognition [12]. It will then give back the top three results.

The server side performs the majority of the functionality. Once it has the OAuth token from the user, it can begin collecting the images using the method described in Section 2. Using OpenCV [13], an open source computer vision library, to perform the face detection, we use two separate front facing face recognition methods provided from OpenCV. After which we check if a tag occurs within the detected face, and if there is more than one we compute the distance, defined in Eq. (1). The face image is then saved and the corresponding user's Hidden Markov Model is updated.

Fig. 3 shows the application in use. A typical scenario involves a user taking a photo of a person (Fig. 3(a)). They then state the gender of the person to be recognized and click ok (Fig. 3(b)). Then they receive back the top 3 recognition results (Fig. 3(c)).

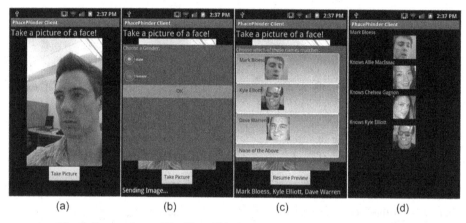

Fig. 3. Screen shots of the PhacePhinder smartphone (mobile) application

3.3   Knowing Who You Know

Since we record friendships and photo occurrences, it is possible to use this data to track connections between people. As a result, not only can we recognize a person, but also suggest to the user how they would know each other. A simple way to do this is to construct a weighted graph, where each user is a vertex and an edge would reflect the connection between two users. Originally we attempted to give a different weight depending on whether an edge is based off a friendship or photo co-occurrence. However, the results from this type of graph did not differ from one where all edges are equal. So an equal edge graph could be used to find a connection between two users.

Although this method would give a social connection, we wanted to find the best social connection. In order to do this we would need to give meaningful weights to the graph which would indicate the strength of a relationship between two users. By using our fusion PLSA to calculate these values, as explained earlier, we are able to do this. Our fusion PLSA will give us a weight for a connection between every possible pair of users, however when building the weighted graph we only want to use real connections. As a result, an edge will still only exist if two users have a friendship on the social network, or appear in an image together. So although the graph is the same, we can give the edge a weight based on our fusion PLSA results. Since our results give a higher value to indicate a stronger connection, and we are using Dijkstra's algorithm to find the shortest path [1], we must use $1 - P(f|u)$ as the weight for the edge between user u and friend f, where $P(f|u)$ is the fusion PLSA result indicating the probability of two users being friends. Fig. 3(d) shows the shortest social path from the current user to the recognized person whom he chose among the top 3 recognition results.

4   Performance Evaluation

4.1   Face Recognition

As for the accuracy of the system, we ran two tests using a subset of collected images. The Hidden Markov Model was used for face recognition. A subset of 50 individual user faces was removed at random from the total database The only constraint being the user must have at least 4 face images. The Hidden Markov Models were then retrained to never have included those images. We then attempted to recognize the removed faces among 1370 users, without stating the gender of the face. The results are as shown in Table 2.

Table 2. Subset test results without gender specified and with gender specified

	without gender	with gender
# of hits in the first place	23	24
# of hits in the second place	4	4
# of hits in the third place	3	4
# of total hits in the top-3 list	30	32

This result shows that the hit-ratio for getting the correct name at the top of the list is 46%. However, we would also like to consider if the correct name was the second or third match. So if we consider the correct name being in the top three, our hit-ratio increases to 60%. The majority of correct identifications are the top match, at 76.67% of all correct identifications being the top suggestion. We performed the same test again, however this time we assumed the gender of the face was known. This test is intended to show whether or not gender information improves the hit-ratio. Gender is the only contextual information we can guarantee we have for the majority of users, since it is public information on Facebook. As shown in Table 2, with gender as a known variable, the hit-ratio for a top match is now 48% and the hit ratio for anywhere in the top 3 is 64%. From this we can make the assumption that if the system is not recognizing a face, it is not likely due to confusion in gender. However, there is a slight improvement; therefore it justifies its addition to the application.

4.2   Friend Recommendation

In order to test friend recommendation, a series of cross-validation tests have been done. Looking at all immediate friends of the primary users who have at least 20 friends, we come up with 490 test users. We then randomly removed 10 users from each of their friends list. We then ran our fusion PLSA with different values for parameter α, as well as the PageRank and Katz algorithms described in [10]. We then retrieved a list of top 10 friend recommendations from each, and checked how many of the recommendations for the test user are in fact their friend. We measured Mean Reciprocal Rank (MRR) to examine the ability of ranking recommended friends.

Table 3. Comparison of Mean Reciprocal Rank at top-10 results

Algorithm	No constraints	< 10	≥ 20	≥ 30
PageRank	0.7727	0.7371	0.8259	0.8563
Katz	0.8604	0.8345	0.8890	0.9266
Fusion PLSA ($\alpha = 0.9$)	0.9280	0.8960	0.9895	1.1035
PLSA ($\alpha = 1$)	0.9724	1.0064	0.9079	0.9704

Table 3 shows the MRR results. When looking at all 490 test users (i.e., no constraints), we saw that the fusion PLSA algorithm outperformed both the PageRank algorithm and Katz algorithm. However, the PLSA based solely on friendship information (i.e. $\alpha = 1$) outperformed the fusion PLSA. This result may be affected by the fact that many test users have very few images. Accordingly, the average MRR results are lower when images are taken into consideration. To gain insight into the impact of photo co-occurrence information on the friend recommendation, we added the constraint of minimum number of photo occurrences to the users in the test group. When the test set was limited to users appearing in less than 10 images (i.e. < 10), the results were similar to the total dataset, with PLSA excelling. However, when we considered test users having at least 20 images (i.e. ≥ 20), the fusion PLSA outperformed the others. When the limit of at least 30 images was added (i.e. ≥ 30), each algorithm's performance improved, but the fusion PLSA had the most drastic improvement.

4.3   Preliminary User Study

To validate our system's functionality, we performed a preliminary user study with real Facebook users. The goal was to test the social path recommendation. To this end, we invited ten volunteers that existed in the database. We then determined the path to "50" random non-friends of each user. By using the Dijkstra's algorithm, we determined the path based on a graph with weights based on our fusion PLSA algorithm, and a graph where all weights are equal.

Table 4 shows the results. The "# of different paths" column indicates how many of the paths from the original 50 actually differed between the two algorithms. The column labeled "Fusion PLSA" indicates how many times the test subject chose the fusion PLSA path as the better result. The column labeled "Unweighted" indicates the number of times the test subject chose the result based on a graph where all edges are equal. From these results we can see the fusion PLSA more often determined a more meaningful path than a graph with equal edges. However, many times there was no difference in the paths between the two algorithms. This is a result of either there is only one path that exists between two users, or that the path the PLSA chose just happened to be the first shortest path found. Only one path existing between users would become less frequent as the database grows and the interconnectivity of users increases. In one case we can observe that the path with equal edges was chosen more often over the PLSA path. However, many times it was the same two paths that were being compared. Although the final connection differed, the intermediary connections were the same. So the user was actually making the same observation several times. Several of the paths, although different, users claimed there to be no social difference. By this they mean that both paths are equally significant. This is because although the system may have determined a stronger relationship, the user does not consider one friend greater than the other.

Table 4. Comparison of user feedback on finding the social path

Subject	# of different paths	Unweighted	Fusion PLSA
S1	7	0	5
S2	7	2	3
S3	28	7	5
S4	9	0	6
S5	33	1	31
S6	32	5	24
S7	32	0	6
S8	20	4	15
S9	29	4	10
S10	19	1	18

5   Conclusions and Future Work

In this paper we have proposed a method of collecting information and images from Facebook. We use this information to build a face recognition database. We then proposed and develop a system in which it can be used by a mobile device to detect people in every day scenarios. We also proposed using the contextual information from Facebook and using a fusion PLSA algorithm to build a weighted graph which can be used to find relations between a user and a recognized face, as well as be used for friend and image recommendation. Our preliminary user study validates the quality of these social connections, and our tests show that the fusion PLSA can outperform other algorithms when sufficient image information is available.

As more primary users grant permission to the PhacePhinder application, the database will grow rapidly. A larger number of users in the database will likely reduce the accuracy of the face recognition; however we will gain additional information on the users in the database. As a result we can use more context than just gender to reduce our search space. For example, "current location" is a piece of information which users can state on Facebook. If we get the current location of most users in the database, then we can use the GPS location of the mobile phone to indicate where the photo is taken to reduce the possible results to users in that geographical area. As for finding connections, using Dijktsra's algorithm could get costly once the database grows extremely large. Instead, a heuristic algorithm could be implemented to improve finding a link between two users. As well, if face recognition accuracy becomes too low, a collaboration of face recognition methods could be used, such as in [3]. The fusion PLSA algorithm also has the potential for image recommendation. We will conduct further user testing to determine the quality of these recommendations. Additionally, user feedback on friend recommendations can also give more insight into the validity of real world recommendations. As the project evolves, it could eventually be adapted to a head mounted display equipped with a camera, such as is proposed in [9].

References

1. Barbehenn, M.: A Note on the Complexity of Dijkstra's Algorithm for Graphs with Weighted Vertices. IEEE Transactions on Computers 47(2), 263 (1998)
2. Becker, B.C., Ortiz, E.G.: Evaluation of Face Recognition Techniques for Application to Facebook. In: 8th IEEE International Conference on Automatic Face and Gesture Recognition (2008)
3. Choi, J.Y.: Collaborative Face Recognition for Improved Face Annotation in Personal Photo Collections Shared on Online Social Networks. IEEE Transactions on Multimedia 13(1), 14–28 (2011)
4. Cohn, D., Hofmann, T.: The Missing Link – A Probabilistic Model of Document Content and Hypertext Connectivity. In: Advances in Neural Information Processing Systems 13 (2001)
5. Davis, M., Smith, M., Canny, J., Good, N., King, S., Janakiraman, R.: Towards Context-Aware Face Recognition. In: 13th ACM International Conference on Multimedia, pp. 483–486 (2005)
6. Hammer-Lahav, E., Recordon, D., Hardt, D.: The OAuth 2.0 Authorization Protocol, Network Working Group Internet-Draft (2011)
7. Hofmann, T.: Probabilistic Latent Semantic Indexing. In: 22nd ACM SIGIR International Conference on Research and Development in Information Retrieval, pp. 50–57 (1999)
8. Kim, H.-N., El Saddik, A., Jung, J.-G.: Leveraging Personal Photos to Inferring Friendships in Social Network Services. Expert Systems with Applications 39(8), 6955–6966 (2012)
9. Kurze, M., Roselius, A.: Smart Glasses Linking Real Live and Social Network's Contacts by Face Recognition. In: 2nd Augmented Human International Conference, vol. 31 (2011)
10. Liben-Nowell, D., Kleinberg, J.: The link-prediction problem for social networks. Journal of the American Society for Information Science and Technology 58(7), 1019–1031 (2007)
11. Mavridis, N., Kazmi, W., Toulis, P.: Friends with Faces How Social Networks Can Enhance Face Recognition and Vice Versa. In: Abraham, A., Hassanien, A.E., Snášel, V. (eds.) Computational Social Network Analysis, pp. 453–482. Springer, Heidelberg (2010)
12. Nefian, A.V.: Face Detection and Recognition Using Hidden Markov Models. In: 1998 International Conference on Image Processing, pp. 141–145 (1998)
13. OpenCV: Open Source Computer Vision Library,
http://opencv.willowgarage.com/wiki/
14. Stone, Z., Zickler, T., Darrell, T.: Autotagging Facebook: Social Network Context Improves Photo Annotation. In: IEEE Computer Society Conference on Computer Vision and Pattern Recognition Workshops (2008)
15. Stone, Z., Zickler, T., Darrell, T.: Toward Large-scale Face Recognition Using Social Network Context. Proceedings of the IEEE 98(8), 1408–1415 (2010)

Hyperspectral Image Classification by Using Pixel Spatial Correlation

Yue Gao and Tat-Seng Chua

School of Computing, National University of Singapore, Singapore
{gaoyue,chuats}@comp.nus.edu.sg

Abstract. This paper introduces a hyperspectral image classification approach by using pixel spatial relationship. In hyperspectral images, the spatial relationship among pixels has been shown to be important in the exploration of pixel labels. To better employ the spatial information, we propose to estimate the correlation among pixels in a hypergraph structure. In the constructed hypergraph, each pixel is denoted by a vertex, and the hyperedge is constructed by using the spatial neighbors of each pixel. Semi-supervised learning on the constructed hypergraph is conducted for hyperspectral image classification. Experiments on two datasets are used to evaluate the performance of the proposed method. Comparisons with the state-of-the-art methods demonstrate that the proposed method can effectively investigate the spatial relationship among pixels and achieve better hyperspectral image classification results.

Keywords: Hyperspectral image classification, spatial correlation, hypergraph learning.

1 Introduction

A hyperspectral image is a spatially sampled image which is gathered from hundreds of contiguous narrow spectral bands (from the visible to the infrared bands) by hyperspectral sensors [3]. Recently, the hyperspectral image data is rapidly increasing collected by hyperspectral instruments such as NASA Airborne Visible Infra-Red Imaging Spectrometer and Reflective Optics System Imaging Spectrometer. Different from general images, hyperspectral image comes with high dimensional feature spaces and many of the cotents are not visible by humans. Hyperspectral imaging has attracted extensive research efforts [11]. One key research in Hyperspectral classification which aims to classify its pixels into different categories. By consider the fact that hyperspectral images contain hundreds of spectral bands and the human labelling is expensive, the main challenges in hyperspectral image classification lie in the need to deal with few training samples with high data dimensionality.

Existing works on hyperspectral image classification mainly focus on either feature dimension reduction or semi-supervised classification. Traditional feature dimension reduction methods, such as Independent Component Analysis and Principal Component Analysis, have been investigated in previous works [17]. A kernel nonparametric weighted feature extraction [10] method has been proposed to extract hyperspectral image feature by using a kernel nonparametric method. For feature dimension reduction,

S. Li et al. (Eds.): MMM 2013, Part I, LNCS 7732, pp. 141–151, 2013.

another approach is to perform band selection which aims to select a group of bands from the original high-dimensional feature space. In [9], the correlation between each two spectral bands is measured by mutual information, and the representative bands are selected by minimizing the distance between the selected bands and the estimated reference map. Then the representative bands are selected by using a clustering-based method [14] in which the bands with the largest similarity to other bands are chosen. For hyperspectral image classifiers, K-Nearest Neighbor classifier (KNN) and Support Vector Machine (SVM) have been employed [13]. A semi-supervised graph-based learning method [4] is introduced to represent the hyperspectral image by using a graph structure, and then a semi-supervised learning process on the graph is conducted for hyperspectral image classification. Gu et al. [8] introduced a representative multiple kernel learningapproach to automatically combine multiple kernels in the learning procedure. Manifold learning [18] has been investigated for hyperspectral image classification in combination with the KNN classifier. In [12], a manifold structure is constructed by the pixels and the local manifold learning is conducted in the manifold structure, and a weighted KNN classifier is employed for hyperspectral image classification. Classification of hyperspectral image with few training data has attracted wide research attention recently [2]. For instance in [15], sparse representation has been investigated in hyperspectral image classification to deal with the few labeled samples.

Similar to traditional image corpus, hyperspectral images contain high spatial correlation among pixels. Nearby pixels in one hyperspectral image are captured from spatially close area, which are likely to share the same labels. This spatial information plays an important role in the understanding and classification of hyperspectral image categories. However, existing works mainly employ spatial information for noise removal or image smoothing. For instance, a spatial preprocessing approach is introduced in [16] to remove noise and smooth the image by enhancing spatial texture information with locally linear embedding in the feature space. However, none of the existing works considers spatial information embedding in terms of hyperspectral image representation, which is one of the main main challenges in the hyperspectral image classification task. In addition, how to better employ the spatial information of the pixels to improve the hyperspectral image classification performance still requires further investigation.

In most of existing hyperspectral image classification works, each pixel is mainly described by the high dimensional feature, which leads to the curse of dimensionality. It is noted that the spatial information can be also used to explore the relationship among pixels. In this work, we propose to employ the pixel spatial correlation for hyperspectral image classification, in which a hypergraph structure is constructed to estimate the relationship among pixels. The use of hypergraph is to alleviate the curse of dimensionality problem as most computations are done at the local region, which has been widely investigated in image search [22,5], object classification [20,21], and 3D object retrieval [6] and recognition [7]. Hypergraph has also been employed in Hyperspectral image classification task [19], in which the constructed hypergraph aims to explore the feature-based pixels relationship. Different from [19], we focus on the pixel spatial correlation in this work. Figure 1 illustrates the flowchart of the proposed method. In this method, the relationship among pixels is formulated in a hypergraph structure, in which each vertex denotes one pixel in the hyperspectral image. To construct the hyperedges, each pixel is connected to its spatial neighbor pixels, which generate one hyperedge for the hypergraph. By using the training data, we conduct semi-supervised learning in the constructed hypergraph for hyperspectral image classification. Experiments on two datasets, i.e., the Indian Pine and the Indian Pine Sub, are used to evaluate the performance of the proposed method.

The advantages of the proposed hypergraph method are two-fold. First, it does not require the high dimensional feature reduction process. The relationship among pixels is constructed by using the spatial correlation. The distance in the feature space is only employed to estimate the weights for each pixel in one hyperedge as introduced in the next section. Second, the employed hypergraph structure is able to capture the complex relationship among different pixels, while leads to superior hyperspectral image classification results.

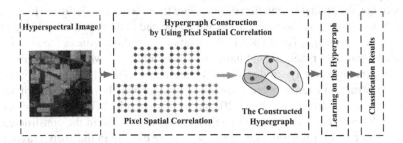

Fig. 1. The flowchart of the proposed hyperspectral image classification method by using pixel spatial correlation

The rest of the paper is organized as follows. Section 2 introduces the proposed hyperspectral image classification method by using the spatial information. Experimental results and comparison with the state-of-the-art methods on two datasets are provided in Section 3. Finally, we conclude the paper in Section 4.

2 Hyperspectral Image Classification by Using Pixel Spatial Correlation

In this section, we introduce the proposed hyperspectral image classification method by using pixel spatial correlation as shown in Fig. 1. First, we introduce the hyperspectral hypergraph construction process by spatial correlation. Next, we describe the learning process on the constructed hypergraph.

2.1 Hypergraph Construction by Using Pixel Spatial Correlation

In the proposed method, the relationship among pixels in the hyperspectral image is formulated in a hypergraph structure. In this part, we introduce the hypergraph

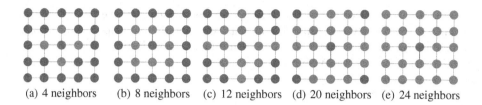

(a) 4 neighbors   (b) 8 neighbors   (c) 12 neighbors   (d) 20 neighbors   (e) 24 neighbors

Fig. 2. The illustration of the spatial-based hyperedge construction procedure

construction procedure by using pixel spatial correlation. In the constructed hypergraph $\mathcal{G} = \{\mathcal{V}, \mathcal{E}, \mathbf{W}\}$, each vertex denotes one pixel in the hyperspectral image $\mathcal{X} = \{x_1, x_2, \cdots, x_n\}$. Therefore, there are n vertices totally in $\mathcal{G}$.

In a hypergraph structure, each hyperedge connects multiple vertices. To construct the hyperedge, the spatial correlation of pixels are taken into consideration. In this process, each pixel is selected as the centroid and connected to its spatial neighbors, which generates one hyperedge. This hyperedge construction method is under the assumption that spatial connected pixels should have large possibility to have the same labels. As each pixel generates one hyperedge, there is a total of n hyperedges. Some spatial hyperedge construction examples are shown in Fig. 2, in which the numbers of spatial neighbors for the centroid are 4, 8, 12, 20, and 24 respectively. In Fig. 2, the red pixel is the centroid, and it connects its spatial neighbors (green pixels) in the constructed hyperedge. Under this formulation, the pixels which are closing in the spatial layout can be connected by hyperedges, and the connection relationship can be further extended through hyperedges.

Let the selected number of spatial neighbors be K, and there are totally $K+1$ vertices in one hyperedge. Each hyperedge $e \in \mathcal{E}$ is given a weight $w(e) = 1$, which reveals that all hyperedges are with equal influence on the constructed hypergraph structure. Though each hyperedge plays an equal role in the whole hypergraph structure, the pixels connected by one hyperedge may be not close enough in the feature space. Therefore, these pixels may have different weights in the corresponding hyperedge. For a hyperedge $e \in \mathcal{E}$, the entry of the incidence matrix $\mathbf{H}$ of the hypergraph $\mathcal{G}$ is generated by:

$$\mathbf{H}(v, e) = \begin{cases} 1 & if \ v = v_c \\ \exp\left(-\frac{d^2(v, v_c)}{2\sigma^2}\right) & otherwise \end{cases} \tag{1}$$

where v_c is the centroid pixel, $d(v, v_c)$ is the distance between one pixel v in $\mathcal{E}$ and v_c, and σ is the mean distance among all pixels. Under this definition, the pixels in one hyperedge which are similar to the centroid pixel in the feature space can be strongly connected by the hyperedge, and other pixels are with weak connection by the hyperedge.

By using the generated incidence matrix $\mathbf{H}$, the vertex degree of a vertex $v \in \mathcal{V}$ and the edge degree of a hyperedge $e \in \mathcal{E}$ are generated by:

$$d(v) = \sum_{e \in \mathcal{E}} \mathbf{H}(v, e) \tag{2}$$

and

$$d(e) = \sum_{v \in \mathcal{V}} \mathbf{H}(v, e) \tag{3}$$

In the above formulation, $\mathbf{D}_v$ and $\mathbf{D}_e$ denote the diagonal matrices of the vertex degrees and the hyperedge degrees respectively, and $\mathbf{W}$ denotes the diagonal matrix of the hyperedge weights, which is an identity matrix.

2.2 Learning on the Constructed Hypergraph

With the constructed hypergraph structure, we conduct a semi-supervised learning for classification by using the training data, which follows the regularization framework proposed in [23] as follows:

$$\arg\min_{\mathbf{F}} \{\Omega(\mathbf{F}) + \lambda R_{emp}(\mathbf{F})\} \tag{4}$$

In the above formulation, $\mathbf{F} = [f_1, f_2, \cdots, f_C]$ is the confidence score matrix for hyperspectral image classification, where $f_s(t)$ is the confidence score to categorize the s-th pixel into the t-th class, and C is the number of pixel categories. This formulation aims to minimize the empirical loss R_{emp} under the constraint of the hypergraph regularizer $\Omega(\mathbf{F})$, and it guarantees that the pixels with strong spatial correlations have large possibilities to share the same labels. R_{emp} is defined by:

$$R_{emp} = \sum_{k=1}^{C} \|f_k - y_k\|^2, \tag{5}$$

where y_k is an $n \times 1$ labeled training vector for the k-th class, and $\mathbf{Y} = [y_1, y_2, \cdots, y_C]$. $\lambda > 0$ is a tradeoff parameter, and $\Omega(\mathbf{F})$ is the hypergraph regularizer on the hyperspectral hypergraph structure, which is defined by Eq. (6),

$$
\begin{aligned}
\Omega(\mathbf{F}) &= \frac{1}{2} \sum_{k=1}^{C} \sum_{e \in \mathcal{E}} \sum_{u,v \in \mathcal{V}} \frac{w(e)\mathbf{H}(u,e)\mathbf{H}(v,e)}{\delta(e)} \left(\frac{\mathbf{F}_{u,k}}{\sqrt{d(u)}} - \frac{\mathbf{F}_{v,k}}{\sqrt{d(v)}} \right)^2 \\
&= \sum_{k=1}^{C} \sum_{e \in \mathcal{E}} \sum_{u,v \in \mathcal{V}} \frac{w(e)\mathbf{H}(u,e)\mathbf{H}(v,e)}{\delta(e)} \left(\frac{\mathbf{F}_{u,k}^2}{d(u)} - \frac{\mathbf{F}_{u,k}\mathbf{F}_{v,k}}{\sqrt{d(u)d(v)}} \right) \\
&= \sum_{k=1}^{C} \sum_{u \in \mathcal{V}} \mathbf{F}_{u,k}^2 \sum_{e \in \mathcal{E}} \frac{w(e)\mathbf{H}(u,e)}{d(u)} \sum_{v \in \mathcal{V}} \frac{\mathbf{H}(v,e)}{\delta(e)} - \sum_{e \in \mathcal{E}} \sum_{u,v \in \mathcal{V}} \frac{\mathbf{F}_{u,k}\mathbf{H}(u,e)w(e)\mathbf{H}(v,e)\mathbf{F}_{v,k}}{\sqrt{d(u)d(v)}\delta(e)} \\
&= \sum_{k=1}^{C} f_k^T (\mathbf{I} - \Theta) f
\end{aligned} \tag{6}
$$

where $\Theta = \mathbf{D}_v^{-\frac{1}{2}} \mathbf{HWD}_e^{-1} \mathbf{H}^T \mathbf{D}_v^{-\frac{1}{2}}$. Here we let $\Delta = \mathbf{I} - \Theta$, and $\Omega(\mathbf{F})$ can be written as:

$$\Omega(\mathbf{F}) = \sum_{k=1}^{C} f_k^T \Delta f_k. \tag{7}$$

Now the objective function can be rewritten as:

$$\arg\min_{\mathbf{F}} \left\{ \sum_{k=1}^{C} f_k^T \mathbf{\Delta} \mathbf{f_k} + \lambda \sum_{k=1}^{C} \| f_k - y_k \|^2 \right\} \tag{8}$$
$$s.t. \quad \lambda > 1$$

According to [23], it can be derived as:

$$\mathbf{F} = \left(\mathbf{I} + \frac{1}{\lambda} \mathbf{\Delta} \right)^{-1} \mathbf{Y} \tag{9}$$

The pixel-category correlation can be obtained after the confidence score matrix $\mathbf{F}$ has been generated. With $\mathbf{F}$, each pixel in the hyperspectral image can be classified to the category with the highest confidence score.

3   Experiments

In this section, we first describe the testing datasets and then discuss the experimental results and the comparison with the state-of-the-art methods.

3.1   The Testing Datasets

In our experiments, two datasets are employed to evaluate the performance of the proposed method. The first dataset is the Airborne Visible/Infrared Imaging Spectrometer (AVIRIS) image taken over NW Indiana's Indian Pine test site, which has been widely employed [1,4]. The Indian Pine dataset is with the resolution of 145×145 pixels and has 220 spectral bands. 20 bands are removed due to the water absorption bands, and finally 200 out of the 220 bands are used in our experiment. There are originally 16 classes in total, ranging in size from 20 to 2455 pixels. Some small classes have been removed and only 9 classes are selected for evaluation. The details information about the selected classes is shown in Table 1.

Table 1. Details of the Indian Pine Dataset

Class	# Pixels	Class	# of Pixels	Class	# of Pixels
Soybeans-no till	972	Corn-no till	1428	Grass/pasture	483
Soybeans-min	2455	Corn-min	830	Grass/trees	730
Soybeans-clean till	593	Woods	1265	Hay-windrowed	478
Total	9134				

We further select a subset scene of the Indian Pine dataset, consisting of the pixels $[27-94] \times [31-116]$ for a size of 68×86 dataset, denoted by Indian Pine Sub. In

Indian Pine Sub, there are 4 labeled classes in total. This dataset [4] aims to evaluate the hyperspectral image classification method when dealing with different classes with similar spectral signatures. 20 bands are removed due to the water absorption bands. The details about the Indian Pine Sub dataset are shown in Table 2.

Table 2. Details of the Indian Pine Sub Dataset

Class	# of Pixels	Class	# of Pixels
Soybeans-clean till	732	Corn-no till	1005
Soybeans-min	730	Grass/trees	1903
Total	5848		

3.2 Compared Methods

To evaluate the effectiveness of the proposed hyperspectral image classification approach, the following methods are employed for comparison.

1. Semi-Supervised Graph Based Method [4]. In semi-supervised graph based method, the hyperspectral image classification is formulated as a graph based semi-supervised learning procedure. All pixels are denoted by the vertices in the graph structure, which is able to exploit the wealth of unlabeled samples by the graph learning procedure. For comparison, the "Cross+Stacked" kernel is chosen which shows the best results in [4]. This method is denoted by "SSG+CS".
2. Representative Multiple Kernel Learning (RMKL) [8]. In RMKL, the multiple kernel-based learning method is employed for hyperspectral image classification. In this method, multiple kernels are selected and they are evaluated according to statistical significance and learned weights for better kernel combination, which is achieved by learning the linear combination of the basis kernels and minimizing the F-norm error.
3. Local Manifold Learning-Based k-Nearest Neighbor (SML+KNN) [12]. SML+KNN combines the local manifold learning and the k-nearest neighbor classifier for hyperspectral image classification. In this method, all pixels are embedded in a manifold, and local manifold learning is conducted to estimate the relationship among pixels. Then, the weighted KNN classifier is employed for pixel classification. The Supervised Locally Linear Embedding (SLLE) method is used as the weighting methods due to its steady performance as introduced in [12].
4. Hypergraph analysis with distance-based hyperedges for hyperspectral image classification (HGD) [19]. HGD is another hypergraph based hypespectral image classification method. In HGD, the hypergraph is constructed by using the neighborhood clustering method, where each pixel is connected to its several neighbor pixels in the feature space.
5. Hypergraph analysis with spatial hyperedges for hyperspectral image classification (HGS), i.e., the proposed method.

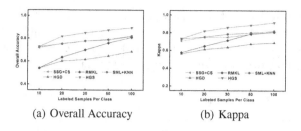

(a) Overall Accuracy   (b) Kappa

Fig. 3. The classification results of compared methods in the Indian Pine dataset

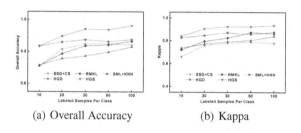

(a) Overall Accuracy   (b) Kappa

Fig. 4. The classification results of compared methods in the Indian Pine Sub dataset

3.3   Experimental Results

In our experiments, the number of labeled training samples for each class varies from 10 to 100, i.e., $\{10, 20, 30, 50, 100\}$. To evaluate the hyperspectral image classification performance, the widely used overall accuracy (OA) and the Kappa statistic are employed [9] as the evaluation metrics. In the following experiments, K is set as 12, and $\lambda = 0.9$.

Experimental comparisons on the two testing datasets are shown in Fig. 3 and Fig. 4. In comparison with the state-of-the-art methods, the proposed method outperforms all compared methods in both of the two testing databases. Here we take the experimental results when 10 samples per class are selected as the training data as an example. In the Indian Pine dataset, the proposed method achieves a gain of 1.23%, 3.50%, 0.03%, and 34.38% in terms of the OA measure and a gain of 3.92%, 27.60%, 0.44%, and 30.52% in terms of the Kappa measure compared with SSG+CS, RMKL, SML-KNN and HGD. In the Indian Pine Sub dataset, the proposed method achieves a gain of 17.47%, 16.82%, 0.17%, and 16.48% in terms of the OA measure and a gain of 27.13%, 16.37%, 1.50%, and 14.01% in terms of the Kappa measure compared with SSG+CS, RMKL, SML-KNN and HGD. Experimental results show that the proposed method achieves the best image classification performance in most of cases in the two testing datasets, which indicates the effectiveness of the proposed method.

Figure 5 and Figure 6 demonstrate the classification map of the proposed method in the testing datasets with different number of selected training samples per class.

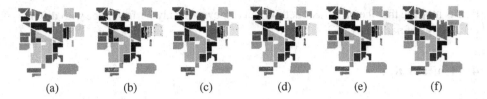

Fig. 5. Classifications maps of the Indian Pine dataset. (a) Groundtruth map with 9 classes (b)-(f) Classifications maps with 10,20,30,50,and 100 labeled training samples for each class.

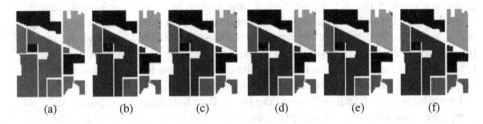

Fig. 6. Classifications maps of the Indian Pine Sub dataset. (a) Groundtruth map with 9 classes (b)-(f) Classifications maps with 10,20,30,50,and 100 labeled training samples for each class.

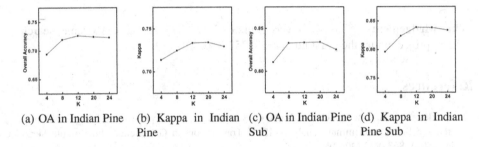

(a) OA in Indian Pine   (b) Kappa in Indian   (c) OA in Indian Pine   (d) Kappa in Indian
                         Pine   Sub   Pine Sub

Fig. 7. Classification performance comparison with different K values by using 10 training sample per class in the Indian Pine dataset and the Indian Pine Sub dataset

3.4 On the Parameter K for Hyperedge Construction

The parameter K determines the number of selected spatial neighbors for each pixel in the hyperedge construction procedure. A larger K value indicates that more pixels are connected by one hyperedge, and a smaller K value means that only a few pixels are linked in one hyperedge. To investigate the influence of different K selection on the hyperspectral image classification performance, we vary the parameter K as $\{4, 8, 12, 20, 24\}$. Figure 7 provides the OA and Kappa performance curves with respect to the variation of K in the two testing databases, where 10 samples per class are selected as the training data.

Experimental results show that the results are stable with the variation of parameter K. When K is large (e.g., 20 or 24) or K is small (e.g., 4), the hyperspectral image classification performance is only a bit lower than that of $K = 8$ and $K = 12$. These results indicate that the proposed method can achieve a steady performance with different settings of parameter K.

4   Conclusion

In this paper, we propose a hyperspectral image classification method by using the spatial correlation of pixels. In the proposed method, the relationship among pixels in the hyperspectral image is formulated in a hypergraph structure. In the constructed hypergraph, each vertex denotes a pixel in the image, and the hyperedge is generated by using the spatial correlation among pixels. Semi-supervised learning on the hypergraph is conducted for hyperspectral image classification. This method employs the spatial information to explore the relationship among pixels, and the high dimensional feature is only used to further enhance the spatial-based correlation in the constructed hypergraph, which is able to avoid the curse of dimensionality.

Experiments on the Indian Pine and the Indian Pine Sub datasets are performed, and comparisons with the state-of-the-art methods are provided to evaluate the effectiveness of the proposed method. Experimental results indicate that the proposed method can achieve better results in comparison with the state-of-the-art methods for hyperspectral image classification.

Acknowledgements. This work was supported by NUS-Tsinghua Extreme Search (NExT) project under the grant number: R-252-300-001-490.

References

1. Bandos, T., Bruzzone, L., Camps-Valls, G.: Classification of hyperspectral images with regularized linear discriminant analysis. IEEE Transaction on Geoscience and Remote Sensing 47(3), 862–873 (2009)
2. Berge, A., Solberg, A.: Structured gaussian components for hyperspectral image classification. IEEE Transaction on Geoscience and Remote Sensing 44(11), 3386–3396 (2006)
3. Bilgin, G., Erturk, S., Yildirim, T.: Unsupervised classification of hyperspectral-image data using fuzzy approaches that spatially exploit membership relations. IEEE Geoscience and Remote Sensing Letters 5(4), 673–677 (2008)
4. Camps-Valls, G., Marsheva, T.B., Zhou, D.: Semi-supervised graph-based hyperspectral image classification. IEEE Transaction on Geoscience and Remote Sensing 45(10), 3044–3054 (2007)
5. Gao, Y., Wang, M., Luan, H., Shen, J., Yan, S., Tao, D.: Tag-based social image search with visual-text joint hypergraph learning. In: ACM Conference on Multimedia, pp. 1517–1520 (2011)
6. Gao, Y., Wang, M., Tao, D., Ji, R., Dai, Q.: 3D object retrieval and recognition with hypergraph analysis. IEEE Transactions on Image Processing 21(9), 4290–4303 (2012)
7. Gao, Y., Wang, M., Zha, Z., Shen, J., Li, X., Wu, X.: Visual-textual joint relevance learning for tag-based social image search. IEEE Transactions on Image Processing (in press)
8. Gu, Y., Wang, C., You, D., Zhang, Y., Wang, S., Zhang, Y.: Representative multiple kernel learning for classification in hyperspectral imagery. IEEE Transaction on Geoscience and Remote Sensing 50(7), 2852–2865 (2012)

9. Guo, B., Damper, S.G.R., Nelson, J.: Band selection for hyperspectral image classification using mutual information. IEEE Geoscience and Remote Sensing Letters 3(4), 522–526 (2006)
10. Kuo, B.-C., Li, C.-H., Yang, J.-M.: Kernel nonparametric weighted feature extraction for hyperspectral image classification. IEEE Transaction on Geoscience and Remote Sensing 47(4), 1139–1155 (2009)
11. Landgrebe, D.: Hyperspectral image data analysis. IEEE Signal Process Magazine 19(1), 17–28 (2002)
12. Ma, L., Crawford, M., Tian, J.: Local manifold learning-based -nearest-neighbor for hyperspectral image classification. IEEE Transaction on Geoscience and Remote Sensing 48(11), 4099–4109 (2010)
13. Marconcini, M., Camps-Valls, G., Bruzzone, L.: A composite semisupervised svm for classification of hyperspectral images. IEEE Geoscience and Remote Sensing Letters 6(2), 234–238 (2009)
14. Martinez-Uso, A., Pla, F., Sotoca, J.M., Garcia-Sevilla, P.: Clustering-based hyperspectral band selection using information measures. IEEE Transaction on Geoscience and Remote Sensing 45(12), 4158–4171 (2007)
15. ul Haq, Q.S., Tao, L., Sun, F., Yang, S.: A fast and robust sparse approach for hyperspectral data classification using a few labeled samples. IEEE Transaction on Geoscience and Remote Sensing 50(6), 2287–2302 (2012)
16. Velasco-Forero, S., Manian, V.: Improving hyperspectral image classification using spatial preprocessing. IEEE Geoscience and Remote Sensing Letters 6(2), 297–301 (2009)
17. Wang, J., Chang, C.-I.: Independent component analysis-based dimensionality reduction with applications in hyperspectral image analysis. IEEE Transaction on Geoscience and Remote Sensing 44(6), 1586–1600 (2006)
18. Wang, J., Zhang, Z., Zha, H.: Adaptive manifold learning. In: Proceedings of Advances in Neural Information Processing Systems (2004)
19. Wen, Y., Gao, Y., Liu, S., Cheng, Q., Ji, R.: Hyperspetral image classification with hypergraph modelling. In: Proceedings of International Conference on Internet Multimedia Computing and Service (2012)
20. Xia, S., Hancock, E.: Learning large scale class specific hyper graphs for object recognition. In: Proceedings of International Conference on Image and Graphics, pp. 366–371 (2008)
21. Yu, J., Tao, D., Wang, M.: Adaptive hypergraph learning and its application in image classification. IEEE Transactions on Image Processing 21(7), 3262–3272 (2012)
22. Zass, R., Shashua, A.: Probabilistic graph and hypergraph matching. In: Proceedings of IEEE International Conference on Computer Vision and Pattern Recognition (2008)
23. Zhou, D., Huang, J., Schokopf, B.: Learning with hypergraphs: Clustering, classification, and embedding. In: NIPS (2007)

Research on Face Recognition under Images Patches and Variable Lighting

Wengang Feng[1,2]

[1] Department of Policing Intelligence, Chinese People's Public Security University,
100038, Beijing, China
[2] Public Security Intelligence Research Center, Chinese People's Public Security University,
100038, Beijing, China
Wengang.feng@gmail.com

Abstract. Many classic and contemporary face recognition algorithms work well on public data sets, but degrade sharply when they are used in variations lighting, expressions and images patches situation. New correlation filter designs have shown to be distortion invariant and the advantages of using images are due to the invariance to visible illumination variations. We propose a conceptually simple face recognition system that achieves a high degree of robustness and stability to illumination variation, image patches based in a simple non-linear correlation filter. The proposed technique is based on the premise that the face is an object composed of facial characteristics. The system can efficiently and effectively recognize faces under a variety of realistic conditions, using only frontal images under the proposed illuminations as training, and the results of detection and identification rate of is 96.3% in face identification, while in verification task reaches 94.6%.

Keywords: Face recognition, correlation filter, Face patches.

1   Introduction

Human face recognition is currently a very active research area [1, 2] with focus on ways to perform robust identification. However, face recognition is a challenging task because of the variability of the appearance of face images even for the same subject as it changes due to expression, occlusion, illumination, pose, aging etc. It is common to hear that we are in the information age, organizations generate large amounts of data that only have been stored and subsequently consulted. That is, the existing technology has advanced to save efficaciously these data but there hasn't been a great advance in technologies that allow automatically analyzing them.

Many biometric sensors output images and thus image processing plays an important role in biometric authentication. Image preprocessing is important since the quality of a biometric input can vary significantly. For example, the quality of a face image depends very much on illumination type, illumination level, detector array resolution, noise levels, etc. Preprocessing methods that take into account sensor characteristics must be employed prior to attempting any matching of the biometric

S. Li et al. (Eds.): MMM 2013, Part I, LNCS 7732, pp. 152–162, 2013.

images. However, this paper will focus on spatial frequency domain image processing technologies that can be used for matching biometric images. Processing in spatial frequency domain is nothing but 2-D filtering and we will refer to this approach as correlation filtering.[3]

Most approaches to face recognition are in the image domain whereas we believe that there are more advantages to work directly in the spatial frequency domain. By going to the spatial frequency domain, image information gets distributed across frequencies providing tolerance to reasonable deviations and also providing graceful degradation against distortions to images in the spatial domain. Correlation filter technology [4] is a basic tool for frequency domain image processing. In correlation filter methods, normal variations in authentic training images can be accommodated by designing a frequency domain array (called a correlation filter) that captures the consistent part of training images while deemphasizing the inconsistent parts (or frequencies). Object recognition is performed by cross-correlating an input image with a designed correlation filter using fast Fourier transforms (FFTs).

Correlation filters have been investigated mostly for automatic target recognition (ATR) [4] applications. The most basic correlation filter is the matched filter (MF), which performs well at detecting a reference image corrupted by additive white noise. But it performs poorly when the reference image appears with distortions (e.g., rotations, scale changes). In biometric verification, the input biometric is bound to have some differences from the reference biometric because of normal variations. Then, one MF will be needed for each appearance of a biometric. Clearly this is computationally impractical. Hester and Casasent [6] addressed this challenge with the introduction of the synthetic discriminant function (SDF) filter. The SDF filter is a linear combination of MFs where the combination weights are chosen so that the correlation outputs corresponding to the training images would yield pre-specified values at the origin. For example, the correlation peak values corresponding to the training images of authentic can be set to 1, and the peak values due to the impostor training images can be set to zero. It is hoped that the resulting correlation filter would yield correlation peak values close to 1 for non-training images from the authentic class and correlation peak values close to zero for non-training images from the impostor class.Several researchers have used correlation pattern recognition for face recognition. This technique is used in target automatic recognition[6], biometric recognition[7], location of objects[8],visual object tracking[9], among others. Savvidesetal[10] used the MACE filter for face verification task. A problem with MACE filter is that it doesn't work well in presence of noise neither with partial occlusion. Santiago-Ramirez et al.[11] evaluated the performance of some correlation filters in verification task. Subsequently, Santiago-Ramirez et al.[12] evaluated how to improve the performance correlation filter using a pre-processing to improve image quality before performing the recognition.

This paper is organized as follow: Section 2 explains the proposed unconstrained nonlinear composite filter. Section 3 describes the face recognition method using the unconstrained non-linear composite filter. Section4 provides the results and discussions. Finally, section 5 presents the conclusions of this work.

2   Composite Correlation Filters

Correlation filters have been successfully applied in biometric recognition. In particular, correlation filters have been demonstrated to work well for fingerprint recognition. Object recognition is performed by cross-correlating an input image with a synthesized template or filter and processing the resulting correlation output. Figure 1 shows schematically how the cross-correlation is obtained using Fast Fourier Transforms (FFTs). The correlation output is searched for peaks, and the relative heights of these peaks are used to determine whether the object of interest is present or not. The locations of the peaks indicate the position of the objects.

A correlation filter takes the form of a two dimensional complex-valued array in frequency and is applied as shown in Fig. 1. The image is converted into the frequency domain with a fast Fourier transform, multiplied by the filter, and converted back to the image domain by an inverse fast Fourier transform; this produces a correlation plane. This process is equivalent to, but much faster than, computing the cross correlation in the original image domain. This process, which produces a correlation plane from a given image, can be modeled as

$$v_i = (DFT)^{-1} C_i DFT x \tag{1}$$

where x is the vectorized form of the n image pixels, DFT is an $n \times n$ matrix containing the basis of a two-dimensional discrete Fourier transform, and C_i is a diagonal matrix containing the correlation filter values (designed for class i) in the Fourier domain along the diagonal.

If the image belongs to the pattern class of the filter, the correlation plane output contains a sharp peak; if not, no such peak exists. We derive a match metric from the correlation plane by measuring the peak-to-correlation energy (PCE), defined as,

$$PCE(v_i) = \frac{\max(|v_i|) - mean(|v_i|)}{stdev(|v_i|)} \tag{2}$$

which yields our cost function,

$$f(v_i) = \frac{1}{PCE(v_i)} \tag{3}$$

Note that correlation filters are shift invariant (i.e., a shift in the input results in a corresponding shift in the correlation plane); the PCE remains constant because it is computed after the correlation peak is located.

UMACE filter is based in a set of training images as the SDF filter, nevertheless this filter doesn't restrict the correlation output in the origin, instead of this it treats to maximize it[20]. This filter is given by the equation:

$$h = D^{-1}(v_i) \tag{4}$$

Where m is a column vector which contains the average of Fourier transform of training images and $D = \dfrac{1}{D \cdot N} \sum_{i=1}^{N} I_i \cdot I_i^*$ is a matrix which diagonal contains the average of spectral energy of the training images and -1 is the inverse of a matrix, d is the number of pixels in the training image, I_i is the Fourier transform of the ith training image and $*$ is the complex conjugate. The proposed filter in this paper is described below. It is based in the idea of a SDF conventional filter, but without the restrictions at the origin 102 and its performance is compared against the UMACE filter.

2.1   Definition of an Unconstrained Nonlinear Composite Filter

Given a set of training images taking values in a d–dimensional image space, the filter design minimizes the average energy of the output correlation planes , corresponding to the images , subject to the constraints that the amplitude value at the origin of each plane is fixed to a specific value.

This result is achieved in the following manner. Let D be a $d \times d$ diagonal matrix carrying the average power spectrum of the training images, i.e.,

$$D(k,k) = \frac{1}{N} \sum_{i=1}^{N} |(v_i)|^2 \qquad (5)$$

where v_i is the 2-D Fourier transform of in vector form. Then, the optimal filter should minimize the average correlation energy (ACE) measure, defined as $ACE = h^+ Dh$ (where $+$ is the transpose conjugate), and meet the constraints $X^+ h = u$. The solution to this problem is

$$h = D^{-1} X (X^+ D^{-1} X)^{-1} U \qquad (6)$$

Traditional SDF filter has a lot of problems. Because it does not consider any input noise, the noise of the input image may be difficult to perfect match, that is, the constant term is not produced origin, even if the test is a training image. Due to be considered only in the origin of the restrictions, which may not comply with the relevant peak. The unconstrained values may be higher than c_i. Only is designated as the limit of the test images of the training image sand. In addition, the purpose of limitation is through the use of multiple exposure techniques [15], the filter is synthesized in optical laboratories.

By minimizing the output variance of the correlation peak, robustness to noise can be achieved. If the noise in the training set is additive zero-mean and stationary, with power spectral density values arranged in a $d \times d$ diagonal matrix P, a measure that can also be minimized through the same process is the output noise variance, that is $ACE = h^+ Dh$.

3 Face Recognition Using Correlation Filter Synthesized with Facial Patches

3.1 Face Patches

The facial expression database is collected from the Yale Face Database, which contains 165 gray scale images in GIF format of 15 individuals. There are 11 images per subject, one per different facial expression or configuration: center-light, w/glasses, happy, left-light, w/no glasses, normal, right-light, sad, sleepy, surprised, and wink. The faces were captured in a video sequence where a face tracker tracked the movement of the user's head and based upon an eye localization routine and extracted registered face images of size 64x64.

The human face is three-dimensional objects, which is recognized the ITS2-D digital image of a tree. Most face recognition method considered face as a single object, they do not consider their own behavior in his left eye, right eye, nose, mouth and other facial features. This behavior will affect the recognizable face. When a man says to his mouth, change its shape. Modifications not only the structure of the function of such a shape changes, but this feature of the entire region. Change shape in order to reduce the face, we recommend the creation of a face region using the biometric template filter. A face image of the face patches, which contains one or more facial features, as shown in Fig. 1.

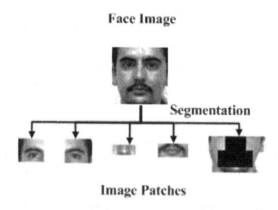

Fig. 1. Main patches of a human face

Each face region is retained in its original position, the correlation output, in the face region of the origin of each matching purposes. One of the advantages of the filter for training in the use of the face region, if an input image contains only a part of the face, then it may be similar to the region, thereby generating a powerful and sharp peaks. This pretend recognize the split plane, with a high degree of reliability. Another advantage is that the face area of the cheeks and chin, as still not too many changes, when a person says, therefore, can be used for tracking.

3.2   Face Recognition by Correlation Filter

In Fig. 2 is shown the block diagram of the proposed method in this paper. First, in the register of a new user, he provides N samples of face images to create their biometric template. Each one, is segmented in r face regions. Be X_i^k the Fourier transform with the k law of the i-th training facial region. The UNCF correlation filter in equation 7 159 is constructed with $N \times r$ facial regions. Second, when we have a test face image, the Fourier transform and the k law are applied, and then it is correlated with the filter. Third, the output correlation plane is examined by the classifier that measures the sharp of the peak and comparing it with a threshold value to determinate if there is a match.

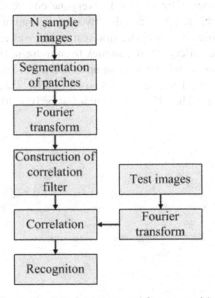

Fig. 2. Block diagram of the proposed face recognition method

4   Experimental Results and Evaluation

In order to evaluate the method of the face image of the raised portion, constructed two correlation filter. First filter constructed five only one training image of the face region, with the complete face image and the second filter structure. Part of the face is artificially generated data sets. The six-sided, with 15.34%, 16.28%, 11.23%, 17.59%, 79.48%, and 100% complete example of the information. The algorithm was tested of the faces of these portions. The sharpness of the peak of the correlation produced was measured by using a measurement known as PSR, the degree of similarity between the test images of the face and it represents the biometric face template [16]. Further, the performance of the filter in the diagram (b) shown in the PSR. As can be seen, and the various regions of the filter built shows better performance than the other tested images 1-4, 5 and 6 of the image is worse. The proposed method, the evaluation of two face database; Yale B [18] and ORL [19]. YALEB database contains 64 different face images, 38 persons. The image is contained in the lighting, shadows, the main change. 40 the ORL database contains10 different face images. The image captured at a different time, there are slight changes in lighting, the expression of a face (open / normally closed eyes, smile / non-smile's) and the face details (glass / non-glasses, and the). Face image of the human eye and a positive (with side some slight movement). From two databases, each face image, cropping and scaling manually in are solution of 64 × 64pixels. The dynamic range of the image gray 0-255.

Three sets of facial images were created, as is described in [17]. First, the gallery $G = g_1, g_2, ..., g_{|G|}$ which contains the biometri c face template of each known person by the system. A biometric template surface is everyone coherency filter, according to the method proposed in this article. The other two are a group of test. A test sample of a biometric face recognition system. The first test group was called the PG contains biological samples in the gallery (these samples from those in the gallery, but belong to the same class can also include training samples). The PG test set contains 38 × 64 = 2432 Yale B face image. The test set is known as PN, which contains biometric sample is not in the gallery. The PN group contains 40 × 10 = 400 ORL face database.

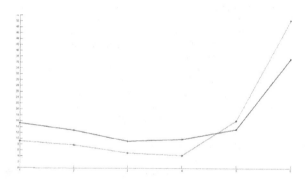

Fig. 3. PSR performances of the correlation filter with test face regions. Continuous line represents the filter performance constructed with facial regions. Dashed line represents the filter performance constructed with complete images.

The filters of the gallery were created with face regions of 4 face images selected in such way that could be recognized 64 face images(6 for training and 58 reminder). The selected training images present dark left half face, normal face illumination, dark right half face, and closed eyes face as is shown in Fig. 4. These images represent the general distortions of the YALE B data set.

Each face image was segmented in five face regions. Therefore, each filter in the G gallery was constructed with 4×5=20 facial regions. Before segmentation, logarithmic transformation is applied to images to improve their quality in case that they present shadows or intense illumination.

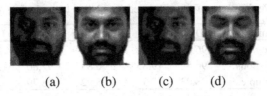

(a)   (b)   (c)   (d)

Fig. 4. Examples of training images. A) Dark left half face B) normal face illumination C) dark right half face D) closed eyes face.

Based on many experiments, we determine using four training images to obtain the minimum acceptable performance. However, as more images are used to construct the filter, better will be the performance. The PSR values(solid line) under the threshold =10belongtoshadowed (partial or total) images and its low value is because do not exist a good matching with the filter. This results aims to include the images A, B, and C of the Fig. 5 into the filter construction, the four images represents the principals variations in shadow.

In Fig. 7 is shown the performance of a correlation synthesized with four images. All those face images that were in the training set and even the most which belong to the same class but that weren't in the training set were recognized correctly. Using the PSR = 10 recognition threshold is correctly rejected all the impostor face images.

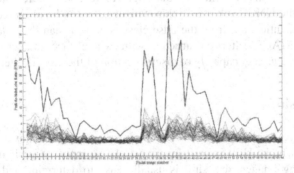

Fig. 5. PSR performance for person 1, the correlation filter was constructed with five face regions obtained from a single face image. Bottom dashed lines correspond to impostors subjects.

This recognition threshold was determinate based on experiments and observations. If a test image produces a PSR value greater or equal to this threshold, then the subject is declared as authentic. With the configuration described before, the performance of the proposed method was calculated with the performance measures described in, i.e., detection and identification rate(DIR) and false alarm rate(FAR) for the identification task. For the verification task, were calculated the statistics of verification rate(VR) and false acceptance rate(FAcR).The statistics obtained with the ORL dataset are showed in Table 1. The precision of the filters for recognizing authentic face images and reject impostors face images are presented too. The test sets PG and PN were used for face identification. For face verification were used only the set PG. In both task, the database G was the same.

Table 1. Performance of unconstrained correlation filters in face recognition

Filter	Training data	DIR	FAR	Precision	VR	FAcR	Precision
UNCF	Regions	96.01	1.27	99.75	95.93	0.25	89.87
UMACE	Regions	96.52	19.83	96.83	97.86	2.18	55.32
UMACE	Complete	98.03	26	95.92	98.79	2.42	52.03

The UNCF filter shows the best performance in terms of precision of the recognition with 98.96% and 89.87% in face identification and face verification, respectively. The proposed method has a very good performance with 96.01% and 95.93% in identification and verification. In identification, the probability that the proposed method declares as authentic an impostor image when it is not even person in the gallery is very low. From 400impostor images, only five were declared as authentic. In verification, the probability 18 of that the method declares as equal two face images that are not is only 0.31%. That is, of the 89994 impostor face images, only 279 were declared incorrectly as authentic. A false acceptance occurs when the algorithm confuses two face images of two persons registered in the databases. The UMACE filter was evaluated with templates constructed with complete face images and with face regions. The best performance was obtained for the whole face images set with 98.03% and 98.79% in identification and verification problems, respectively. A problem with UMACE filter is the high FAR and FAcR rates, while the performance of the UNCF is better in this statistics. As can be observed, the gap that separates the authentic class from the false class is greater than the obtained with the best case of UMACE filter constructed with whole face images. This gap of separation means a greater capacity of discrimination of the UNCF filter.

5   Conclusions

As we will see, correlation filter offer several advantages over model-based approaches. First is the built-in shift-invariance. If the input image is translated with respect to training images, that shift is usually easy to determine and correct when correlation filters are used. Second, correlation filters are based on integration operation and thus offer graceful degradation in that impairments to the test image cause only gradual degradation in the quality of the output. Third, correlation filters can be designed to exhibit attributes such as noise tolerance, high discrimination, etc. Finally, correlation filter designs offer closed form expressions.

It is a non-linear transformation by the Fourier transform. Then, it is associated with the filter. Output plane in search of peak detection. The PSR metric to measure the sharpness of the peak, then it is more perception threshold. Two databases used in the experimental section, the method is a valid verification and identification. It reaches the identification and detection rate of 96.3% and 94.6%, respectively. This method is effective part of the face image, the whole face image. Using the logarithmic transformation in the face image, to improve the filter performance of a shadow or strong lighting.

Acknowledgments. This work was financially supported by the Chinese People's Public Security University Natural Science Foundation (2011LG08).

References

1. John, H.: The national biometrics challenge. Tech. rep., NSTC Subcommittee on Biometrics and Identity Management (2011)
2. Sirovich, L., Kirby, M.: Low-dimensional procedure for the characterization of human faces. Optical Society of America, 519–3087 (1987)
3. Kumar, B., Savvides, M.: Spatial frequency domain image processing for biometric recognition. In: IEEE ICIP, pp. 53–58 (2002)
4. Lu, J., Plataniotis, K.N., Vanetsanopoulus, A.N.: Regularized discriminant analysis for the small samples size problem in face recognition. Pattern Recognition Letters, 3079–3087 (2003)
5. Jun, B., Lee, J., Kim, D.: A novel illumination-robust face recognition using statical and nonstatical method. Pattern Recognition Letters, 329–336 (2010)
6. Vijaya-Kumar, B.V.K.: Tutorial survey of composite filter design for optical correlators. Applied Optics, 4773–4801 (1992)
7. Vijaya-Kumar, B.V.K., Savvides, M., Xie, C., Venkataraman, K., Thornton, J., Mahalanobis, A.: Biometric verification with correlation filters. Optical Society of America, 391–402 (2004)
8. Bolme, D.S., Beveridge, J.R., Draper, B.A., Lui, Y.M.: Visual object tracking using adaptive corre283 lation filters. In: IEEE Conference on Computer Vision and Pattern Recognition, pp. 2544–2550 (2010)
9. Bolme, D.S., Draper, A., Beveridge, J.R.: Average of synthetic exact filters. In: IEEE Conference 285 on Computer Vision and Pattern Recognition, pp. 2105–2112 (2009)
10. Savvides, M., Kumar, B.V., Khosla, P.: Face verification using correlation filters. Automatic Identification Advanced Technologies, 56–62 (2002)
11. Santiago-Ramirez, E., Gonzalez-Fraga, J.A., Ascencio-Lopez, J.I.: Performance of Correlation Filters in Facial Recognition. Springer, Heidelberg (2011)
12. Santiago-Ramirez, E., Gonzlez-Fraga, J.A., Ascencio-Lopez, J.I., Buenrostro, O.: Performance of composite correlation filters for object recognition. In: Proc. SPIE, vol. 8011, pp. 174–178 (2011)
13. VanderLugt, A.: Signal detection by complex spatial filtering. IEEE Transactions on Information Theory, 139–145 (1964)
14. Casasent, D., Chang, W.-T.: Correlation synthetic discriminant functions. Appl. Opt., 2343–2350 (1986)
15. Vijaya-Kumar, B.V.K., Mahalanobis, A., Juday, R.: Correlation pattern recognition. Cambridge University Press (2005)
16. Javidi, B., Wang, W., Zhang, G.: Composite fourier-plane nonlinear filter for distortion-invariant pattern recognition. Society of Photo-Optical Instrumentation Engineers, 2690–2696 (2005)
17. Phillips, P.J., Grother, P., Micheals, R.: Handbook of face recognition. Springer Science Business (2005)
18. Georghiades, A.S., Belhumeur, P.N.: From few to many: Illumination cone models for face recognition under variable lighting and pose. IEEE Transactions on Pattern Analysis and Machine Intelligence, 643–660 (2001)

19. Samaria, F., Harter, A.: Parameterization of a stochastic model for human face identification. In: Proceedings of the Second IEEE Workshop on Applications of Computer Vision, pp. 138–142 (1994)
20. Mahalanobis, A., Kumar, B.V.K.V., Song, S., Sims, S.R.F., Epperson, J.F.: Unconstrained correlation filters. Applied Optics, 3751–3759 (1994)
21. Wang, M., Hua, X., Hong, R.: Unified Video Annotation Via Multi-Graph Learning. IEEE Transactions on Circuits and Systems for Video Technology, 733–746 (2009)
22. Wang, M., Hua, X., Tao, M., Hong, R.: Semi-Supervised Kernel Density Estimation for Video Annotation. Computer Vision and Image Understanding, 384–396 (2009)
23. Wang, M., Hong, R., Li, G.: Event Driven Web Video Summarization by Tag Localization and Key-Shot Identification. IEEE Transactions on Multimedia, 975–985 (2012)
24. Wang, M., Hong, R., Yuan, X., Yan, S., Chua, T.: Movie2Comics: Towards a Lively Video Content Presentation. IEEE Transactions on Multimedia, 858–870 (2012)

A New Network-Based Algorithm
for Human Group Activity Recognition
in Videos

Gaojian Li[1], Weiyao Lin[2,*], Sheng Zhang[2], Jianxin Wu[3],
Yuanzhe Chen[2], and Hui Wei[1]

[1] School of Computer Science and Technology, Fudan University, China
[2] Department of Electronic Engineering, Shanghai Jiao Tong University, China
[3] School of Computer Engineering, Nanyang Technological University, Singapore

Abstract. In this paper, a new network-based (NB) algorithm is proposed for human group activity recognition in videos. The proposed NB algorithm introduces three different networks for modeling the correlation among people as well as the correlation between people and the surrounding scene. With the proposed network models, human group activities can be modeled as the package transmission process in the network. Thus, by analyzing the energy consumption situation in these specific "package transmission" processes, various group activities can be effectively detected. Experimental results demonstrate the effectiveness of our proposed algorithm.

Keywords: Network model, Package transmission, Group activity recognition, Energy consumption.

1 Introduction

Detecting human group activities or human interactions has attracted increasing research interests in many applications such as video surveillance and human-computer interaction [3-7]. Some typical group activities of interest include people being followed, people gathering together, and person leaving a group in a party.

There have been many researches on group activity recognition. Zhou et al. [6] detect pair-activities by extracting the causality features from bi-trajectories. Ni et al. [7] further extend the causality features into three types including individuals, pairs and groups. Cheng et al. [4] use the Group Activity Pattern for representing and differentiating group activities where Gaussian parameters from trajectories are calculated from multiple people. Lin et al. [5] use group representative to represent each group of people for detecting the interaction of people groups such that the number of people can vary in the group activity. However, while these methods suitably handle the interaction among people, many of them neglect the relationship between people and their surrounding scene. Thus, they may have limitations when

---

[*] Corresponding author.

S. Li et al. (Eds.): MMM 2013, Part I, LNCS 7732, pp. 163–173, 2013.
© Springer-Verlag Berlin Heidelberg 2013

detecting the scene-related activities. Furthermore, their abilities for detecting complex activities (such as people first approach and then split) are also limited.

Although some methods [1, 3, 9] can recognize the group activity as well as the scene-related activity by using some pre-designed graphical models such as the layered Hidden Markov Model (HMM) [3], they often require large amount of training data in order for working well. Besides, the restricted graphical structure used in these methods may also limit their ability to handle various unexpected cases.

In this paper, we propose a new network-based (NB) algorithm for recognizing human group activities. The proposed framework first introduces three different networks for modeling the correlation among people as well as the correlation between people and the surrounding scene. With the proposed network models, human group activities can be modeled as the package transmission process in the network. Thus, by analyzing the energy consumption situation in these specific "package transmission" processes, various group activities can be effectively detected. Our NB algorithm is flexible and capable of handling both the interactions among people and the interaction between people and the scene (e.g., differentiating whether a person is moving or following irregular paths in the scene). Experimental results demonstrate the effectiveness of our proposed algorithm.

The rest of the paper is organized as follows: Section 2 describes the framework of our proposed NB algorithm. Section 3 describes the detailed implementations of our NB algorithm for group activity recognition. The experimental results are shown in Section 4 and Section 5 concludes the paper.

2   Framework of the NB Algorithm

In this section, we will first describe the basic idea of our proposed network-based (NB) group activity recognition algorithm, and then describe the framework of NB algorithm.

2.1   Basic Idea of the Algorithm

The basic idea of our NB algorithm can be described by Fig. 1 and Fig. 2. In order for detecting the interaction between people and the surrounding scene, the NB algorithm first divides the entire scene into patches where each patch is modeled as a "node" in the network (as in Fig. 1). Based on this network, the process of people moving in the scene can be modeled as the package transmission process in the network (i.e., a person moving from one patch to another can be modeled as a 'package' transmitted from one node to another, as the red trajectory in Fig. 1). By this way, various human activity recognition problems can be transferred into the package transmission analysis problem in the network.

Furthermore, when detecting the interactions among people, a similar "relative" network can be modeled for handling the task. For example, as in Fig. 2, the relative network can be constructed where one person is always located in the center of the network and the movement of another person can be modeled as the package transmission process in this "relative" network based on his relative movement to the network-center person. By this way, the interaction among people can also be effectively modeled. recognized by evaluating different transmission energies in our network-based model.

With these network-based models, one key problem is how to use this model for recognizing activities. We further observe that if we model the process of person moving among patches as the 'energy' consumed to transmit a package, the activities can then be differentiated with these 'transmission energy' features. For example, we can differentiate whether a person is moving in the scene based on his consumption energy value. Thus, by carefully modeling and analyzing the networks, various group activities can be the effectively detected.

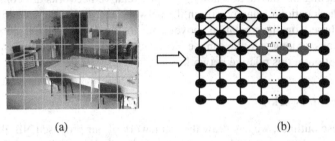

(a)   (b)

Fig. 1. (a) Divide the scene into patches. (b) Model each patch in (a) as a node in the network and the edges between nodes are modeled as the activity correlation between the corresponding patches. The red trajectory $R(u, q)$ in (a) is modeled by the red package transmission route in (b). (Note that (b) can be a fully connected network (i.e., each node has edges with all the other nodes in the network). In order to ease the description, we only draw the four neighboring edges for each node in the rest of the paper) (best view in color).

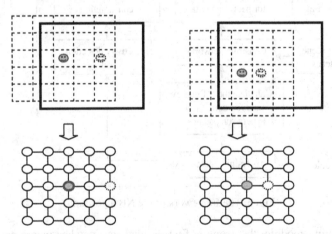

Fig. 2. Constructing relative networks for modeling people interactions. Upper: the locations of the two people in two different frames (the dashed patches are divided by making the red-circled grey person at the network center). Down: the transferred networks of the upper frames (the red-circled grey node and the blue-circled dotted node are the locations of the two people in the network). The location of the red-circled grey person is fixed in the network while the location of blue-circled dotted person in the network is decided by his relative location to the red-circled grey person.

From the above discussions, we can outline the basic idea of our proposed NB algorithm as follows:

(1) The entire scene is divided into patches for constructing a network. In this network, each node represents a patch and each edge represents the transmission energy when moving between the corresponding nodes. The transmission energy can be modeled by the activity correlation between the corresponding patches.

(2) For modeling the interaction between people and their surrounding scene, the scene-related network is constructed where each patch is fixed in the scene and a package transmitted in the network can represent a person moving in the scene.

(3) For modeling the interaction among people, relative networks are constructed by fixing the location of one person in the network and derive the locations of other people based on their relative movements to the location-fixed person. Thus, interactions among people can be detected based on the transmission energy consumption in this relative network.

2.2   The Framework

Based on these outlines, we can draw the framework of our proposed NB algorithm as in Fig. 3. In Fig. 3, the part in the dashed rectangular is the training module while the four blocks on the top are the testing process.

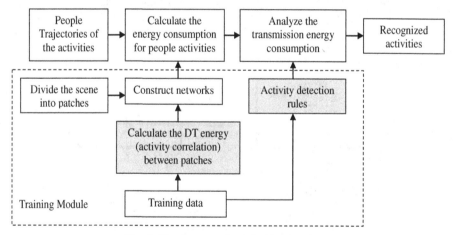

Fig. 3. The framework of the NB algorithm

In the training module, the scene is first divided into patches where each patch is modeled as a node in the network. Then the activity correlations between patches are estimated based on the training data and these activity correlations will be used as the edge values in the network. With these patches and edges, the transmission networks can be constructed. At the same time, the activity detection rules are also derived from the training data for detecting activities of interest during the testing process.

In the testing process, after extracting the activity trajectories of the people, their corresponding transmission energies are first calculated based on the constructed network. Then, these transmission energies are analyzed and the activity detection rules will be applied for detecting the activities.

Furthermore, several things need to be mentioned about our NB algorithm. They are described in the following:

(1) Although there are other works [1, 9] trying to segment the scene into parts for activity recognition, our NB algorithm is different from them in: (a) our NB algorithm construct a package transmission network over the patches while other works [1, 9] use graphical models for recognition. While the fixed structures of the graphical models [1, 9] may limit their ability to handle various unexpected cases, our fully-connected transmission network is more generalized and flexible for handling various scenarios; (b) With the transmission network model, our NB algorithm is robust to the patch segmentation styles (e.g., in this paper, we just simply segment the scene into identical rectangular blocks as shown in Fig. 1). Comparatively, the graphical model-based methods normally require careful segmentation of the scene [1, 9]; (3) We also propose to construct relative networks for modeling the interaction among people, which has not been used in the previous works [1, 9].

(2) From Fig. 3, it is clear that the steps of "calculate the energy between patches" and "activity detection rules" are the key parts of our NB algorithm. Therefore, in the next section, we will describe the detailed implementations of our algorithm in group activity recognition.

3   The Implementation of the NB Algorithm in Group Activity Recognition

In the group activity recognition scenarios, we want to recognize various group activities such as people approach each other, one person leaves another, and people walk together. As mentioned, when recognizing the interaction among people, the relative networks can be constructed as in Fig. 2. At the same time, since some group activities also include the relationship between people and their surrounding scene (e.g., we need to recognize whether a person is moving or standing still in the scene in order to differentiate activities such as both people walk to "meet" or one person stand still and another one "approaches" him), a scene-related network is also necessary. Therefore, in this section, we propose to use two types of networks for representing group activities. The detailed implementation of the key parts in Fig. 3 for group activity recognition can be described in the following.

3.1   Construct Networks

In this paper, we construct three networks for recognizing group activities: the scene-related network, the normal relative network, and the weighted relative network. The scene-related network is used to model the correlation between people and the scene and it can be constructed as in Fig. 1. The normal relative network and the weighted relative network are used for modeling the interaction among people and they can be constructed by fixing the location of one person in the network and derive the locations of other people based on their relative movements to the location-fixed person, as in Fig. 2. Besides, the following two points need to be mentioned about the networks.

(1) The structures of the normal relative network and the weighted relative network are the same. They only differ in the edge values (i.e., the energy when moving between the corresponding nodes).

(2) Note that the scene-related network is a non-directional network (i.e., the energy consumption when moving from patch i to j is the same as moving from j to i). However, the normal relative network and the weighted relative network are directional networks (i.e., the energy from i to j is different from j to i). This point will be further described in detail in the following sub-sections.

3.2   Calculate the Energy Consumption for People Activities

In this paper, we propose to calculate a set of transmission energies from the three networks for describing group activities. For the ease of description, we use two people group activity as an example to describe our algorithm. Multiple people scenarios can be easily extended from our description. The total transmission energy set for two people group activity can be calculated by:

$$[E_1(u_1,q_1),\ E_2(u_2,q_2),\ ENR(u_2\text{-}u_1,q_2\text{-}q_1),\ EWR(u_2\text{-}u_1,q_2\text{-}q_1)] \tag{1}$$

where $E_1(u_1,q_1)$ and $E_2(u_2,q_2)$ are the total transmission consumption for person 1 and person 2 in the "scene-related" network, respectively. And they can be calculated by Eqn. (2).

$$E_c\left(u_m,q_m\right)=\sum_{(i,j)\in R_c\left(u_m,q_m\right)}e\left(i,j\right) \tag{2}$$

where $R_c(u_m,q_m)$ is the current trajectory for person m with u_m being the starting patch and q_m being person m's current patch. Also $e(i,j)$ is the Direct Transmission (DT) energy for the edge between patches i and j (i.e., the energy used by directly transmitting a package from patch i to j without passing through other patches).

Furthermore, $ENR(u_2\text{-}u_1,q_2\text{-}q_1)$ in Eqn. (1) is the total transmission consumption in the "normal relative" network where $R(u_2\text{-}u_1,q_2\text{-}q_1)$ is the relative trajectory of person 2 with respect to person 1. And $EWR(u_2\text{-}u_1,q_2\text{-}q_1)$ is the total transmission consumption in the "weighted relative" network. $ENR(u_2\text{-}u_1,q_2\text{-}q_1)$ and $EWR(u_2\text{-}u_1,q_2\text{-}q_1)$ can be calculated by Eqn. (3).

$$\begin{cases} ENR\left(u_2-u_1, q_2-q_1\right)= \sum_{(i,j)\in R\left(u_2-u_1,q_2-q_1\right)} enr\left(i,j\right) \\[3em] EWR\left(u_2-u_1, q_2-q_1\right)= \sum_{(i,j)\in R\left(u_2-u_1,q_2-q_1\right)} ewr\left(i,j\right) \end{cases} \qquad (3)$$

where $enr(i,j)$ and $ewr(i,j)$ are the Direct Transmission (DT) energy from patch i to j in the normal relative network and weighted relative network, respectively. The calculation of $e(i,j)$, $enr(i,j)$ and $ewr(i,j)$ will be described in detail in the next sub-section.

3.3   Calculate the DT Energy (Edge Value) between Patches

The DT energy (i.e., edge) between patches for the three networks is shown by Fig. 4.

For the scene-related network, since we only need it to detect the movement of the person in our scenario, we simply set all the DT energies to be 1, as in Fig. 4 (a). Note that we can also extend the DT energy calculation method for detecting more complicated scene-related group activities (e.g., we can estimate different DT energy values according to the activity trajectories in the training data for detecting people following irregular paths).

For the normal relative network, three DT energy values are used as shown in Fig. 4 (b). For edges pointing toward the center node, their DT energy values $enr(i,j)$ will be 1 (as the red dashed arrows in Fig. 4 (b)). For edges pointing outward the center node, their DT energy values will be -1 (as the blue dash-dot arrows in Fig. 4 (b)). And the DT energy values will be 0 for edges between nodes having the same distance to the center node (as the black solid arrows in Fig. 4 (b)). Since in the normal relative network, person 1 is fixed at the center node, the normal relative energy $ENR(u_2-u_1, q_2-q_1)$ is mainly calculated by the movement of person 2 with respect to person 1. Based on our DT energy definition, when person 2 is moving close to the center node (i.e., moving toward person 1), $ENR(u_2-u_1, q_2-q_1)$ will be increased. On the contrary, when person 2 is leaving the center node, $ENR(u_2-u_1, q_2-q_1)$ will be decreased. By this way, the relative movement between people can be effectively modeled by the transmission energy.

The structure of the weighted relative network is the same as the normal relative network. However, the DT energy values $ewr(i,j)$ are "weighted" as shown in Fig. 4 (c). For edges either pointing toward or outward the center node, the DT energy values will become larger when they are closer to the center node. However, these DT energy values differ in that edges pointing toward the center node are positive while edges pointing outward the center node are negative. With this weighted relative network, we can extract the "history" or "temporal" information of the relative movement between people. For example, when person 2 moves from the red node in Fig. 4 (c) toward person 1 and moves back, the corresponding total weighted relative transmission energy $EWR(u_2-u_1, q_2-q_1)$ will be a positive value. On the contrary, EWR will be a negative value when person 2 leaves person 1 from the red node and then comes back.

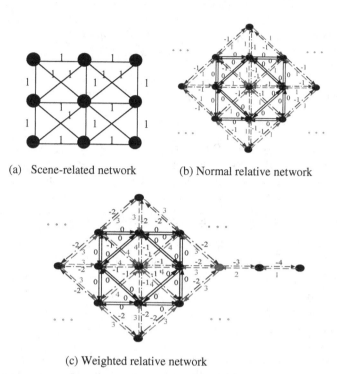

(a) Scene-related network   (b) Normal relative network

(c) Weighted relative network

Fig. 4. The DT energy values for the three networks (note that (a) is an undirected network while (b) and (c) are directed networks)

If we take a more careful look at the three networks in Fig. 4, we can see that since the scene-related network is constructed based on the scene without being affected by the people movements, it can be viewed as an identical field where packages need to consume energy to move and its moving distance is proportional to its consumed energy. Comparatively, since the two relative networks in Fig. 4 (b)-(c) are constructed based on person 1, they can be viewed as the repulsive fields where person 1 in the network center is creating "repulsive" forces. Thus, packages need to consume energy in order to approach person 1 while "gain" energy when leaving person 1. At the same time, no energy will be consumed or gained when packages are revolving around person 1.

3.4   Activity Detection Rules

With the three networks and their corresponding DT energies, we can calculate the total transmission energy set for the input group activity trajectories, as in Eqn. (1). Then when recognizing group activities, we can view the total transmission energy set in Eqn. (1) as a feature vector and train classifiers for automatically achieving the detection rules. In this paper, we use Support Vector Machine (SVM) [2] to learn the detection rules from the training set and use it for group activity recognition. Experimental results demonstrate that our NB algorithm can effectively recognize various group activities.

4   Experimental Results

In this section, we perform experiments for the group activity recognition. The experiments are performed on the public BEHAVE dataset [10] where 800 activity clips are selected for recognition and people trajectories are extracted by a particle-filter-based tracking method [8]. Eight group activities are recognized as shown in Table 1. Some frames are shown in Fig. 5.

Table 2 compares the results of the four methods:

(1) The group-representative-based algorithm that extracting group representatives for detection group activities [5] (GRAD in Table 2).

(2) The pair-activity classification algorithm based on bi-trajectories analysis which using causality ratio and feedback ratio as features [6] (PAC in Table 2).

(3) The localized-causality-based algorithm using individual, pair, and group causalities for group activity detection [7] (LCC in Table 2).

(4) Our proposed NB algorithm with transmission energy sets from three networks (NB in Table 2).

Table 1. The group activities recognized in our experiments

(I) meet: two people walk toward each other.
(II) follow: two people are walking. One people follow another.
(III) approach: one people stand and another walk toward the first people.
(IV) separate: two people escape from each other.
(V) leave: one people stand and another leave the first people.
(VI) together: two people are walking together.
(VII) exchange: two people first gathered and then leave each other.
(VIII) return: two people first separate and then meet.

In Table 2, three rates are compared: the miss detection rate (Miss), the false alarm rate (FA) [5], and the total error rate (TER) [5]. The TER rate is calculated by N_{t_miss}/N_{t_f} where N_{t_miss} is the total number of misdetection activities for both normal and abnormal activities and N_{t_f} is the total number of activity sequences in the test set. TER reflects the overall performance of the algorithm in detecting both the normal and the abnormal activities [5]. From Table 2, we can see that our proposed NB algorithm can achieve obviously better performance than the other three state-of-art algorithms. This demonstrates that our NB algorithm with the transmission energy features can precisely catch the inter-person spatial interaction and the activity temporal history characteristics of the group activities. Specifically, our NB is obviously effective in recognizing complex activities (i.e., exchange and return). In Fig. 6, (a) shows two example trajectories of the complex activities, (b) shows the values of the major features in the PAC algorithm [6], and (c) shows the transmission energy (EWR) from the weighted relative network in our NB algorithm. From Fig. 6 (b), we can see that the features in the PAC algorithm [6] cannot show much differences between the two complex activities. Compared to (b), our EWR energy in (c) are obviously more distinguishable by effectively catching the activity history information.

(a) Leave   (b) Follow

Fig. 5. Example frames of the BEHAVE dataset

Table 2. Miss, False Alarm, and TER rates of the group recognition algorithms under 75% training and 25% testing

		GRAD	PAC	LCC	**NB**
Meet	Miss(%)	11.4	12.1	18.5	**0.0**
	FA(%)	1.6	1.3	2.4	**0.0**
Follow	Miss(%)	12.2	8.3	11.7	**2.5**
	FA(%)	0.9	0.8	1.4	**0.2**
Approach	Miss(%)	10.3	9.0	12.8	**0.0**
	FA(%)	1.8	2.6	2.6	**0.5**
Separate	Miss(%)	6.3	5.0	5.8	**2.5**
	FA(%)	1.2	0.7	1.9	**0.6**
Leave	Miss(%)	5.2	4.9	9.7	**5.6**
	FA(%)	1.8	1.1	1.4	**1.0**
Together	Miss(%)	2.7	2.9	6.6	**8.8**
	FA(%)	0.7	0.2	0.9	**0.9**
Exchange	Miss(%)	24.7	28.9	26.3	**2.6**
	FA(%)	2.4	3.2	2.7	**0.5**
Return	Miss(%)	28.6	36.8	31.6	**5.3**
	FA(%)	2.6	2.5	2.3	**0.2**
TER(%)		10.7	11.1	13.8	**3.4**

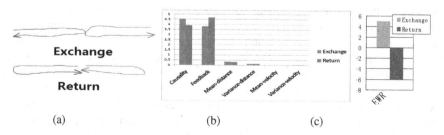

(a)   (b)   (c)

Fig. 6. (a) Example trajectories for complex group activities; (b) The major feature values for PAC algorithm; (c) The EWR energy values by our NB algorithm

5   Conclusion

In this paper, a new network-based algorithm is proposed for human group activity recognition in videos. The proposed algorithm models the entire scene as a network. Based on this network, we further model people in the scene as packages. Thus, various human activities can be modeled as the process of package transmission in the network. By analyzing the transmission process, various activities such as abnormal activities and group activities can be effectively recognized. Experimental results demonstrate the effectiveness of our algorithm.

Acknowledgement. This work was supported in part by the following grants: National Science Foundation of China grants (61001146, 61025005), Chinese national 973 project grants (2010CB731401), the Open Project Program of the National Laboratory of Pattern Recognition (NLPR), the SMC grant of SJTU, Shanghai Pujiang Program (12PJ1404300), and the China National Key Technology R&D Program (2012BAH07B01).

References

1. Loy, C.C., Xiang, T., Gong, S.: Modelling activity global temporal dependencies using time delayed probabilistic graphical model. In: Int'l Conf. Computer Vision (ICCV), pp. 120–127 (2009)
2. Chang, C.-C., Lin, C.-J.: LIBSVM: a library for support vector machines. ACM Trans. Intelligent Systems and Technology 2(3), 1–27 (2011)
3. Zhang, D., Gatica-Perez, D., Bengio, S., McCowan, I.: Modeling individual and group actions in meetings with layered HMMs. IEEE Trans. Multimedia 8(3), 509–520 (2006)
4. Cheng, Z., Qin, L., Huang, Q., Jiang, S., Tian, Q.: Group activity recognition by Gaussian process estimation. In: Int'l Conf. Pattern Recognition, pp. 3228–3231 (2010)
5. Lin, W., Sun, M.-T., Poovendran, R., Zhang, Z.: Group event detection with a varying number of group members for video surveillance. In: IEEE Trans. Circuits and Systems for Video Technology, pp. 1057–1067 (2010)
6. Zhou, Y., Yan, S., Huang, T.: Pair-activity classification by bi-trajectory analysis. In: IEEE Conf. Computer Vision Pattern Recognition, pp. 1–8 (2008)
7. Ni, B., Yan, S., Kassim, A.: Recognizing human group activities with localized causalities. In: IEEE Conf. Computer Vision and Pattern Recognition, pp. 1470–1477 (2009)
8. Hess, R., Fern, A.: Discriminatively Trained Particle Filters for Complex Multi-Object Tracking. In: IEEE Conf. Computer Vision and Pattern Recognition, pp. 240–247 (2009)
9. Li, J., Gong, S., Xiang, T.: Discovering multi-camera behaviour correlations for on-the-fly global prediction and anomaly detection. In: Int'l Workshop. Visual Surveillance, pp. 1330–1337 (2009)
10. BEHAVE set,
    http://groups.inf.ed.ac.uk/vision/behavedata/interactions/

Exploit Spatial Relationships among Pixels for Saliency Region Detection Using Topic Model

Guang Jiang[1,2], Xi Liu[3], JinPeng Yue[1,2], and Zhongzhi Shi[1,*]

[1] The Key Laboratory of Intelligent Information Processing,
Institute of Computing Technology, Chinese Academy of Sciences,
Beijing 100190, China
[2] Graduate University of Chinese Academy of Sciences, Beijing 100049, China
[3] Fujitsu Research & Development Center Co., LTD, Beijing 100025, China

Abstract. In this paper, we describe an approach to saliency detection as a two-category (salient or not) soft clustering using topic model. In order to simulate human's paralleled visual neural perception, many sub-regions are sampling from an image, where each one is considered as a set of colors from a codebook, which is a color palette for the image. We assume salient pixels would appear spatial adjacent more possibly, therefore in a same sub-region, while less salient pixels would either. Consequently, all the sub-regions are clustered into two assumed topics with probabilities: "salient"/"non-salient", while "salient" one is decided to give saliency value of each pixel according to its posterior conditional probability. Our method will give a global saliency map with full resolution, and experiments illustrate it is competitive with the state-of-art methods.

Keywords: Probabilistic Latent Semantic Analysis, Salient region detection.

1 Introduction

Saliency region detection dedicates to find which part attracts attention mostly when human sees an image. It is significant to identify the "proto objects", and extract as a saliency map, which would facilitate many subsequent image or video analysis, such as object recognition[1], image retrieval[2], target tracking[3] etc.

Saliency was firstly proposed in psychology and cognition fields, which found that, when looking at an image, our visual system would first quickly focus on one or several "interesting" regions of the image before further exploring the contents[4]. Therefore, there are two sequential stages in visual perception of human visual system: firstly, pre-attentive (bottom-up), which process all the

* Supported by the National Natural Science Foundation of China (No. 61072085, 60933004, 61035003, 60903141), National Program on Key Basic Research Project (973 Program) (No. 2013CB329502), National High-tech R&D Program of China (863 Program) (No.2012AA011003), National Science and Technology Support Program(2012BA107B02).

S. Li et al. (Eds.): MMM 2013, Part I, LNCS 7732, pp. 174–184, 2013.

information available fast but coarsely; secondly, focused attention (top-down), which process only parts of the input information with more intensive efforts of exploration[5], however, it often depends on specific application. Therefore, many efforts devoted to saliency detection in a bottom-up way.

Intuitively, saliency is determined by its visual uniqueness or rarity which appears in image with attributes such as color, gradient, etc. Fig.1 gives some examples of saliency map. Ideal saliency map is to obtain a "global" and full resolution saliency map, as in Fig. 1, which means to compute saliency for each pixel by comparing with all the others for an image. However, it is time-consuming usually, for example, a 1024*768 image, the comparison of all two pixels would achieve $6.2 * 10^{11}$. To alleviate intensive computation, kinds of approaches are proposed.

Fig. 1. Some examples of our method for salient region detection. Top: source images; Down: saliency map.

Some researchers exploit frequency domain of images for saliency detection. Hou et al. [6] propose to analyze the log-spectrum of an input image, and extract the spectral residual in spectral domain to construct saliency map. Furthermore, Guo et al. [7] suggested that the phase spectrum of Fourier transform alone is sufficient to get the salient map. Nonetheless, these methods are usually computational efficient, but usually focus on the boundary of salient object.

Saliency is also considered in color space directly, however, a "local" saliency map is often obtained: computing saliency of a pixel from its rarity with its neighbor pixels. One of fundamental works is Itti et al. [8], which define saliency according to the difference of fine scale and coarse scale from Gaussian pyramid. Ma and Zhang [9] provide three-level attention analysis, and propose a fuzzy growing method for local contrast. Harel et al. [10] extract feature vectors to form "activation map", and define Markov chains to combine several maps into a normalized one. Liu et al. [11] combine multi-scale contrast, center-surround histogram and color spatial distribution using Conditional Random Field (CRF) for salient object detection. However, these works often downsize the input image, and use limited range of spatial frequencies, therefore, the saliency results often are ill-defined on object boundary, or cannot uniformly map the entire salient region[12]. An exception is Achanta et al. [12], which proposes a frequency-tuned saliency detection method, and choose Difference of Gaussians (DoG) for a pixel among large range of frequencies to give a full-resolution saliency.

Though color space is huge, many colors don't appear in a specific image, hence could be reduced. Zhai and Shah [13] reduce the number of colors by only using luminance, while Cheng et al. [14] construct histogram of colors to reduce color space and give saliency map using global contrast. In addition, patch is considered instead of pixel for efficiency, several methods employ information theory to study information of a patch to analyze its saliency: Li et al.[15] learn sparse coding bases for random sampled patches, and used incremental coding length (ICL) to measure the saliency of a patch. Moreover, Han et al [16] considered both residual and sparseness of sparse coding to define the saliency. Recently, Duan et al.[17] regard a patch as a vector of all pixels involved and reduce the dimension by PCA(Principal Component Analysis), lastly a global saliency for each patch is defined by comparing with all others. However, patch-based methods cannot give a full-resolution saliency map either.

In conclusion, all the above methods mentioned try to define the saliency of a pixel/patch by comparing itself with others (neighbors or global ones) directly. Among them, [13], [12], [14] try to give a full-resolution saliency map; while [14] and [17] try a global contrast to give a "global" saliency map.

In this paper, we try to define a "global" saliency map with full-resolution using Probabilistic Latent Semantic Analysis(PLSA). Actually, topic model has been utilized in object categorization, image annotation and many other computer vision missions [18]. However, there are little work for saliency detection, Li et al. [19] combine latent topic model with existing saliency detection method to categorize images. As far as we know, the only work of employing topic model to detect saliency is Li et al. [20], in their work, proto-objects and saliency map are computed in an iterative way: proto-objects are detected based on saliency map using latent topic model, while saliency map is improved by the proto-objects. Therefore, topic model is considered as a top-down view. However, as an initialization, their method depends on multiple segmentation of image, which is still known as an open problem.

We argue that latent topic model could be used directly as a bottom-up way to model the saliency map, and propose several assumptions for this aim. The saliency of a pixel is modeled in a cluster way, concretely, as the conditional probability of the "salient" topic. we assume that salient pixels would exist with high probability under "salient" topic, while vice versa for less-salient pixels under "non-salient" topic. Our contributions are as follows:

1. We propose to give saliency map of an image using topic model, in which, topic(or latent state) is usually hidden, however, we judge their meaning in this work.

2. The saliency map is full resolution: saliency of each pixel is given by its posterior probability of "salient" topic.

3. The saliency map is "global" because two "global" latent topics are shared with all the pixels.

2   Proposed Methods

2.1   Protocols

We use typical terms in PLSA for ease, consider color palette as a color codebook, denoted as C, and a color in C as term "word", while a sub-region from an image as term "document". It should be noticed that, in topic model, a document is considered as "bag of words" (BOW), i.e., the appearance number of all words from C, which means that words are exchangeable in a document, therefore "words" in the "document" are unordered, not the same situation as pixels in a patch, has their own spatial position.

We propose four assumptions as the basis of our model:

Assumption 1. Two pixels with similar color incline to same saliency.

Assumption 2. Two pixels are more spatial adjacent, then incline to same saliency because of higher possiblility of belonging to a same appearing object.

Assumption 3. There are two latent topics: "salient" and "non-salient", a pixel is salient only when it has higher probability of the "salient" topic, but lower probability of the "non-salient" topic.

Assumption 4. The saliency values of all pixels are described by a distribution under the "salient" topic, and the posterior probability of the "salient" topic gives a saliency map of the image.

In the following, our methods try to utilize these assumptions to compute saliency map for a given image.

2.2   Reduce Color

As mentioned above, color space is too huge, for a colorful image, RGB space is $256 * 256 * 256 = 2^{24}$, which is size of color codebook C, is tremendous for PLSA. Therefore, we must reduce color space.

For a color image, Cheng et al.[14] suggested that 85 colors on average can still give a good illustration for an image. Quantizing color is not our key point, so we adopt a quantization method[1] with variance minimization, which use greedy orthogonal bipartition of RGB space aided by inclusion-exclusion tricks to quantize the color space, then result in the color codebook C.

Therefore, according to C, each pixel in an image is mapped to a nearest-neighbor word $w \in C$. It is obvious that similar color would be mapped to same word w, which leads to same saliency as in follows; this fact conforms to our assumption 1.

2.3   Sample Patches and Construct Documents

We use assumption 3 to model the saliency via the latent topic: C is the dictionary, therefore, many documents is essential to learn a distribution of the latent topic using PLSA. For this purpose, we simulate human's paralleled visual neural perception as sampling many patches of arbitrary size at arbitrary position

---

[1] http://www.ece.mcmaster.ca/~xwu/cq.c

within an image, for each, construct a sub-region as a document d, meanwhile all the pixels within it are mapped to the words in C. Therefore, all the documents are collected, and denoted as D.

When sampling a patch p from an image I, there are two choices for constructing a document d:

Strategy 1. Treat p as d directly;

Strategy 2. Subtract p from I, and residue region in I is regarded as d;

Obviously, d is a sub region of I in any strategy, but we prefer choice strategy 2 than 1, and explaining the reason in the following primary theory of topic model, also illustrated in the experiments. Considering two pixels p_1 and p_2 more spatial adjacent in an image, therefore are more possible being a a part of an object involved in I, likewise, more possible appear in different documents d many times, which is just the assumption 2.

2.4   Model Saliency via Topic Model

There are many topic models and variants could be employed for our method. In the paper, we use PLSA[21], which is a generative model to model co-occurrence data by latent variables which generate all words in documents. Given a set of M documents $D = \{d_i, 1 \leq i \leq M\}$, and a word dictionary $C = \{w_j, 1 \leq j \leq N\}$, which $N = |C|$ is the size of C. Consequently, PLSA introduces K latent topics $Z = \{z_k, 1 \leq k \leq K\}$: a document is a mixture of K topics, while a topic is a mixture of all words from C. The joint probability of a document and a word is as follows:

$$P(d_i, w_j) = \sum_{k=1}^{K} P(z_k)P(d_i|z_k)P(w_j|z_k) \tag{1}$$

In PLSA, topics are latent, so the meaning depends on specific application: they would be "outdoor/indoor" for scene recognition, or "fruit/bear" for image annotation. For word w, $P(w|z_k)$ represent its relevance with topic z_k. For instance, word "apple" and "banana" would have higher probability under topic "fruit". Generally, Bayesian rules can be used to compute the posterior probability of latent topic z_k for a word w_j:

$$P(z_k|w_j) = \frac{P(w_j|z_k)P(z_k)}{\sum_k^K P(w_j|z_k)P(z_k)} \tag{2}$$

We called w_j belongs to topic z_k when satisfying (3); for a pixel p_j is mapped to w_j, we also called p_j belongs to z_k.

$$z_k = max_{p=1}^{K} P(z_p|w_j) \tag{3}$$

In our method, we assume there are only two topics: "salient" topic, which is corresponding to the salient region in the image; "non-salient" topic, which can be seen as background of the salient region. Fig.2 illustrates to explain our method: the flower is the salient object O in the image, there are two pixels p_{s1} and p_{s2} in O, and several less salient pixels: four pixels p_1, p_2, p_3, and p_4, which

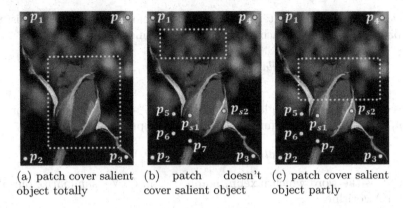

(a) patch cover salient object totally

(b) patch doesn't cover salient object

(c) patch cover salient object partly

Fig. 2. Three situations when sampling a patch

are adjacent to four boundary edges and corners of the image, and three pixels near O: p_5, p_6, p_7.

Basically, there are three typical situations when we sample a patch p as (a)-(c) in Fig. 2: the rectangle with dotted line represented the sampled patch P, therefore subtract it from the image, and the sub region considered as d according to strategy 2 in section 2.3. Therefore:

- For Fig. 2(a), (b) and (c), p_1, p_2, p_3, and p_4 co-occur in same documents, therefore would incline to same topic, because of accord with the fact in PLSA: for two words w and w' share both higher condition probability under specific topic z_k if and only if they co-occur in many documents. In this sense, it is obvious that p_1 to p_4 would appear seldom in any document under strategy 1 mentioned in section 2.3, hence cannot contribute any.
- For Fig. 2(b) and (c), salient pixels p_{s1} and p_{s2} co-occur with their neighbor less-salient pixels p_5, p_6 and p_7; however, due to saliency of p_{s1}, its color differs much with its less-salient neighbor pixels, which lead to that p_5, p_6 and p_7 are more possible mapped to same word in C, while p_{s1} and p_{s2} to other words. Therefore, p_5, p_6 and p_7 would belong to one same topic, while p_{s1} and p_{s2} belong to another, because all the words in a document are clustered into only two topic in our model.

We can see that all the words will compete for the "salient" topic among many documents via PLSA, therefore the convergence states of the model give the answer, as proposed in assumption 4. So far, we didn't know which topic is "salient" one, therefore should identify the "salient" one from the two topics.

2.5 Predicate "Salient"/"Non-Salient" Topic

Researchers suggest the center of image would attract people's attention more likely[22]. Therefore, most algorithms consider a strong bias to the center of the image. Inspired by this fact, we assume the pixels belong to the "salient" topic apt to locate around the center of the image. Therefore we examine the distance of all pixels p_j belongs to each topic z_k, with the image center:

$$SalPossibility(z_k) = \sum_{p_j} Dist(p_j, center) \tag{4}$$

We select the topic with higher SalPossibility as "salient" topic z_{sal}; for each pixel p_j in the image, we define its saliency as the probability of z_{sal}:

$$saliency(p_j) = P(z_{sal}|w_j) \tag{5}$$

Where w_j is the word which pixel p_j mapped to in C. We summarize our algorithm in Algorithm 1.

Algorithm 1. Bottom-up saliency detection by topic model

Input: Image I

Output: Salience map of I

1: Create color codebook C, and reduce the number of colors of I to $|C|$;
2: Map each pixel p_j in I to a word w_j in C;
3: Sample a patch, and construct a document according to strategy 2 in section 2.3;
4: Repeat Step 3 for M times to prepare M documents for training a PLSA model with two topics;
5: Predicate "salient" topic using (4);
6: For all the pixels, compute saliency using (5) and compose the final salience map.

---

3   Experiments

We select a publicly available benchmark data set which is also used in Achanta et al.[12] to evaluate our algorithm. The database contains 1000 images (average size is 300*400 pixels), with ground truth labeled by human.

We implement our algorithm on the win7 64 bit platform, using RGB color space, parameters setting are: $|C| = 80$ and $M = 500$ according to the following parameter selecting experiment. We use the algorithm supplied by J Sivic [18] to train the PLSA model. In order to judge "salient" topic, all pixels of the image belong to only one topic according to (3), and should be considered; however, for efficiency, only 200 pixels for each topic are sampled randomly to compute the distance with image center. Lastly, the pixel values in saliency map are normalized from 0 to 255. As a whole, our algorithm will spend 1.8 seconds per image on average to create the saliency map.

Fig.3 illustrates the parameters varying of our method: Fig. 3(a) shows the AUC(Area under curve) of recall-precision when $|C|$ varying from 10 to 200, we also try if larger patch will perform better, therefore limit the minimum of patch size from 0.2 to 0.8 times of image size, however, all of these are inferior to no limit setting of patch size(green curve), which maybe explained as simulating

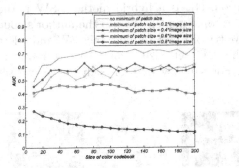
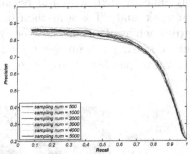

Fig. 3. Performance of our method with different parameters: (a) AUC performance with $|C|$ from 10 to 200, and different patch size limitations; (b) Recall-precision curve with M from 500 to 5000

different scale of humans paralleled visual neural perception. We can see that for $|C|$ varying from 20 to 80, the AUC value increase because more colors would discriminate more different pixels; however, colors more than 80 couldn't give prominent improvement any more, this fact is also in line with the result of Cheng et al.[14]. Fig.3(b) shows recall-precision curve when M varying from 500 to 5000, their performance show small differences, therefore we select $M = 500$ for efficiency. It can be concluded that our method is not sensitive with parameters $|C|$ and M.

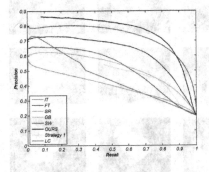
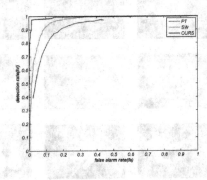

Fig. 4. Performance comparison with other methods: (a) precision-recall curves with saliency threshold varying from 0 to 255; (b) ROC of detection rate(hr)-false alarm rate(fa) performance comparison with PT and SW

We compare our algorithm with seven algorithms: IT(Itti et al.[8]), FT(Achanta et al.[12]), SR(Hou et al.[6]), GB(Harel et al.[10]) SW(Duan et al. [17]), LC(Zhai et al.[13]),and PT(Li et al. [20]). In which, IT is the early and widely referenced work, SR is according to frequency statistics, FT and LC supply saliency map with full resolution, GB is a hybrid method, SW is the recent work, and PT is a method using topic model. We use the author's code directly when available on the Web. We try our best to reproduce SW and select parameters using 100 images from the data set: size of patch is 10, reducing dimension of patch to 8 using PCA. For saliency map of each image, we set

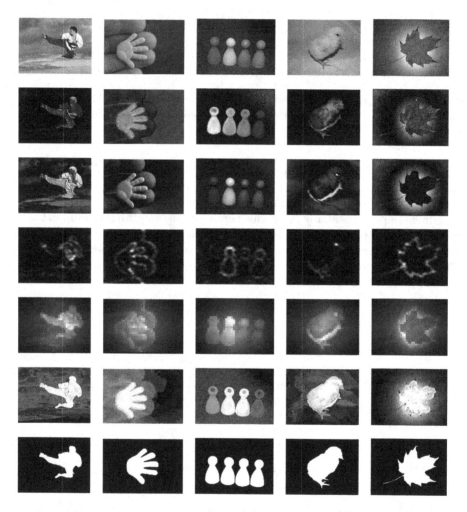

Fig. 5. Saliency map examples of some methods, rows from top to down: source images, FT, LC, SR, SW, OURS, Ground Truth

saliency threshold from 0 to 255. For each threshold, pixel in the saliency map is set to 1 if it is bigger than the threshold, then obtain recall and precision by comparing with the saliency ground truth. Finally average over all images are plotted as precision-recall curves as shown in Fig. 4(a). It can be seen that our method is competitive with SW, when recall is lower than about 0.88, however, outperforms all the other methods in both recall and precision. We also give ROC(receiver operator characteristic curve) of our method and SW, and compare with result of PT published in their work: with saliency threshold varying from 0 to 255, the higher detection rate(hr), and lower false alarm rate(fa), better the method is. As shown in Fig. 4(b), our method outperform both SW and PT, we attribute this fact to the four reasonable assumptions. We also implement strategy 1, as shown in fig4.(a), it is unsatisfactory, mainly due to deficiency of absorbing sufficient pixels far from the image center, which often belong to "non-salient" topic.

Fig.5 gives some examples using these methods. We can see that our method gives fine saliency map for both single object(other columns) or multi-objects(3rd column), what's more, the saliency map gives a holistic map and uniform on the saliency object, which may attribute to assumption 2 we proposed and the fact that all the pixels of the image can share the two global topics in our model. Recall p_1 to p_4 in Fig. 2, they are almost dark in all salience maps of Fig. 5 because they belong to "non-salient" topic, therefore with lower probabilities under "salient" topic.

4   Conclusions

In this paper, we proposed to model the saliency of each pixel by the probability of the assumed "salient" topic, and experiments illustrate effectivity of our method. In our experiment, PLSA is employed, however, it could be replaced with any other topic model(Latent Dirichlet allocation, LDA). On the other hand, many paralleled versions could accelerate implementation of topic model.

In our model, we quantize the pixels by a color codebook directly, without considering the closeness among the colors; in the future, we consider to construct a continuous color space or continuous function, because many topic models support continuous variables. On the other hand, we give an empirical way to judge "salient topic" in this paper, which should be studied in depth.

References

1. Rutishauser, U., Walther, D., Koch, C., Perona, P.: Is bottom-up attention useful for object recognition? In: IEEE Computer Society Conference on Computer Vision and Pattern Recognition (CVPR), vol. 2, p. II–37 (2004)
2. Chen, T., Cheng, M., Tan, P., Shamir, A., Hu, S.: Sketch2photo: internet image montage. ACM Transactions on Graphics (TOG) 28, 124 (2009)
3. Zhang, G., Yuan, Z., Zheng, N., Sheng, X., Liu, T.: Visual Saliency Based Object Tracking. In: Zha, H., Taniguchi, R.-i., Maybank, S. (eds.) ACCV 2009, Part II. LNCS, vol. 5995, pp. 193–203. Springer, Heidelberg (2010)
4. Leavers, V.: Preattentive computer vision towards a two-stage computer vision system for the extraction of qualitative descriptors and the cues for focus of attention. Image and Vision Computing 12(9), 583–599 (1994)

5. van der Heijden, A.: Two stages in visual information processing and visual perception? Visual Cognition 3(4), 325–362 (1996)
6. Hou, X., Zhang, L.: Saliency detection: A spectral residual approach. In: IEEE Conference on Computer Vision and Pattern Recognition (CVPR), pp. 1–8 (2007)
7. Guo, C., Ma, Q., Zhang, L.: Spatio-temporal saliency detection using phase spectrum of quaternion fourier transform. In: IEEE Conference on Computer Vision and Pattern Recognition(CVPR), pp. 1–8 (2008)
8. Itti, L., Koch, C., Niebur, E.: A model of saliency-based visual attention for rapid scene analysis. IEEE Transactions on Pattern Analysis and Machine Intelligence 20(11), 1254–1259 (1998)
9. Ma, Y.F., Zhang, H.J.: Contrast-based image attention analysis by using fuzzy growing. In: Proceedings of the Eleventh ACM International Conference on Multimedia (MM), pp. 374–381. ACM, New York (2003)
10. Harel, J., Koch, C., Perona, P.: Graph-based visual saliency. Advances in Neural Information Processing Systems 19, 545 (2007)
11. Liu, T., Yuan, Z., Sun, J., Wang, J., Zheng, N., Tang, X., Shum, H.: Learning to detect a salient object. IEEE Transactions on Pattern Analysis and Machine Intelligence 33(2), 353–367 (2011)
12. Achanta, R., Hemami, S., Estrada, F., Susstrunk, S.: Frequency-tuned salient region detection. In: IEEE Conference on Computer Vision and Pattern Recognition (CVPR), pp. 1597–1604 (2009)
13. Zhai, Y., Shah, M.: Visual attention detection in video sequences using spatiotemporal cues. In: Proceedings of the 14th Annual ACM International Conference on Multimedia (MM), pp. 815–824 (2006)
14. Cheng, M., Zhang, G., Mitra, N., Huang, X., Hu, S.: Global contrast based salient region detection. In: IEEE Conference on Computer Vision and Pattern Recognition (CVPR), pp. 409–416 (2011)
15. Li, Y., Zhou, Y., Xu, L., Yang, X., Yang, J.: Incremental sparse saliency detection. In: IEEE International Conference on Image Processing (ICIP), pp. 3093–3096 (2009)
16. Han, B., Zhu, H., Ding, Y.: Bottom-up saliency based on weighted sparse coding residual. In: Proceedings of the 19th ACM International Conference on Multimedia (MM), pp. 1117–1120 (2011)
17. Duan, L., Wu, C., Miao, J., Qing, L., Fu, Y.: Visual saliency detection by spatially weighted dissimilarity. In: IEEE Conference on Computer Vision and Pattern Recognition (CVPR), pp. 473–480 (2011)
18. Sivic, J., Russell, B., Efros, A., Zisserman, A., Freeman, W.: Discovering objects and their location in images. In: IEEE International Conference on Computer Vision, vol. 1, pp. 370–377 (2005)
19. Li, Z., Wang, Y., Chen, J., Xu, J., Larid, J.: Image topic discovery with saliency detection. In: Proceedings of the British Machine Vision Conference (BMVC), pp. 33–31 (2010)
20. Li, Z., Xu, J., Wang, Y., Geers, G., Yang, J.: Saliency detection based on proto-objects and topic model. In: IEEE Workshop on Applications of Computer Vision (WACV), pp. 125–131 (2011)
21. Hofmann, T.: Unsupervised learning by probabilistic latent semantic analysis. Machine Learning 42(1), 177–196 (2001)
22. Tatler, B.: The central fixation bias in scene viewing: Selecting an optimal viewing position independently of motor biases and image feature distributions. Journal of Vision 7(14) (2007)

Mining People's Appearances to Improve Recognition in Photo Collections

Markus Brenner and Ebroul Izquierdo

Queen Mary University of London, UK
{markus.brenner,ebroul.izquierdo}@eecs.qmul.ac.uk

Abstract. We show how to recognize people in Consumer Photo Collections by employing a graphical model together with a distance-based face description method. To further improve recognition performance, we incorporate context in the form of social semantics. We devise an approach that has a data mining technique at its core to discover and incorporate patterns of groups of people frequently appearing together in photos. We demonstrate the effect of our probabilistic approach through experiments on a dataset that spans nearly ten years.

1 Introduction

Faces, along with the people's identities behind them, are an effective element in organizing a personal collection of photos, as they represent *who* was involved. However, along with the fact that current face recognition approaches require a notable amount of user interaction for training a classification model, the accurate recognition of faces itself is still very challenging. Particularly in the uncontrolled environments of consumer photos, faces are usually neither perfectly lit nor captured. Wide variations in pose, expression or makeup are common and difficult to handle as well.

One way to address this issue is to keep improving face recognition techniques. Another way is to incorporate *context*; in other words, to consider additional information aside from just faces within photos and across entire collections. As a wide variety of literature shows, such contextual information might include time, location or scene. The people appearing in photos can also provide further information. For instance, a person's demographics such as gender and age are often easier to surmise than his or her identity. The same applies to ethnic indicators, including skin tone and traditional costumes. Finally, clothing in general can provide useful contextual information.

Since consumer photos are often very personal in nature, they involve a highly social aspect. Recognition may also be improved by considering social semantics, especially when multiple and related people are involved.

2 Background and Related Work

There are several previous works considering social semantics as a contextual cue. One apparent constraint modeled by many, including [1,2], is the exclusivity

S. Li et al. (Eds.): MMM 2013, Part I, LNCS 7732, pp. 185–195, 2013.

constraint wherein multiple faces in a single photo cannot belong to the same individual person. [3] takes into account the collective arrangement (relative and absolute) of people, or the fact that they often group in an apparently structured way (e.g. men on the edges and children centered in the first row). However, the authors primarily exploit this information to infer demographics like gender and age or the activity of a group (e.g. having dinner) rather than to recognize people.

More generally, one can study and exploit the collective appearance of people. Such statistics can help improve recognition as they give an indication about someone's identity in case the person repeatedly appears together with other people. [1] calls this *group prior* and shows improved recognition performance. However, the work depends on photos that depict multiple people. A key aspect of our proposed work is to extend this idea to cover *events* (temporal clusters) consisting of multiple related photos. In that case, groups of people do not (always) need to appear together in a single photo (e.g. when group members alternate between subject and photographer). Another work considering people's co-occurrences is [4], but it models only pair-wise occurrences of people.

As [5] shows, it is also possible to gather such social semantics from other, external sources rather than from the photo collection itself. The authors incorporate social context directly provided (e.g. a person's network of friends) by social online networks like Facebook.

Some related works such as [6,7] study the reverse problem. They aim to discover social relationships (e.g. identifying parents, their children, their relatives, ...) from photos without utilizing them for people recognition. Two works that actually incorporate such social relationships into a recognition framework are [8,9]. In summary, the following sets our work apart from previous studies:

- We consider people's appearances on an event rather than only on a photo level. This is more difficult, as the effectiveness of most previous works depends on photos depicting groups of people together at the same time (basically, in the same photo). However, this is often not the case and is thus a limitation.
- Previous works are limited in that they only model the pairwise appearances between two people at a time. We, however, intend to jointly capture appearances of higher degrees (e.g. family gatherings with always the same four or five people but never certain other people).
- To our knowledge, we are the first to utilize a data mining technique to discover and exploit social semantics for improving people recognition.

3   Objective and Approach

We present a flexible framework in three stages to recognize people across a Consumer Photo Collection. Compared to traditional approaches where a classifier is typically trained to recognize a single random face at a time, we intend to consider and thus recognize all people's appearances within an entire dataset simultaneously. To accomplish this, we first lay out a probabilistic graphical model with a similarity or distance-based description technique at its core.

Next, to further improve recognition performance, we aim to incorporate context in the form of social semantics that is usually only implicitly at hand within photo collections. For example, family relations are usually not labeled, but it is often possible to infer this information by looking at multiple photos spanning a longer time period. Thus, we propose a method that utilizes a data mining technique at its core to discover patterns of groups of people who frequently appear together. In order to discover such patterns even when the training set is sparse, we devise our overall approach in an iterative fashion. Lastly, we extend our initial recognition framework by incorporating the gained additional information in an effective way.

We are interested in improving the performance of people recognition based solely on the information contained – explicitly or implicitly – within a photo collection. By doing so, we are not relying on information from external sources that are not always available.

In the next two sections, we summarize how we detect and discriminate faces and reiterate the basics of our graph-based recognition approach, both of which we detail in [2]. Thereafter, we set forth the techniques we employ to mine people's appearances, process them and incorporate them into a unified framework. We demonstrate the effectiveness of the proposed approach with experiments on a new dataset and present our conclusions.

4   Face Detection and Basic Recognition

We choose to utilize the seminal work of Viola and Jones included in the OpenCV package to detect faces. Their detection framework builds upon Haar-like features and an AdaBoost-like learning technique. The face recognition technique we introduce next provides some leeway for minor misalignment and thus scaling the patches identified as faces to a common size and converting them to a gray-scale representation is the only *normalization* we perform.

Compared to holistic face recognition approaches that typically require training, we turn to a feature-based method using histograms of Local Binary Patterns [10] that allows us to directly compute face descriptors and subsequently compare these with each other based on a distance measure (e.g. utilizing χ^2 Statistics). To actually recognize faces, the most straightforward approach is then nearest-neighbor matching against a set of known face descriptors.

5   Graph-Based Recognition

Like in our previous work [2], we employ a graphical model to further improve recognition of people (e.g. by incorporating constraints). Such models are factored representations of probability distributions where nodes represent random variables and edges probabilistic relationships. In particular, we choose a pairwise Conditional (Markov) Random Field (CRF). In our proposed approach, people's appearances (e.g. as represented by their faces) correspond to the nodes in an undirected graphical model. We set up one graph with nodes for both a testing

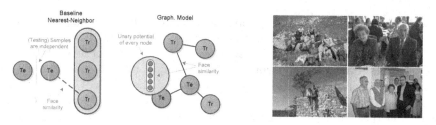

Fig. 1. Left: Face and people recognition by traditional nearest-neighbor matching (each testing sample Te is independently matched against all training samples Tr) and our graph-based framework that considers all people's appearances within an entire dataset simultaneously (faces are represented by interconnected nodes, where each node encodes a list of state likelihoods based on face similarities). Right: Exemplary photos of the employed dataset.

and a training set (signified by samples Tr and Te in the left side of Figure 1), where we condition on the *observed* training samples with known label classes. The states of the nodes reflect the people's identities.

We use the graph's unary node potentials to express how likely people's appearances belong to particular individuals (in the training set). We define the unary term as in Equation 1, where w signifies the states, s a face similarity function (the distances among faces), and Z_s an optional normalization constant. Note that for each state we retain only the closest match among several possible training samples of the same label class.

$$u\left(w\right) = \frac{1}{Z_s} s(w) \qquad (1)$$

As there is one unary potential for each node, there is one pairwise potential for each edge (connecting two nodes) in a pairwise CRF. They allow us to model the combinations of the states two connected nodes can take, and thus, to encourage a spatial smoothness among neighboring nodes in terms of their states. They also allow us to enforce a *uniqueness* constraint that no individual person can appear more than once in any given photo:

$$p\left(w_n, w_m\right) = \begin{cases} \tau, & \text{if } w_n = w_m \wedge i_n \neq i_m \\ 0, & \text{if } w_n = w_m \wedge i_n = i_m \\ 1, & \text{otherwise} \end{cases} \qquad (2)$$

The larger the spatial smoothness τ in Equation 2, the more we encourage neighboring nodes to take on the same state, therefore leading to fewer but larger clusters. Note that for the smoothness and constraint to be effective, we need to establish edges reflecting direct dependencies among nodes – in the latter case, simply among all nodes that share the same photos. In general, however, we only connect nodes representing the most similar appearances of people with each other for our aim of a sparse but effective graph representation. In other words, we connect each node with its η closest matches among the combined testing and training sets. We control η based upon a threshold such that $\check{s} - s_x < \varepsilon\check{s}$, where $\check{s}$ denotes the closest match (with the smallest distance and thus highest similarity), s_x the next x closest matches, and $\varepsilon \in [0, 1]$ a constant.

Ultimately, we wish to infer the states - as in a discrete model - of the random variables. An approach is to compute the individual marginal distributions, and then the largest marginal values would reflect the *world states* that are individually most probable. However, it is also possible to find the maximum a posteriori (MAP) solution of such a model. To deal with eventual cycles (*loops*) in our graph, we employ Loopy Belief Propagation as our method for inference.

6 Mining People's Appearances

Our aim is to discover and exploit patterns of how people appear together in a photo collection. We do so primarily by finding repeated appearances of groups of people. While multiple people may appear together in a single photo, it is evident that people typically do not congregate for the sole purpose of taking one photo. Instead, people usually meet for a longer time period (e.g. numerous relatives attending the same family birthday party), over which multiple photos might get captured. Thus, it is also necessary to model the case of an individual person who only appears by himself or with different members of the group in all photos captured throughout such a period, which we refer to as an event. Note that for simplicity we are only interested in consecutive, non-overlapping events. Finding such patterns based solely on the information given by the training set (some labeled people) is usually not a viable option because of the often low number of training samples. Thus, we devise the following iterative approach:

1. First, we use our basic graph-based approach (as described in Section 5) to initially recognize people.
2. Then, based on these preliminary recognition results, we attempt to discover appearance patterns.
3. Lastly, we perform inference - as in Step 1 - a second time while also considering the information gained through pattern mining in Step 2 to refine the initial recognition results.

Note that in Step 1 we also compute a measure of confidence $\nu \in [0, 1]$ for a person's face appearance based on probabilities (corresponding to the nodes' states) provided by the graph's inference method. We utilize this confidence measure later.

6.1 Temporal Clustering

Our first step towards finding patterns of people's appearances within a photo collection is to split the photo collection into several sets of photos according to events, or simply temporal clusters (e.g. photos falling on the same day).

Based on the photos' capture timestamps, we find temporal clusters e that span a maximum duration $\hat{\delta}$ of several hours. We do so by employing Mean-Shift, a non-parametric, iterative clustering algorithm based on density estimation, thus linking the clustering criteria to the cluster duration.

6.2   Frequent Itemset Mining

In the area of data mining and knowledge discovery, Association Rule Mining is a method for discovering interesting relations (among variables) in a dataset. Part of this is Frequent Itemset Mining (FIM), which attempts to find sets of items frequently occurring together; for example, products that are often purchased together in a store.

More formally, given a set of transactions T (each containing several items that have occurred together) over an item base B (e.g. the different people as in our case), the goal of FIM is to generate subsets $I \subseteq B$ that are common to at least a minimum threshold s_{min} (this number is also referred to as minimum support) of the transactions T. In Equation 3 below, $s_T(I)$ signifies the absolute support of I w.r.t. T.

$$F_T(s_{min}) = \{I \subseteq B \mid s_T(I) \geq s_{min}\} \tag{3}$$

One of the earlier and more popular FIM algorithms is Apriori. Later, Eclat and FP-growth were presented – each one outperforming previous works. We employ RElim, a simpler relative to the latter. The frequent mining parameter important to us is the minimum support s_{min}.

Recall that we run our graph-based recognition approach twice. After the initial run, we have a preliminary recognition result R (that is, a label along with a confidence value for each face). Based on R, we then compile the set of transactions T (essentially, a set of individual people for each event) to be fed to the FIM method. Note that we also apply the following filtering while compiling the transactions with the aim of omitting appearances that we may have incorrectly recognized in the initial run:

1. We discard all appearances with associated confidence values below a threshold ν_{min}^m. Then: $R^m \subseteq R$.
2. For each event e, we discard individuals from R^m (in fact, the parts corresponding to the event e) who appear less than a certain number of times n_{min} (in this event). Note, however, that we always include people who are part of the training set D^{Tr}.

The FIM's main result F_T is again a set of transactions (or sets of people). In addition, we book-keep the frequency f (specifying the number of times a transaction or pattern occurs) for each event e.

7   Incorporating Mining Results

At this point we have numerous appearance patterns; however, we do not know to which events they apply. Thus, we show next how to match both. Like in the previous section, we first discard all *uncertain* appearances from R. These are appearances with associated confidence values ν below a certain threshold ν_{min}^i. Then $R^i \subseteq R$. Next, we compile an intermediate result H_e for each event e:

1. First, we form a set of individual people P_e over R^i that is associated with the event e.
2. We then include people in $P_e' \supseteq P_e$, who are part of the training set D^{Tr} and are associated with the event e, but are not contained within P_e (because of possible recognition errors during our initial run).
3. Next, we match the given event with any appearance pattern based on it's set of associated people. To do so, we form a list L_{F_T} by iterating through all transactions in the FIM's result F_T and including any transaction I that matches the following criteria:
   - The number of distinct individuals who are in P_e' and I is above a threshold $\max(1, n_e m_{min})$ with n_e being the number of distinct people in P_e'.
   - The number of distinct individuals who are either *only* in P_e' or I is below a threshold $\max(2, n_e m_{max})$.
4. Then, we transform the sets of people within L_{F_T} into stacked vectors with a vector length equal to the number of total individuals. If an individual appears in a set, we store the transaction's frequency f at each individual's position in the vector.
5. Finally, we sum up all vectors of the given event e into a histogram vector H_e, thus aggregating the frequencies.

All histogram vectors H_e together then make up the final intermediate result H, which we now incorporate into our graph-based recognition framework with the aim of improving recognition on its subsequent run. Again, we perform several steps for each event e as follows:

1. We identify *applicable* people in the corresponding histogram vector H_e:
   (a) First, we omit people with a low frequency (below f_{min}). We define the vector corresponding to the remaining people as H_e'.
   (b) Then, we compute the mean average $\bar{f}_e$ of H_e', and omit people from H_e' whose frequency is less than $\bar{f}_e \alpha$. H_e'' then defines the remaining people.
2. If H_e'' corresponds to less than two distinct people, the mining result is not useful for the given event and we continue with the next event.
3. Next, we include people into $H_e''' \supseteq H_e''$, who are part of the training set D^{Tr} and are associated with the event e, but are not *contained* within H_e'' (again, we do this because of possible recognition errors during our initial run). We initialize their frequencies with $\bar{f}_e$.

At this point, H_e''' tells us which people *should* be present in a given event e according to the previously discovered patterns. Moreover, the aggregated frequency values indicate how likely this is true for particular people. We propose two complementary ways to incorporate this information.

Recall that the graph's unary node potentials (Equation 1) express how likely people's appearances belong to particular individuals. We propose to *adjust* the potentials according to H_e''' based on following exponential regularization, where λ is used to dampen large frequency values:

$$u'(w) = u(w)H_e'''(w)^{\frac{1}{\lambda}} \tag{4}$$

Note that *adjusting* the unary potentials primarily affects only the recognition of appearances associated with an event e. Thus, we also propose to influence how we establish the graph's edges (recall that these reflect dependencies among nodes and thus appearances). Let $s(a)$ be a vector signifying an appearance's similarities with every other appearance (as used when we find the η closest matches among all appearances as outlined in Section 5). Since we obtain a preliminary recognition result R in our initial step, we are able to *adjust* $s(a)$ w.r.t. the label prediction w of the appearances. We perform this step for every appearance vector $s(a)$ associated with a given event e. Similarly to before, we also incorporate a regularization parameter λ:

$$s'(a_w) = s(a_w)H_e'''(w)^{\frac{1}{\lambda}} \; \forall \; w, a_w \in a \tag{5}$$

8   Experiments

8.1   Dataset

Typical face datasets like *FERET* or *LFW* are not suitable for our aim of recognition in Consumer Photo Collections because they usually only contain single faces stripped of their contexts. The public *Gallagher Dataset* does include family photos (personal photos centering around a single person or family); however, it contains only 589 photos captured over a shorter time span.

To better demonstrate our data mining driven approach, we compile a more challenging dataset that spans a significantly longer time period and contains notably more photos and face appearances. Similarly to the *Gallagher Dataset*, all photos are shot in an uncontrolled environment with a typical consumer camera. Many of the photos depict the main subjects, a couple and their friends and family, in a broad variety of settings and scenes both indoors and outdoors (see Figure 1, right side).

The dataset depicts a total of 56 different individual people over a period of nearly ten years. Altogether, there are roughly 3000 face appearances spread over approximately 2200 photos. Most individuals appear at least ten times in total. Note that infants and children appear quite seldom; most people are grown-ups. The number of distinct people appearing together throughout the events (e.g. when we consider an event to last around one day) seems to be mostly under ten, but is in several cases more than 15.

The ground truth specifies the true face boundaries along a unique class label. All photos include EXIF metadata as embedded by the camera, and it is thus possible to extract the time of capture.

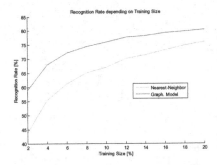

Configuration	Events	Incorr.	Correct
Default	93.85	109.98	80.41
$s_{min} = 3$	79.81	107.39	81.60
$n_{min} = 2$	15.19	23.90	15.76
$n_{min} = 2$, relax.	89.42	88.96	77.27
relaxed	99.62	103.70	79.11
$\nu_{min}^m = 0.1$	81.15	115.15	82.68
$\nu_{min}^i = 0.1$	96.92	111.87	78.02
$\alpha = 1.4$	69.23	82.46	84.75
$\alpha = 1.8$	64.04	62.51	87.75

Fig. 2. Left: Basic recognition performance (without using mined appearances) depending on training set size. Right: Evaluation of the intermediate result H depending on different parameter configurations (all values are in percent).

8.2   Implementation Details and Setup

We refer the reader to our previous work [2] for details on Sections 4 and 5. Note, however, that in this work we do not consider any of the social semantics that [2] introduces except people's uniqueness within photos. We are interested in evaluating the effectiveness of our graph-based approach that considers social semantics gained through a data mining technique. Thus, we first compare our basic graph-based approach against the traditional nearest-neighbor method. Then, we evaluate the intermediate outcome of the proposed data mining technique (separately) against the ground truth as well as the impact of the overall approach when incorporating the gained data mining results. Lastly, to avoid being influenced by the face detection method, we only consider correctly detected faces as verified against the ground truth; then: $D' \subseteq D$. Our primary measure is then the 1-hit recognition rate.

For all experiments, we split the dataset D' into a training set D^{Tr} and a testing set D^{Te}, such that D^{Tr} represents a small random but stratified subset of D'. By default, we use a rather small training set size of 3% (with at least three samples of each label class). We repeat all experiments five times and average the results. If not otherwise mentioned, we base all experiments on the following default parameters and configuration: MAP-based graph inference with $\tau = 1.2$; $\eta = 3$; $\varepsilon = 0.05$; $\hat{\delta} = 10$ hours; $s_{min} = 2$; $\nu_{min}^m = \nu_{min}^i = 0$; $n_{min} = 1$; $m_{min} = 0.67$; $m_{max} = 0.33$; $f_{min} = 2$; $\alpha = 1.0$; and $\lambda = 25$.

8.3   Results and Evaluation

Face Detection and Recognition. Given D, the Viola-Jones face detection method we utilize correctly detects 2498 faces (*true positives*) with respect to the ground truth. Recall that for D' we retain only correctly detected faces. The left plot of Figure 2 illustrates the 1-hit recognition rates for varying training set sizes for the baseline nearest-neighbor matching approach as well as our graph-based approach. We notice notably better results for the latter, especially when the training set size is small (in that case, up to 15% gain).

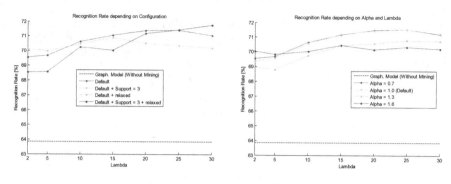

Fig. 3. Final recognition rate depending on λ, configuration (left) and α (right)

Mining People's Appearances. We find 104 temporal cluster sets (or events) by applying Mean-Shift on the photos' capture timestamps. Based on our preliminary recognition results (as outlined in Section 6, Step 1), we are able to mine between 287 ($s_{min} = 3$, $n_{min} = 2$, $\nu_{min}^m = 0.1$) and 98781 ($s_{min} = 2$, $n_{min} = 1$, $\nu_{min}^m = 0$) appearance patterns on average (recall that we average the results over five runs) depending on the various parameter combinations. Using either $s_{min} = 3$ or $n_{min} = 2$ greatly reduces the number of patterns to below 6500. Note that we limit our experiments to following *suitable* values: $s_{min} = \{2, 3\}$; $n_{min} = \{1, 2\}$; and $\nu_{min}^m = \{0, 0.05, 0.1\}$.

Incorporating Mining Results. Next, we evaluate the intermediate result H. In the right table of Figure 2, we list for how many events we can find a matching appearance pattern (where 100% reflects the optimum result). Moreover, we compare applicable patterns (that match an event) against the ground truth w.r.t. the amount of individuals correctly predicted (we desire high precision values) as well as the amount of incorrectly predicted individuals (we desire low values). Due to lack of space, we only list some parameter combinations where *relaxed* denotes the combination ($m_{min} = m_{max} = 0.5$).

We notice that higher values for α and s_{min} lead to more correctly predicted individuals (and, likewise, less incorrectly predicted individuals), but they also lead to fewer matching events. For example, we are able to find at least one matching pattern for about 94% of all events using our default configuration, but only for about 80% when using a higher support s_{min} of 3. In both cases, we predict about 80% of the people collectively appearing in the events. We also see that ν^{min} and n_{min} do not show much positive effects.

The two plots of Figure 3 show the final recognition performance when incorporating the mined appearance patterns. We achieve the best performance (about 71%) when α is between 0.7 and 1.2 and λ between 15 and 25. We notice that as long as many events are matched to an appearance pattern, the mentioned parameters are quite insensitive w.r.t. the performance. Looking only at matched events, we notice recognition gains of nearly 20% (e.g. with α of 1.8, where we match only 64% of the events, but these with a higher *precision*). Overall, we are able to improve recognition performance by around 8% and 22% when compared to our basic graph-based approach and a traditional nearest-neighbor method, respectively.

Lastly, without detailing complexity, we list the following final single-core measures (on a AMD X6 2.6 GHz): Viola-Jones face detection takes about 0.2 seconds per photo; RElim-based frequent itemset mining takes 1-2 seconds; non-optimized event matching against the mining results (as described in Section 7) takes 2-3 seconds; and inference over the entire dataset finishes in 2-3 seconds.

9   Conclusion

We present a framework to detect and recognize people in Consumer Photo Collections. We incorporate context in the form of social semantics to further improve recognition performance. To do so, we devise a method that utilizes a data mining technique to discover and incorporate patterns of groups of people frequently appearing together in photos. We conclude that by considering such social semantics, we are able to notably improve recognition performance. For future experiments, we intend to additionally discover people's appearance patterns with respect to time (e.g. a family that meets every Christmas) and possibly location (e.g. given geo-tagged photos).

Acknowledgements. This work is partially supported by EU project CUbRIK (grant FP7-287704).

References

1. Gallagher, A.C., Chen, T.: Using group prior to identify people in consumer images. In: CVPR, pp. 1–8 (2007)
2. Brenner, M., Izquierdo, E.: Graph-based Recognition in Photo Collections using Social Semantics. In: MM SBNMA, pp. 47–52 (2011)
3. Gallagher, A.C., Chen, T.: Understanding images of groups of people. In: CVPR, pp. 256–263 (2009)
4. Lin, D., Kapoor, A., Hua, G., Baker, S.: Joint People, Event, and Location Recognition in Personal Photo Collections Using Cross-Domain Context. In: Daniilidis, K., Maragos, P., Paragios, N. (eds.) ECCV 2010, Part I. LNCS, vol. 6311, pp. 243–256. Springer, Heidelberg (2010)
5. Stone, Z., Zickler, T., Darrell, T.: Toward Large-Scale Face Recognition Using Social Network Context. IEEE 98(8), 1408–1415 (2010)
6. Singla, P., Kautz, H., Luo, J., Gallagher, A.: Discovery of social relationships in consumer photo collections using markov logic. In: CVPR Workshops, pp. 1–7 (2008)
7. Zhang, T., Chao, H., Tretter, D.: Dynamic estimation of family relations from photos. In: AMM, pp. 65–76 (2011)
8. Zhang, T., Chao, H., Willis, C., Tretter, D.: Consumer image retrieval by estimating relation tree from family photo collections. In: ICMR, pp. 143–150 (2010)
9. Wang, G., Gallagher, A., Luo, J., Forsyth, D.: Seeing People in Social Context: Recognizing People and Social Relationships. In: Daniilidis, K., Maragos, P., Paragios, N. (eds.) ECCV 2010, Part V. LNCS, vol. 6315, pp. 169–182. Springer, Heidelberg (2010)
10. Ahonen, T., Hadid, A.: Face description with local binary patterns: Application to face recognition. PAMI 28(12), 2037–2041 (2006)

Person Re-identification by Local Feature Based on Super Pixel

Cheng Liu[1] and Zhicheng Zhao[1,2]

[1] Multimedia Communication and Pattern Recognition Labs
[2] Beijing Key Laboratory of Network System and Network Culture,
Xitucheng Road No. 10, Haidian District, Beijing, China
{chengcic,zhaozc}@bupt.edu.cn

Abstract. In many multi-camera surveillance systems, there is a need to identify whether a captured person have emerged before over the network of cameras. This is the person re-identification problem. In this paper, we propose a novel re-identification method based on super pixel feature. Firstly, local C-SIFT features based on super pixel are extracted as visual words, and appearance details are used to describe detecting objects. Secondly, a TF-IDF vocabulary index tree is built to speed up person search. Finally, an image-retrieval way is adopted to implement person re-identification. Experimental results on ETHZ dataset show that our method is better than the approach proposed by Schwartz et.al and two machine learning methods based on SVM and PCA.

Keywords: person re-identification, super pixel, visual word, vocabulary tree.

1 Introduction

With widely application of multi-camera video surveillance system, a need of matching the same person in different camera view fields emerged. This matching problem, called person re-identification which aims to identify person separated in time and locations, has attracted great interest in recent years. However basically person re-identification is still a difficult open question. It faces several challenges below: 1) Low resolution of surveillance videos which makes re-identification through biometric clue, for example, face detection and recognition, intractable. 2) Varying illumination conditions and different camera parameters leading to inconstant representation of one same individual. 3) Pose variability of persons, together with camera viewpoint differences, resulting in partial consistence of captures of the same person, aggravates the difficulty of re-identification task.

To address these challenges, a lot of works [1, 2, 3, 4, 5, 8, 13, 17] have been done in the past years. Most of studies turn to resort appearance modeling of a person instead of biometric detection. Most of them follow similar frame-work: firstly, extract features to represent person's appearance, and then build a

S. Li et al. (Eds.): MMM 2013, Part I, LNCS 7732, pp. 196–205, 2013.
© Springer-Verlag Berlin Heidelberg 2013

similarity metrics to measure the distance of two objects, finally get the rank of each compared person. Therefore, state-of-art works can be roughly divided into two groups: the first group emphasizes the invariable and stable feature representation of persons, and tries to get kinds of features invariant to illumination, viewpoint and pose variety [1, 2, 3, 4, 8, 13], while another group concentrates their efforts on exploring similarity metrics so as to obtain a higher similarity score for the same person than the others [5, 17].

By examining the methods of researches regarding invariant feature representation, we find that they are mostly dependent on global appearance of a person, such as weighted color histogram [13], dominant color descriptor [18]. color-position histogram [5]. The dominant characteristic in person appearance would be efficiently represented by using such features, while some details of appearance, such as a patch of logo on the clothes, can not be effectively described.

Local feature method was used in limited literatures, although there are a lot of works applying it in image searching and representing. In [10], Harris corner detector is adopted to localize the local feature point, and then person re-identification methods based on SIFT [9] were proposed. Jüngling et.al proposed a re-identification approach by combining SIFT feature and appearance codebook scheme. SIFT feature is used for three tasks: person detection, tracking and re-identification. With the help of appearance codebook, similarity measurement is implemented in a linear weighed sum manner of each codebook entry. In [8], Gheissari et.al used Hessian affine invariant interest operator to extract local feature. Firstly, the locations of key points are determined by Hessian Affine detector, and then circular support regions centered at the locations are generated. Finally, the paper incorporated a point to point corresponding match method to measure the similarity of two images.

However, different poses of person and viewpoints of cameras are easy to generate interest-point-location drift, as a result, the mismatching often appears. To avoid the drawback of global and local feature, we propose a novel re-identification method based on local feature of super pixel [11].

In this paper, we formulate the person re-identification problem as an image searching problem using local feature based on super pixel as visual words. According to the super pixel algorithm, a person can be segmented into patches where local features are extracted. Inspired by Sivic and Zisserman [7], who perform image search in terms of text retrieval, we regard these local features as words of a person's appearance. By incorporating vocabulary trees schemes [6], our person re-identification method is fulfilled in a more efficient way.

Our contributions can be summarized as following aspects: 1) Instead of using local points detector, we incorporate super pixel segmentation to generate visual patches as interest regions for local feature, alleviating the influence of interest point shift mentioned before under the situation that person's gesture and illumination condition changed. 2) Visual word TF-IDF scheme are adopted, making our re-identification procedure a text-retrieval-like way. 3) Vocabulary tree's using speeds up the person search process.

The paper is organized as follows. In Sec. 2 the detail of our approach is described. The result of our approach is represented in Sec 3. In Sec. 4, we sum up our approach and envisage the future perspectives.

2   The Approach

As we use vocabulary tree as our matching scheme, our approach can be divided into two stages, offline training stage, online indexing and searching stage. The whole process of the approach is: we firstly implement the training stage, involving the feature extraction and vocabulary tree building process, to form the basic structure of our searching system - a tree which contains the visual words of the person appearance; then by using the vocabulary tree, we index the image of person to the tree, in a scheme similar to text retrieval inverted file regarding each local feature as a visual word; and when we meet a query to reidentificate a person, a TF-IDF scheme is incorporated to fulfill the scoring and matching task. In sum, these stages are made up of three processes: feature extraction, vocabulary tree building and person matching. The implementation of them would be detailed below.

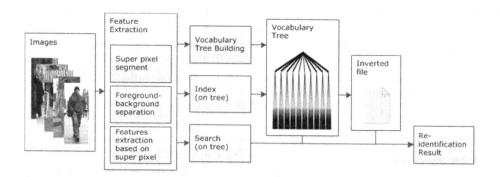

Fig. 1. An overview of our framework

2.1   Feature Extraction

The super pixel algorithm is first used to segment the image of person into several patches. Then combining the super pixel segmentation and the result of STEL component analysis [12], we get the foreground of a person separated from the scenario background noise. Local feature based the super pixel patches can be extracted finally.

Super Pixel Segment: Super pixel is an over-segmentation preprocess for segmenting a specific item task. Because it uses contour and texture cues, the boundary of the person's silhouette and the edge of the person's different parts are well detected (See Fig. 2) and the whole appearance of the person is segmented into visual patches. In such case, the same individual's different images would share most amounts of similar visual patches, making the super pixel segment feature a reasonable cue for matching person. To alleviate the influence of different image scales, the size of super pixel is adjusted according to the resolution of images to process - large scale image generates large size super pixels, small generates small - ensuring the number of the super pixels is stable regardless of the scale of the image. In our experiment the number of super pixels in each image is stable around 30.

Fig. 2. An example of super pixel segmentation result. From left to right: original image, image segmented, visual patches generated by segmentation.

Foreground-Background Separation: Without doubt, the visual patches in the background generated in the super pixel segmentation process would certainly influence the matching procedure. Background should be discarded. In [13], STEL component analysis is used to generate silhouette mask. However the masks generated are coarse, which incorporate some background scenario and discard some foreground details. To refine the mask, the results of STEL component analysis are mapped to the super pixel segmentation structure and the proportion of the masked pixels in each super pixel is calculated. Regarding every super pixel as basic element of foreground and background, we classify super pixels into two groups, foreground and background, according to the masked pixel proportion mentioned before (See Fig. 3). Because super pixel segmentation performs well in the silhouette-background separation, the extracted foreground mask is improved and more exact compared with STEL mask. We set an empirical threshold 0.8, which means that if more than 80% pixels in a super pixel are masked pixels, the super pixel would be classified as the foreground.

Features Extraction Based on Super Pixel: After segmenting a person into visual patches, low-level visual features are extracted to represent these patches. Usually, two visual cues are used to represent the appearance of one person: color and texture. Thus color features such as HSV histogram [13], dominant color descriptor [18] and texture features like Haar, LBP can be used to represent the visual patches. In our approach, we adopt color SIFT descriptor, to represent appearance. In [14], several color SIFT descriptors are introduced

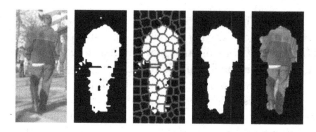

Fig. 3. An illustration of foreground-background separation, from left to right the images are original person image, mask generated by STEL analysis, result of mapping the STEL mask to super pixel segmentation, refined mask and separation result image

and their performances for object and scene recognition are evaluated. C-SIFT outperforms other descriptors for its color invariance to light intensity changes, making it a preferred descriptor candidate for describing person appearance in different illumination conditions. Ensuring in every super pixel region C-SIFT points can be extracted, we adopt dense sampling method rather Harris-Laplace detector for keypoint detection. By averaging all C-SIFT points in a super pixel, the feature based super pixel can be got. As mentioned before, every image of person generates about 30 super pixels, thus there would be about 30 local features extracted for each person image.

2.2   Off-line Vocabulary Tree Building

Recently encouraging results have been achieved in the field of searching large-scale image database by using TF-IDF (term frequency-inverse document frequency) schemes and inverse voting mechanism similar to text retrieval [6, 7]. In this paper, we can utilize this text-retrieval-like scheme by regarding each local feature as visual word.

Fig. 4. A three-level vocabulary tree with factor 10

Vocabulary tree is a tree-like structure whose nodes are visual words of vocabulary. Two attributions define a vocabulary tree: branch factor (number of children of each node) and depth. By organizing the visual words in such tree structure, visual words' search procedure becomes more efficient compared with non-hierarchical visual word vocabulary.

In order to train a vocabulary tree, we extract a large amount of features from training data. The selection of training data is important, since visual words in vocabulary tree ought to occupy the possible feature space of appearance. Whether the training images populate the feature space will determine the representativeness and the performance of the vocabulary tree. On the basis of extracted features, we should determine the branch factor and depth of the tree, and then, a vocabulary tree can be trained by implementing the hierarchical K-means clustering algorithm. Supposing a tree's depth is L and branch factor is K, it's leaf node number would be K^L, which means the number of visual words in the vocabulary is K^L. The vocabulary tree plays a core role in our re-identification task. This offline training process is implemented only once in the whole procedure.

2.3   On-line Person Matching

Once the vocabulary tree is built, we can implement the person matching procedure. Our matching procedure adopts TF-IDF scoring scheme and can be divided into two parts: index and search.

TF-IDF Scheme: TF-IDF is often used as a weighting factor in text retrieval. TF is the term count in the given document which refers to the number of a specific term or word in such document. IDF is an indicator whether a term is common or rare across all documents.

In our approach, we regard the leafage of vocabulary tree as visual words similar to the term in the TF-IDF scheme. Through propagating the local feature down to the leafage of tree, each image of person would be represented by the visual words. By propagating each local feature in the image database to the leafage, we get the number of local features attach to each leaf node, thus the entropy weight of the node can be calculated by Eq. (1)

$$w_i = \log(N/n_i) . \tag{1}$$

where N is the number of visual features in the database, n_i is the number of occurrences of term i in the whole database. In order to ensure the equation valid, n_i would be equal to the number of occurrences of term plus one in case that terms doesn't exist in the database.

Index: Index procedure aims to form an inverted file attach to the leafage of the vocabulary tree. For each image, we propagate the features to the leafage to get the visual words. In the inverted file, the image would be added to every visual word corresponding to it. Repeat the procedure above to every image in the database, the inverted file of the whole database would be formed. After setting up index, the weight of the node, visual word, can be calculated by Eq. (1).

Search: If we want to reidentificate a person, search procedure will be implemented. Firstly, the features of the query image would be propagated to the leafage of the tree to get the visual words. Secondly, every visual word would be mapped to a list of images sharing the same word with the query. Thirdly, by multiplying the TF, the occurrence frequency of one specific visual word in the query image, and IDF, the weight of that visual word, the contributing score of that visual word and corresponding images of the visual word can be got. Fourthly, add all the scores to form a image-score list to the query image. Finally, the search result would be got by sorting the images in decreasing order of their scores.

The index and search procedure can be implemented online. An additive or removal action of image of the database is feasible under the framework of our approach. All you need to do is altering the inverted file and the weight of visual word.

3   Experiments and Analysis

In this section, we evaluate our approach by implementing the re-identification task on public ETHZ [15, 16] dataset. The result of re-identification is showed in terms of Cumulative Matching Characteristic (CMC) curve, which represents the expecting position of correct match in the searching result. To evaluate the influence of different vocabulary sizes on re-identification performance, we compare the experiment results under different vocabulary trees.

3.1   Dataset

The ETHZ dataset contains three video sequences captured from moving cameras. These three sets of images are named as ETHZ1, ETHZ2 and EHTZ3 in our experiment. Because the camera is moving when capturing video, the images representing the same person have a wide range of variation in person appearance. ETHZ1 contains 83 pedestrians with 4875 images; ETHZ2 contains 35 pedestrians with 1936 images; ETHZ3 contains 28 pedestrians with 1762 images.

3.2   Vocabulary Tree

In [6], experimental results show that larger vocabulary would generate better performance. Therefore, we train several vocabulary trees with different vocabulary sizes to compare the re-identification results. For each dataset, we train three vocabulary trees. The numbers of tree leaf nodes are 100, 1000 and 10000 respectively.

Fig. 5 shows the result of our experiment on ETHZ dataset. It clearly shows that with the growth of vocabulary size, the CMC curve also rises up, indicating a better re-identification performance. This may attribute to the fact that compared with the larger vocabulary tree, the small one is coarsely quantified in

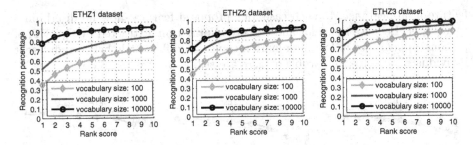

Fig. 5. Performance of ETHZ dataset. The experiment is carried out using vocabulary trees with different vocabulary sizes. With growth of vocabulary size, the better re-identification result is got.

visual words, leading the visual words indistinctive. In a small vocabulary tree, unrelated images sharing the same visual words with the query image easily get involved into searching procedure acting as noises, influencing the whole re-identification procedure. However too large vocabulary size would generate too small quantization cell of visual words, and meanwhile, some visual words that should have be matched are assigned into different quantization cells resulting in mismatch. In our experiment, when the vocabulary tree grow to 10000-leafage scale, some image queries get no corresponding match in the retrieval results while the CMC curve shows a better performance than any other vocabulary tree. This is a tradeoff between performances (high accuracy) versus reliability (low miss rate).

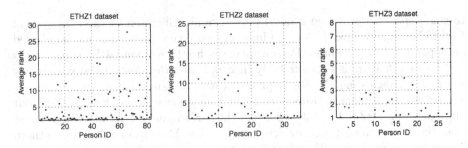

Fig. 6. Average rank of each individual's search result in the ETHZ dataset (using result of 10000-vocabulary-size)

An overall examination on the CMC curves of ETHZ dataset reveals that about 90% correct matches of person rank top 10 in the search result. However, the performance differs in different person model. We average the rank of search result of each person. As illustrated in Fig. 6, person 1, 2, 12 in ETHZ1 enjoy a high re-identification performance, while 4, 81 in ETHZ1 perform poorly in the task. By examining the images of 4, 81 and their search results, we find that the images are captured in low illumination condition and have little discriminative feature in appearance. In such case, our proposed feature show no discriminatory power. Because the color and texture information, extracted by C-SIFT, of persons' appearance in low illumination condition differs little between different individuals. To solve this problem there are still a lot of work to do.

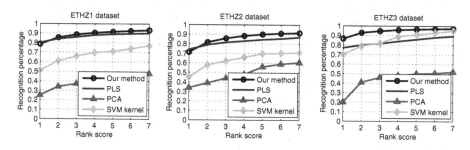

Fig. 7. Performance comparison considering our proposed method (using 10000-vocabulary-size), PCA, SVM kernel and PLS [15]

Fig. 7 shows a performance comparison among our method, PLS method [15] and two machine learning methods of PCA, SVM also reported in [15]. We can see that the recognition rate of our super-pixel-based approach is higher than the other methods, indicating that feature based super pixel has discriminative ability and is effective in person re-identification task.

4   Conclusion

In this paper, we proposed a novel person re-identification method using local feature based super pixel. A novel local appearance representing method based on super pixel is presented. Person image is firstly segmented by super pixels and C-SIFT feature is extracted. By incorporating the visual word vocabulary scheme and regarding the super pixel feature as visual words, the person re-identification problem is transformed as an image search problem. We test our approach on ETHZ dataset showing that the performance of our method is better than the approach proposed by Schwartz et.al, SVM method and PCA method. These experiments show the effectiveness of our proposed method.

Acknowledgments. This work is supported by National Natural Science Foundation of China under Projects 90920001, 61101212, and by Fundamental Research Funds for the Central Universities.

References

1. Gray, D., Tao, H.: Viewpoint Invariant Pedestrian Recognition with an Ensemble of Localized Features. In: Forsyth, D., Torr, P., Zisserman, A. (eds.) ECCV 2008, Part I. LNCS, vol. 5302, pp. 262–275. Springer, Heidelberg (2008)

2. Madden, C., Cheng, E., Piccardi, M.: Tracking people across disjoint camera views by an illumination-tolerant appearance representation. Machine Vision and Applications 18(3), 233–247 (2007)
3. Park, U., Jain, A.K., Kitahara, I., Kogure, K., Hagita, N.: ViSE: Visual Search Engine Using Multiple Networked Cameras. In: 18th International Conference on Pattern Recognition, ICPR 2006, pp. 1204–1207 (2006)
4. Cai, Y., Pietikäinen, M.: Person Re-identification Based on Global Color Context. In: Koch, R., Huang, F. (eds.) ACCV Workshops 2010, Part I. LNCS, vol. 6468, pp. 205–215. Springer, Heidelberg (2011)
5. Truong Cong, D.N., Khoudour, L., Achard, C., Meurie, C., Lezoray, O.: People re-identification by spectral classification of silhouettes. Signal Processing 90(8), 2362–2374 (2010)
6. Nister, D., Stewenius, H.: Scalable Recognition with a Vocabulary Tree. In: 2006 IEEE Computer Society Conference on Computer Vision and Pattern Recognition, pp. 2161–2168 (2006)
7. Sivic, J., Zisserman, A.: Video Google: a text retrieval approach to object matching in videos. In: Proceedings of the Ninth IEEE International Conference on Computer Vision, pp. 1470–1477 (2003)
8. Gheissari, N., Sebastian, T.B., Hartley, R.: Person Reidentification Using Spatiotemporal Appearance. In: 2006 IEEE Computer Society Conference on Computer Vision and Pattern Recognition, pp. 1528–1535 (2006)
9. Lowe, D.G.: Distinctive Image Features from Scale-Invariant Keypoints. International Journal of Computer Vision 60(2), 91–110 (2004)
10. Jüngling, K., Arens, M.: Local Feature Based Person Reidentification in Infrared Image Sequences. In: 2010 Seventh IEEE International Conference on Advanced Video and Signal Based Surveillance (AVSS), pp. 448–455 (2010)
11. Ren, X., Malik, J.: Learning a classification model for segmentation. In: Proceedings of the Ninth IEEE International Conference on Computer Vision, pp. 10–17 (2003)
12. Jojic, N., Perina, A., Cristani, M., Murino, V., Frey, B.: Stel component analysis: Modeling spatial correlations in image class structure. In: IEEE Conference on Computer Vision and Pattern Recognition, CVPR 2009, pp. 2044–2051 (2009)
13. Farenzena, M., Bazzani, L., Perina, A., Murino, V., Cristani, M.: Person re-identification by symmetry-driven accumulation of local features. In: 2010 IEEE Conference on Computer Vision and Pattern Recognition, pp. 2360–2367 (2010)
14. van de Sande, K.E.A., Gevers, T., Snoek, C.G.M.: Evaluating Color Descriptors for Object and Scene Recognition. IEEE Transactions on Pattern Analysis and Machine Intelligence 32(9), 1582–1596 (2010)
15. Schwartz, W.R., Davis, L.S.: Learning Discriminative Appearance-Based Models Using Partial Least Squares. In: 2009 XXII Brazilian Symposium on Computer Graphics and Image Processing (SIBGRAPI), pp. 322–329 (2009)
16. Ess, A., Leibe, B., Van Gool, L.: Depth and Appearance for Mobile Scene Analysis. In: IEEE 11th International Conference on Computer Vision, ICCV 2007, pp. 1–8 (2007)
17. Wei-Shi, Z., Shaogang, G., Tao, X.: Person re-identification by probabilistic relative distance comparison. In: 2011 IEEE Conference on Computer Vision and Pattern Recognition, pp. 649–656 (2011)
18. Ba, S., et al.: Person Re-identification Using Haar-based and DCD-based Signature. In: Advanced Video and Signal Based Surveillance (AVSS), pp. 1–8 (2010)

An Effective Tracking System
for Multiple Object Tracking in Occlusion Scenes

Weizhi Nie[1], Anan Liu[1,*], Yuting Su[1], and Zan Gao[2]

[1] School of Electronic Information Engineering,
Tianjin University, Tianjin, China 300072
[2] Key Laboratory of Computer Vision and System, Ministry of Education,
Tianjin University of Technology, Tianjin, China 300072
{truman.nie,zangaonsh4522}@gmail.com, {liuanan,ytsu}@tju.edu.cn

Abstract. In this paper, we propose an effective multi-object tracking system which can handle the partial occlusion in the tracking process. First, this method employs the part-based model to localize the person and body parts in every frame. Then it leverages the motion characteristics of both parts and the entire body to generate the trajectories of individuals. To overcome the difficulty in partial occlusion, we propose to formulate the task of multi-object tracking into multi-object matching with body part cues. The large scale comparison experiment on the popular tracking datasets demonstrates the superiority of the proposed method.

Keywords: multiple object matching, partial occlusion, part-based model, linear programming.

1 Introduction

The task of multiple object tracking is to automatically recognize and localize persons in the dynamic scene and simultaneously record the history of individual's movement. This technique plays an important role in the field of video surveillance. Several factors make this problem rather challenging, such as the irregular changes of human shape, partial occlusion, and the variation of illumination in the surveillance scene. Many algorithms on this problem have been proposed in the last decades [1,2]. The state-of-the-art methods can be roughly divided into three categories. First, several multi-person tracking algorithms [3-5] implement the stereo or color blob representations for the tracking targets. The recovered blobs can then be statistically characterized by shape and color features for tracking. Although this method has good performance in some special scenes, these features are too weak to provide reliable tracking results under complex conditions. Second, several methods [6-11] adopts a tracking-by-detection framework by associating the detection candidates with spatiotemporal constraints. The object detectors are firstly implemented to localize individuals in each frame or generate tracklets. Then the detected candidates are

---

* Corresponding author.

S. Li et al. (Eds.): MMM 2013, Part I, LNCS 7732, pp. 206–216, 2013.

linked within the spatial and temporal neighborhoods. Huang et al. proposed to generate tracklets by conservative grouping the detection results [7]. Zhang [11] proposed a minimum-cost flow network to resolve the global data association approaches of multiple objects. Although this kind of methods can overcome the negative influence by template drift, it does not work well enough when facing the highly dense crowds because of and the existence of partial occlusion. Third, some tracking systems design filter models to predict statues of human between frames [12-14]. Although the filter models can leverage the prior knowledge of trajectory for prediction, they usually fail when drifting or occlusion happens due to the incapability of adaptive correction of the tracking results as the tracking-by-detection framework does.

To tackle this problem, we propose an effective multi-object tracking system which can handle the partial occlusion. First, the proposed method employs the part-based model to localize the person and body parts in every frame. Then it leverages the motion characteristics of both parts and the entire body to generate the trajectories of individuals. To overcome the difficulty in partial occlusion, we propose to formulate the task of multi-object tracking into multi-object matching with body part cues. At last, it updates the trajectory with both results of detection and tracking. Comparing with the current track-by-detection methods using final detection windows to represent the feature of human body and using online learning to track the corresponded object, the proposed approach employs a set of detected parts to represent the feature of global object and looks for the corresponding object in the consecutive frames to generate the final trajectory of the individual. The major contribution lies in two-folds: 1) the set of detected parts can provide a better description of the articulated body than the global detection window does because the part-based detection excludes most of the background in the final detection window and avoids the influence of the background changes. Moreover, the part-base representation of human can facilitate to handle the occlusion which always brings in much difficulty for multiple object tracking. 2) the proposed multi-object matching algorithm can automatically links the corresponding objects with part cues and spatiotemporal constraints for the entire trajectory of one object in one shot when facing partial occlusion while skipping the tracklet generation and association in the traditional tracking-by-detection framework.

The remainder of this paper is organized as follows. Section 2 will introduce the system framework. Section 3 will detail the key techniques of the system. The experimental results are illustrated in Section 4. We will conclude this paper in Section 5.

2   System Framework

The proposed multiple person tracking system includes five modules as show in Figure 1 : (1) Detector, which detects and localizes candidate persons in the input image sequences; (2) Predictor, which inputs the candidate person regions by the detector in the T frame and outputs the predicted region in the T+1 frame; (3)

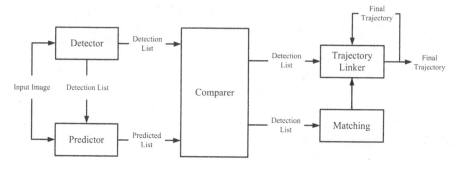

Fig. 1. System overview

Comparer, which compares the detection person regions by the detector in the T+1 frame with the predicted regions by the predictor in the T+1 frame and then outputs the overlapping score. The existence of confusing objects would be decided with the overlap score. If confusing objects appear in the current frame, comparer will output the detection list to matching module. Otherwise, the detection list will be outputted to trajectory linker. (4) Matching, which inputs the detection list by comparer module, and outputs the reliable trajectory segments to the trajectory linker; (5) Trajectory linker, which oversees the entire tracking history and updates individual trajectory.

This tracking system starts by processing the input image sequences. The output of this system is a complete trajectory for individual person. For every tracking object, the system will record the edge, color and texture features for every person in each frame. We would like to provide an overview of the system workflow as follow.

Step1: Initializing the trajectory list for every detected person in the first frame with the detected results by the part-based detector. System utilizes the bounding box to represent the region and location for the entire body of each person while maintains the region and location information of eight body parts. Therefore, 9 regions and 18 coordinates are used to represent every detected person. At the same time, every person is initialized by a unique track ID.

Step2: Detector module detects the persons in every frame and simultaneously records region and position for every person. The output is also 9 regions and 18 coordinates to represent every detected person in the current frame.

Step3: Predictor module predicts the corresponding regions for every track ID in the next frame by the well-known mean-shift tracking algorithm. Especially it will not only predict the region of the entire body in the next frame but also predict the part regions in the next frame. Therefore, the output of this module also outputs 9 regions and 18 coordinates for each person.

Step4: Comparer module receives the results of both the predictor and the detector to compute the overlap rate for every tracking object. The score function can be formulated as:

$$p = \sum_{i=1}^{9} \frac{A_D(i) \bigcap A_P(i)}{A_D(i) \bigcup A_P(i)})$$ (1)

where i is the part region ID; A_D is the region by detector; A_P is the region by predictor; p is the overlap rate for individual person in the current frame. If p is higher than the preset threshold T, system regards the detected region as the same person and then outputs the detection list to trajectory linker module. Otherwise, detection list will be outputted to the matching module especially in the partial occlusion case.

Step5: Matching module receives the detection list from the comparer and extracts features of every object in the current frame to match them to the candidates in the continuous frames within a temporal window until the matching score is above T. The output is a set of reliable linkages between the candidates.

Step6: Trajectory linker connects the detection list from comparer module or trajectory segments from matching module with the original trajectory if both are within a preset spatiotemporal window and the missing trajectory segments in-between caused by spatiotemporal matching can be reconstructed by interpolation. The new trajectory will also be fed back to trajectory linker for updating if necessary.

The next section will elaborate the key techniques of the system, detector module and matching module.

3   Method

3.1   Part-Based Human Detection

Part-based human detection can localize the body parts of one person by dynamic programming with the visual features [15]. In the train processing, we firstly train a root filter for the entire body region and then train the part filters within the region of the root filter with the latent-SVM.

Latent-SVM is a special SVM^{struct}[15], predicts complex objects y like trees, sequence, or sets. The SVM^{struct} algorithm can also be used for linear-time training for binary and multi-class SVMs under the linear kernel. Consider a classifier that scores an example x with a function of the form,

$$f_\beta(x) = \max_{z \in Z(x)} \beta \cdot \Phi(x, z) \tag{2}$$

where β is a vector of model parameters and z are latent values. The set $Z(x)$ defines the possible latent values for an example x, a binary label for x can be obtained by thresholding its score.

In analogy to classical SVMs we train β from labeled examples $D = ((x_1, y_1), (x_2, y_2), \ldots\ldots(x_n, y_n))$, where $y_i \in \{-1, 1\}$, by minimizing the objective function,

$$L_D(\beta) = \frac{1}{2} \| \beta \|^2 + C \sum_{i=1}^{n} \max(0, 1 - y_i f_\beta(x_i)) \tag{3}$$

where $\max(0, 1 - y_i f_\beta(x_i))$ is the standard hinge loss and the constant C controls the relative weight of the regularization term.

If there is a single possible latent value for each example $(|z(x_i)|=1)$ and f_β is linear in β, we will own an linear SVMs. Therefore, we said that latent SVMs is a special SVMs algorithm.

In the detection step, part-based detectors are implemented on the detected region by root filter to localize each human part. The final score of object is computed by the responses of all the part filters and the root filter. The detection score at location (x_0, y_0) can be formulated as follows:

$$score(x_0, y_0) = b + \sum_{i=1}^{i=n} s(p_i) \tag{4}$$

where b is the bias, n is the number of parts, $s(p_i)$ is the score of part i and can be computed as:

$$s(p_i) = F_{pi} \cdot \phi(H, p_i) - d_{pi} \cdot \phi_d(d_x, d_y) \tag{5}$$

where F_{pi} is the part filter, and $\phi(H, p_i)$ denotes the vector obtained by concatenating the feature vectors from H at the sub-window of the part p_i. (dx, dy) is the displacement of the part with respect to its anchor position, $\phi_d(dx; dy) = (dx; dy, d^2x, d^2y)$ represents the deformation features, and d_{pi} specifies the coefficients of the deformation features.

Every part filter can be used to localize the specific body part whenever one person can be detected by root filter and devotes its score to the final decision score. Part-based detection is a process of sliding-windows searching, which has some problems such as speed and precision. For these reasons, we employ a temporal differencing detection algorithm to find moving regions, which are used as potential human regions. This change will improve the speed and precision of part-based detection algorithm. Some human detection results are shown in Figure.2.

3.2   Spatiotemporal Multiple Object Matching Algorithm

In this section, matching algorithm is performed when several objects exist closely, especially when partial occlusion happens, which is named as confusing objects. We use the result of object matching to handle the occlusion problem as shown in the Figure 3.

Let $H = \{h_i\}_{i=1}^{N_i}$ be a hypotheses set of trajectories when occlusion happens and each h_i denotes one candidate trajectory linking by associating the corresponding objects in the continuous detection list to the one in the old detected list. Given the detected object lists, $D = \{d_j\}_{j=1}^{N_j}$, the task of matching can be formulated in an optimization manner to obtain the optimal hypothesis H^* by:

$$H^* = \arg\max_H P(H \mid D) \tag{6}$$

where $P(H \mid D)$ denotes the conditional probability of the trajectory H given the list D.

Fig. 2. Samples of part-based human detection results. Left : human detection result using HOG [18]. Right: human detection result using the part-based human detector. The red boxes show the human detected as full bodies and green boxes show the human detected as part bodies. It is clear that HOG failed to detect occluded human and accurately localize person while our approach achieves significant improvements.

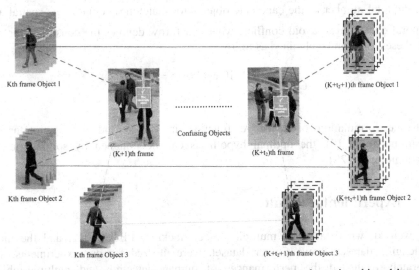

Fig. 3. Sample of matching algorithm: the multi-object correspondence is achieved between (K)th frame to (k+t_2+1)th frame by skipping the confusing objects from (K+1)th frame to (k+t_2)th frame. The trajectory segments in-between can be reconstructed by interpolation.

With the Bayesian rule, the optimization problem in Eq. (6) can be factorized as (in logarithm format):

$$H^* = \arg\max_H \log P(D\,|\,H)P(H)$$
$$= \arg\max_H (\sum_{i=1}^{N_i} \log P(d_j\,|\,H) + \sum_{i=1}^{N_i} \log P(h_i)) \tag{7}$$

In Eq.(7), $P(d_j\,|\,H)$ denotes the probability of the detected object list d_j in hypothesis H and can be computed by referring to [28]. $P(h_i)$ means the probability of one hypothesis trajectory segment. For each hypothesis, h_i consists of two object lists, $d_{h_i}^1$ (the old detection list) and $d_{h_i}^k$ (the continuous detection list). Therefore the key factor of $P(h_i)$ lies in the probability of state translation and can be computed as:

$$P(h_i) \propto \max_i(P(d_{h_i}^k\,|\,d_{h_i}^1)\cdot e^{-|k-l|}) \tag{8}$$

where $P(d_{h_j}^1\,|\,d_{h_j}^k)$ means the transition probability and can be computed with both body and parts cues in Eq.(1); $e^{-|k-l|}$ denotes the decay factor which means that the earlier the matching score reach the preset threshold T, the less decay would affect.

For the tracking problem in partial occlusion, the spatiotemporal constraints can be formulated as follows: Considering that one old list can be only linked to one new list, we reasonably supposing C ($C \in R^{N_i \times 2N_j}$, N_i: the total number of hypotheses h_j; N_j: the total number of the candidate objects for matching) contains the spatial and temporal constraint to avoid conflicts where each row denotes the constraint for one hypotheses. C can be obtained as follows:

$$C(h_j, n) = \begin{cases} 1, & \text{if } n = l \text{ or } n = N_j + k \\ 0, & \text{otherwise} \end{cases}. \tag{9}$$

With the formulation of objective function in Eq.(7) and the spatiotemporal constraints in Eq.(9), the optimal hypothesis can be obtained by solving integer programming as [28].

4   Experimental Results

The well known dataset for multiple object tracking, PETS 2012, and the more challenging dataset, Town Center dataset, were utilized for our experiments. We separately evaluated the performances of human detection and multiple object tracking.

4.1   Evaluation of Human Detection

To improve the accuracy of human detection, we first implemented *Gaussian Mixture Model* [17] to get motion regions in every frame and then apply part-based human detector to detect human body and parts in the foreground. We evaluated the performance of human detection by part-based model and got 98.2% precision and 82.3% recall on the PETS 2012 dataset and 93.44% precision and 71.42% recall on the Town Center dataset.

4.2   Evaluation of Multiple Person Tracking

Referring to [24-27], we extracted the visual features of each part region with 36-D HoG feature [18] and 16-D intensity histogram. We evaluated the proposed method using the standard CLEAR MOT metrics [19], TA (tracking accuracy), DP (detection precision) and DA (detection accuracy).

We tested the proposed method on the well known dataset, S2_L1_View001 in PETS 2012 dataset. The experiment datasets is challenging due to the existence of occlusion, crowded scenes, and cluttered background. We got 72.6% MOTA, 72.8% MODA, and 75.8 MODP.

We also tested the proposed method on the Town Center dataset. The resolution of each frame is 1920×1080 and the frame rate is 25 fps. This dataset contains the street scene with long-term occlusions. We got 72.2% MOTA, 72.6% MODA, and 72.1% MODP. We compared the results with the recently proposed methods [10, 20-23]. With the same experimental setting, the performance of our match is better than others in MOTA and MODA. The experiment results are shown in the Table I. The improvement by our match is due to two factors: First, the part models could improve the recall of detection. Second, the proposed matching algorithm could generate the reliable trajectory for each object.

Table 1. Comparison of Tracking Results on the Town Center Dataset

Criteria	MOTA (%)	MODA (%)	MODP (%)
Benfold et al. [20]	64.8	64.9	**80.5**
Zhang et al. [10]	65.7	66.1	71.5
Pellegrini et al. [21]	63.4	64.1	70.8
Yamaguchi et al. [22]	63.3	64.0	71.1
Leal-Taixe et al. [23]	67.3	67.6	71.6
Our Track Match	**72.2**	**72.6**	72.1

5   Conclusion

In this paper, we present an effective multi-person tracking system which can handle the partial occlusion in the tracking process. The proposed method employs the part-based model to localize the person and body parts in every frame. The motion characteristics of both parts and the entire body are leveraged to generate the

Fig. 4. Example results on the PETS 2012 S2_L1_View001 dataset

Fig. 5. Example results on the Town Center dataset

trajectories of individuals. Especially, the paper originally proposes a spatiotemporal multiple object matching algorithm to handle the partial occlusion. The comparison experiment on PETS 2012 datasets and the Town Center datasets demonstrates the superiority of the proposed method.

Acknowledgement. This work was supported in part by the National Natural Science Foundation of China (61170239, 61100124, 21106095, 61202168, 61172121), in part by the Tianjin Research Program of Application Foundation and Advanced Technology (10JCYBJC25500), and in part by the 2010/2011 Innovation Foundation of Tianjin University.

References

1. Watada, J., Musa, Z.B.: Tracking human motions for security system. In: SICE Annual Conference (2008)
2. Sungmin, K., Chang-Beom, P., Seong-Whan, L.: Tracking 3D Human Body using Partile Filter in Moving Monocular Camera. In: ICPR (2006)

3. Wren, C., Azarbayejani, A., Darrell, T., Pentland, A.: Real-time tracking of the huma body. IEEE Trans. Pattern Analysis and Machine Intelligence 19(7), 780–785 (1997)
4. Intille, S.S., Davis, J.W., Bobick, A.F.: Real-time closed-world tracking. In: Proc. of the IEEE Conf. on Computer Vision and Pattern Recognition, pp. 697–703. IEEE Computer Society Press, Los Alamitos (1997)
5. Krumm, J., Meyers, B., Brumitt, B., Hale, M., Shafer, S.: Multi-camera multi-person tracking for EasyLiving. In: Proc. of the 3rd IEEE Int. Work. on Visual Surveillance (July 2000)
6. Guang, S., Afshin, D., Omar, O., Hand, E., Shah, M.: Part-based Multiple-Person Tracking with Partial Occlusion Handling. In: CVPR (2012)
7. Huang, C., Wu, B., Nevatia, R.: Robust Object Tracking by Hierarchical Association of Detection Responses. In: Forsyth, D., Torr, P., Zisserman, A. (eds.) ECCV 2008, Part II. LNCS, vol. 5303, pp. 788–801. Springer, Heidelberg (2008)
8. Zheng, W., Thangali, A., Sclaroff, C., Betke, M.: Coupling detection and data association for multiple object tracking. In: CVPR 2012 (2012)
9. Breitenstein, M.D., Reichlin, F., Leibe, B., Koller-Meier, E., Van Gool, L.: Online multiperson tracking by detection from a single uncalibrated camera. IEEE Transactions on Pattern Analysis and Machine Intelligence 33(9), 1820–1833 (2011)
10. Fragkiadaki, K., Jianbo, S.: Detection free tracking: Exploiting motion and topology for segmenting and tracking under entanglement. In: CVPR 2011 (2011)
11. Zhang, L., Li, Y., Nevatia, R.: Global data association for multi-object tracking using network flows. In: CVPR (2008)
12. Comaniciu, D., Ramesh, V., Meer, P.: Real-time Tracking of Non-rigid Objects using Mean Shift. In: Proc. of the IEEE Conf. on Computer Vision and Pattern Recognition (2000)
13. Grimson, W.E.L., Stauffer, C., Romano, R., Lee, L.: Using Adaptive Tracking to Classify and Monitor Activities in a site. In: Proc. Computer Vision and Pattern Recognition, pp. 22–29 (1998)
14. Beymer, D., Konolige, K.: Real-time tracking of multiple people using stereo. In: Proc. of the IEEE Frame Rate Workship, Corfu, Greece (1999)
15. Felzenszwalb, P., Girshick, R., McAllester, D., Ramanan, D.: Object detection with discriminatively trained part based models. In: PAMI (2010)
16. Joachims, T., Schölkopf, B., Burges, C., Smola, A. (eds.): Making Large-Scale SVM Learning Practical. Advances in Kernel Methods - Support Vector Learning, pp. 169–184 (1999)
17. Haque, M., Murshed, M., Paul, M.: On stable dynamic background generation technique using gaussian mixture models for robust object detection. In: AVSS 2008 (2008)
18. Dalal, N., Triggs, B.: Histogram of Oriented Gradients for Human Detection. In: CVPR 2005 (2005)
19. Kasturi, R., Goldgof, D., Soundararajan, P., Manohar, V., Garofolo, J., Boonstra, M., Korzhova, V., Zhang, J.: Framework for performance evaluation for face, text and vehicle detection and tracking in video: data, metrics, and protocol. In: PAMI (2009)
20. Benfold, B., Reid, I.: Stable multi-target tracking in realtime surveillance video. In: CVPR (2011)
21. Pellegrini, S., Ess, A., Van Gool, L.: Improving Data Association by Joint Modeling of Pedestrian Trajectories and Groupings. In: Daniilidis, K., Maragos, P., Paragios, N. (eds.) ECCV 2010, Part I. LNCS, vol. 6311, pp. 452–465. Springer, Heidelberg (2010)
22. Yamaguchi, K., Berg, A., Ortiz, L., Berg, T.: Who are you with and where are you going. In: CVPR (2011)

23. Leal-Taixe, L., Pons-Moll, G., Rosenhahn, B.: Everybody needs somebody: Modeling social and grouping behavior on a linear programming multiple people tracker. In: ICCV Workshop on Modeling, Simulation and Visual Analysis of Large Crowds (2011)

24. Wang, M., Hong, R., Li, G., Zha, Z.-J., Yan, S., Chua, T.-S.: Event Driven Web Video Summarization by Tag Localization and Key-Shot Identification. IEEE Transactions on Multimedia 14(4), 975–985 (2012)

25. Wang, M., Hong, R., Yuan, X.-T., Yan, S., Chua, T.-S.: Movie2Comics: Towards a Lively Video Content Presentation. IEEE Transactions on Multimedia 14(3), 858–870 (2012)

26. Wang, M., Hua, X.-S., Tang, J., Hong, R.: Beyond Distance Measurement: Constructing Neighborhood Similarity for Video Annotation. IEEE Transactions on Multimedia 11(3), 465–476 (2009)

27. Wang, M., Hua, X.-S., Hong, R., Tang, J., Qi, G.-J., Song, Y.: Unified Video Annotation Via Multi-Graph Learning. IEEE Transactions on Circuits and Systems for Video Technology 19(5), 733–746 (2009)

28. Bise, R., Li, K., Eom, S., Kanade, T.: Reliably Tracking Partially Overlapping Neural Stem Cells in DIC Microscopy Image Sequences. In: Proceedings of the MICCAI Workshop on Optical Tissue Image analysis in Microscopy, Histopathology and Endoscopy (OPTIMHisE), London, UK, pp. 67–77 (September 2009)

Image Search Reranking
with Semi-supervised LPP and Ranking SVM

Zhong Ji[1,2], Yanru Yu[1], Yuting Su[1], and Yanwei Pang[1]

[1] School of Electronic Information Engineering
Tianjin University, Tianjin, 300072, China
[2] State Key Laboratory for Novel Software Technology
Nanjing University, Nanjing, 210093, China
{jizhong,yanruyu,ytsu,pyw}@tju.edu.cn

Abstract. Learning to rank is one of the most popular ranking methods used in image retrieval and search reranking. However, the high-dimension of the visual features usually causes the problem of "curse of dimensionality". Dimensionality reduction is one of the key steps to overcome these problems. However, existing dimensionality reduction methods are typically designed for classification, but not for ranking tasks. Since they do not utilize ranking information such as relevance degree labels, direct utilization of conventional dimensionality reduction methods in ranking applications generally cannot achieve the best performance. In this paper, we study the task of image search reranking, and propose a novel system scheme based on Locality Preserving Projections (LPP) and RankingSVM. And further, in the proposed scheme, we improve LPP by incorporating the relevance degree information into it. Since this kind of method can use the information of labeled and unlabeled data, we name it as semi-supervised LPP (Semi-LPP). Experiments on the popular MSRA-MM dataset demonstrate the superiority of the proposed scheme and Semi-LPP method in image search reranking application.

Keywords: Image search reranking, dimensionality reduction, learning to rank, Locality Preserving Projections.

1 Introduction

The overwhelming of multimedia data makes an urgent requirement to accurately and rapidly find the concerned images and videos [1]-[3]. Nowadays, most popular image search engines still depend on text-based search techniques, which employ the metadata utilize the metadata associated with media contents as features, and then use some well-known information retrieval methods, such as tf-idf and okapi BM25 to rank the images. However, since the visual contents are entirely neglected, irrelevant results are often returned in the top position, which severely decreases the users' search experience.

To address this problem, visual search reranking technique is proposed and has attracted substantial research interest. It applies visual information to reorder the

S. Li et al. (Eds.): MMM 2013, Part I, LNCS 7732, pp. 217–227, 2013.

text-based search results to get search performance improvement [4]-[7], which is a new paradigm followed by content-based image retrieval (CBIR) and concept detection in the domain of image content analysis and retrieval. Many research works have been carried out on this topic, which can be broadly classified into three categories, i.e., graph-based approaches [4][7], classification-based approaches [8]-[9] and learning to rank based approaches [5][10]-[14].

Among them, learning to rank based approaches have shown promising results and have received increasing attention. Learning to rank is a kind of newly proposed machine learning technology used to build a ranking model, and have shown promising performance in information retrieval domain. It can be divided into three categories [15]: the pointwise approach, the pairwise approach, and the listwise approach. Popular algorithms include RankingSVM [16], ListNet [17], and so on. Generally speaking, learning to rank based approaches first extract the image visual features from the initial search results, and then build a ranking function with the labeled training data, finally reorder the images with the ranking function. For example, the work proposed by Yang et al. [10] was the first one to employ the learning to rank technique in visual search reranking task. The authors utilized RankingSVM and ListNet algorithms to build the ranking function respectively and then rerank the initial search results, which had proved to be very effective and efficient through extensive experiments. Based on the observation that the ranks of different documents for a query are interdependent, Geng et al. [11] viewed the ranking as a structure output problem. To this end, based on learning to rank, they proposed a ranking model with large margin structured output learning, in which both textual and visual information are simultaneously leveraged in the ranking learning process. Liu et al. [12] discussed the differences of classification and ranking in reranking detailed, and presented two novel pairwise reranking models by formulating reranking as an optimization problem. They first converted the individual documents to "document pairs," and then found the optimal document pairs and their relevant relation with the proposed pairwise reranking models, at last adopted a round robin criterion to recover the final ranked list. More recently, Yang and Hanjalic [13] proposed a prototype-based reranking framework, in which the RankingSVM algorithm was also used as the ranking model.

However, the dimensionality reduction methods are not taken into account in these algorithms. The high-dimensional visual features not only bring the high computational and storage burden, but also decrease the generalization ability of the learning algorithms. Unfortunately, existing dimensionality reduction methods are typically designed for classification, clustering, and visualization, but not for ranking tasks. Direct utilization of them in ranking applications generally cannot achieve the best performance. Therefore, in this paper, we first propose a novel system scheme based on dimensionality reduction and learning to rank for image search reranking task. And then, we improve LPP by incorporating the ranking information into it.

The main contributions of our work can be summarized as follows. 1) We present an effective image search reranking scheme based on dimensionality reduction and learning to rank. Specifically, LPP and RankingSVM algorithms are adopted in the scheme. 2) We propose a semi-supervised dimensionality reduction algorithm called semi-supervised LPP (Semi-LPP), which incorporates the ranking information of relevance degree into LPP.

The rest of the paper is organized as follows. In Section 2, we first give a brief overview the LPP algorithm, and then detailedly introduce the proposed image search reranking scheme and Semi-LPP dimensionality reduction algorithm. Experimental results are presented and analyzed in Section 3. Section 4 concludes the paper.

2   The Proposed Image Search Reranking Scheme and Semi-LPP Algorithm

2.1   Locality Preserving Projections (LPP)

Locality Preserving Projections (LPP) [18] is a popular manifold learning based dimensionality reduction technique, which finds the optimal linear approximations to the eigenfuctions of the Laplace Betrami operator on the manifold. Although it is a linear approximation of the nonlinear Laplacian Eigenmap algorithm [19], it can well recover important aspects of the intrinsic nonlinear manifold structure by preserving local structure.

Given a training matrix $\mathbf{X} = [\mathbf{x}_1, ..., \mathbf{x}_n] \in \mathbb{R}^{D \times n}$, LPP uses the obtained transformation matrix $\mathbf{W} = [\mathbf{w}_1, ..., \mathbf{w}_d] \in \mathbb{R}^{D \times d}$ with the basis vector $\mathbf{w}_i \in \mathbb{R}^{D \times 1}$ to map the high-dimensional samples $\mathbf{x}_i \in \mathbb{R}^{D \times 1}$ to low-dimensional samples $\mathbf{y}_i \in \mathbb{R}^{d \times 1} (d \ll D)$:

$$\mathbf{y}_i = \mathbf{W}^\mathrm{T} \mathbf{x}_i. \tag{1}$$

Then, the objective function of LPP is defined as:

$$\arg\min_{\mathbf{W}} \sum_{i,j}^{n} (\mathbf{w}^\mathrm{T}\mathbf{x}_i - \mathbf{w}^\mathrm{T}\mathbf{x}_j)^2 a_{ij}, \tag{2}$$

$$\mathrm{s.t.} \mathbf{w}^\mathrm{T}\mathbf{X}\mathbf{D}(\mathbf{w}^\mathrm{T}\mathbf{X})^\mathrm{T} = \mathbf{I},$$

where a_{ij} measures the similarity of $\mathbf{x}_i$ and $\mathbf{x}_j$, $\mathbf{D}$ is a diagonal matrix with its element $D_{ii} = \sum_j a_{ij}$, $\mathbf{I}$ is unit matrix. A commonly used similarity is heat kernel. Let a_{ij} constitutes a weight matrix $\mathbf{S}$. The Laplacian matrix $\mathbf{L}$ is then formed by:

$$\mathbf{L} = \mathbf{D} - \mathbf{S}. \tag{3}$$

The optimization problem of (2) can be reduced to a generalized eigen-decomposition problem:

$$\mathbf{X}\mathbf{L}\mathbf{X}^\mathrm{T}\mathbf{w} = \lambda \mathbf{X}\mathbf{D}\mathbf{X}^\mathrm{T}\mathbf{w}. \tag{4}$$

2.2   The Proposed Image Search Reranking Scheme

Although LPP has been widely used in classification oriented applications, such as face recognition, handwritten digit recognition, image classification, and so on, little attempt has been made to use it in multimedia learning to rank applications. In this paper, we proposed an image search reranking scheme based on LPP [18] and RankingSVM [16]. We named this algorithm as LPPR, and the system scheme is shown in Fig. 1.

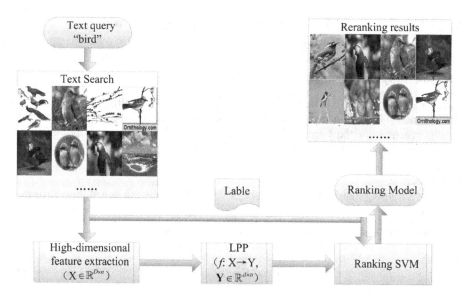

Fig. 1. The flowchart of the proposed LPPR scheme

Take the query term "bird" as an example. When "bird" is submitted to the web image search engine, an initial text-based search result is returned to the user. The result is unsatisfactory because some irrelevant images are retrieved as top results. To rerank these images, multimodal features are first extracted to represent their visual contents. Since the features' dimension is usually very high, which may cause the problem of "curse of dimensionality", and moreover, the high computational and storage burden. Therefore, dimensionality reduction is necessary. We choose LPP as the dimensionality reduction method for its excellent performance in multimedia analysis. Meanwhile, since there is generally no explicit training data, a manual labeling or pseudo-relevance feedback mechanism is adopted to label some data with relevance degrees to train the RankingSVM learning model. Finally, all the images are reranked with the reduced low dimensional features and ranking model.

2.3   The Proposed Semi-LPP Algorithm

From Fig. 1, we can observe that there are some labeled samples for the RankingSVM to train the learning model. However, they are not utilized for the dimensionality reduction stage. Since the labeled samples have a kind of useful relevance degree information, they have better been used in LPP to accurately uncover the intrinsical feature representation. However, to the best of our knowledge, there are few algorithms that designed for LPP or other manifold learning methods to incorporate the relevance degree information.

Graph construction is one of the most key steps in manifold learning methods, which determines the local manifold structure. Therefore, we further improve the LPP by incorporating the relevance degree information into the graph construction step of LPP, and call this algorithm as Semi-LPP, since the labeled and unlabeled data are both utilized. The idea of the proposed Semi-LPP algorithm is as follows: First, the information of labeled samples is used to automatically label the pseudo relevance degree to the unlabeled samples. Second, the edge weights are computed with these pseudo-labeled samples and labeled samples. Finally, this kind of weight edges is utilized in LPP to reduce the feature's dimension. In this way, we can adopt Semi-LPP in the image search reranking scheme illustrated in Fig. 1. In the following, we will introduce the Semi-LPPR algorithm in detail.

Let $\mathbf{X}_L = [\mathbf{x}_1, ..., \mathbf{x}_l] \in \mathbb{R}^{D \times l}$ be a set of l labeled samples, and $z_i \in \{0, ..., r-1\}$ be the corresponding relevance degree label. In addition to the labeled samples, let $\mathbf{X}_U = [\mathbf{x}_{l+1}, ..., \mathbf{x}_n] \in \mathbb{R}^{D \times (n-l)}$ be a set of $(n-l)$ unlabeled samples. The aim of the Semi-LPP algorithm is to find a transformation matrix $\mathbf{W}$ that can map $\mathbf{X} = [\mathbf{x}_1, ..., \mathbf{x}_n] \in \mathbb{R}^{D \times n}$ to low-dimensional vectors $\mathbf{Y} = [\mathbf{y}_1, ..., \mathbf{y}_n] \in \mathbb{R}^{d \times n} (d \ll D)$ with both the labeled and unlabeled samples, that is $\mathbf{Y} = \mathbf{W}^\mathrm{T} \mathbf{X}$. The algorithm procedure of the proposed Semi-LPP is stated below:

1) *k -nearest neighbors selection*: For unlabeled data samples $\mathbf{x}_i$ in set $\mathbf{X}_U$, we take their k -nearest neighbors $\mathbf{x}_{ij}$ from labeled set $\mathbf{X}_L$ to construct the index set $N(i)$.

2) *Reconstruction coefficients calculation*: In this step, we reconstruct the unlabeled samples with the labeled samples with the idea borrowed from locally linear embedding (LLE) [4]. The data samples $\mathbf{x}_i$ in set $\mathbf{X}_U$ are reconstructed with the samples in set $N(i)$ in original high-dimensional space:

$$\min_{\mathbf{x}_i \in \mathbf{X}_U, \mathbf{x}_{ij} \in N(i)} \left\| \mathbf{x}_i - \sum_{j=1}^{k} c_{ij} \mathbf{x}_{ij} \right\|^2, \tag{5}$$

where c_{ij} is the reconstructive coefficient, and $\sum_{j=1}^{k} c_{ij} = 1$.

By some simple algebra derivations, we can get $c_{ij} = \sum_{t=1}^{k} G_{jt}^{-1} \bigg/ \left(\sum_{p=1}^{k} \sum_{q=1}^{k} G_{pq}^{-1} \right)$, where

$G_{jt} = \left(\mathbf{x}_i - \mathbf{x}_{ij} \right)^\mathrm{T} \left(\mathbf{x}_i - \mathbf{x}_{it} \right)$ is a Gram matrix, $\mathbf{x}_{ij}, \mathbf{x}_{it} \in N(i)$.

3) *Pseudo relevance degree label*: All the unlabeled samples are assigned a pseudo relevance degree label automatically with the reconstruction coefficients:

$$z_i = \sum_{j=1}^{k} c_{ij} z_{ij}, \tag{6}$$

where z_{ij} is the corresponding label of $\mathbf{x}_{ij}$, $l+1 \leq i \leq n$.

4) *Adjacency graph construction*: Let **G** denote a graph with n nodes. We put an edge between nodes i and j if samples $\mathbf{x}_i$ and $\mathbf{x}_j$ are neighborhood. The heat kernel is adopted to assign the weight to the edge:

$$A_{ij} = \begin{cases} e^{-\frac{d(\mathbf{x}_i, \mathbf{x}_j)^2}{2\sigma}}, & \text{if } \mathbf{x}_i \in N(j) \text{ or } \mathbf{x}_j \in N(i), \\ 0 & , \text{ otherwise} \end{cases} \tag{7}$$

$$\sigma = \frac{1}{n^2} \sum_{i,j=1}^{n} d\left(\mathbf{x}_i, \mathbf{x}_j\right)^2, \tag{8}$$

where $d(\mathbf{x}_i, \mathbf{x}_j)$ denotes the distance between $\mathbf{x}_i$ and $\mathbf{x}_j$.

To incorporate the relevance degree information into the adjacency graph, we adopt the method in [21], i.e., keeping the distance between data examples within the same relevance degree unchanged, while that from different relevance degrees enlarged. The distance $d(\mathbf{x}_i, \mathbf{x}_j)$ is defined as:

$$d(\mathbf{x}_i, \mathbf{x}_j) = (|z_i - z_j| + 1) \|\mathbf{x}_i - \mathbf{x}_j\|_2, \tag{9}$$

where $|\cdot|$ represents the absolute value operator and $\|\cdot\|_2$ represents the L_2-norm operator. For the labeled samples, z_i denotes the true label, while for those unlabeled samples, z_i denotes the pseudo relevance degree obtained from the step 3). In this way, the relevance degree difference is reflected by the extent of enlargement.

5) Eigenmaps: Compute the eigenvectors and eigenvalues for (4). The embedding reduced features are $\mathbf{Y} = \mathbf{W}^T \mathbf{X}$.

3   Experimental Results

In this section, we demonstrate the effectiveness of the proposed LPPR and Semi-LPPR algorithms with MSRA-MM image dataset [22], which is very popular in image visual search reranking domain. We first introduce the dataset and methodologies, and then show the effectiveness of the proposed algorithms.

3.1   Experimental Setup

MSRA-MM image dataset consists of 68 popular queries collected from the Microsoft Live image Search engine. These queries cover a wide variety of categories, including objects, people, event, entertainments, and location, as illustrated in Fig. 2. For each query, about top 1000 images along with the surrounding texts are collected. Thus, there is totally 65 443 images in the dataset. The rank orders of these images are obtained as the initial ranked lists. The original images are not provided due the copyright issue, however, their Webpage URLs are provided to enable the users to obtain the images.

Fig. 2. Example images from MSRA-MM dataset. From left to the right, the columns are images of categories "baby", "fish", "fruits", "bird" and "car". From top to the bottom, the rows are images with relevance label "2", "1" and "0".

In the dataset, each image to the corresponding query was manually assigned with a relevance level: "irrelevant," "relevant," and "very relevant." The three levels are indicated by scores 0, 1, and 2, respectively. We adopt the features provided by the dataset to make the results reproducible and comparable. They are totally 899D features, including: (1) 225D block-wise color moment; (2) 64D HSV color histogram; (3) 256D RGB color histogram; (4) 144D color correlogram; (5) 75D edge distribution histogram; (6) 128D wavelet texture; and (7) 7D face features. In addition, we used top 500 images in the initial search results for reranking in our experiments, since it is typical that there are very few relevant images after the top 500 search results [7].

NDCG (Normalized Discounted Cumulative Gain) is a commonly adopted metric for evaluating a search engine's performance, especially when there are more than two relevance levels [23].Therefore, we adopt NDCG criterion to evaluate the ranking performance. Given a query q, the NDCG score at the depth d in the ranked documents is defined by:

$$NDCG @ d = Z_d \sum\nolimits_{j=1}^{d} \frac{2^{r_j} - 1}{\log(1 + j)}, \tag{10}$$

where r_j is the rating of the j-th document, Z_d is a normalization constant and is chosen so that a perfect ranking's $NDCG @ d$ value is 1. We obtain the final performance by averaging NDCG from 68 queries.

3.2   Evaluations

The scheme with RankingSVM and Semi-LPP is named as Semi-LPPR in this paper. To evaluate the effectiveness of the proposed LPPR and Semi-LPPR methods, we compare them with two state-of-the-art image search reranking algorithms and three dimensionality reduction algorithms. In these algorithms, we selected the parameters achieving the best performance.

The two state-of-the-art image search reranking algorithms are: 1) Bayesian reranking (referred as "Bayesian") [4], a representative graph-based reranking method, which used local learning regularizer to model the multiple-wise consistency. The method made an improvement to the assumption of ranking score consistency, and had been proved to be superior to the traditional graph-based method [3]. 2) Supervised learning-to-rerank (referred as "Letorr") [5], a recently proposed typical learning to rank based reranking method, which designed 11 lightweight reranking features and utilized RankingSVM model for image search reranking. And the three dimensionality reduction algorithms used for comparison in the proposed scheme are PCA, LDA and SELF [24], which are unsupervised, supervised and semi-supervised algorithms respectively. Note that for all the statistical experiments, we repeat them for three times by randomly selecting the different samples and report the average results. In addition, we use "Text" to denote the performance of initial search result, and "Baseline" to denote the performance using the provided 899D original image features. For PCA, LPP, and Semi-LPP, the reduced dimension d is 150. As for LDA, the dimension is set to be 2 since the three relevance degrees are regarded as three classes.

We randomly label $m = 3$ images from each relevance degree group. Fig. 3 illustrates the performance comparison at the depth of $\{1, 2, 3, 4, 5, 6, 7, 8, 9, 10\}$ using the same training data. And we can observe that: 1) The proposed LPPR and Semi-LPPR methods are very effective and outperform the Baseline, Text, and other methods significantly. 2) The performances of PCA and LDA are lower than the Baseline at most depths, which shows that it is not all dimensionality reduction methods is effective in the proposed scheme. 3) The relative low performances of Bayesian and Lettor mainly lie in the fact that no human label in their algorithms. In addition, the impact of human labeled numbers is evaluated in Fig. 4, from which we can observe that the performance increases with the growth of label number, and the performance is still very good even when $m = 2$.

4   Conclusion

In this paper, we propose a learning to rank based image search reranking scheme, in which the learning to rank algorithm (Ranking SVM) and dimensionality reduction algorithm (LPP) are two key components. Moreover, we also propose a novel dimensionality reduction algorithm, which incorporates the ranking information into the adjacency graph construction step of LPP, and uses the information of both labeled and unlabeled data. The experimental results demonstrate the effectiveness of the proposed scheme and algorithm.

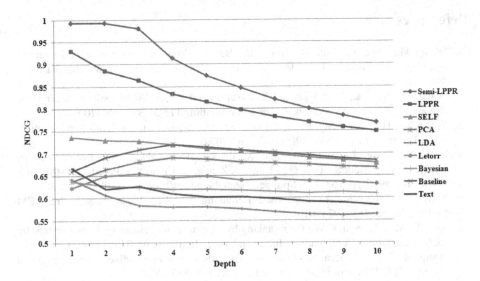

Fig. 3. Performance comparisons of the proposed methods with other methods

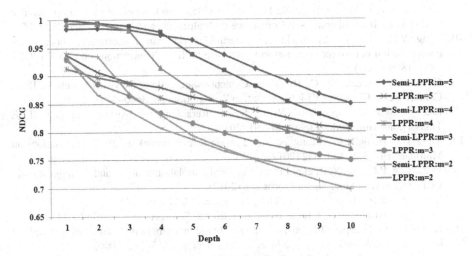

Fig. 4. Performance comparisons of different labeled number *m*

Acknowledgments. This work was supported by the National Natural Science Foundation of China (Nos. 61271325, 60975001, 61172121, 61222109, 61271412), the Tianjin Research Program of Application Foundation and Advanced Technology (No. 10JCYBJC07700), the Specialized Research Fund for the Doctoral Program of Higher Education (No. 20090032110028), the Program for New Century Excellent Talents in University (No. NCET-10-0620), and the Open Project Program of the State Key Laboratory for Novel Software Technology (No. KFKT2012B35), Nanjing University.

References

1. Wang, M., Hua, X., Tang, J., Hong, R.: Beyond distance measurement: constructing neighborhood similarity for video annotation. IEEE Transactions on Multimedia 11(3), 465–476 (2009)
2. Wang, M., Hua, X., Mei, T., et al.: Semi-supervised kernel density estimation for video annotation. Computer Vision and Image Understanding 113(3), 384–396 (2009)
3. Wang, M., Hua, X., Hong, R., et al.: Unified video annotation via multi-graph learning. IEEE Transactions on CSVT 19(5), 733–746 (2009)
4. Tian, X., Yang, L., Wu, X., Hua, X.-S.: Visual Reranking with Local Learning Consistency. In: Boll, S., Tian, Q., Zhang, L., Zhang, Z., Chen, Y.-P.P. (eds.) MMM 2010. LNCS, vol. 5916, pp. 163–173. Springer, Heidelberg (2010)
5. Yang, L.J., Hanjalic, A.: Supervised Reranking for Web Image Search. In: ACM Multimedia, pp. 183–192 (2010)
6. Yao, T., Mei, T., Ngo, C.W.: Co-reranking by mutual reinforcement for image search. In: ACM International Conference on Image and Video Retrieval, pp. 34–41 (2010)
7. Wang, M., Yang, K., Hua, X., Zhang, H.: Towards a relevant and diverse search of social images. IEEE Transactions on Multimedia 12(8), 829–842 (2010)
8. Wei, S.K., Zhao, Y., Zhu, Z.F., Liu, N.: Multimodal fusion for video search reranking. IEEE Transactions on KDE 22(8), 1191–1199 (2010)
9. Liu, Y., Mei, T., Hua, X.S., et al.: Learning to video search rerank via pseudo preference feedback. In: IEEE International Conference on Multimedia and Expo, pp. 207–210 (2008)
10. Yang, Y.H., Hsu, W., Chen, H.: Online reranking via ordinal informative concepts for context fusion in concept detection and video search. IEEE Transactions on CSVT 19(12), 1880–1890 (2009)
11. Geng, B., Yang, L.J., Xu, C., Hua, X.S.: Content-aware ranking for visual search. In: IEEE Conference on Computer Vision and Pattern Recognition, pp. 3400–3407 (2010)
12. Liu, Y., Mei, T.: Optimizing Visual Search Reranking via pairwise learning. IEEE Transactions Multimedia 13(2), 280–291 (2011)
13. Yang, L.J., Hanjalic, A.: Prototype-Based Image Search Reranking. IEEE Transactions on Multimedia 14(3-2), 871–882 (2012)
14. Ji, Z., Jing, P.G., Su, Y.T., et al.: Rank canonical correlation analysis and its application in visual search reranking. Signal Processing (2012), http://dx.doi.org/10.1016/j.sigpro.2012.05.006
15. Liu, T.Y.: Learning to rank for information retrieval. Springer Press, Berlin (2011)
16. Herbrich, R., Graepel, T., Obermayer, K.: Large margin rank boundaries for ordinal regression. In: Advances in Large Margin Classifiers, pp. 115–132 (2000)
17. Cao, Z., Qin, T., Liu, T.Y., Tsai, M.F., Li, H.: Learning to rank: from pairwise approach to listwise approach. In: ICML, pp. 129–136 (2007)
18. He, X.F., Yan, S.C., Hu, Y.X., Niyogi, P., Zhang, H.J.: Face recognition using Laplacianfaces. IEEE Transactions on PAMI 27(3), 328–340 (2005)
19. Belkin, M., Niyogi, P.: Laplacian eigenmaps and spectral techniques for embedding and clustering. In: Advances in Neural Information Processing System, pp. 585–591 (2001)
20. Roweis, S., Saul, L.: Nonlinear dimensionality reduction by locally linear embedding. Science 290(22), 2323–2326 (2000)
21. Liu, Y., Liu, Y., Zhong, S.H., Chan, K.: Semi-supervised manifold ordinal regression for image ranking. In: ACM Multimedia, 1393–1396 (2011)

22. Wang, M., Yang, L.J., Hua, X.S.: MSRA-MM: bridging research and industrial societies for multimedia information retrieval. Microsoft Technical Report, Beijing, MSR-TR-2009-30 (2009)
23. Järvelin, K., Kekäläinen, J.: IR evaluation methods for retrieving highly relevant documents. In: ACM SIGIR Conference on Research and Development in Information Retrieval, pp. 41–48 (2000)
24. Sugiyama, M., Ide, T., Nakajima, S., Sese, J.: Semi-supervised local fisher discriminant analysis for dimensionality reduction. Machine Learning 78(1-2), 35–61 (2010)

Co-ranking Images and Tags via Random Walks on a Heterogeneous Graph

Lin Wu, Yang Wang, and John Shepherd

School of Computer Science and Engineering
The University of New South Wales, Sydney, Australia
{linw,wangy,jas}@cse.unsw.edu.au

Abstract. Ranking on image search results has attracted considerable attentions. Despite many graph-based ranking algorithms have demonstrated remarkable success, most of their applications are limited to single image-networks such as the network of tags associated with images. In this paper, we investigate the problem of co-ranking images and tags attached in a heterogeneous network, which consists of three graphs: the image graph connecting images, the tag graph connecting tags attached to the images, as well as the image-tag graph connecting the above two graphs together. Observing that existing ranking approaches do not consider images and tags simultaneously, a novel co-ranking method via random walks on all three graphs is proposed to significantly improve the ranking effectiveness on both images and tags. Experimental results conducted on three benchmark data sets show that our approach outperforms the state-of-the-art local ranking approaches for image ranking and tag ranking and scales well on large scale data sets.

Keywords: Co-ranking, Random walk, Heterogeneous network.

1 Introduction

The explosion of online community-contributed multimedia data results in great focus on image retrieval. Most of social media sharing websites like Flickr allow users to upload personal images and annotate content with descriptive keywords called tags. Many ranking algorithms specialized to image on such social media repositories have been proposed to help organize the shared media data [8] or to facilitate the image ranking process[11]. For instance, Jing *et al.*[11] propose to rank images by utilizing visual similarity among images. Although a lot of encouraging results have been reported from these approaches focusing on centrality measures on image content, the evaluations of the relative importance of images have been carried independently, which fails to take advantage of useful metadata including tags, manual labels, etc. In a similar way, the well-studied treatments on tag recommendation [8] and tag ranking [12] are solely conducted on the tag graph, However, the natural connections between images and tags are still not fully leveraged. Another approach proposed in [8] helps users in tagging process by suggesting relevant tags. Admittedly, these methods [12,8,11] have

S. Li et al. (Eds.): MMM 2013, Part I, LNCS 7732, pp. 228–238, 2013.

achieved better ranking performance over previous approaches, however, they do not consider the reinforcing dependency between images and tags, which is beneficial to further improving ranking results. For instance, the tag ranking list provided by Liu *et al.*[12] simply relies on the tag graph built upon a given image, while discarding the additional ranking information coming from the image graph, which would be potential to improve the accuracy of tag ranking greatly. Therefore, it is desirable to develop a novel algorithm to handle dual-relational data over a combined graph for co-ranking images and tags simultaneously. This paper aims to design a co-ranking scheme for images and their associated tags in a heterogeneous graph. Our contributions are summarized as follows.

- We explore the mutually reinforcing relationship between image and tag graphs by constructing three graphs with image graphs, tag graphs and the bridging graph combining the above two graphs. For image graph, we model the relationship between images using a probabilistic hypergraph, in which a hyperedge corresponds to a particular tag and the probability of images that belong to hyperedges effectively describes a higher-order relationship.
- The importance of the images and tags are obtained by random walks on image graphs and tag graphs, the importance of tags are effectively utilized to enhance the image ranking through bridging graph and vice versa for tag ranking by combining with the importance of images.
- The extensive experiments are designed to show the effectiveness and efficiency of our ranking approach over the existing local ranking approaches for image ranking and tag ranking.

The rest of the paper is organized as follows. In Section 2, we describe the construction of image graph, tag graph as well as image-tag graph connecting them together. Section 3 presents the co-ranking algorithm. We conducted experiments in Section 4 and conclusions are given in Section 5.

2   Graph Constructions

2.1   The Design of Image Hypergraph

The existing methods to model the relationship between images is to construct the pairwise image graph, in which images are taken as vertices and two similar images are connected by an edge whose weight is computed as image-to-image affinities [11,16]. However, the simple pairwise graphs cannot describe the high order relationship of more than two vertices as pointed out in [1,10,9,17]. Due to this fact, we consider a weighted hypergraph $G_M = (V, T, W)$, which consists of a set of vertices $V = \{v_1, v_2, \ldots, v_{|V|}\}$, a set of nonempty subsets of V, referred to as the hyperedges $T = \{t_1, t_2, \ldots, t_{|T|}\}$, and a weight matrix W in which $w(t_i)$ is the hyperedge weight of t_i. A toy example to show the rationality of image hypergraph is illustrated in Fig. 1.

We establish the relationship between hyperedges and vertices in a probabilistic incidence matrix H with the size of $|V| \times |T|$, which can be defined as:

$$h(v_i, t_j) = \begin{cases} p(v_i|t_j), & \text{if } v_i \in t_j; \\ 0, & \text{otherwise.} \end{cases} \qquad (1)$$

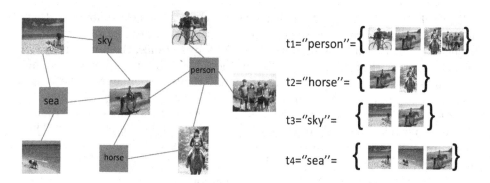

Fig. 1. A toy example for image hypergraph. We consider each tag t_i as a hyperedge containing an arbitrary number of images which are associated with the tag t_i.

H induces the definition of a vertex-degree matrix D_v with diagonal elements $d(v) = \sum_{t \in T} h(v,t)w(t)$, and a hyperedge-degree matrix D_t with diagonal elements $\delta(t) = \sum_{v \in V} h(v,t)$. To estimate the probability function of $p(v_i|t_j)$, we employ the kernel density estimation approach, which can be formulated as

$$p(v_i|t_j) = \frac{1}{|Y_i|} \sum_{y_k \in Y_i} K_\sigma(y - y_k) \qquad (2)$$

where Y_i is the set of images that contain the tag t_i and K_σ is the Gaussian kernel function with radius parameter σ.

Hyperedge Weight. In order to quantize the similarity between images, we use two feature descriptors: the SURF-based appearance feature descriptor [2] and the pyramid histogram of oriented gradients-based shape feature descriptor[3]. SURF features are extracted from images and then create a 128-bin codebook of SURF features by employing k means algorithm, which are further quantized into a histogram by soft assignment, as suggested in [14,10]. For the PHOG descriptor, we discretize gradient orientation into 8 bins to build corresponding histograms. Thus, the distance between two images (v_i, v_j) can be calculated by a spatial pyramid matching (SPM) approach formulated as Eq. 3

$$\phi(v_i, v_j) = \sum_{l=0}^{L} \frac{1}{2^{L-l}} \sum_{p=1}^{m(l)} \beta_p^l \chi^2(His_p^l(v_i), His_p^l(v_j)) \qquad (3)$$

where $His_p^l(\cdot)$ denotes local histograms at the pth position of level l; β is the weighting parameter; $\chi(\cdot, \cdot)$ is the chi-square distance function to measure the distance between two histograms. Both SURF and PHOG descriptors are based on 3-level image pyramids. Specifically, three levels of spatial pyramids are:

- 1×1: whole image, $l=0, m(0)=1, \beta_1^0=1$;
- 2×2: image quarters, $l=1, m(1)=4, \beta_1^1 \sim \beta_4^1 = \frac{1}{4}$;
- 4×4: 16-divided image, $l=2, m(2)=16, \beta_1^2 \sim \beta_{16}^2 = \frac{1}{16}$.

Therefore, the similarity between two images can be computed using a following kernel function:

$$A(v_i, v_j) = \exp(-\frac{1}{\Delta}\phi(v_i, v_j)) \tag{4}$$

where Δ is the standard deviation of $\phi(v_i, v_j)$ over all the data.

To determine the weight of a hyperedge, we employ the homogeneity of appearance and shape features at all images constituting the hyperedge. In particular, we define the weight as a function of the variances of features at images inside a hyperedge:

$$w(t) = \exp(-\frac{\sum_{v_i, v_j \in t} \phi(v_i, v_j)}{|t|(|t| - 1)\mu^2}) \tag{5}$$

where parameter μ characterizes the homogeneity of the region and $|t|$ denotes the number of images belonging to the hyperdege t. It can be seen that a large value of μ allows for an object of potentially different shape or appearance, while a small value of μ prohibits the difference.

2.2   The Design of Tag Graph

In the tag graph construction, we consider a simple pairwise graph to model the pairwise constraints between tags.

We remark that it constructs the simple graph rather than a hypergraph lies in the difficulties to define the important concept of hyperedge due to the fact that tags are more difficult to be categorized in a particular image if an image is defined as a hyperedge. Moreover, the determination of hyperedges is equivalent to manually assigning tags to images, which is too laborious.

An effective way to model the relationship between tags is to build the models using visual cues that come from nearest neighbor images associated with the tags. The introduce of nearest neighbor strategy could avoid the noises caused by polysemy [12]. Unlike the most popular method [12] that considers nonflexible number of nearest neighbors and equal contribution offered by different neighbors, we learn the distances between tags more accurately by introducing the locally specific distance metrics, which is further combined with the popular Google distance [5] to measure the similarity between tags. Suppose that we have two sets of U_i and U_j to be the representative image collections of tag t_i and t_j, respectively. Then the distance between t_i and t_j is defined as follows:

$$d(t_i, t_j) = \sum_{u_i \in U_i, u_j \in U_j} \| u_i - u_j \|^2 \tag{6}$$

The distance defined in Eq. (6) is Euclidean distance, which is independent of the input data. To truly reflect the similarity among tags and capture ranking information from image graph, we consider to use the Mahalanobis distance with an appropriate distance metric.[1] Essentially, instead of using Eq. (6), we compute the tag distances using the Mahalanobis distance as:

$$d_m(t_i, t_j) = \sum_{u_i \in U_i, u_j \in U_j} [(u_i - u_j)^T M_{ij}(u_i - u_j)] \qquad (7)$$

where M_{ij} is the element in the Mahalanobis matrix M, which captures the ranking information of representative images. The Google distance[5] is also incorporated into the metric for the tags, which is motivated by [12]. Specifically, the Google distance between t_i and t_j by the concurrence similarity is as follows:

$$d_g(t_i, t_j) = \frac{\max(\log g(t_i), \log g(t_j)) - \log g(t_i, t_j)}{\log G - \min(g(t_i), \log g(t_j))} \qquad (8)$$

where G is the total number of images, $g(t_i)$ and $g(t_j)$ denotes the number of images containing tag t_i and t_j respectively, and $g(t_i, t_j)$ is the number of images containing both t_i and t_j. Note that these numbers can be obtained by performing search on social media websites, e.g., Flickr, with tags as the keywords. Based on the above two distances defined as Eq.(7) and Eq.(8), we formalize the affinity value between tags t_i and t_j as:

$$S(t_i, t_j) = \alpha \cdot \exp(-d_m(t_i, t_j)) + (1 - \alpha) \cdot \exp(-d_g(t_i, t_j)) \qquad (9)$$

where $0 < \alpha < 1$. In our experiments, we set $\alpha = 0.8$ to be consistent with the baseline method [12].

2.3   The Combined Heterogeneous Graph

The intuitive definition of the heterogeneous graph G^* is that the entries in its adjacency matrix W^* are the probabilistic values of tags associated with the images, i.e.

$$w^*(i, j) = p(t_i|v_j) \qquad (10)$$

To perform the transition of a random walk moving from image i (tag j) to tag j (image i) via the graph G^*, we define the conditional transition matrix $\langle MT \rangle$ (with the entry $\langle MT \rangle_{ij}$) and $\langle TM \rangle$ (with the entry $\langle TM \rangle_{ji}$), containing the transition probabilities from image graph G_M to tag graph G_T and vice versa. Hence given the next step takes place in the graph G^*, we have

$$\langle MT \rangle_{i,j} = \mathcal{P}(t_j|v_i) = \frac{w^*(i,j)}{\sum_k w^*(i,k)}; \quad \langle TM \rangle_{j,i} = \mathcal{P}(v_i|t_j) = \frac{w^*(i,j)}{\sum_k w^*(k,i)} \qquad (11)$$

The above matrices $\langle MT \rangle$ and $\langle TM \rangle$ reflect the asymmetric relationship between images and tags, which implies that it is desirable for a tag to be associated with

---

[1] As each tag does not equally contribute to the labeling of image, it is desirable to learn multiple weights for each tag. The weights reflects the relative importance of the tag with respect to an image.

many related images, while for an image it is better to have tightly correlated tags, but not necessarily more tags. Moreover, the significance of a particular tag depends on the degree of such a tag, which has the underlying meaning to reveal that the more popular the tag is, the more connections there will be among the tag and others.

3   Co-ranking Framework

In this section, we present the co-ranking mechanism involving the random walks on the heterogeneous network described in section 2. The framework is illustrated in Fig. 2.

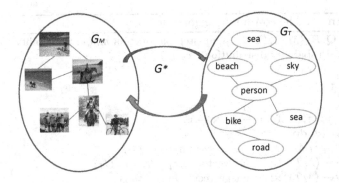

Fig. 2. The framework for co-ranking images and tags. G_M and G_T are image graph and tag graph, respectively. G^* is the image-tag graph derived from images and their associated tags.

3.1   Random Walks on Image and Tag Graphs

Consider the transition matrix $\mathbf{Q}$ derived from the weighted image hypergraph G_M, and $r_k(i)$ denotes the relevance score of image v_i with respect to the query image at iteration k. Consequently, the relevance scores of all the images in the hypergraph at iteration k form a column vector $\mathbf{r}_k = [r_k(v_i)]_{n \times 1}$. The element q_{ij} in matrix $\mathbf{Q}$ indicates the probability of the transition from image v_i to image v_j, which can be formulated as

$$q_{ij} = \sum_{v_i, v_j \in t} w(t) A(v_i, v_j) \frac{h(v_i, t)h(v_j, t)}{d(v_i)d(v_j)} \tag{12}$$

Accordingly, the random walk process is formulated as

$$r_k(j) = \rho \sum_i r_{k-1}(i) q_{ij} + (1 - \rho) c_j \tag{13}$$

where c_j is the initial relevance score of image v_j, and ρ is a damping factor between [0,1]. The unique solution of Eq. (13) is $\mathbf{r}_\pi = (1 - \rho)(\mathbf{I} - \rho \mathbf{Q})^{-1} \mathbf{c}$ [4].

Similarly, given the transition matrix $\mathbf{R}$ induced by the tag graph G_T, the random walk over G_T converges to a stationary probability distribution of $\mathbf{f}_\pi$.

3.2   Combined Random Walks on a Heterogenous Network

We present the combined random walk on the heterogeneous graph in terms of a random surfer who is capable of jumping over images and their tags as well. Thus, in the process of coupling two random walks, a probability distribution will have the form $(\mathbf{r}_\pi, \mathbf{f}_\pi)$ [18], satisfying $\parallel \mathbf{r}_\pi \parallel_1 + \parallel \mathbf{f}_\pi \parallel_1 = 1$. Furthermore, we parameterize the coupling process by using four parameters, i.e., m, n, k and λ, which are elaborated in the following procedures: For the current state of random surfer $v \in V_{G_M}$ ($v \in V_{G_T}$), it has probability λ to take $2k+1$ steps on G^*, while with probability 1-λ to take m (n) steps on G_M (G_T). We summarize this process in Algorithm 1.

Algorithm 1. The co-ranking algorithm.

Input: $\mathbf{Q}, \mathbf{R}, \langle MT \rangle, \langle TM \rangle$; jump-step parameters m,n,k; probability value λ.
Output: Probability distribution ($\mathbf{r}_\pi, \mathbf{f}_\pi$).
$\mathbf{r}_\pi \leftarrow \frac{1}{n_Q}\mathbf{1}_{n_Q \times 1}$;
$\mathbf{f}_\pi \leftarrow \frac{1}{n_R}\mathbf{1}_{n_R \times 1}$;
repeat
    $\mathbf{r}'_\pi \leftarrow \mathbf{r}_\pi$; $\mathbf{f}'_\pi \leftarrow \mathbf{f}_\pi$; $\mathbf{a} \leftarrow \mathbf{f}'_\pi$; $\mathbf{x} \leftarrow \mathbf{r}'_\pi$;
    for $i=1{:}k$ **do**
        $\mathbf{b} \leftarrow \langle TM \rangle^T \mathbf{a}$; // tag vector jumps to G_M;
        $\mathbf{a} \leftarrow \langle MT \rangle^T \mathbf{b}$; // tag vector jumps back to G_T;
        $\mathbf{y} \leftarrow \langle MT \rangle^T \mathbf{x}$; // image vector jumps to G_T;
        $\mathbf{x} \leftarrow \langle TM \rangle^T \mathbf{y}$; // image vector jumps back to G_M;
    $\mathbf{b} \leftarrow \langle TM \rangle^T \mathbf{a}$;//the tag vector takes the $(2k+1)th$ step on G^*;
    $\mathbf{y} \leftarrow \langle MT \rangle^T \mathbf{x}$;//the image vector takes the $(2k+1)th$ step on G^*;
    $\mathbf{r}_\pi \leftarrow (1-\lambda)(\mathbf{Q}^T)^m \mathbf{r}'_\pi + \lambda \mathbf{b}$;
    $\mathbf{f}_\pi \leftarrow (1-\lambda)(\mathbf{R}^T)^n \mathbf{f}'_\pi + \lambda \mathbf{y}$;
until $\mid \mathbf{r}_\pi - \mathbf{r}'_\pi \mid \leq \varepsilon$ *or* $\mid \mathbf{f}_\pi - \mathbf{f}'_\pi \mid \leq \varepsilon$;
return ($\mathbf{r}_\pi, \mathbf{f}_\pi$);

---

4   Experiments

In this section, we experimentally evaluate the co-ranking algorithm in the tasks of both image ranking and tag ranking on three benchmark data sets: LabelMe[13], which is a large collection of annotated and unlabeled images; Corel5K data set [6] and PASCAL2010 data set[7].

4.1   Experimental Settings

- For image ranking, we compare our algorithm with two popular approaches: PageRank for image search [11] and Efficient Manifold Ranking (EMR)[16].
- For tag ranking, we conduct the comparison on another two state-of-the-art methods: tag ranking [12] and the algorithm of learning to tag[15].

To evaluate the performance of the co-ranking method as well as the baseline algorithms, we adapt the following evaluation metric, Mean Average Precision (MAP) to measure the quality of returned results. For a single query, Average Precision is obtained for the set of top k items existing after each relevant item is retrieved, and this value is then averaged over all queries. If the set of relevant items for a query $b_i \in B$ is $\{d_1, \ldots, d_{m_j}\}$ and R_{jk} is the set of ranked retrieved results from the top result until we get to item d_k, then we have $MAP(B) = \frac{1}{|B|} \sum_{j=1}^{|B|} \frac{1}{m_j} \sum_{k=1}^{m_j} Precision(R_{jk})$.

4.2   Co-ranking Results: A Study Case

To evaluate the performance of co-ranking on images and tags, we randomly select five images from the Flickr collection [12] as the query set, as illustrated in the first column of Fig.3. For each image query, top 10 returns are retrieved according to their relevance scores to the query. At the same time, orderless tags attached to the query image are ordered by considering their intrinsically semantic similarity as well as the reinforcing impact generated by the image graph. As illustrated in Fig.3, each row shows the original query followed by the top 10 most relevant items, and the ranking list of tags is tailed simultaneously.

Fig. 3. Top image-returns and corresponding tags w.r.t various image queries

4.3   Image Ranking

Note that all the above methods except our algorithm perform ranking procedure over the image graph individually, while overlooking the mutual reinforcement between images and tags. The MAP values are reported in Fig.4 in which we present three groups of bars corresponding to 15 popular topics collected from three benchmarks. This evaluation shows that the co-ranking method outperforms the other two algorithms, e.g., achieving an average improvement of 33.11%, 20.82%, over PageRank and EMR in terms of LabelMe database. We can draw the conclusion that owing to fully leveraging tag ranking scores provided by random walks on the tag graph, our co-ranking approach shows superiority over baseline methods, which are limited on the simple image graph, leading to the failure of utilizing additional tag ranking information.

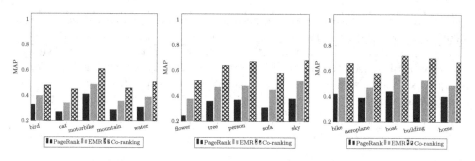

Fig. 4. Comparisons on image ranking with respect to three benchmarks. From left to right: LabelMe, Corel5K, and PASCAL 2010.

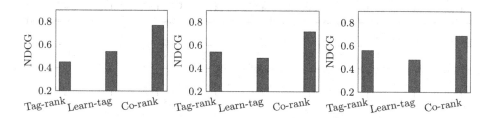

Fig. 5. Performance of different tag ranking strategies. Baseline algorithms include: tag ranking in [12], and learning to tag proposed in [15].

4.4 Tag Ranking

We use NDCG as the metric to evaluate the performance of tag ranking. For a query image, each of its tags is labeled as one of five levels: most relevant(score 5), relevant(score 4), partially relevant(score 3), marginally relevant(score 2), and irrelevant(score 1). Assume that an image associated with a tag ranking list $L = t_1, t_2, \ldots, t_n$, the NDCG value of the list is $NDCG(n) = Z_n \sum_{j=1}^{n} \frac{2^{r(j)}-1}{\log_2(j+1)}$, where $r(j)$ is the relevance score of the jth tag and Z_n is a normalization constant. For a particular database, e.g., Corel5K, given the NDCG values of each image's tag list, we average them to obtain an overall performance of the tag ranking algorithm with respect to such a data set. The experimental results in terms of NDCG are shown in Fig. (5). The following conclusions can be drawn.

- Our co-ranking algorithm outperforms baselines greatly by leveraging the ranking information from the image graph that is incorporated in the Mahalanobis matrix M in Eq.(7), which is the main difference between [12] that considers the weight equally.
- Existing tag ranking methods fall short in using ranking scores of images since only tag graph is considered, which is not comparable with our co-ranking method.

4.5   Parameters Learning

Considering that the probability value λ balances the random walks over the heterogeneous network, we report the learning process of parameter λ with different values ranging from 0.1 to 0. In Fig.6, MAP metric is used to evaluate the performance of co-ranking w.r.t varying λ in terms of three benchmarks. We can observe that the selection of λ is critical to the algorithm of co-ranking and parameter learning is necessary in different image databases.

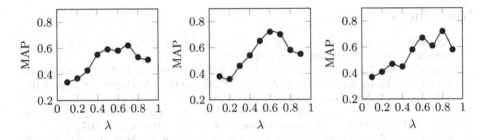

Fig. 6. Learning of papameter λ ranging from 0.1 to 0.9 on three benchmark databases. From left to right: LabelMe, Corel5K, and PASCAL 2010

4.6   Running Time Evaluations

We also show the effect of m and n on the number of iterations as well as CPU running time before convergence. The evaluated results are reported in Fig.7, from which we can observe that as m and n increase, the number of iterations and CPU running time (sec) decrease slowly. This is due to the fact that the random walks on the individual graph have sufficient steps to become locally stationary before taking the next step on the combined graph.

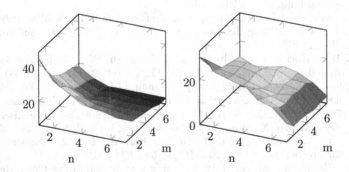

Fig. 7. Effect of $[m, n]$ on number of iterations (left) and CPU running time (right)

5   Conclusion

In this paper, we propose a novel co-ranking method that couples two random walk simultaneously by exploiting the mutually reinforcing relationship between images and tags. The ranking information upon individual graphs on images and tags are complementary, and hence the leveraging mechanism incorporated in our algorithm can achieve more satisfactory performance than current algorithms for image ranking and tag ranking. Extensive experiments on various image databases validate the effectiveness and efficiency of our method.

References

1. Agarwal, S., Lim, J., Zelnik-Manor, L., Perona, P., Kriegman, D., Belongie, S.: Beyond pairwise clustering. In: CVPR (2005)
2. Bay, H., Tuytelaars, T., Van Gool, L.: SURF: Speeded Up Robust Features. In: Leonardis, A., Bischof, H., Pinz, A. (eds.) ECCV 2006. LNCS, vol. 3951, pp. 404–417. Springer, Heidelberg (2006)
3. Bay, H., Tuytelaars, T., Gool, L.V.: Presenting shape with a spatial pyramid kernels. In: International Conference on Image and Video Retrieval (2007)
4. Berkhin, P.: A survey on pagerank computing. Internet Mathematics 2(1), 73–120 (2005)
5. Cilibrasi, R., Vitányi, P.M.B.: The google similarity distance. IEEE Transactions on Knowledge and Data Engineering 19(3), 370–383 (2007)
6. Duygulu, P., Barnard, K., de Freitas, J.F.G., Forsyth, D.A.: Object Recognition as Machine Translation: Learning a Lexicon for a Fixed Image Vocabulary. In: Heyden, A., Sparr, G., Nielsen, M., Johansen, P. (eds.) ECCV 2002, Part IV. LNCS, vol. 2353, pp. 97–112. Springer, Heidelberg (2002)
7. Everingham, M., Gool, L.V., Williams, C., Winn, J., Zisserman, A.: The pascal visual object classes challenge (VOC 2010) results (2010), http://pascallin.ecs.soton.ac.uk/challenges/VOC/voc2010/
8. Guan, Z., Bu, J., Mei, Q., Chen, C., Wang, C.: Personalized tag recommendation using graph-based ranking on multi-type interrelated objects. In: SIGIR (2009)
9. Huang, Y., Liu, Q., Lv, F., Gong, Y., Metaxas, D.N.: Unsupervised image categorization by hypergraph partition. TPAMI 33(8), 1266–1273 (2011)
10. Huang, Y., Liu, Q., Zhang, S., Metaxas, D.N.: Image retrieval via probabilistic hypergraph ranking. In: CVPR (2010)
11. Jing, Y., Baluja, S.: Pagerank for product image search. In: WWW (2008)
12. Liu, D., Hua, X.-S., Yang, L., Wang, M., Zhang, H.-J.: Tag ranking. In: WWW, pp. 351–360 (2009)
13. Russell, B.C., Torralba, A., Murphy, K.P., Freeman, W.T.: Labelme: A database and web-based tool for image annotation. IJCV 77(1-3), 157–173 (2008)
14. van Gemert, J.C., Geusebroek, J.-M., Veenman, C.J., Smeulders, A.W.M.: Kernel Codebooks for Scene Categorization. In: Forsyth, D., Torr, P., Zisserman, A. (eds.) ECCV 2008, Part III. LNCS, vol. 5304, pp. 696–709. Springer, Heidelberg (2008)
15. Wu, L., Yang, L., Yu, N., Hua, X.: Learning to tag. In: WWW, pp. 361–370 (2009)
16. Xu, B., Bu, J., Chen, C., Cai, D., He, X., Liu, W., Luo, J.: Efficient manifold ranking for image retrieval. In: SIGIR, pp. 525–534 (2011)
17. Zhou, D., Huang, J., Schölkopf, B.: Learning with hypergraphs: Clustering, classification, and embedding. In: NIPS (2006)
18. Zhou, D., Orshanskiy, S.A., Zha, H., Giles, C.L.: Co-ranking authors and documents in a heterogeneous network. In: ICDM, pp. 739–744 (2007)

Social Visual Image Ranking
for Web Image Search

Shaowei Liu[1,2], Peng Cui[1,2], Huanbo Luan[1,4], Wenwu Zhu[1,2],
Shiqiang Yang[1,2], and Qi Tian[3]

[1] Department of Computer Science and Technology, Tsinghua University, China
[2] Beijing Key Laboratory of Networked Multimedia, China
[3] Computer Science Department, University of Texas at San Antonio, USA
[4] School of Computing, National University of Sinapore
liu-sw11@mails.tsinghua.edu.cn, cuip@mail.tsinghua.edu.cn,
luanhuanbo@gmail.com, {wwzhu,yangshq}@tsinghua.edu.cn, qitian@cs.utsa.edu

Abstract. Many research have been focusing on how to match the textual query with visual images and their surrounding texts or tags for Web image search. The returned results are often unsatisfactory due to their deviation from user intentions. In this paper, we propose a novel image ranking approach to web image search, in which we use social data from social media platform jointly with visual data to improve the relevance between returned images and user intentions (i.e., social relevance). Specifically, we propose a community-specific Social-Visual Ranking(SVR) algorithm to rerank the Web images by taking social relevance into account. Through extensive experiments, we demonstrated the importance of both visual factors and social factors, and the effectiveness and superiority of the social-visual ranking algorithm for Web image search.

Keywords: Social image search, Image reranking, Social relevance.

1   Introduction

Fundamentally, user intention plays an important role in image search. Most of traditional image search engines represent user intentions with textual query. Thus, a lot of existing research work focuses on improving the relevance between the textual query and visual images. However, there exists semantic gap between user intention and textual query. Let's take the query "jaguar" as an example, as shown in Fig.1. Different users have different intentions when inputting the query "jaguar". Some are expecting leopard images, while others are expecting automobile images. This scenario is quite common, particularly for queries with heterogeneous concepts or general (non-specific) concepts. This raises a fundamental but yet unsolved problem in Web image search: how to understand user intentions when users conducting image search?

Today user interests is mostly used to understand user intentions. For the instance in last paragraph, if we have the knowledge that the user is interested

S. Li et al. (Eds.): MMM 2013, Part I, LNCS 7732, pp. 239–249, 2013.

in animals, we can infer that he is likely to want the images about leopards when he searches "jaguar". In the past years, interest analysis is very difficult due to the lack of personal data. With the development of social media platforms, such as Flickr and Facebook, the way people can get social data has been changed: users' profiles, interests and their favorite images are exposed online and open to public, which are crucial information sources to implicitly understand user interests.In this paper, we exploit social data to assist image search, aiming to improve the relevance between returned images and user interests, which is termed as *Social Relevance*.

By considering social relevance and visual relevance comprehensively, we can understand user intention better, thereby improving the performance of our image ranking approach. However, the combination faces the following challenges:

(1) **Social data sparseness.** In social media platform, most users only possess a small number of favored images, from which it is difficult to discover user intentions. With the hypothesis that users in the same community share similar interests, a community-specific method is more practical and effective than a user-specific method.

(2) **The tradeoff between social relevance and visual relevance.** Although social relevance may guarantee the interest of returned images for the user, the quality and representativeness of images, cannot be ignored. Both of which are necessary for good search results. Thus, both social relevance and visual relevance are needed to be addressed and subtly balanced.

(3) **Complex factors.** To generate the final image ranking, one needs to consider the user query, returned images from current search engines, and many complex social factors derived from social media platforms. How to integrate these heterogeneous factors in an effective and efficient way is quite challenging.

Fig. 1. The results returned by Flickr for the query "jaguar", recorded on April, 10th, 2012

To address the above problems, we propose a novel community-specific Social-Visual Ranking (SVR) algorithm to rerank the Web images returned by current image search engines. More specifically, in SVR, given the preliminary image search results and the user's Flickr ID, we will use group information in social platform and visual contents of the images to rerank the Web images for a group that the user belongs to, which is termed as *the user's membership group*. In SVR, user interests and textual query are both utilized to predict user intention. The SVR algorithm is implemented by PageRank over a hybrid image link graph, which is the combination of an image social-link graph and an image visual-link graph. Through SVR, the Web images are reranked according to their interests to the users while maintaining high visual quality and representativeness for the query.

The contributions of our proposed approach are highlighted as follows:

1) We propose a novel image ranking method for by combining the information in social media platforms and traditional image search engines to address the user intention understanding problem in Web image search, which is of ample significance to improve image search performances.

2) We propose a community-specific Social-Visual Ranking algorithm to rerank Web images according to their social relevance and visual relevance. In this algorithm, complex social and visual factors are effectively and efficiently incorporated by hybrid image link graph, and more factors can be naturally enriched.

3) We have conducted intensive experiments, indicated the importance of both visual factors and social factors, and demonstrated the advantages of social-visual ranking algorithms for Web image search. Except image search, our algorithm can also be straightforwardly applied in other related areas, such as product recommendation and personalized advertisement.

The rest of the paper is organized as follows. We introduce some related works in Section 2. Image link graph generation and image ranking is presented in Section 3. Section 4 presents the details and analysis of our experiments. Finally, Section 5 concludes the paper.

2   Related Work

Aiming at improving the visual relevance, a series of methods are proposed based on incorporating visual factors into image ranking [16,7]. An essential problem in these methods is to measure the visual similarity[4]. As an effective approach, VisualRank[3] determines the visual similarity by the number of shared SIFT features[1]. After a similarity based image link graph was generated, an iterative computation similar to PageRank[10,15] is utilized to rerank the images. VisualRank obtains a better performance than text-based image search in the measurement of relevance for queries with homogeneous visual concepts. However, for queries with heterogeneous visual concepts, VisualRank does not work well[8].

With the development of social media platform, the concept of social image retrieval was proposed, which brings more information and challenges to us[13]. Most of works in social image search focus on tags [14,6,5]. However, the quality of recommendation is based on the technique of tag annotation[12], which is not mature enough. Overall, understanding user intention is significant but challengeable in social media platform. Many social media sites such as Flickr offer millions of groups for users to share images with others. There are tons of works based on improving the user experience [11]. Group information is an efficient way to estimate user interests.

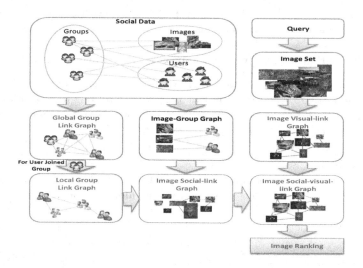

Fig. 2. The framework of our approach. Major intermediate results and final target are marked in red

3   Social-Visual Reranking

Fig.2 illustrates the framework of our Social-Visual ranking algorithm. In our approach, a random walk model based on PageRank[10] is utilized for image ranking. The weight $p(I_i, I_j)$ of the link from image I_i to image I_j represents the probability that a user will jump to I_j after viewing I_i. This procedure can be considered in both social factor and visual factor. From the social point of view, if I_i's group G_p is similar to I_j's group G_q, the probability of user's jump from I_i to I_j will be high. From the visual point of view, a user may be attracted by some visual contents of I_i and then decide to view I_j which also contains these contents. As a result, these two factors will both have significant effects in image ranking. Thus, we define our image link graph as the linear combination of the visual-link graph and the social-link graph. i.e.,

$$P_G = \alpha \cdot P_G^S + (1 - \alpha) \cdot P^V \tag{1}$$

where P_G is the adjacency matrix of the hybrid image link graph. P_G^S is the matrix for image social-link and P^V is the matrix of image visual-link graph. α is a parameter to balance these factors. The estimation of α will be discussed in Section 4. In this equation, P_G and P_G^S are relevant to the user's membership group G. Therefore they have a subscript as 'G'. The symbols with the subscript 'G' in our algorithm have the same meaning.

3.1 Image Social-Link Graph

Global Group Ranking. First, a global group link graph is generated in preprocessing phase of our algorithm based on group similarity. Group similarity in user interests can be measured by the overlap of user sets and data sets, which is defined as:

$$S(G_u, G_v) = \lambda \cdot overlap(\mathcal{M}_u, \mathcal{M}_v) + (1 - \lambda)overlap(\mathcal{I}_u, \mathcal{I}_v) \qquad (2)$$

where $\mathcal{M}_i$ is the user set of group G_u and $\mathcal{I}_u$ is the image set of group G_u. λ is a parameter to balance the user factor and the image factor. The overlap of $\mathcal{M}_i$ and $\mathcal{M}_j$ can be described as the Jaccard distance:

$$overlap(\mathcal{M}_u, \mathcal{M}_v) = \frac{\mathcal{M}_u \cap \mathcal{M}_v}{\mathcal{M}_u \cup \mathcal{M}_v} \qquad (3)$$

so is the overlap of $\mathcal{I}_u$ and $\mathcal{I}_v$.

After the pair-wised group similarities are computed, the iterative computation based on PageRank can be utilized to evaluate the centrality of the groups:

$$gr = d \cdot S \cdot gr + (1 - d)e_0, e_0 = \left[\frac{1}{N_G}\right]_{N_G \times 1} \qquad (4)$$

where S is a column-normalized matrix constructed by $S(G_u, G_v)$. N_G is the number of groups. d is the probability for user to visit the images along the graph links rather than randomly.

Local Group Link Graph. Local group link graph can be generated based on social strength of pairwise groups. Social strength of group G_u and group G_v for the given membership group G, represented as $T_G(G_u, G_v)$, describes the correlation between G_u and G_v with respect to G's interests. In other words, $T_G(G_u, G_v)$ denotes the probability that an user in G will jump to the images of G_v after viewing the images of G_u.

The group similarity $S(G_u, G_v)$ can represent the degree that G_u recommend G_v to G. If users in G are interested in images in G_u and G_u recommend G_v to G, then users in G may also be interested in images in G_v. Therefore, we can formulate the social strength $T_G(G_u, G_v)$:

$$T_G(G_u, G_v) = (S(G, G_u) + S(G, G_v)) \cdot S(G_u, G_v) \cdot f(gr(G_u)) \cdot f(gr(G_v)) \quad (5)$$

where $f(gr(G_u))$ is a function of the group rank value of G_u in the rank vector calculated in Eq.4. It denotes the weight of group importance. In this paper, we just consider the basic form of power function, which is proved to be valid[17], i.e.:

$$f(x) = x^r \qquad (6)$$

where r is a parameter which will be estimated by experimental study.

Image Social-Link Graph. For images and groups, we first construct a basic image-group graph. The edge from an image to a group denotes the image belonging to the group:

$$A(I_i, G_u) = \begin{cases} 1 & I_i \text{ belongs to } G_u \\ 0 & \text{otherwise} \end{cases} \tag{7}$$

Based on local group link graph and image-group graph, we can define the weight of the edge in image social-link graph as:

$$p_G^S(I_i, I_j) = \frac{Z_1}{(\sum_{u=1}^{N_G} A(I_i, G_u))(\sum_{u=1}^{N_G} A(I_j, G_u))} \cdot$$
$$\sum_{u=1}^{N_G} \sum_{v=1}^{N_G} A(I_i, G_u) \cdot A(I_j, G_v) \cdot T(G_u, G_v) \tag{8}$$

where Z_1 is a column-normalization factor to normalize $\sum_j p_G^S(I_i, I_j)$ to 1. $p_G^S(I_i, I_j)$ denotes the probability that group G will visit I_j after viewing I_i in social factor.

3.2   Image Visual-Link Graph and Social-Visual Ranking

SIFT descriptors of the images are clustered into some visual words by a hierarchical visual vocabulary tree[9]. Then, an image can be regarded as a document including some words. The weight of the edge in visual image link graph can be defined as:

$$p^V(I_i, I_j) = \frac{C(I_i, I_j)}{\sum_i C(I_i, I_j)} \tag{9}$$

where $C(I_i, I_j)$ is the count of co-occurrence of visual words in image I_i and I_j.

After two image link graphs are generated, hybrid image link graph can be constructed by Eq.1. Then, the iteration procedure based on PageRank can be formulated as:

$$r_G = d \cdot P_G \cdot r_G + (1 - d)e \tag{10}$$

where $d = 0.8$ as in Eq.4. e is a parameter to describe the probability a user jumps to another image without links when he is tired of surfing by links. In our experiments, we have two choices of e:

$$e_1(i) = \frac{1}{N_I} \tag{11}$$

where N_I is the number of images, and

$$e_G(i) = Z_2 \frac{\sum_{u=1}^{N_G} A(I_i, G_u) \cdot S(G, G_u)}{\sum_{u=1}^{N_G} A(I_i, G_u)} \tag{12}$$

where Z_2 is the factor to normalize the sum of $\sum e_G(i)$ to 1. These two cases of e will be compared in our experiments.

4   Experiments

4.1   Dataset and Settings

In this paper, we conduct experiments with data including images, groups, users, group-user relations and group-image relations from Flickr.com. 30 queries are collected and 1000 images are downloaded for each query by Flickr API. The selected queries includes:(1)Daily articles with no less than two different meanings, such as "apple", "jaguar" and "golf";(2) Natural scenery photos with multiple visual categories, such as "landscape", "scenery" and "hotel";(3)Living facilities with indoor and outdoor views, such as "restaurant" and "hotel";(4)Fashion products with different product types, such as "smart phone" and "dress".

In our experiment, we compare our algorithm SVR with other three image ranking methods: VisualRank(VR), SocialRank(SR) and Flickr search engine by relevance(FR) as baseline. Among them, VR is the special case for SVR when $\alpha = 0$, and SR is the special case for $\alpha = 1$.

4.2   Measurements

Social Relevance. Defined as the relevance to user intention, social relevance is an important measurement in our experiments. For a query, we randomly select n testing pairs (I_i, G_u) from the dataset, which means a group G_u and an image I_i belongs to this group. When a user in G_u inputs a query, I_i should be one of the images he wants to find. In another word, I_i should get a high rank order in our algorithm. Therefore, we define a measurement called Average Rank(AR) to reflect the degree to which we can capture user intentions:

$$AR = \frac{1}{|T|} \sum_{I_i \in T} rank(I_i) \qquad (13)$$

where T is the set of testing pairs. $rank(I_i)$ is the image I_i's ranking order. In our experiments, we select 20 testing pairs for each query. The smaller the AR value, the better the algorithm performance.

Visual Relevance. All images in our dataset are labeled according to their visual relevance in 4 levels, 0:irrelevant, 1:so-so, 2:good, 3:excellent. Normalized Discounted Cumulative Gain ($NDCG$) is adopted to measure the visual relevance[2]. Giving a ranking list, the score $NDCG@n$ is defined as

$$NDCG@n = Z_n \Sigma_{i=1}^{n} \frac{2^{r(i)} - 1}{log(1 + i)} \qquad (14)$$

$r(i)$ is the score of the image in the ith rank order. Z_n is the normalization factor to normalize the perfect rank to 1.

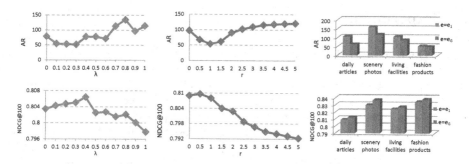

Fig. 3. Parameter settings for λ, r and e with $\alpha = 0.3$. Best performance is obtained with $\lambda = 0.4$, $r = 0.5$ and $e = e_G$

4.3   Parameter Settings

In our approach, there are four parameters: λ in Eq.(2), r in Eq.(6), α in Eq.(1) and e in Eq.(10). To study the effect of one parameter, we fix three other parameters as constants. Iteratively, we can find the optimal values for all the parameters to achieve the best performance.

From the Fig.3 we can find that our approach obtains the best performance when $\lambda = 0.4$, $r = 0.5$, $\alpha = 0.3$ and $e = e_G$. As the parameter representing the trade-off between the users' overlap and the images' overlap to determine group similarity, the value of λ shows users are more likely to be interested in a group because of its images rather than users. The value of r indicates that the importance of a group has small impact on visual relevance. In other words, an important group may also share some low-quality images. Besides, it can be observed that the algorithm with $e = e_G$ is significantly better than $e = e_1$ for all categories of queries. Thus, personalized vector e_G can indeed improve the performance of our approach.

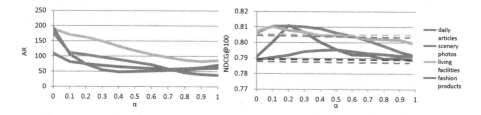

Fig. 4. Performance for different values of α with $\lambda = 0.4$, $r = 0.5$ and $e = e_G$. Best performance is obtained with $\alpha = 0.3$.

α is an important parameter to balance social factor and visual factor. We estimate the setting of α for each of the four categories. From the results in Fig.4, we can observe following: (1)For any category, best α is not small. i.e., social factor is helpful in image search.(2)Measured by AR, larger α produces better performance. Therefore, social factor can improve the images' relevance to user interests.(3)The curve of $NDCG$ indicates that, as the weight of social factor growing after a critical point, more images with low visual relevance are ranked to the front. Based on these observation, α is determined to be 0.3 in our approach, which can guarantee a reasonable balance between social relevance and visual relevance.

4.4 Results and Performance

To prove the results of SVR can really reflect the user intentions, we select 2 queries "jaguar" and "hotel" to show cases of our results. For each query, we select 2 groups that we can obviously estimate the interests by their group names. Fig.5 shows the results. The content in the bracket after SVR is the group name. It can be observed that our approach really knows what the users want and the results are mostly of high quality. For the query "jaguar", which has obvious different concepts, SVR can find the images fit for the group names fairly well. In contrast, the top-10 results of VisualRank for "jaguar" are all about leopards.

Fig. 5. Top-10 reranking results of our approach for two different groups compared to FlickrRank and VisualRank for two typical queries .

For the quantitative evaluation of the performance, we compare our approach with other three ranking methods. Fig.6 shows the comparison results. It can be observed that our approach achieves the best performance in $NDCG$ and

Fig. 6. The performance of our approach compared to other two methods FlickrRank and Visual Rank by the measurements AR and $NDCG@100$ for four categories of queries

has great improvement in AR compared to VR. Although AR of SR is the best, $NDCG$ of SR is much worse than VR. Under the comprehensive consideration, our approach performs the best in these four ranking methods.

5   Conclusions

In this paper, we propose a novel framework of community-specific Social-Visual image Ranking for Web image search. We explore to combine the social factor and visual factor together based on image link graph to improve the performance of social relevance under the premise of visual relevance. Comprehensive experiment shows effectiveness of our approach. In that, it is significantly better than VisualRank and Flickr search engine in social relevance as well as visual relevance. Besides, the importance of both social factor and visual factor is discussed in details.

Acknowledgements. This work is supported by National Natural Science Foundation of China, No. 60933013 and No. 61003097; National Program on Key Basic Research Project, No. 2011CB302206 and National Significant Science and Technology Projects of China under Grant, No. 2011ZX01042-001-002. Thanks for the support of Tsinghua-Tencent Associated Laboratory.This work was supported in part to Dr. Qi Tian by ARO grant W911NF-12-1-0057, NSF IIS 1052851, Faculty Research Awards by Google, NEC Laboratories of America and FXPAL, respectively.

References

1. Broder, A., Kumar, R., Maghoul, F., Raghavan, P., Rajagopalan, S., Stata, R., Tomkins, A., Wiener, J.: Graph structure in the web. Computer Networks 33(1-6), 309–320 (2000)
2. Järvelin, K., Kekäläinen, J.: Ir evaluation methods for retrieving highly relevant documents. In: Proceedings of the 23rd Annual International ACM SIGIR, SIGIR 2000, pp. 41–48. ACM, New York (2000)

3. Jing, Y., Baluja, S.: Visualrank: Applying pagerank to large-scale image search. IEEE Transactions on Pattern Analysis and Machine Intelligence 30(11), 1877–1890 (2008)
4. Kondor, R.I., Lafferty, J.: Diffusion kernels on graphs and other discrete structures. In: Proceedings of the ICML, pp. 315–322 (2002)
5. Larson, M., Kofler, C., Hanjalic, A.: Reading between the tags to predict real-world size-class for visually depicted objects in images. In: Proceedings of the 19th ACM Multimedia, MM 2011, pp. 273–282. ACM, New York (2011)
6. Li, X., Snoek, C.G., Worring, M.: Learning tag relevance by neighbor voting for social image retrieval. In: Proceedings of the 1st ACM Multimedia Information Retrieval, MIR 2008, pp. 180–187. ACM, New York (2008)
7. Liu, J., Lai, W., Hua, X.-S., Huang, Y., Li, S.: Video search re-ranking via multi-graph propagation. In: Proceedings of the 15th Multimedia, MULTIMEDIA 2007, pp. 208–217. ACM, New York (2007)
8. Liu, Y., Mei, T., Hua, X.-S.: Crowdreranking: exploring multiple search engines for visual search reranking. In: Proceedings of the 32nd International ACM SIGIR, SIGIR 2009, pp. 500–507. ACM, New York (2009)
9. Nister, D., Stewenius, H.: Scalable recognition with a vocabulary tree. In: Computer Vision and Pattern Recognition, vol. 2, pp. 2161–2168 (2006)
10. Page, L., Brin, S., Motwani, R., Winograd, T.: The pagerank citation ranking: Bringing order to the web. Technical Report 1999-66, Stanford InfoLab (November 1999) (previous)
11. Park, L.A.F., Ramamohanarao, K.: Mining web multi-resolution community-based popularity for information retrieval. In: Proceedings of the Sixteenth ACM CIKM, CIKM 2007, pp. 545–554. ACM, New York (2007)
12. Sang, J., Liu, J., Xu, C.: Exploiting user information for image tag refinement. In: Proceedings of the 19th ACM Multimedia, MM 2011, pp. 1129–1132. ACM, New York (2011)
13. Wang, M., Ni, B., Hua, X.-S., Chua, T.-S.: Assistive tagging: A survey of multimedia tagging with human-computer joint exploration. ACM Comput. Surv. 44(4), 25:1–25:24 (2012)
14. Wang, M., Yang, K., Hua, X.-S., Zhang, H.-J.: Towards a relevant and diverse search of social images. IEEE Transactions on Multimedia 12(8), 829–842 (2010)
15. Zhang, S., Huang, Q., Hua, G., Jiang, S., Gao, W., Tian, Q.: Building contextual visual vocabulary for large-scale image applications. In: Proceedings of Multimedia, MM 2010, pp. 501–510. ACM, New York (2010)
16. Zhou, W., Lu, Y., Li, H., Song, Y., Tian, Q.: Spatial coding for large scale partial-duplicate web image search. In: Proceedings of Multimedia, MM 2010, pp. 511–520. ACM, New York (2010)
17. Zhou, W., Tian, Q., Lu, Y., Yang, L., Li, H.: Latent visual context learning for web image applications. Pattern Recognition 44(10-11), 2263–2273 (2011); Semi-Supervised Learning for Visual Content Analysis and Understanding

Fusion of Audio-Visual Features and Statistical Property for Commercial Segmentation

Bo Zhang*, Bailan Feng, and Bo Xu

Institute of Automation
Chinese Academy of Sciences
Beijing 100190, China
{bo.zhang,bailan.feng,xubo}@ia.ac.cn

Abstract. Commercial segmentation is a primary step of commercial management which is an emerging technology. Relative to general video scene segmentation, commercial segmentation is particular because of dramatic changes in acoustic effect and chromatic composition. Conventional algorithms emphasize on utilizing new audio and visual features to adapt with change over time. In this paper, we have proposed a novel scheme to fuse audio-visual characteristics and statistical property of commercial length to find individual commercial boundaries. First, mid-level descriptors such as Static Shot with Product Information (SSPI) are used to predict the likelihoods of commercial boundary for every shot boundary. And then, Dynamic Programming (DP) refiner with Distribution of Individual Commercial Length (DICL) constraint is applied to find the optimal path of a Markov Chain of these shot boundaries. Experiments on simulated and real datasets show promising results.

Keywords: Commercial Segmentation, Scene Segmentation, Support Vector Machine, Video Analysis, Dynamic Programming.

1 Introduction

TV commercials have great influence upon our lives, and millions of people's living and working habits are affected by them. The management of commercials which concludes detection, identification, retrieval and recommendation has multiple potential applications, such as monitoring broadcast time of target commercial and pushing suitable commercial to target customer. Using audio and visual features to detect and identify known commercials [1] [9] is an efficient and reliable method for commercial management, which relies on a database of known individual commercials. Automatic new commercial detection and segmentation are preliminary stages to establish the known commercial database. Previous approaches focus on detecting new commercial blocks [5] [6] [12] and get

* This work was supported by the National Natural Science Foundation of China (Grant No.90820303) and the National Natural Science Foundation of China (61202326).

S. Li et al. (Eds.): MMM 2013, Part I, LNCS 7732, pp. 250–260, 2013.

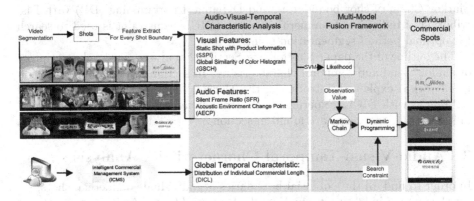

Fig. 1. The illustration of the proposed scheme

good performance while ignoring the segmentation of commercial blocks into individual commercial spots which are the main components of known commercial database.

Unlike traditional scene segmentation problem in video analysis [10] [13] which groups semantically related shots to a scene, the visual and acoustic change are much more drastic than that in general program such as movies. But there are also some unique sematic descriptors for commercial segmentation which are generated from produce and broadcast process. Hauptmann [2] utilized the occurrence of black frames among the individual commercials to segment. However, for some countries, especially in China, there is seldom this indicator to tell where to locate the actual commercial boundaries. Duan [7] have proposed an intermediate feature named Image Frames Marked with Product Information (FMPI) to segment the commercial blocks into individual commercial spots. Liu [12] propose an enhanced FMPI representation by exploiting the intrinsic visual and textual characteristics to reinforce the discrimination ability. Most of the previous works emphasize on utilizing audio-visual features and ignoring statistical information of individual commercials which is very important to refine the segmentation result.

In this paper, we focus our research on commercial segmentation and propose a novel scheme based on the collaborative exploitation of audio-visual characteristics and statistical information of individual commercial length. As illustrated in Fig.1, the video stream is first segmented into a sequence of shots, and four kinds of mid-level audio-visual descriptors [Static Shot with Product Information (SSPI), Global Similarity of Color Histogram (GSCH), Silent Frame Ratio (SFR), and Acoustic Environment Change Point (AECP)] are extracted from each shot boundary to describe the intrinsic characteristics of intra-commercial and inter-commercial. Then, to determine whether or not these shot boundaries belong to individual commercial boundaries, Support Vector Machine (SVM) classifier is utilized to exploit the audio and visual cues to form a consolidated fuse of them, and gets the likelihood of commercial boundary for every shot boundary. At last, the output likelihoods are regarded as observations of a

Markov Chain of shot boundaries, and Dynamic Programming (DP) with Distribution of Individual Commercial Length (DICL) constraint is used to search the optimal path of segmentation point.

The outline of the paper is as follows. Section 2 analyzes mid-level descriptors of audio, visual and temporal characteristics. In section 3, the proposed fusion framework is explained, introducing the Markov Chain of shot boundaries and DP refiner with DICL. Experiments on commercial segmentation are presented in section 4. Conclusions are found in section 5.

2   Audio-Visual-Temporal Characteristics Analysis

In order to find all the potential boundaries of individual commercial shots, we use the method in [8] to get a high recall of cuts and fade-in/-out to cover most of the potential commercial boundaries. The problem of commercial segmentation is transformed to commercial boundary detection. Five unique sematic descriptors are introduced to exploit the intrinsic audio, visual and temporal semantic cues.

2.1   Static Shot with Product Information (SSPI)

As shown in Figure 1, most commercials end with a relative static shot which highlights the promoted product information in commercials to reinforce viewer's appeal of the names and features of their products. To utilize this kind of semantic information, Duan [7] resorted to the combination of texture, edge, and color features to represent FMPI. Liu [12] proposed an enhanced FMPI representation by exploiting the intrinsic visual and textual characteristics. Although their method achieved promising detection results, it only concentrated on visual features about image. As one of the most distinct characteristics, the temporal semantic information was not considered in their work. Accordingly, in order to reinforce the discrimination ability of this characteristic, we proposed a descriptor named Static Shot with Product Information (SSPI) by exploiting the intrinsic global temporal visual and action attributes.

Edge Change Ratio and Frame Difference which describe structural changes in scene such as entering, exiting and moving objects are always low at the end of commercial blocks to express the information of product. While those cues change much more rapidly in the middle of commercial blocks to attract the viewer's appeal. In this paper, we use the average and variance of Edge Change Ratio and Frame Difference to describe the temporal cue of SSPI (4 dimensions). More details about those two features can be find in [1] and [5]. The key-frame of SSPI which contains the information of product is same as FMPI. As described in [7], a 141-d visual feature vector comprising 128-d local features (48-d local color features, 16-d density features and 64-d Gabor texture features) and 13-d global features (9-d global color features and 4-d edge direction features) is constructed. The likelihood of FMPI for every key-frame is obtained from SVM classifier with radial basis function (RBF) kernel (1 dimension).

Indubitably, the presence of SSPI, which generally appears around the end of individual commercials, can be reasonably taken to form an effective descriptor for commercial segmentation, which indicates a series of potential positions for the boundaries of individual commercials. The descriptors of SSPI from left shot and right shot around the candidate shot boundary are described with Edge Change Ratio, Frame Difference and FMPI likelihood (10 dimensions).

2.2   Global Similarity of Color Histogram (GSCH)

It can be seen from Fig.1, shots which belong to one commercial often have similar dominated color attribute, and we consider the global similarity of color histogram around shot boundary rather than the individual shot pairs to describe this characteristic. A 16 bin HSV color histogram [4] is used to describe the color information for each shot's key frame. The value of these histograms are summarized with their shots' length weights to generate a final histogram. Then, the Global Similarity of Color Histogram (GSCH) $s_{ch} = \sum_{h \in bins} \min(H_L(h), H_R(h))$ is constructed for each shot boundary, where H_L and H_R are the HSV color histogram of left window and right window (10s-long), respectively (1 dimension).

2.3   Silent Frame Ratio (SFR)

Silent audio frames always appear at the end of commercials in most cases owing to the audio-visual asynchrony. Thus, the occurrence of silent frames around the commercial boundaries can be reasonably taken as an essential characteristic for commercial segmentation. The audio signals in our studies are digitized at 22050Hz. For each visual shot, the audio stream is segmented into a sequence of 15-ms-long non-overlapping frames. Silence point detection is performed on these audio frames by comparing the zero crossing rate and short time energy of each frame with the pre-defined thresholds. The Silent Frame Ratio (SFR) is set as $R_{SF} = N_{SF}/N_S$, where N_{SF} is the number of silent frames, and N_S is the total number of frames of shot. Then, the SFRs of the left shot and right shot around the shot boundary are used to describe this characteristic (2 dimensions).

2.4   Acoustic Environment Change Point (AECP)

In general, different commercials have different acoustic environment such as background music and speaker. A proper modeling of Acoustic Environment Change Point (AECP) detection can facilitate the identification of commercial boundaries. For each shot boundary, every 100 audio frames are combined into 1.5s-length clips to detect the AECP. Two hypothesis have been formulated, H_0 considers that the left side x^L and the right side x^R of candidate shot boundary share the same acoustic environment, and H_1 considers that they belong to different acoustic environments. Each hypothesis is modeled by a GMM following the BIC (Bayesian Information Criterion) like algorithm [3] with MFCC features. H_0 is modeled by θ_0, and H_1 is model by GMM per side θ_1^L and θ_1^R. The proposed criterion is then simply the log likelihood ratio (with constant number of parameters)

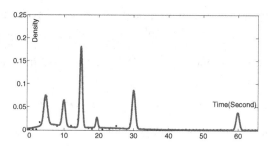

Fig. 2. Distributions of individual commercial length

$$d_{ac} = L_1(x^L|\theta_1^L) + L_1(x^R|\theta_1^R) - L_0(x^L, x^R|\theta_0) - \frac{\lambda}{2} \cdot \Delta k \cdot \log N \qquad (1)$$

where Δk is the difference of the number of the model parameters, N is the number of data points are modeling, and λ is the penalty weight. If d_{ac} is greater than zero, it is considered to be an AECP, then the descriptor of AECP for it will set to be 1; otherwise, is equal to 0 (1 dimension).

2.5   Distribution of Individual Commercial Length (DICL)

Commercials are organized videos of communicating information about products which advertisers want to promote to viewers, and commercial spots do not occur randomly in the television video stream. The timing of the insertion of a commercial segment is both affected by the content planning strategies of broadcasters and the budget of advertisers. The standard TV commercial length was 60-second in the 1950s, and 30-second got eventual dominance in 1970s. Quickly following the media inflation, 15-second spots and even shorter films emerge and became popular. The cost of creating, producing, and airing TV commercial is very expensive. Newstead [11] compared the proportionately attention of 15-s and 30-s commercials, and pointed out that the shorter advertising can get three-quarters of the memorable impact with only half the film time, which costs roughly half as much in airtime. For advertisement awareness can be achieved simply by increasing the length and frequency of commercial, advertisers generally use longer commercial to introduce a new product, and broadcast the shorter one frequently to reinforce the impact of audience. Today, TV commercials are produced for 5, 10, 15, 30, and 60 seconds for different purpose and strategy. The statistical information of individual commercial length from an Intelligent Commercial Management System (ICMS) which has monitored over 100 Chinese television channels every day shows some interesting characteristic. As shown in Fig.2, we get the real broadcast list of individual commercials from ICMS and model the density distribution of different lengths with 8 components Gaussian Mixture Model (GMM). The density function $P_{GMM}(t) = \sum_{k=1}^{K} \pi_k N(t|\mu_k, \Sigma_k)$ ($K = 8$) is considered as the temporal descriptor named Distribution of Individual Commercial Length (DICL).

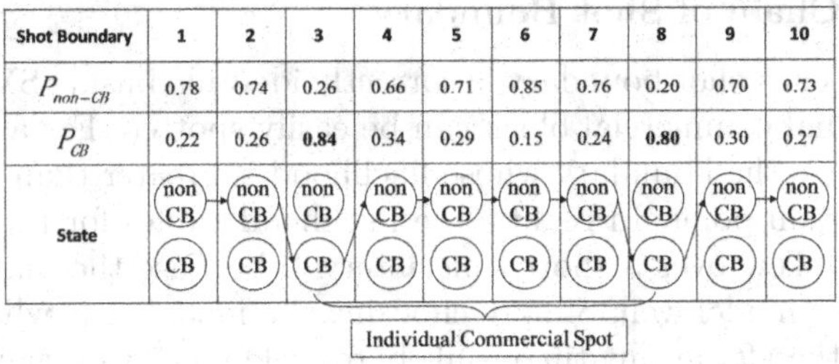

Shot Boundary	1	2	3	4	5	6	7	8	9	10
P_{non-CB}	0.78	0.74	0.26	0.66	0.71	0.85	0.76	0.20	0.70	0.73
P_{CB}	0.22	0.26	**0.84**	0.34	0.29	0.15	0.24	**0.80**	0.30	0.27
State	non CB / CB	non CB / CB	non CB / CB	non CB / CB	non CB / CB	non CB / CB	non CB / CB	non CB / CB	non CB / CB	non CB / CB

Individual Commercial Spot

Fig. 3. The illustration of a Markov Chain of shot boundaries

3 Multi Model Fusion Framework

As shown in Fig.3, a commercial video is constituted by commercial boundary (CB) and non-commercial boundary (non-CB). An individual commercial spot is always composed of a start CB, some non-CBs and an end CB. Different commercial spots are divided by CB. The start CB of current commercial spot is always the end CB of previous commercial spot, and the end CB is always the start CB of the next commercial spot. Then, we present a SVM-DP fusion framework to collaboratively exploit these audio-visual-temporal characteristics. First, SVM is used to fuse audio-visual cues to get the likelihood of commercial boundary for every shot boundary. Then, those likelihoods are treated as observations of a Markov Chain of shot boundaries. At last, DP refiner with DICL constraint is used to find the optimal segmentation point.

3.1 SVM-Based Fusion of Audio-Visual Features

Now, we get 14-dimensional basic audio-visual features from section 2. As well known, the neighborhoods of current video shot boundary are also very important for distinguishing intra-commercial from inter-commercial. A symmetrical sliding window is used (5 seconds). Besides utilizing 14-dimensional basic audio-visual features, the mean and variance of these basic features of its neighborhood shot boundaries are contained in both the left and right sides of the window, respectively. A 70-dimensional combined feature vector is obtained. Then, LIB-SVM [14] is employed to learn a classifier with the RBF kernel, and the optimal parameters are obtained by cross-validation. Since we use RBF kernels, we have to tune two parameters, C (the cost parameter in soft-margin SVMs), and γ (the width of the RBF function). Through analysis the broadcast video, we found that the numbers of commercial boundary and non-commercial boundary are highly unbalanced. Thus, we should consider three model parameters: C^{CB} (the cost parameter for the commercial boundary examples), C^{non-CB} (the cost parameter for the non-commercial boundary examples), and γ. In practical implementation, we assigned $C^{CB} = \frac{N^{CB}}{N^{CB}+N^{non-CB}} * C$ and $C^{non-CB} = \frac{N^{non-CB}}{N^{CB}+N^{non-CB}} * C$, where N^{CB} and N^{non-CB} are the number of individual commercial boundaries and non-commercial boundaries in training set, respectively.

3.2   Markov Chain of Shot Boundary

In an ideal case, each shot boundary is correctly identified using SVM classifiers, then the individual commercial block can be easily spotted. For a binary classifier, we can choose the boundary whose likelihood is greater than T_p ($T_p = 0.5$) as the segmentation point. In reality, the likelihood values for the correct class can temporarily skew over a short time interval, because the audio-visual features can't fit the model well. Simply choosing the model that whose likelihood value is greater than T_p at any time is likely to yield a noisy classification result. By considering the statistic information of individual commercial length, the misclassified results can be refined. As shown in Figure 3, likelihoods from SVM classifiers are considered as observations of a two states markov chain about CB and non-CB. The transition probability of moving from one state to other state is modeled by DICL with GMM in our experiment. The problem of commercial segmentation is therefore transformed into that of finding the optimal sequence of the hidden states of this time duration dependent model. Dynamic Programming approaches with DICL constraint are introduced to solve the problem.

3.3   Dynamic Programming Refiner Using DICL

One way to refine the short misclassified results is to search optimal state path based on an accumulative likelihood that a shot boundary belongs to a particular state from the starting point, with a penalty assigned to any transition from one state to another to suppress false transitions. The dynamic programming tries to find the optimum state transition path, which maximizes the score $L_i(d) = \max_k\{L_k(d-1) + T_p(k,i)\} + P_i(d)$ at every point d, $L_i(d)$ is the accumulated likelihood for the most likely path ending in state i at point d, $T_p(k,i)$ is the probability for transition from state k to state i, $P_i(d)$ is the likelihood value for state i at point d, $i = 1, 2$. Here, we introduce some variables: $B_i(d)$ is the state backtrack pointer, specifying the state at point $d-1$ for the most likely path ending in state i at point d; $C^*(d)$ is the class index at point d; $H_i(d)$ which record the state change point is the start point of the current state; D is the total number of candidate points. The search algorithm is outlined as follows:

Initialization:

$$L_i(1) = P_i(1), \quad B_i(1) = 0, \quad H_i(1) = 1, \quad i = 1, 2 \tag{2}$$

Recursion:

$$L_i(d) = \max_{1 \le k \le 2} \{L_k(d-1) + T_p(k,i)\} + P_i(d) \tag{3}$$

$$B_i(d) = \arg \max_{1 \le k \le 2} \{L_k(d-1) + T_p(k,i)\}, \quad i = 1, 2 \tag{4}$$

$$H_i(d) = \begin{cases} d & , \quad if \quad i = CB, \\ H_i(d-1) & , \quad else; \end{cases} \tag{5}$$

Termination:

$$L^* = \max_{1 \le i \le 2} (L_i(D)) \tag{6}$$

Table 1. The database of individual commercials

Commercial Length	5s	10s	15s	30s	60s	Total
Segment	95	130	470	230	80	1005
Duration (hours)	0.13	0.36	1.96	1.92	1.33	5.70

$$C^*(D) = \arg \max_{1 \leq i \leq 2} (L_i(D)) \tag{7}$$

Class Backtracking:

$$C^*(d) = B_{C^*(d+1)}(d+1), \quad d = D-1, D-2, \cdots, 1 \tag{8}$$

Ideally, the penalty value $T_p(k, i)$ should be set to the transition probability from state k to state i, measured from training data. Accurate measurement of this probability requires training data that contain a sufficient number of state transition events. But such data is difficult to obtain. In our experiment, this value is assigned based on DICL which is simulated by GMM. Then the transition probability is defined as:

$$T_p(k, j) = \begin{cases} \alpha * P_{GMM}(t_d - t_{H_i(d)}) & , \quad if \quad i = CB, \\ 1 - \alpha * P_{GMM}(t_d - t_{H_i(d)}) & , \quad else; \end{cases} \tag{9}$$

where t_d is the time of point d, and $t_{H_i(d)}$ is the time of point $H_i(d)$, α is an amplifier which makes the median line of the GMM to around 0.5. $P_{GMM}(t)$ is the density function of GMM which model the DICL.

4 Experiments and Analysis

The goal of the experiments are three aspects. First, study the performances of different fusion strategies. Second, compare our proposed method with the traditional fusion framework, and analyze the effect of using Dynamic Programming refiner with DICL constraint. Third, test the performance of our proposed scheme on real TV data. Therefore, we organized an extensive set of experiments on simulated and real datasets.

4.1 Simulated Dataset and Performance Evaluation

As no publicly available dataset exists for such task, TV individual commercial spots are obtained from known commercial database of ICMS. As shown in Table 1, the total duration of simulated dataset which includes 1005 different individual commercial spots are 5.70 hours. The dataset is equally divided into 5 sub-sets, and each sub-set is tested with the other four beings used for training. The training and testing data are generated from these basic individual commercial spots randomly. The evaluation measures based on *precision*, *recall* and $F1$ are utilized to evaluate the performance of the proposed scheme.

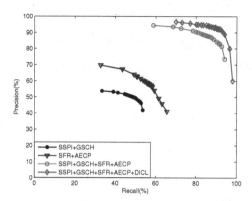

Fig. 4. Precision-Recall curves of different fusion strategies

A temporal boundary error tolerance of n shots is used when calculating *recall* and *precision*. That is to say, a detected boundary which is less than n shots apart from a real commercial boundary can be regarded as a true positive.

4.2 Multi-Model Fusion Performance

To analyze the detection results, we compare *precision-recall* curves yielded from different fusion strategies ($n = 2$). As shown in Fig.4, we can find that the visual features (SSPI + GSCH) does not achieve desirable results as expected. A possible explanation is that the main visual indicator SSPI occurs not only at the end of commercials but also irregularly interposed in the course of commercials to timely highlight the product information. Audio features (SFR + AECP) gets a good result, that maybe the silence and acoustic environment change always occur between different commercials. Noticeably, the combinations of all descriptors (SSPI + GSCH + SFR + AECP + DICL) which can gain the benefit from the collaborative exploitation of the intrinsic audio-visual-temporal characteristics achieves the best performance.

4.3 Comparison with Traditional Fusion Framework

Then, we conduct a comparison between our proposed scheme and other methods proposed in the literatures. Besides the proposed algorithm, we implemented those in [7] [12] for comparison. The F_1 results of different methods with different error tolerance are shown in Fig.5. The traditional fusion method which utilizes SVM-based classifier fused audio-visual features (SSPI + GSCH + SFR + AECP) without post-proceeding gets better results than Duan's method [7] and Liu's method [12]. That means the mid-level descriptors introduced in this paper is suitable for commercial segmentation. All methods get good results while error tolerance n is equal to 2, and our proposed method which fuses the audio-visual characteristics and statistical property of individual commercials with SVM-DP fusion framework introduced in this paper gets more promising

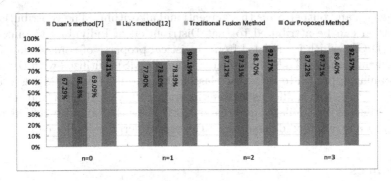

Fig. 5. The F1 results of different commercial segmentation methods

result than other methods while error tolerance n is equal to 0. That is because, the audio-visual descriptors of SSPI, GSCH, SFR and AECP might not simultaneously appear in the last shot for an individual commercial, but may gradually occur in a certain range around the end of commercial. Our proposed method which exploits the statistic information DICL can refine the classification result.

4.4   Experiment on Real Broadcast Data

For verifying the practical applicability of our proposed method on establishing and updating the known commercial database of ICMS, we have done an experiment on real data which is collected from five Chinese broadcast channels. The real dataset contains 30 commercial blocks which include 164 different individual commercial spots, about 0.93 hours. We obtain the ground truth by manually labeling the individual commercial boundary. The evaluation of our proposed method with different error tolerance is shown in Table 2. N_{gt} is the number of individual commercial boundary of Ground Truth, N_d is the number of detected commercial boundaries, and N_r is the number of detected right commercial boundaries. It can be clearly seen that, the proposed method gets promising results on the real data, and is very close to the performance on the simulated data.

Table 2. The evaluation of different Error Tolerance on real data

	N_{gt}	N_d	N_r	$Recall$	$Precision$	F_1
n=3	164	165	152	92.68%	92.12%	92.40%
n=2	164	165	151	92.07%	91.52%	91.80%
n=1	164	165	148	90.24%	89.70%	89.97%
n=0	164	165	144	87.80%	87.27%	87.54%

5   Conclusion

In this paper, we propose an effective SVM-DP scheme to fuse audio-visual features and statistical information of commercial length for commercial segmentation. This method can be used in automatic and interactive commercial management, and outperforms the conventional algorithms. The contribution of this method can be attributed to that Distribution of Individual Commercial Length (DICL) which is modeled from statistical property of commercials from an Intelligent Commercial Management System (ICMS) is utilized with Dynamic Programming approach to refine the classification results from SVM-based classifier which fuses four audio-visual descriptors (SSPI, GSCH, SFR and AECP). Our experiments on simulated and real datasets show that about 0-20% increase in F-measure is observed depending on the error tolerance is allowed.

References

1. Lienhart, R., Kuhmunch, C., Effelsberg, W.: On the Detection and Recognition of Television Commercials. In: Proc. of IEEE Conf. on Multimedia Computing and Systems (1997)
2. Hauptmann, A.G., Witbrock, M.J.: Story segmentation and detection of commercials in broadcast news video. In: Proc. Conf. ADL (1998)
3. Ajmera, J., McCowan, I., Bourlard, H.: Robust speaker change detection. IEEE Signal Processing Letters 11, 649–651 (2004)
4. Rasheed, Z., Shah, M.: Detection and representation of scenes in videos. IEEE Transactions on Multimedia 7, 1097–1105 (2005)
5. Hua, X.S., Lu, L., Zhang, H.J.: Robust Learning-based TV Commercial Detection. In: Proc. ICME (2005)
6. Mizutani, M., Ebadollahi, S., Chang, S.F.: Commercial Detection in Heterogeneous Video Streams Using Fused Multi-Modal and Temporal Features. In: Pro. ICASSP (2005)
7. Duan, L.Y., Wang, J.Q., Zheng, Y., Jin, J.S., Lu, H.Q., Xu, C.S.: Segmentation, categorization, and identification of commercials from tv streams using multimodal analysis. In: Proc. ACM MM 2006, pp. 202–210 (2006)
8. Yuan, J.H., Wang, H.Y., Xiao, L., Zheng, W.J., Li, J.M., Lin, F.Z., Zhang, B.: A Formal Study of Shot Boundary Detection. IEEE Trans. on Circuits and Systems for Video Technology 17, 168–186 (2007)
9. Duxans, H., Conejero, D., Anguera, X.: Audio-based automatic management of TV commercials. In: Pro. ICASSP (2009)
10. Chasanis, V.T., Likas, A.C., Galatsanos, N.P.: Scene Detection in Videos Using Shot Clustering and Sequence Alignment. IEEE Transactions on Multimedia 11, 89–100 (2009)
11. Newstead, K., Romaniuk, J.: Cost Per Second: The Relative Effectiveness of 15- and 30-Second Television Advertisements. Journal of Advertising Research 50, 68–76 (2009)
12. Liu, N., Zhao, Y., Zhu, Z.F., Lu, H.Q.: Exploiting Visual-Audio-Textual Characteristics for Automatic TV Commercial Block Detection and Segmentation. IEEE Transactions on Multimedia 13, 961–973 (2011)
13. Sidiropoulos, P., Mezaris, V., Kompatsiaris, I., Meinedo, H., Bugalho, M., Trancoso, I.: Temporal video segmentation to scenes using high-level audiovisual features. IEEE Transactions on Circuits and Systems for Video Technology 21, 1163–1177 (2011)
14. Libsvm, http://www.csie.ntu.edu.tw/cjlin/libsvm/

Learning Affine Robust Binary Codes
Based on Locality Preserving Hash

Wei Zhang[1,2], Ke Gao[1], Dongming Zhang[1], and Jintao Li[1]

[1] Advanced Computing Research Laboratory,
Beijing Key Laboratory of Mobile Computing and Pervasive Device,
Institute of Computing Technology,
Chinese Academy of Sciences, 100190 Beijing, China
[2] University of Chinese Academy of Sciences, 100049 Beijing, China
{zhangwei,kegao,dmzhang,jtli}@ict.ac.cn

Abstract. In large scale vision applications, high-dimensional descriptors extracted from image patches are in large quantities. Thus hashing methods that compact descriptors to binary codes have been proposed to achieve practical resource consumption. Among these methods, unsupervised hashing aims to preserve Euclidean distances, which do not correlate well with the similarity of image patches. Supervised hashing methods exploit labeled data to learn binary codes based on visual or semantic similarity, which are usually slow to train and consider global structure of data. When data lie on a sub-manifold, global structure can not reflect the inherent structure well and may lead to incompact codes. We propose locality preserving hash (LPH) to learn affine robust binary codes. LPH preserves local structure by embedding data into a sub-manifold, and performing binarization that minimize false classification ratio while keeping partition balanced. Experiments on two datasets show that LPH is easy to train and performs better than state-of-the-art methods with more compact binary codes.

Keywords: Similarity Search, Binary Code Learning, Locality Preserving Hash, Locality Preserving Projection.

1 Introduction

Descriptor extracting is fundamental to computer vision applications such as image retrieval, copy detection, object recognition, etc. The descriptors are always high-dimensional and in large quantities. To allow for fast matching with practical memory consumption, there have been many recent attempts that compact descriptors to binary codes in Hamming space based on hashing techniques [1-11].

Compacting methods such as locality sensitive hashing (LSH) [1] and spectral hashing (SH) [2] are unsupervised and aim to preserve the Euclidean distances of original vectors, but do not take the visual similarity of patches into account. As can be seen from Fig.1, descriptors of visually similar patches are not always adjacent to each other in vector space. The Euclidean distance cannot reflect similarity relationships

S. Li et al. (Eds.): MMM 2013, Part I, LNCS 7732, pp. 261–271, 2013.

properly in such situations. Christoph Sander et al. proposed supervised LDAHash [9], which performs linear discriminant analysis (LDA) to preserve similarity relationships of a training set. However, LDA considers the global structure of data and has little to do with the manifold or local structure [10]. When data lie on a sub-manifold, it is hard to handle the inherent structure well and may lead to incompact binary codes[10].

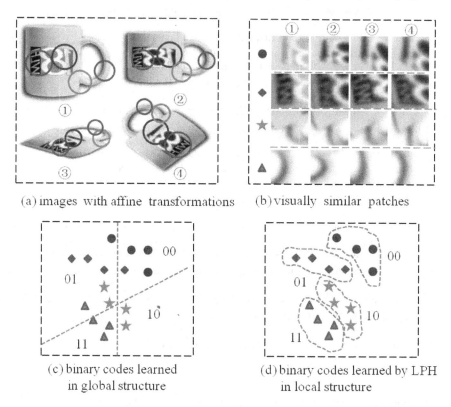

(a) images with affine transformations   (b) visually similar patches

(c) binary codes learned   (d) binary codes learned by LPH
in global structure   in local structure

Fig. 1. An illustration of affine robust binary codes learned by locality preserving hash (LPH) in local structure (d), compared with binary codes learned in global structure (c). Visually similar patches (b) are detected and normalized from images with affine transformations (a) and consistent in visual contents and geometric information.

To overcome the drawbacks in existing methods mentioned above, we propose locality preserving hash (LPH), as illustrated in Fig.1. LPH generates affine robust and compact binary codes by preserving the local structure of data. Specifically, LPH has the following new characteristics:

(1)LPH embeds data into a sub-manifold by locality preserving projection (LPP) [12] with a patch similarity matrix. Locality preserving projection (LPP) is a linear algorithm and has shown its effectiveness in exploring the intrinsic manifold structure in face recognition. However, the original LPP is unsupervised. To exploit the training samples, we define a new similarity matrix following the supervised learning scheme [10].

The matrix is based on similarity of image patches, which are automatically labeled by mining stable patches in affine transformed images [13]. In this way, binary codes learned are affine robust. Supervised learning methods are usually prone to over fitting when labeled data are of small amount or noisy [5]. This can be avoided in our case, for training samples are automatically generated and labeled.

(2)After the embedding, a binarization process is performed which partitions the data space and represents data as binary codes. The traditional methods mainly focus on the partition balance of data space [2, 14-15] for efficiency. We propose a novel binarization method that minimizes false classification ratio while keeping partition balanced. Therefore the local structure is preserved as much as possible in the partitioned space while an efficient retrieval is achieved.

The rest of the paper is organized as follows: section 2 presents the related work. In section 3, we present our LPH by formulating the hashing problem and giving our solutions. In section 4, we introduce how to conduct off-line training and then evaluate our method on two datasets. Finally, we conclude the paper in section 5 and give research perspectives for the future of this work.

2   Related Work

In large-scale computer vision applications, high-dimensional descriptors are extracted in huge quantities to represent image contents, which challenge the limited computational and storage resources. There have been many attempts at compacting descriptors to allow for fast matching with practical memory consumption. One class is based on quantization such as bag of visual words [16] and product quantization [17]. However, these approaches are not sufficient to produce short codes without loss of performance. Another class is to compact descriptors into binary codes. These approaches are promising, for binary codes can retain the desired similarity information with much reduced memory. And various methods have been proposed [1-11] in recent years.

A simple idea is to embed data to a new data space using a matrix, whose elements are random variables sampled from Gaussian distributions as in LSH [1]. Since the distribution of high dimensional data in vision applications is far from uniform [18], binary codes generated by machine learning approaches [2-5, 9-11] are always more compact than random methods. Weiss et al. propose Spectral Hashing [2] that attempts to preserve the given similarities defined by Euclidean metric and form codes with each bit carrying more information, which achieves significant improvement over LSH. Transform coding [19] and integrated binary codes [8] utilize principle component analysis (PCA) to embed data which aim to maximally preserve the variance.

The methods mentioned above aim to preserve the neighborhood relationships in Euclidean space of descriptors. Unfortunately, the descriptors of visually similar image patches may be far away in the vector space and thus those methods are not sufficient to reflect the underlying relationships of such patches.

By contrast, many supervised methods take advantage of similarity information contained in training data, and thus can map similar descriptors closer and dissimilar ones far apart. There are various forms of supervision to define similarity, such as learned Mahalanobis metric [20], Restricted Boltzmann Machines (RBMs) [4], and Binary Reconstruction Embeddings [3]. However, as discussed in [21], these methods are suffered from the difficulty of optimization and slow training mechanisms. LDAHash performs supervised optimization easily and tries to capture the global structure of data [9].

We propose LPH to learn affine robust binary codes by exploring the underlying manifold structure of training samples, with effective and efficient optimization. We define image patches as visually similar if they are detected from transformed images but consistent in visual contents and geometric information, as is shown in Fig.1. We aim to preserve the local structure of visual similar patches when compacting descriptors to binary codes. This is a relaxation of the similarity preserving problem, compared with methods [4, 11] that define patches with quite different visual contents as similar. Besides, our training method is simple and easy to implement.

3   Locality Preserving Hash

Compacting descriptors to binary codes is usually implemented by hashing, which include two key steps: (1) embed data to a new space represented by a projection matrix and (2) binarize the embedded or projected values with thresholds. The two steps boil down to the generation of projection matrix and selection of optimal thresholds. In this section, we formulate the problem and present our solution of the generation of projection matrix and the selection of binarization thresholds in LPH.

3.1   Problem Formulation

Given a training set of descriptors containing groups of visually similar patches, the aim of LPH is to map descriptors into affine robust and compact binary codes, by preserving the local structure of similar patches. Formally, let $X=\{x_1,...,x_n\}$ be a set of d-dimensional descriptors and $h(x_i)$ be a hash function, we seek a binary feature vector $y_i=[h_1(x_i),...,h_k(x_i)]$ that preserves the visual similarity relationships using a Hamming distance.

Suppose we want to get a k-bit code y_i of x_i, then k hash functions leading to k Hamming embeddings are needed. We use linear projection and threshold as the hash function, and then binary code is computed using the following equation:

$$y_i = \mathrm{sgn}(Px_i^T + t) . \tag{1}$$

where P is a $k*d$ matrix and t is a $k*1$ vector.

The projection matrix P is the key of embedding and threshold t is the key of binarization. The existing methods of generating P and t, randomly or supervised, have been discussed in the previous section. Ours are discussed in the next section.

3.2   Projection Matrix Generation

Given a matrix W with weights characterizing the similarity of two image patches, we want to learn a P to preserve the local structure, which satisfies the following:

$$\text{minimize:} \quad \sum_{ij} \left\| y'_i - y'_j \right\|^2 W_{ij} . \tag{2}$$

where y' is the data projected by P, i.e. $y'=Px^T$ and W_{ij} is defined as follows:

$$W_{ij} = \begin{cases} 1, & \text{if image patches of } x_i \text{ and } x_j \text{ are visually similar} \\ 0, & \text{otherwise} \end{cases} . \tag{3}$$

To simplify the similarity computation between different points, we just set the weight equal to 1. The similarity of image patches is automatically obtained when collecting descriptors. And descriptors are collected by simulating affine transformations of images, extracting descriptors from each transformed image and mining stable ones that are consistent in visual contents and geometric information. The details can be found in [13]. Then we define stable descriptors as visually similar to construct W. The similarity learning method is the spirit of LPP and P can be obtained by solving a generalized Eigen value problem as in [12].

3.3   Optimal Thresholds Selection

The threshold selection is a separate problem from projection matrix generation. In this section, $y'=P_i x^T$ denotes the i-th element of the projected x, and $y=\text{sgn}(P_i x^T + t_i)$ will generate the i-th bit of the final code. Usually, thresholds are selected to partition data space in balance for efficiency considerations. Here, we consider both effectiveness and efficiency aspects of a partition. On one hand, we want to minimize the ratio of false partitions that divide in-class data into two parts. In this way, the local structure will be preserved as much as possible in the partitioned space. The false ratio (FR) is also the probability that the value of a threshold t between the projected in-class data, which is given by:

$$FR(t) = \Pr\left(\min\{Y'\} < t < \max\{Y'\}\right) . \tag{4}$$

where Y' is projected data from a set of similar patches.

On the other hand, we want to generate a balanced partition for further efficient retrieval of binary codes. A balanced partition means that data are distributed uniformly in each sub-parts and indicates that less noises are accessed during searching. Since entropy can be used to measure the distribution of data, a balanced partition is achieved by maximizing the following partition entropy (PE):

$$PE(t) = - p_1 \log p_1 - p_2 \log p_2 , \quad p_1 = \Pr(y' \leq t) \\ p_2 = \Pr(y' > t) . \tag{5}$$

where y' denotes any vectors x projected by P.

Suppose the value of α and β represent the importance of the two sub functions *FR* and *PE*, then the final optimization problem is as follows:

$$\text{minimize:} \quad \alpha FR(t) - \beta PE(t) . \tag{6}$$

If $\alpha=0$, each sub-partition is of the same size and the most balanced partition is obtained. If $\beta=0$, data remains unpartitioned when *FR(t)* is maximized, which is meaningless. In the following experiments, we set α and β equal to 1.We solve the problem approximately by computing *FR(t)* and *PE(t)* with different thresholds and select the threshold when the optimal trade-off is gained.

4   Experimental Results

We perform our experiments on two datasets: (1) image patches extracted from INRIA Holiday dataset [15] and (2) the Ukbench dataset [22] with a distractor dataset downloaded from Internet. For evaluation we compute the average recall of each query and then take the mean value over the set of queries. All experiments are run on a workstation with 16GB memory and 4 CPUs of 2.13GHz.

Image Patches Extracted from Holiday Dataset (*91,400 image patches, 500 queries*). We collect images of different scene types from Holiday dataset and simulate affine transformations like ASIFT [23]. Then descriptors are obtained by DoG detector and SIFT descriptor [24]. Visually similar patches are obtained by mining descriptors that are stable [13] in various transformations. We define a descriptor as stable if it appears in more than 5 transformed images with slightly changed geometric information. And severely changed descriptors in vector space are filtered to reduce the gap between feature space and vector space during learning. Then the patches of stable descriptors are collected as the database. The queries are randomly chosen from groups of visually similar patches and the correct retrieval results are the other ones of the corresponding groups.

Ukbench Dataset with Distractor Images (*10,200 images, 10,000 distractor images, 7.8million descriptors*). We apply our method to object retrieval on Ukbench dataset. The dataset contains groups of 4 objects shot in different viewpoints and illuminations. To evaluate in a larger database, we also introduce distractor images downloaded from Internet with various objects. The descriptors are obtained by DoG detector and SIFT descriptor. There are 7.8 million descriptors in total. The queries are randomly chosen from 500 groups and the correct retrieval results are the 4 images of the corresponding groups.

4.1   Off-line Training

The training dataset contains 10,000 groups of patches randomly chosen from our first dataset. The training process includes two steps, i.e. projection matrix generation and threshold selection. When visual similarity relationships are available, the similarity matrix W in (3) is constructed. The supervised LPP that generate the projection matrix is performed using Deng's source codes [25].

Then we select thresholds on each separate dimension after projecting data to the matrix. On each dimension of the projected data, we compute FR in (4) and PE in (5) with different thresholds. And then following (6), a threshold is selected when an optimal trade-off between false partitioned ratio and partition balance is achieved.

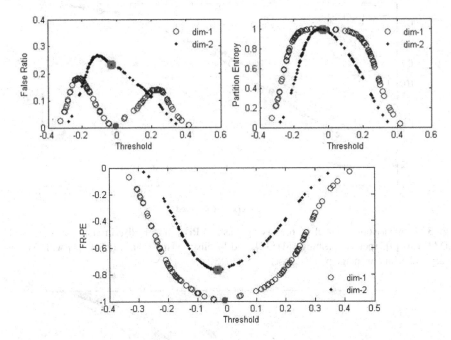

Fig. 2. Histograms of *false ratio* (*FR*), *partition entropy* (*PE*) and *FR-PE* in (6) computed with different threshold settings on two sample dimensions of projected data. The threshold is marked when optimal value is gained.

Fig.2 shows the selection of thresholds on separate dimensions. As can been seen, false ratio is minimized while partition entropy is maximized on the first dimension, which is the perfect case. This means that data are partitioned uniformly and only 0.65% data are false partitioned. The false ratio is higher on the second dimension for the partition balance is emphasized. If the false ratio is emphasized, α in (6) should be enlarged. Besides, the false ratio can be compensated by exploring a Hamming ball during query process.

4.2   Image Patch Retrieval

In this section, our proposed LPH is evaluated on the image patch database, for which the ground truth of visually similar patches are available. The average recall as a function of the rate of codes retrieved is used as the evaluation metric for this retrieval task. A higher recall with a lower rate of codes retrieved indicates that more correct codes are found with fewer noises returned. After the off-line training process, we compact database descriptors into binary codes. Then we adopt linear scan for codes retrieval. In addition, the efficiency can be improved further by using multi-index [26]. By setting Hamming radius of the search from 0 to 32, comparisons of LPH with the state-of-the-art methods are demonstrated in Fig.3 and Fig.4.

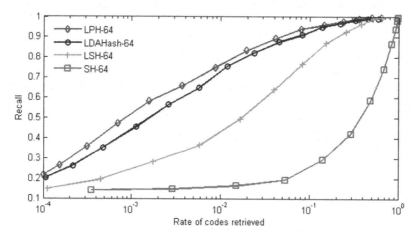

Fig. 3. Comparison of locality preserving hash (LPH), linear discriminant analysis hash (LDAHash) [9], spectral hashing (SH) [2], and locality sensitive hash (LSH) [1] with 64-bits binary codes on our image patch dataset

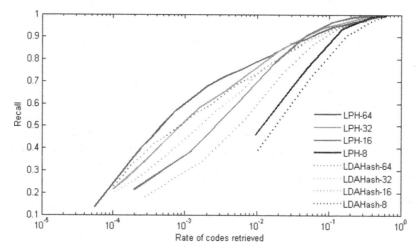

Fig. 4. Comparison of locality preserving hash (LPH) and linear discriminant analysis hash (LDAHash) [9] with different number of bits on our image patch dataset

As can been seen from Fig.3, LPH and LDAHash using supervised information show significant improvements over SH and LSH. And LPH performs the best. This is because LPH generates more robust binary codes according to the local structure preserving hash functions. An untypical observation is that SH performs worse than LSH on this dataset. Since SH aims to preserve Euclidean distances between data points, it performs well when the ground truth data are nearest neighbors in Euclidean space. However, in this experiment the ground truth data are based on visually similar patches. Hence SH degrades. This also proves that Euclidean distance can not reflect visual patch similarity properly. Besides, since we filter severely changed descriptors in the ground truth data of visually similar patch groups, LSH can compact similar points to similar codes in some extent and thus shows a higher recall than SH.

We further assess the impacts of the number of bits, as is shown in Fig.4. As the number of bits increases, recall is improved for both LPH and LDAHash but the improvement becomes smaller. With the same number of bits, LPH always perform better than LDAHash. LPH with 32 bits shows comparable performance to LDAHash with 64 bits and even better when the recall is larger than 0.5. This indicates that our binary codes are more compact, i.e. preserve more similarity information with the same number of bits.

4.3   Object Retrieval

To evaluate our approach on a more challenged dataset, we perform object retrieval on the Ukbench dataset which consists of groups of objects under different viewpoint, rotation, scale and lighting conditions. We set Hamming radius to 2 and the length of binary codes to 32 bits for all approaches. Fig 5 gives the average recall with a varying number of retrieved images.

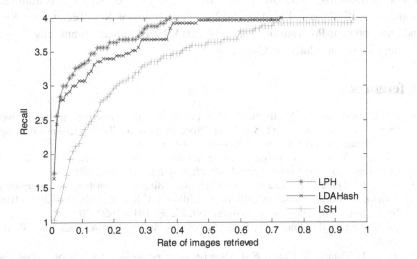

Fig. 5. Performance of object retrieval with different number of retrieved images for locality preserving hash (LPH), linear discriminant analysis hash (LDAHash) [9], and locality sensitive hash (LSH) [1]. The dataset is Ukbench with distractors.

For a dataset of 20,200 images and 7.8 million descriptors, the recall of LPH is 6.03% higher than LDAHash on average, and 45.61% higher than LSH. LPH achieves much better performance than unsupervised LSH and supervised LDAHash on the dataset. This proves that binary codes generated by LPH are more robust to complex transformations than the baseline methods.

The original SIFT descriptors needs 7.89G for storage while LPH only needs 78.2M. The memory cost of binary codes is one hundredth of original descriptors, which is critical for large scale applications. In this experiment we want to compare descriptor compacting approaches independently and thus adopt a simple vote scheme to rank the final results. Actually, the results can be improved further by geometry consistency verification [27-28].

5   Conclusion

This paper describes a new method named locality preserving hash (LPH) to compact descriptors into binary codes. The binary codes are affine robust and compact, for LPH has the following characteristics: (1) the embedding preserves manifold structure of descriptors extracted from visually similar patches; (2) automatic threshold selection achieves a good trade-off between positive classification ratio and partition balance. Besides, the off-line training of LPH is easy to implement. The experimental results on two datasets show that LPH often performs better than the baseline methods. Our future work would focus on the indexing methods in Hamming space to improve efficiency. Furthermore, there have been various improved versions of locality preserving projection and it will be interesting to explore them to improve LPH further.

Acknowledgments. This work was supported by the National Nature Science Foundation of China (61271428, 61273247), National Key Technology Research and Development Program of China (2012BAH39B02) and Co-building Program of Beijing Municipal Education Commission.

References

1. Datar, M., Immorlica, N., Indyk, P., Mirrokni, V.: Locality-sensitive hashing scheme based on p-stable distributions. In: SCG 2004: Proceedings of the Twentieth Annual Symposium on Computational Geometry, pp. 253–262. ACM (2004)
2. Weiss, Y., Torralba, A., Fergus, R.: Spectral hashing. In: Proceedings of the 22nd Annual Conference on Neural Information Processing Systems, NIPS (2008)
3. Kulis, B., Darrell, T.: Learning to Hash with Binary Reconstructive Embeddings. In: Bengio, Y., Schuurmans, D., Lafferty, J., Williams, C.K.I., Culotta, A. (eds.) Advances in Neural Information Processing Systems, vol. 22, pp. 1042–1050 (2009)
4. Salakhutdinov, R., Hinton, G.: Semantic hashing. Int. J. Approx. Reasoning 50, 969–978 (2009)
5. Wang, J., Kumar, S., Chang, S.-F.: Semi-supervised hashing for scalable image retrieval. In: Computer Vision and Pattern Recognition (CVPR), pp. 3424–3431 (2010)
6. Min, K., Yang, L., Wright, J., Wu, L., Hua, X.-S., Ma, Y.: Compact Projection: Simple and Efficient Near Neighbor Search with Practical Memory Requirements. In: IEEE Conference on Computer Vision and Pattern Recognition (CVPR), San Francisco, USA (2010)
7. He, J., Radhakrishnan, R., Chang, S.-F., Bauer, C.: Compact hashing with joint optimization of search accuracy and time. In: 2011 IEEE Conference on Computer Vision and Pattern Recognition (CVPR), pp. 753–760 (2011)

8. Zhang, W., Gao, K., Zhang, Y., Li, J.: Efficient approximate nearest neighbor search with integrated binary codes. In: Proceedings of the 19th ACM International Conference on Multimedia, pp. 1189–1192. ACM, Scottsdale (2011)
9. Strecha, C., Bronstein, A., Bronstein, M., Fua, P.: LDAHash: Improved Matching with Smaller Descriptors. IEEE T. PAMI 34, 1 (2012)
10. Wong, W.K., Zhao, H.T.: Supervised optimal locality preserving projection. Pattern Recogn. 45, 186–197 (2012)
11. Mu, Y., Shen, J., Yan, S.: Weakly-supervised hashing in kernel space. In: 2010 IEEE Conference on Computer Vision and Pattern Recognition (CVPR), pp. 3344–3351 (2010)
12. He, X., Niyogi, P.: Locality Preserving Projections. In: Neural Information Processing Systems. MIT Press (2003)
13. Gao, K., Zhang, Y., Luo, P., Zhang, W., Xia, J., Lin, S.: Visual stem mapping and geometric tense coding for augmented visual vocabulary. In: IEEE Conference on Computer Vision and Pattern Recognition, CVPR (2012)
14. Arya, S., Mount, D.M., Netanyahu, N.S., Silverman, R., Wu, A.Y.: An optimal algorithm for approximate nearest neighbor searching fixed dimensions. J. ACM 45, 891–923 (1998)
15. Jegou, H., Douze, M., Schmid, C.: Hamming Embedding and Weak Geometric Consistency for Large Scale Image Search. In: Forsyth, D., Torr, P., Zisserman, A. (eds.) ECCV 2008, Part I. LNCS, vol. 5302, pp. 304–317. Springer, Heidelberg (2008)
16. Sivic, J., Zisserman, A.: Video Google: A Text Retrieval Approach to Object Matching in Videos. In: Proceedings of the Ninth IEEE International Conference on Computer Vision (ICCV), vol. 2, p. 1470 (2003)
17. Jegou, H.: Product Quantization for Nearest Neighbor Search. IEEE Transactions on Pattern Analysis and Machine Intelligence 33, 117–128 (2011)
18. Poullot, S., Buisson, O., Crucianu, M.: Z-grid-based probabilistic retrieval for scaling up content-based copy detection. In: Proceedings of the 6th ACM International Conference on Image and Video Retrieval, CIVR (2007)
19. Brandt, J.: Transform coding for fast approximate nearest neighbor search in high dimensions. In: IEEE Conference on Computer Vision and Pattern Recognition (CVPR), pp. 1815–1822 (2010)
20. Jain, P., Kulis, B., Grauman, K.: Fast image search for learned metrics. In: Computer Vision and Pattern Recognition (CVPR), pp. 1–8 (2008)
21. Liu, W., Wang, J., Ji, R., Jiang, Y.-G., Chang, S.-F.: Supervised Hashing with Kernels. In: IEEE Conference on Computer Vision and Pattern Recognition, CVPR (2012)
22. Nister, D., Stewenius, H.: Scalable Recognition with a Vocabulary Tree. In: Proceedings of the 2006 IEEE Computer Society Conference on Computer Vision and Pattern Recognition, vol. 2, pp. 2161–2168 (2006)
23. Morel, J.M., Yu, G.: ASIFT: A New Framework for Fully Affine Invariant Image Comparison. SIAM Journal on Imaging Sciences 2, 438–469 (2009)
24. Lowe, D.G.: Object recognition from local scale-invariant features. In: The Proceedings of the IEEE International Conference on Computer Vision, vol. 1152, pp. 1150–1157 (1999)
25. http://www.cad.zju.edu.cn/home/dengcai/Data/ReproduceExp.html
26. Norouzi, M., Punjani, A., Fleet, D.J.: Fast Search in Hamming Space with Multi-Index Hashing. In: IEEE Conference on Computer Vision and Pattern Recognition, CVPR (2012)
27. Wang, W., Zhang, D., Zhang, Y., et al.: Robust Spatial Matching for Object Retrieval and Its Parallel Implementation on GPU. IEEE Trans. on Multimedia 13(6), 1308–1318 (2011)
28. Xie, H., Gao, K., Zhang, Y., Tang, S., et al.: Efficient Feature Detection and Effective Post-Verification for Large Scale Near-Duplicate Image Search. IEEE Trans. on Multimedia 13(6), 1319–1332 (2011)

A Novel Segmentation-Based Video Denoising Method with Noise Level Estimation

Shijie Zhang[1], Jing Zhang[1], Zhe Yuan[1], Shuai Fang[2], and Yang Cao[1]

[1] University of Science and Technology of China
[2] Hefei University of Technology
{sjz1901,zjwinner,zheyuan}@mail.ustc.edu.cn, fangshuai@hfut.edu.cn,
forrest@ustc.edu.cn

Abstract. Most of the state of the art video denoising algorithms consider additive noise model, which is often violated in practice. In this paper, two main issues are addressed, namely, segmentation-based block matching and the estimation of noise level. Different with the previous block matching methods, we present an efficient algorithm to perform the block matching in spatially-consistent segmentations of each image frame. To estimate the noise level function (NLF), which describes the noise level as a function of image brightness, we propose a fast bilateral medial filter based method. Under the assumption of short-term coherence, this estimation method is consequently extended from single frame to multi-frames. Coupling these two techniques together, we propose a segmentation-based customized BM3D method to remove colored multiplicative noise for videos. Experimental results on benchmark data sets and real videos show that our method significantly outperforms the state of the art in removing the colored multiplicative noise.

1 Introduction

Noise reduction (denoising) is of crucial importance in multimedia applications, as digital images and videos are often contaminated by noise during acquisition, compression, and storage. Accordingly, it is desirable to reduce noise for visual improvement or as a preprocessing step for subsequent processing tasks.

A number of previous video denoising methods are directly extended from image denoising, such as block matching and 3D filtering [1], wavelet shrinkage [2], PDE based methods [3] and non-local means (NLM) [4]. Several methods integrate motion estimation with spatial filtering. For example, a NLM framework integrating with robust optical flow is introduced in [5]. Besides, the idea of sparse coding in a patch dictionary has also been applied on video denoising (e.g. [6,7]), where the denoised image patches are found by seeking for the sparsest solution in a patch dictionary. In [8], the problem of denoising patch stacks is converted to the problem of recovering a complete low rank matrix from its noisy and incomplete observation.

The performance of a denoising method is highly dependent on how close the real noise fits the noise model assumed by the method. Most of the state of the

S. Li et al. (Eds.): MMM 2013, Part I, LNCS 7732, pp. 272–282, 2013.

art video denoising works consider additive noise model. However, this is often violated in practice. According to [9], there are five primary noise sources, fixed pattern noise, dark current noise, shot noise, amplifier noise, and quantization noise. The practical image noise is very likely to be the colored multiplicative noises, which are not independent of the pixel value. Therefore, most current denoising approaches cannot truly effectively estimate and remove the real noise. This fact prevents the noise removal techniques from being practically applied to multimedia applications.

The main contribution of this paper is that we propose a segmentation-based customized BM3D method that can effectively reduce colored multiplicative noise introduced by today's digital cameras. Two main issues are addressed in this paper: segmentation based block matching and the estimation of noise level. Different with the previous block matching methods which search similar blocks in a fixed-size neighborhood, we present an efficient method to perform the block matching in spatial-temporally consistent segmentations for each image frame. The estimation of the noise level function (NLF), which describes the noise level as a function of image brightness, is the key to ensure the removing of the colored multiplicative noise. Here in our method, a bilateral median filter is exploited to estimate NLF by fitting a lower envelope to the discrete samples measured from image segmentations. To obtain a more reliable NLF, we further extend the estimation method from single frame to multi-frames, under the assumption of short-term coherence of segmentations.

2   Related Work

Plenty of video denoising methods have been proposed in the recent few years. As it is beyond this paper to provide a comprehensive and detailed review, we will just focus on reviewing the work closest to ours.

In the past several decades, a variety of denoising methods have been developed in the image area. The significant part of denoising processing is the way to exploit image sparsity [10]. In frequency domain, when a natural image is decomposed into multiscale-oriented subbands [11], we can observe the highly kurtotic marginal distributions [12]. With this case, image sparsity leads to coring algorithms [13] to suppress low-amplitude values while retaining high-amplitude values. In spatial domain, the image sparsity is formulated as image self-similarity, namely patches in an image are similar to one another. NLF methods was proposed by finding the similar patterns to a query patch and take the mean or other statistics to estimate the true pixel value in [4]. Other denoising algorithms, such as PDE [3] and region based denoising [14] also implicitly formulate sparsity in their representation. Sparsity also resides in videos. A number of previous video denoising methods are directly extended from image denoising. Since video sequences usually have very high temporal redundancy, a new frame can be well predicted from previous frames by motion estimation. Therefore, several methods integrate motion estimation with spatial filtering for better performance [5]. Besides, the advances in the sparse representations have achieved outstanding denoising results [6,7,15]. The frequency and spatial forms of the image sparsity are introduced by BM3D method [1]. This method produces high-quality results and has been extended for video denoising by exploiting the much more redundancy information in video.

Although many video denoising methods achieve the state of the art results (e.g. [1,4,8]), they mainly consider additive noise model which is often violated in practice. As stated in [9], there are five primary noise sources, fixed pattern noise, dark current noise, shot noise, amplifier noise, and quantization noise. The noise model of a CCD camera is proposed in [14] as:

$$I = f(L + n_s + n_c) + n_q \tag{1}$$

where I is the observed image luminance, f is camera response function, n_s denotes all the noise components that are dependent on irradiance L, n_c is the independent noise before gamma correction, and n_q is additional quantization and amplification noise.

Therefore, the practical image noise is very likely to be the colored multiplicative noises, which are not independent of the pixel value. Most existing denoising methods are sensitive to the noise model violation. The existence of other type of noises will severely degrade the denoising performance. This inspires us to develop a robust denoising algorithm capable of removing color multiplicative noise from the given video data. In this paper, we propose a segmentation-based customized BM3D method to remove colored multiplicative noise in real world videos. Firstly, to improve the efficiency of the grouping process in BM3D, we propose a segmentation-based k-nearest neighbors searching approach. Secondly, we propose a fast bilateral median filter based method to estimate the noise level function (NLF). Once the NLF is properly estimated, the noise parameters in those collaborate filtering steps of BM3D method can be adaptively selected.

3   A Segmentation-Based Customized BM3D Method for Video Denoising

3.1   Overview

Figure 1 is the flowchart of our proposed method. This novel method consists of three modules. The first module aims at establishing the spato-temporal consistent segmentation [16,17]. Based on the segmentation, we can group similar blocks together efficiently by using our proposed k-nearest neighbors searching approach. In addition, we can estimate the NLF through a large number of discrete samples under the assumption of short-term coherence between consistent segmentations. Functionally, this module is designed as an auxiliary part to improve the performance of the classical BM3D algorithm when dealing with the colored multiplicative noise in real world videos. Thus, we have our customized BM3D algorithm shown as the second module. The third module is the output module which is to separate the filtered segmentations and stitch them together to form the final denoising result.

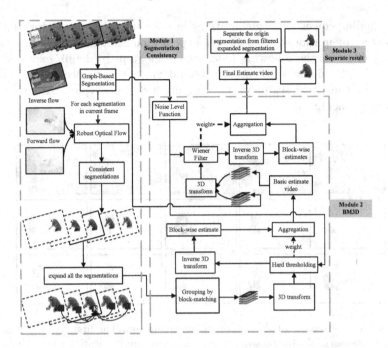

Fig. 1. Flowchart of our proposed segmentation-based customized BM3D method. The operations surrounded by dashed lines are repeated for each segmentation.

3.2 Segmentation Consistency

To achieve high quality video denoising result, we should fully utilize of the spatial-temporal consistency of videos to exploiting the redundancy information properly. Firstly we use Graph-based segment method [18] to get piece-wise smooth regions. Then we adopt robust optical flow [19] to find consistent segmentations across adjacent frames.

The segmentation size affects the performance of the following BM3D algorithm a lot. On one hand, the smaller the area is, the more accurate the noise level estimate will be. On the other hand, the bigger the area is, the more similar patches to the reference one we may find. In this paper, we set the maximum size as 0.04 times of image size, and set the minimum size as 0.002 times of image size. In this way, we make a trade-off between the maximum size and the minimum size about segmentation.

We use the robust optical flow proposed by C. Liu et al. [5] to estimate motion. We estimate forward flow $w^b(z) = [v_x, v_y, 1]$ from frame I_t to I_{t+1} and backward flow $w^b(z) = [v_x, v_y, -1]$ from frame I_t to frame I_{t-1} to establish bidirectional temporal correspondence. For each pixel in one segment, we use forward and backward flow to find the corresponding points in adjacent frames. In addition, if a pixel moves out of the image boundary, we just ignore it. In this way, all the points in one segment are mapped from one frame to its neighboring frames, and the temporal consistent segmentation is established.

After finding consistent segmentations in adjacent frames, we expand all the consistent segmentations with a morphological expansion factor. This is to avoid boundary effect in the following block-matching step. In detail, expansion enables the block on the segmentation boundary higher chance to find similar blocks inside consistent segmentations. Figure 2 shows a typical example of the consistent segmentation.

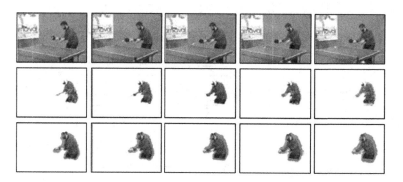

Fig. 2. A typical example of the consistent segmentation. The first row shows several consecutive frames of the tennis sequences with colored multiplicative noise. The second row shows the consistent segmentations established by robust optical flow. The third row shows the consistent segmentations after expansion.

3.3   Segmentation-Based Block-Matching

In the original BM3D algorithm, the searching process is performed in a fixed-size window. In addition, no motion information is exploited to guide the searching process. In contrast, we implement our searching process in the consistent segmentations. This implement has superiority in two aspects: the motion information helps us to improve the searching accuracy in the temporal domain and the segmentation helps us to improve the searching accuracy in the spacial domain. Consequently, we can fully utilize the spatial-temporal redundancy information contained in the videos.

As shown in Fig. 3, for each block centered at pixel z, we want to obtain a set of segmentation-based k-nearest neighbors. For efficiency, we use the priority queue data structure to store the k-nearest neighbors. The priority queue is initialized by the k-nearest blocks to the reference one inside the segmentation in the current frame, which is shown as green boxes in Fig. 3. For each block in the priority queue, we can determine their corresponding blocks in the adjacent frames after we estimate the motion in the video. For example, for block $P(z)$ in current frame t, we find its corresponding block $P\left(z+\mathrm{w}^{f}(z)\right)$ in the next frame $t+1$. If $z+\mathrm{w}^{f}(z)$ is out of segmentation boundary, shown as the blue boxes in Fig. 3, this block $P\left(z+\mathrm{w}^{f}(z)\right)$ will be discarded. Otherwise, this block $P\left(z+\mathrm{w}^{f}(z)\right)$ will be inserted into the priority queue and the last one in the queue is discarded. Then we can determine a $N_{FR} \times N_{FR}$ searching neighborhood of

$z+w^f(z)$, shown as the pink translucent regions in Fig. 3. And we try to insert more blocks similar to the reference one into the priority queue and update it. Consequently, we use this method to complete the priority queue after searching among those 2H adjacent frames.

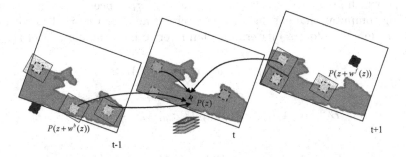

Fig. 3. Illustrations of the supporting patches in spatial-temporal domain for a block $P(z)$

3.4   Noise Level Estimation

The noise parameter controls the behavior of the collaborating filtering steps in BM3D. Therefore, it is important to set the noise parameter appropriately. C. Liu et al firstly introduced Noise level function (NLF) in [14] that describes the noise level as a continuous function of image intensity. They also proposed a method to rough estimate the upper bound of the noise based on Bayesian MAP method. However, the estimation method is computational expensive and needs a large number of statistical data. In this paper, we propose an easier but effective method that apply bilateral median filter to estimate an upper bound of NLF. Furthermore, we extend this method to the multi-frame situations.

Firstly, we calculate the means I_i and standard deviation σ_i of each consistent segmentations for all the three color channels (Here, the consistent segmentations are obtained according to the method shown in Section 3.2). Thus, we obtain the discrete point sets $\Omega_k = \{(I_{ik}, \sigma_{ik}), i = 1, 2, ...\}$, where i is the index of segmentations and k represents the color channel (r, g and b). In the next, the point sets Ω_k are plotting in Cartesian coordinates. The problem of estimation of NLF is then transformed into fitting the lower envelope of the point sets Ω_k by a smooth curve that is subject to some constraints. The fitting problem can be formalized as follows:

$$\arg\max_i \int_i \tau(I_i) - \alpha(\|\nabla\tau(I_i)\|^2) \tag{2}$$

with the constraints:

$$0 \leq \tau(I_i) \leq \sigma_{\min}(I) \tag{3}$$

$$b_{\min}(I) \leq T(I_i) \leq b_{\max}(I) \tag{4}$$

Here $\tau(I_i)$ is the noise level at index i and the factor α controls the smoothness of the solution. $T(I_i)$ is the first-order derivatives of $\tau(I_i)$. We discretize the range of brightness $[0,1]$ into uniform intervals Θ. The lower bound of the standard deviation is denoted as $\sigma_{\min} = \min_\Omega\{\sigma_i\}, i \in \Theta$. Similarly, $b_{\min}(I)$ and $b_{\max}(I)$ represent the lower and upper bound of the first-order derivatives on $\tau(I_i)$, which are calculated by the derivative of σ_i. Since the optimization of (2) being computational expensive, we search for another way to deal with this problem. Here we adopt a bilateral median filter to infer the noise level $\tau(I_i)$ as follows:

$$A(I_i) = median_{S_w}(\sigma_{\min}(I_i)) \tag{5}$$

$$B(I_i) = A(I_i) - median_{S_w}(|\sigma_{min}(I_i) - A(I_i)|) \tag{6}$$

$$C(I_i) = B(I_{i-1}) + \max(\min(B(I_i) - B(I_{i-1}), b_{\max}(I_i)), b_{\min}(I_i)) \tag{7}$$

$$\tau(I_i) = \max(\min(C(I_i), \sigma_{\min}(I_i)), 0) \tag{8}$$

Here S_w is the size of window used in the median filter. Figure 4 shows the example of the estimation of NLF using the above method with $S_w=15 \times 15$.

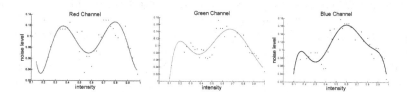

Fig. 4. Figure for noise estimation inside one frame. The red, green, blue curves stand for three channels respectively. X axis stands for intensity and Y axis stands for noise level.

3.5   Segmentation-Based Customized BM3D Algorithm

Coupling the above two techniques together, we propose a segmentation-based customized BM3D method. Figure 1 shows the overview of this method. It can be summarized as: firstly, we obtain consistent segmentations across adjacent frames using the method in Section 3.2. Then, we group the similar blocks by using k-nearest neighbors searching approach and estimate the NLF based on the consistent segmentation results. In the next, for the two collaborate filtering steps in BM3D method, Hard-thresholding filtering and Wiener filtering, the noise parameters are adaptively selected according to the estimated NLF. Finally, we separate the filtered segmentations and stitch them together to form the final denoising result.

4   Experimental Results

In this section we present and discuss the experimental results obtained by our method. In our experiments, we use 8×8 patches and 5 adjacent frames. A graph-based image segmentation algorithm is used to partition each image frame into regions. More details of the segmentation algorithm can be found in [18] and the source code is also available online[1]. For segmentation expansion, we use a hexagon morphological operator with the radius of 9 pixels. For the optical flow, we use a robust optical estimation algorithm [19] and the source code is available online[2]. We download BM3D code package from BM3D website[3] for comparison. In this package, the function VBM3D is designed to deal with gray videos, and CVBM3D is designed to deal with color videos. The gray benchmark videos are also downloaded from BM3D website, and the color benchmark videos are downloaded from website[4]. We adopt the synthetic method shown in [14] to add the colored multiplicative noise into the benchmark videos. As shown in (1), there is a camera response function (CRF) f and two noise variances σ_s and σ_c which control the noise level of n_s and n_c in the synthesized video. We download 190 CRFs from the website[5]. To achieve better synthesis results, in our experiment, we select the 60th function to synthesize the colored multiplicative noise.

Table 1. We conduct two groups of experiments on different color videos and compare the PSNR results of our proposed method and the one of the best denoising algorithm: classical CVBM3D (with $\sigma = 5\sqrt{2}$ and $\sigma = 10\sqrt{2}$ respectively for the following two groups experiments)

Method	CVBM3D		Our method	
Parameters	$\sigma_s = 5, \sigma_c = 5$	$\sigma_s = 10, \sigma_c = 10$	$\sigma_s = 5, \sigma_c = 5$	$\sigma_s = 10, \sigma_c = 10$
Foreman	30.53	28.67	30.36	30.43
Newsreport	31.92	29.35	32.46	31.5
Tennis	24.02	23.36	24.21	23.91
Hall-monitor	28.51	27.14	28.49	27.71
Flower	20.82	20.43	20.34	20.15
Mother-daughter	30.71	28.27	32.39	31.74

Firstly, We apply our method on the Foreman sequences with synthetically colored multiplicative noise (with $\sigma_s = 10$, $\sigma_c = 10$) to show its performance on gray video. We use the ground-truth video and VBM3D results (with noise parameters $\sigma = 10\sqrt{2}$) for comparison. The PSNR values of VBM3D method

---

[1] http://www.cs.brown.edu/~pff/segment/
[2] http://people.csail.mit.edu/celiu/OpticalFlow/
[3] http://www.cs.tut.fi/~foi/GCF-BM3D/
[4] http://media.xiph.org/video/derf/
[5] http://www.cs.columbia.edu/CAVE

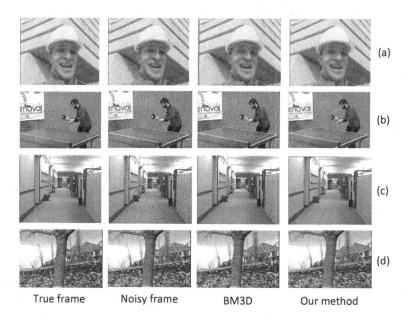

| True frame | Noisy frame | BM3D | Our method |

Fig. 5. Comparison results on a series of video sequences. (a) results of Foreman sequence, with $\sigma_s= 10$, $\sigma_c= 10$ and noise parameters $\sigma= 10\sqrt{2}$ for classical VBM3D. (b) results of Tennis sequence, with $\sigma_s= 10$, $\sigma_c= 10$ and noise parameters $\sigma= 10\sqrt{2}$ for classical CVBM3D. (c) results of Hall-monitor sequence, with $\sigma_s= 10$, $\sigma_c= 10$ and noise parameters $\sigma= 10\sqrt{2}$ for classical CVBM3D. (d) results of Flower sequence, with $\sigma_s= 5$, $\sigma_c= 5$ and noise parameters $\sigma= 5\sqrt{2}$ for classical CVBM3D. Please see color version of this figure in electronic edition.

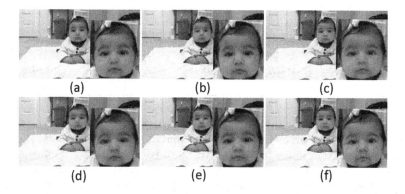

Fig. 6. Comparison results on real video sequences Baby. (a) One frame of real video. (b) CVBM3D with noise parameter $\sigma=20$. (c) CVBM3D with noise parameter $\sigma=30$. (d) CVBM3D with noise parameter $\sigma=40$. (e) Result of C. Liu et al's method. (f) result of our proposed method. Please see color version of this figure in electronic edition.

and our proposed method are 27.39dB and 28.32dB, respectively. As shown in the first row of Fig. 5, in terms of visual quality, the VBM3D result is highly affected by the presence of colored multiplicative noise, while our method can achieve better result no matter in the smooth region or the boundary area.

Then, we move on to the color benchmark videos. We select 6 videos and synthesize the colored multiplicative noised videos with different colored multiplicative noise level (with $\sigma_s = 5$, $\sigma_c = 5$ and $\sigma_s = 10$, $\sigma_c = 10$). The contrastive PSNR results of our method with that of CVBM3D on these videos are included in Table 1. We can see that, for most case, our proposed method can achieve better results in term of PSNR. The visual comparison results on color Tennis sequence, Hall-monitor sequence and Flower sequence are shown in the 2nd to 4th row of Fig. 5. It can be seen that some structured noises are still remains in the CVBM3D's results. Our method, however, can smooth out flat regions, as well as keep subtle texture details. In Fig. 5b, our method can smooth out the noise in the background and preserve the textures of table as well. In Fig. 5c and Fig. 5d, though our method performs slightly worse in terms of PSNR, our method completely remove the noise in the flat region and achieve more visually pleasing results.

Finally, we deal with the real video sequences Baby download from website[6], a 720P HD video clip catured by SONY HDR-XR150. One input frame and denoised frame are shown in Fig. 6(a). We compare our method with CVBM3D and C. Liu et al's denoising algorithm. For CVBM3D method, we try three noise parameters $\sigma = 20$, $\sigma = 30$ and $\sigma = 40$. The contrastive denoising results are shown in the Fig. 6(b-f). The close-up view is shown at the lower right corner. We can see that some structured noises are still remains in the CVBM3D's results. C. Liu et al's method and our method both can remove the structured noise and preserve details. Relatively, our method also reduce other types of multiplicative noise and achieve a more visually pleasing result.

5   Conclusion

In this paper, we propose an adaptive segmentation-based video denoising method to remove colored multiplicative noise from video data. We perform block-matching in spatial-temporally consistent segmentations, and integrate 3D filtering with the estimation of noise level function. The effectiveness of our proposed method is validated in various experiments on benchmark datasets and real videos. There is still a limitation of our approach: when the motion is large, the segmentation consistency becomes weak. We will investigate the use of stronger motion estimate method to compensate this case. In addition, we will try to embed our proposed method in other video denoising frameworks for further study.

Acknowledgements. This paper is supported by the Fundamental Research Funds for the Central Universities of China. (No. WK2100100009), NSFC (No.61175033) and NSFY of Anhui (No.BJ2100100018) and STP (No.11010202192) of Anhui.

---

[6] http://research.microsoft.com/en-us/um/people/celiu/ECCV2010/

References

1. Dabov, K., Foi, A., et al.: Image denoising by sparse 3-d transform domain collaborative filtering. IEEE Trans. on Image Processing 16, 2080–2095 (2007)
2. Sendur, L., Selesnick, I.W.: Bivariate shrinkage functions for wavelet-based denoising exploiting interscale dependency. IEEE Trans. on Signal Processing 50, 2744–2756 (2002)
3. Tschumperle, D.: Fast anisotropic smoothing of multi-valued images using curvature-preserving pdes. International Journal of Computer Vision (IJCV) 68, 65–82 (2006)
4. Buades, A., Coll, B., Morel, J.M.: Nonlocal image and movie denoising. International Journal of Computer Vision (IJCV) 76, 123–139 (2008)
5. Liu, C., Freeman, W.T.: A High-Quality Video Denoising Algorithm Based on Reliable Motion Estimation. In: Daniilidis, K., Maragos, P., Paragios, N. (eds.) ECCV 2010, Part III. LNCS, vol. 6313, pp. 706–719. Springer, Heidelberg (2010)
6. Mairal, J., Bach, F., et al.: Non-local sparse models for image restoration. In: International Conference on Computer Vision, ICCV (2009)
7. Protter, M., Elad, M.: Image sequence denoising via sparse and redundant representations. IEEE Transactions on Image Processing 18, 27–35 (2009)
8. Jiy, H., Liuz, C., Sheny, Z., Xuz, Y.: Robust video denoising using low rank matrix completion. In: Computer Vision and Pattern Recognition, CVPR (2006)
9. Healey, G., Kondepudy, R.: Radiometric ccd camera calibration and noise estimation. IEEE Trans. on Pattern Analysis and Machine Intelligence 16, 267–276 (1994)
10. Yang, J., Wright, J., Huang, T., Ma, Y.: Image super-resolution as sparse representation of raw image patches. In: Computer Vision and Pattern Recognition(CVPR) (2008)
11. Mallat, S.: A theory for multiresolution signal decomposition: the wavelet representation. IEEE Trans. on Pattern Analysis and Machine Intelligence 11, 674–693 (1989)
12. Field, D.: Relations between the statistics of natural images and the response properties of cortical cells. Journal of the Optical Society of America (JOSAA) 4, 2379–2394 (1987)
13. Simoncelli, E., Adelson, E.: Noise removal via bayesian wavelet coring. In: IEEE Int. Conf. on Image Processing, ICIP (1996)
14. Liu, C., Szeliski, R., et al.: Automatic estimation and removal of noise from a single image. IEEE Trans. on Pattern Analysis and Machine Intelligence (TPAMI) 30, 299–314 (2008)
15. Wang, M., Hua, X., et al.: Beyond distance measurement: Constructing neighborhood similarity for video annotation. IEEE Trans. on Multimedia 11, 465–476 (2009)
16. Zitnick, C., Jojic, N., et al.: Consistent segmentation for optical flow estimation. In: International Conference on Computer Vision, ICCV (2005)
17. Klappstein, J., Vaudrey, T., et al.: Moving object segmentation using optical flow and depth information. In: Advances in Image and Video Technology (2008)
18. Felzenszwalb, P.F., Huttenlocher, D.P.: Efficient graph-based image segmentation. International Journal of Computer Vision (IJCV) 59, 167–181 (2004)
19. Liu, C.: Beyond pixels: exploring new representations and applications for motion analysis. PhD thesis, Massachusetts Institute of Technology (2009)

Blocking Artifact Reduction in DIBR
Using an Overcomplete 3D Dictionary

Cheolkon Jung, Licheng Jiao, and Hongtao Qi

Key Laboratory of Intelligent Perception and Image Understanding of Ministry of Education,
Xidian University, Xi'an 710071, China
zhengzk@xidian.edu.cn

Abstract. It has been reported that the image quality and depth perception rates
are undesirably decreased by compression in DIBR. This is because high-
frequency components are filtered by compression, and thus several
compression artifacts occur. Blocking artifacts are the most representative ones
which seriously degrade the picture quality, and are annoying to viewers. The
fundamental cause of the coding artifacts is that the 3D contents are delivered to
users over the existing 2D broadcast infrastructures which employ BDCT
coding as image compression techniques. In this paper, we propose a new
deblocking method in 3D images, i.e., video-plus-depth, using an overcomplete
3D dictionary. We generate the 3D dictionary from natural and depth images
using the k-singular value decomposition (K-SVD) algorithm, and estimate an
error threshold to utilize the 3D dictionary using a compression factor of the
compressed images. Experimental results demonstrate that the proposed method
is very effective in reducing undesirable blocking artifacts in 3D images.

Keywords: 3D dictionary, blocking artifact, depth-image-based rendering
(DIBR), sparse representation.

1 Introduction

By recent advances in the multimedia processing technologies, 3DTV is expected as a
new broadcasting paradigm that provides a more natural and life-like visual home
entertainment experience to viewers. To realize realistic 3D services in 3DTV, depth-
image-based rendering (DIBR) is one of the essential technologies. DIBR is the
process of generating virtual left and right views of a real-world scene which provide
3D effects from reference image sequences and its corresponding depth sequences. In
3DTV, the DIBR technique was first introduced by the advanced three dimensional
television system technologies (ATTEST) project, which started in March 2002 as
part of the European Information Society Technologies (IST). Since depth
information can be compressed and transmitted efficiently through the additional data
channel (bandwidth: 64kbps), DIBR has received much attention in the broadcast
research society as a promising technology for 3DTV systems [1-7]. However, we
found that the image quality and 3D effects in DIBR are remarkably reduced after
compression through the research.

S. Li et al. (Eds.): MMM 2013, Part I, LNCS 7732, pp. 283–294, 2013.

Fig. 1. Example of compression artifacts in the *Ballet* sequence when a compression factor, q, is 10. Top left: original reference image. Top right: compressed reference image. Bottom left: original depth image. Bottom right: compressed depth image.

Thus, it is necessary to improve the image quality and 3D effects in 3DTV. It has been reported that the image quality and depth perception rates are undesirably decreased by compression in DIBR [1]. This is because high-frequency components are filtered by compression and thus several compression artifacts occur. As shown in Fig. 1, representative artifacts are blocking ones. They seriously degrade the picture quality and depth perception rates, and are annoying to viewers of the reconstructed 3D images. The fundamental cause of the compression artifacts is that the 3D contents are delivered to users over the conventional 2D infrastructures which employ a block-based Discrete Cosine Transform (BDCT) coding as image compression techniques. Block-based Discrete Cosine Transform (BDCT) has been widely used in image and video compression, and thus it has been adopted in most image/video compression standards including JPEG, MPEG, H.264, and so on. However, BDCT has major drawback at low bit rate, which is usually referred to as compression artifacts such as blocking, ringing, and shine ones. Conventional methods for blocking artifact reduction are classified into two main groups: 1) image enhancement based methods using adaptive or low pass filtering, and 2) image restoration based methods using the projection onto convex sets (POCS) or maximum a posteriori (MAP) estimation. Image enhancement based methods mainly use some filters in the spatial domain or in the transform domain to reduce the compression artifacts [8-12]. Above all, they assume that the blocking artifacts are artificial high frequency components around block boundaries. To remove the artifacts, several filtering methods including adaptive, loop, and low pass filters have been proposed. Image restoration based methods assume that the compression operation is a distortion process [13-18]. To recover the original image from distorted images, image restoration techniques such as POCS and MAP estimation are used. Image restoration from compressed images is one of representative inverse problems in image processing.

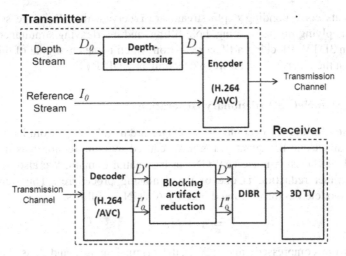

Fig. 2. Block diagram of the 3D broadcasting system based on DIBR. The main contribution of this work lies in the blocking artifact reduction module.

Recently, sparse representation has been actively studied to solve several inverse problems in image processing. Some researchers have made significant contributions to image denoising, restoration, and super-resolution using sparse representation [19-23]. Sparse representation is that original signals can be accurately recovered by several elementary signals called atoms. Our approach is inspired by the recent results based on sparse representation. In this paper, we design an overcomplete 3D dictionary based on sparse representation to reduce blocking artifacts in DIBR. Both reference and depth images in DIBR are called 3D images, i.e., video-plus-depth. We generate the 3D dictionary from 3D images using the K-SVD algorithm. Thus, the 3D dictionary includes two dictionaries for reference and depth images. Above all, we consider the deblocking process as a denoising procedure. Thus, as in the denoising procedure, error thresholds for orthogonal matching pursuit (OMP) are also estimated to use the 3D deblocking dictionary. Experimental results demonstrate that the overcomplete 3D dictionary effectively reduce blocking artifacts in DIBR even at a low bit rate, and reconstruct high-quality 3D images.

2   Proposed Method

2.1   3D Broadcasting System Based on DIBR

As shown in Fig. 2, the 3D broadcasting system based on DIBR consists of two main parts: transmitter and receiver ones. In the transmitter part, a depth sequence is preprocessed for taking advantages of both reduction of disocclusion area (holes) and distortion minimization when virtual views are created. The preprocessed depth sequence and the reference sequence are coded by the H.264/AVC baseline encoder and transmitted through T-DMB channel. In the receiver part, once both the reference stream and its corresponding depth stream are received, the 3D image sequence is created by applying image warping, hole filling, and interleaving simultaneously, and displayed in 3DTV. Blocking artifact reduction which is the chief goal of this study is conducted in the receiver part between decoding and DIBR.

2.2   Overcomplete 3D Dictionary Generation

The distortions of the JPEG compressed images are mainly made from DCT. Meier et al. [18] assumed that the quantization error caused by compression was modeled by additive white Gaussian noise (AWGN) in the spatial domain. We also assume that blocking artifact reduction is one of the denoising procedures. Thus, compressed images are modeled as the corrupted ones by AWGN as follows:

$$Y = X + N \tag{1}$$

where Y is the compressed image; X is the original image; and N is the AWGN. Sparse representation can be used to reconstruct X from Y by reducing N. That is, by sparse representation, each image patch can be represented as a linear combination of a small subset of atoms, taken from an overcomplete dictionary. Based on this assumption, the deblocked results can be obtained by an energy minimization method. The overcomplete 3D dictionary, D_{3d}, for deblocking is comprised of reference and depth dictionaries as follows:

$$D_{3d} = [D_r, D_d] \tag{2}$$

where D_r is the reference dictionary generated from natural images while D_d is the depth dictionary obtained from depth images. Both of the dictionaries are created by the K-SVD algorithm and used for blocking artifact reduction in reference images and depth images, respectively. Here, the K-SVD algorithm is an iterative method to generate an overcomplete dictionary that fits the training examples well [19, 20]. It alternatively performs sparse coding and dictionary updating. To generate the deblocking dictionary D, the objective function of the K-SVD algorithm is defined as follows:

$$\min_{D,\Theta} \left\| \bar{X} - D \cdot \Theta \right\|_F^2 \quad s.t. \quad \forall i, \|\theta_i\|_0 \leq S \tag{3}$$

where $\bar{X} = [x_1, x_2, ..., x_P]$ is a matrix of training square patches; $D = [d_1, d_2, ..., d_K]$ is a dictionary; $\Theta = [\theta_1, \theta_2, ..., \theta_P] = [\varphi_1, \varphi_2, ..., \varphi_P]^T$ is a sparse representation matrix of $\bar{X}$; θ_i is the i-th column of Θ; φ_i^T is the i-th row of Θ; and S is the given sparisty level. Then, the error term in (3) can be represented as $\left\| \bar{X} - D \cdot \Theta \right\|_F^2 = \left\| \bar{X} - \sum_{i \neq j} d_i \varphi_i^T - d_j \varphi_j^T \right\|_F^2$. Let $H_j = \bar{X} - \sum_{i \neq j} d_i \varphi_i^T$, then (3) can be rewritten as follows:

$$\min_{D,\Theta} \left\| H_j - d_j \varphi_j^T \right\|_F^2 \quad s.t. \ \forall i, \left\| \theta_i \right\|_0 \leq S$$
(4)

Let ϕ_j be the set of all indices corresponding to the training patches that use the atom d_j for the current dictionary, i.e., it can be determined by using OMP. In addition, let Φ_j be a $P \times \phi_j$ matrix with ones on the entries $(\phi_j(i), i)$ and zero elsewhere. Thus, (4) is equivalent to:

$$\min_{D,\Theta} \left\| H_j \Phi_j - d_j \varphi_j^T \Phi_j \right\|_F^2$$
(5)

Let $H_j^R = H_j \Phi_j$, $\varphi_{j,R}^T = \varphi_j^T \Phi_j$. Then, let $U \Lambda V^T$ be the singular value decomposition (SVD) of H_j^R. Thus, (5) is rewritten as:

$$\min_{D,\Theta} \left\| U \Lambda V^T - d_j \varphi_{j,R}^T \right\|_F^2$$
(6)

The highest error component is eliminated by defining as follows:

$$d_j = u_1, \varphi_{j,R} = \delta_1 v_1$$
(7)

where δ_1 is the largest singular value of H_j^R, u_1 and v_1 are the corresponding left and right singular vectors. This procedure improves the dictionary atom d_j based on the patches that have used it when considering the temporary dictionary. This procedure continues in the same way for all the other atoms.
Then, we obtain the trained dictionary. For the error-based minimization, the objective function is as follows:

$$\min_{\Theta} \left\| \Theta \right\|_0 \quad s.t. \ \left\| Y - D_{3d} \cdot \Theta \right\|_2 \leq T$$
(8)

where X_i is the i-th training signal; and T is the error threshold. In this paper, we use (3) to generate the 3D dictionary and (8) to remove the blocking artifacts.

2.3   Automatic Error Threshold Estimation

To remove the blocking artifacts effectively, we provide an effective method which can be used to automatically estimate the error threshold T in (8). First, we find the boundaries of the 8×8 blocks in the BDCT compressed images. Second, we compute the differences C_i between the pixels on both sides of the boundaries where the blocking artifacts occur. We consider half of the absolute difference value as a noise value, i.e., $N_i = 0.5 \times |C_i|$. Then, we select the noise values $N_i \geq 2$ to compute the standard deviation σ:

$$\sigma = \sqrt{\frac{\sum_{i=1}^{m}(N_i - M)^2}{m}} \quad s.t. \ N_i \geq 2 \tag{9}$$

where M is the mean value of all the selected noise values, m is the number of the selected noise values. Because we consider the blocking artifacts as AWGN, the blocks in the image have the same noise level. Thus, we can compute the threshold based on σ as follows:

$$T_{old} = C \cdot \sigma \tag{10}$$

where C is noise gain. Thus, T_{old} of (1) is the same as the error threshold used in the KSVD denoising method [19]. However, (1) is not suitable for BDCT compressed images because blocking artifacts depends on a compression factor instead of σ for denoising.

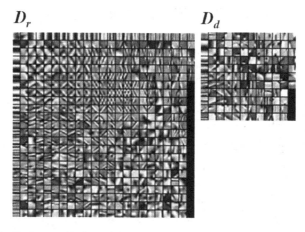

Fig. 3. The 3D dictionary. Left: reference dictionary (64×512). Right: depth dictionary (64×128).

The compression factor q contains a value between 0 and 100, and the higher q is the less image degradation due to compression. Different values of q correspond to different quantization tables. By experiments, we found that the optimal threshold T_{new} of T in (8) for image deblocking is somewhat different from T_{old} for image denoising by experiments. That is, T_{old} fits well when it is about 12, but in other cases, T does not follow actual distribution of the error threshold according to a compression factor q as shown in the figure. Here, the red line is a distribution of the error. To get more accurate results, we use the third-order polynomial fitting, and thus T_{new} is determined by the following equation:

$$T_{new} = T_{old} \cdot f(q) = C \cdot \sigma \cdot (c_1 + c_2 q + c_3 q^2 + c_4 q^3), \quad q \leq 30 \tag{11}$$

where $f(q)$ is a polynomial function of q with polynomial coefficients $c_1,...,c_4$. The polynomial coefficients were determined on the training data set, and c_1, c_2, c_3, and c_4 were set to be -0.0001, 0.0073, -0.1710, and 2.1580, respectively. We employ T_{new} of (11) to optimize the objective function of (8).

3   Experimental Results

To generate the reference dictionary D_r, we use the dataset provided by the Yang et al.'s work [21], which includes 91 natural images. First, we select 100000 image patches from the images for training, and then obtain D_r by the K-SVD algorithm. Total 512 atoms are learned with each atom of the 8×8 size. Thus, the size of the training data is 64×100000, and the sparsity level is 6. The number of training iterations is 20. The depth dictionary D_d is obtained from depth images in 3D image database. The training procedure to obtain D_d is the same as that of D_r. The overcomplete 3D dictionary is shown in Fig. 3. To demonstrate the effectiveness of the overcomplete 3D dictionary, we perform experiments on five 3D image sequences including *Interview*, *Orbi*, *EtriCG*, *BreakDancer*, and *Ballet*. The reference images are 2D color images.

Fig. 4. Deblocking Results in the *Ballet* sequence when q=5, 10, 15, 20. First row: The compressed *Ballet* reference image when q=5, 10, 15, 20 from left to right. Second row: Our deblocking results. Third row: The compressed *Ballet* depth image when q=5, 10, 15, 20 from left to right. Fourth row: Our deblocking results.

We firstly transform RGB format to YCbCr format for the reference images. Then, we compress the Y component according to each compression factor. Because depth images have only one component, we also compress exactly like the Y component of reference images according to each compression factor. Fig. 4 shows some deblocking results in the *Ballet* sequence. In the figure, the first row is the compressed reference images of *Ballet*, when q=5, 10, 15, 20 from left to right; the second row is the deblocking results by the proposed method; the third row is the compressed depth images when quality=5, 10, 15, 20 from left to right; the fourth row is the deblocking results by the proposed method. It can be observed that blocking artifacts are remarkably reduced by the overcomplete 3D dictionary. To provide more quantitative evaluation results, we evaluate the performance of the proposed method in terms of PSNR and SSIM. We evaluate the deblocking performances in three types of the dictionaries: 64×128, 64×256 and 64×512 for D_r; and 64×64, 64×128, and 64×256 for D_d. The PSNR results of reference and depth images are listed in Tables 1 and 2 while the SSIM results are listed in Tables 3 and 4. In the tables, the bold numbers represent the best PSNR and SSIM value at each q. It can be observed from Table 1 that there are little differences in performances among the three dictionaries. This means that image patches can be represented well even by the atoms of the smallest size dictionary. As can be seen in Table 2, the effects of the three depth dictionaries generally are similar to those of the reference dictionaries.

Table 1. The PSNR comparison results for reference images with different dictionary size and SA-DCT [24] when q =1, 5, 10, 15, 20. Bold numbers represent the best PSNR value at each q.

Reference	Method	1	5	10	15	20
Interview	JPEG	22.7242	25.2289	27.8474	29.3846	30.4757
	64×128	23.8503	26.2072	28.8535	30.3392	31.3582
	64×256	23.8848	26.2610	28.8733	30.3650	31.3828
	64×512	23.9073	26.2244	28.8445	30.3380	31.3947
	SA-DCT	**23.9390**	**26.3733**	**28.9192**	**30.4440**	**31.5234**
Orbi	JPEG	25.5000	27.6903	30.6352	32.3443	33.4009
	64×128	**26.8076**	**28.8783**	31.7633	33.3330	34.2844
	64×256	26.7965	28.8682	31.7805	33.3631	34.3085
	64×512	26.7591	28.8514	**31.7867**	**33.3768**	34.3348
	SA-DCT	26.6920	28.7040	31.7356	33.3355	**34.3491**
EtriCG	JPEG	22.6488	25.5129	28.4706	30.0193	31.0745
	64×128	23.6344	26.6490	29.7003	31.2769	32.1597
	64×256	23.6724	26.7025	29.7264	31.2950	32.1819
	64×512	23.7082	26.7329	29.7677	31.3200	32.2185
	SA-DCT	**23.8826**	**26.7736**	**29.7725**	**31.4029**	**32.4396**
BreakDancer	JPEG	25.7393	29.2527	32.5902	34.3822	35.5736
	64×128	**27.2705**	30.9893	34.1697	35.7020	36.8553
	64×256	27.2630	**31.0122**	34.1791	35.7162	36.8583
	64×512	27.2098	30.9543	34.1427	35.6917	36.8581
	SA-DCT	27.1767	30.9687	**34.2353**	**35.8085**	**36.9458**
Ballet	JPEG	25.4662	28.8742	31.7802	33.6675	34.9753
	64×128	**26.7282**	30.0765	32.9331	34.7718	36.1352
	64×256	26.7128	30.1144	32.9297	34.7548	36.1220
	64×512	26.7128	**30.1249**	32.8656	34.7481	36.0967
	SA-DCT	26.5822	30.0639	**33.0730**	**35.0793**	**36.4737**

Table 2. The PSNR comparison results for depth images with different dictionary size and SA-DCT [24] when q =1, 5, 10, 15, 20. Bold numbers represent the best PSNR value at each q.

Depth	Method	1	5	10	15	20
	JPEG	27.3541	28.8340	33.8483	35.2523	36.7267
	64×64	29.2411	**32.3420**	35.6943	37.1491	38.0978
Interview	64×128	29.3721	32.2921	35.6962	37.2029	38.2187
	64×256	**29.4375**	32.3269	35.698	**37.2177**	38.2367
	SA-DCT	29.2390	30.0856	**36.1066**	37.1909	**38.8565**
	JPEG	28.6113	30.1850	35.7064	37.4753	38.9792
	64×64	30.5925	**33.0227**	37.2584	**39.3848**	40.4012
Orbi	64×128	**30.6809**	33.0164	37.2485	39.3808	40.4093
	64×256	30.6430	32.9874	37.2314	39.3655	40.3989
	SA-DCT	30.5914	31.6097	**37.6551**	39.0149	**40.4883**
	JPEG	25.7716	28.3883	31.6472	33.1207	34.2699
	64×64	26.6473	29.9760	32.7221	34.0513	34.9657
EtriCG	64×128	26.5571	30.1089	33.0843	34.5711	35.5946
	64×256	26.7349	**30.2604**	33.2745	34.7868	35.8775
	SA-DCT	**27.6205**	30.2011	**34.0280**	**35.4138**	**36.5840**
	JPEG	27.2756	30.7495	34.8009	36.8703	38.3717
	64×64	28.2235	31.8085	36.0561	38.1417	39.7975
BreakDancer	64×128	28.1856	31.7731	36.0235	38.0877	39.7283
	64×256	28.1416	31.7368	35.9670	38.0448	39.7674
	SA-DCT	**28.5489**	**32.4316**	**36.8928**	**39.0978**	**40.2460**
	JPEG	25.5307	28.6018	32.1959	33.8267	35.0332
	64×64	27.1393	30.5512	34.0755	35.5130	36.7152
Ballet	64×128	27.3075	30.5729	34.0557	35.4622	36.7281
	64×256	**27.3121**	30.5944	34.0177	35.4677	36.7499
	SA-DCT	27.0325	**30.6598**	**34.8106**	**36.4294**	**37.7190**

Table 3. The SSIM comparison results for reference images with different dictionary size and SA-DCT [24] when q =1, 5, 10, 15, 20. Bold numbers represent the best SSIM value at each q.

Reference	Method	1	5	10	15	20
	JPEG	0.571	0.6873	0.7924	0.8384	0.8698
	64×128	0.6374	0.7333	0.8265	0.8667	0.8891
Interview	64×256	0.6385	0.736	0.8279	0.8682	0.8902
	64×512	0.638	**0.7359**	0.8284	0.8689	0.891
	SA-DCT	**0.639**	0.7452	**0.8328**	**0.8699**	**0.8962**
	JPEG	0.6614	0.7329	0.8218	0.8691	0.8951
	64×128	0.7304	0.8005	0.8737	0.9064	0.923
Orbi	64×256	**0.7313**	**0.8012**	**0.8747**	**0.9072**	0.9239
	64×512	0.7304	0.8008	0.8743	0.9072	**0.9242**
	SA-DCT	0.7236	0.7929	0.8694	0.9026	0.9209
	JPEG	0.7312	0.8081	0.8785	0.9053	0.925
	64×128	0.7645	0.8632	0.9264	0.9489	0.9579
EtriCG	64×256	0.7668	0.8657	0.9273	0.9495	0.9584
	64×512	0.7691	**0.8667**	**0.9282**	**0.9499**	0.9589
	SA-DCT	**0.7843**	0.8664	0.9271	0.9484	**0.959**

Table 3. (*Continued*)

BreakDancer	JPEG	0.7351	0.7972	0.8582	0.8946	0.9139
	64×128	0.805	0.8675	0.9092	0.928	0.9394
	64×256	**0.8047**	**0.8676**	**0.9095**	**0.9282**	0.9397
	64×512	0.8029	0.8663	0.909	0.928	**0.9398**
	SA-DCT	0.804	0.8646	0.9084	0.9271	0.9387
Ballet	JPEG	0.7315	0.8122	0.8664	0.9003	0.9213
	64×128	0.7957	0.8625	0.9067	0.9306	0.9449
	64×256	**0.7958**	**0.8637**	0.907	0.9305	0.9448
	64×512	0.7948	0.8635	0.9063	0.9305	0.9448
	SA-DCT	0.7898	0.8624	**0.9082**	**0.9332**	**0.9470**

Table 4. The SSIM comparison results for depth images with different dictionary size and SA-DCT [24] when q =1, 5, 10, 15, 20. Bold numbers represent the best SSIM value at each q.

Depth	Method	1	5	10	15	20
Interview	JPEG	0.7707	0.4991	0.9013	0.7783	0.9374
	64×64	0.8800	0.9008	0.9476	0.9563	0.9638
	64×128	0.8905	0.9044	0.9480	0.9565	0.9643
	64×256	**0.8933**	**0.9058**	0.9483	**0.9566**	0.9644
	SA-DCT	0.8809	0.5514	**0.9505**	0.8370	**0.9673**
Orbi	JPEG	0.7833	0.5987	0.9092	0.8285	0.9428
	64×64	0.8743	**0.8789**	0.9399	0.9542	0.9621
	64×128	**0.8774**	0.8785	0.9416	0.9545	0.9626
	64×256	0.8759	0.8781	0.9410	**0.9546**	0.9624
	SA-DCT	0.8737	0.6516	**0.9473**	0.8711	**0.9632**
EtriCG	JPEG	0.8564	0.7579	0.9264	0.8825	0.9512
	64×64	0.9062	0.9343	0.9612	0.9699	0.9744
	64×128	0.9077	0.9357	0.963	0.9708	0.9755
	64×256	0.9096	**0.9405**	0.9644	**0.9723**	0.9766
	SA-DCT	**0.9256**	0.7962	**0.9679**	0.9219	**0.9809**
BreakDancer	JPEG	0.8462	0.8714	0.9146	0.9337	0.9490
	64×64	0.8911	0.9158	0.9490	0.9599	0.9706
	64×128	0.8912	0.9158	0.9485	0.9597	0.9702
	64×256	0.8886	0.9139	0.9480	0.9593	0.9702
	SA-DCT	**0.9089**	**0.9360**	**0.9614**	**0.9698**	**0.9742**
Ballet	JPEG	0.8023	0.8359	0.8955	0.9196	0.9359
	64×64	0.8905	0.9282	0.9541	0.9619	0.9697
	64×128	0.8948	**0.9292**	0.9544	0.9619	0.9698
	64×256	**0.8954**	0.9291	0.9544	0.9619	0.9701
	SA-DCT	0.8804	0.9227	**0.9579**	**0.9669**	**0.9739**

However, in *EtriCG*, there are big differences in performances between them when q=20. The results show that the 64×64 dictionary is sometimes not enough to represent the depth images, and the 64×128 dictionary is more suitable for reducing the blocking artifacts in depth images. In the four tables, we also report the performance of pointwise shape adaptive DCT (SA-DCT) [24], which has been known to be one of the-state-of-the art methods. As can be seen, the proposed method can achieve consistently good PSNR gains, and the SSIM results are improved a lot after deblocking. Our method achieves better performances than SA-DCT in most cases except in PSNR of reference images. Above all, our method produces higher PSNR and SSIM gains than SA-DCT at low bit rates.

4   Conclusions

In this paper, we have proposed a new deblocking method in 3D images based on an overcomplete 3D dictionary. The proposed method is based on sparse representation which image patches can be represented by the linear combination of several atoms in an overcomplete dictionary. First, we have generated the 3D dictionary by using the K-SVD algorithm.

Then, to make use of the 3D dictionary in reducing blocking artifacts, we have estimated an error threshold for OMP using the third-order polynomial fitting. Experimental results show that the proposed method consistently achieves good performance in terms of PSNR and SSIM. We believe that the proposed method can be effectively employed for enhancing image quality and 3D effects in 3DTV.

Acknowledgement. This work was supported by the National Natural Science Foundation of China (Nos. 61271298, 61050110144, 61173092, 61072106, 60972148, 60971128, 60970066, 61003198, 61001206, 61077009), the Fund for Foreign Scholars in University Research and Teaching Programs (the 111 Project) (No. B07048), and the Program for Cheung Kong Scholars and Innovative Research Team in University (No. IRT1170).

References

[1] Park, Y.K., Jung, K., Oh, Y., Lee, S., Kim, J.K., Lee, G., Lee, H., Yun, K., Hur, N., Kim, J.: Depth-image-based rendering for 3DTV Service over T-DMB. Signal Processing: Image Communication 24, 122–136 (2009)

[2] Choi, S., Jung, C., Lee, S., Kim, J.K., Jung, K., Lee, G., Hur, N., Kim, J.: 3D DMB Player and its realistic 3D services over T-DMB. In: Proc. IEEE ISM 2009, pp. 440–441 (2009)

[3] Fehn, C.: Depth-image-based rendering (DIBR), compression and transmission for a new approach on 3D-TV. In: Proc. SPIE Conf. Stereoscopic Displays and Virtual Reality Systems, pp. 93–104 (2004)

[4] Zhang, L., Tam, W.J.: Stereoscopic image generation based on depth images for 3DTV. IEEE Transactions on Broadcasting 51, 191–199 (2005)

[5] Jung, C., Jiao, L.C.: Disparity-map-based rendering for mobile 3D TVs. IEEE Transactions on Consumer Electronics 57, 1171–1175 (2011)

[6] Jung, C., Jiao, L.C., Oh, Y., Kim, J.K.: Depth-preserving DIBR based on disparity map over T-DMB. Electronics Letters 46, 628–629 (2010)

[7] Merkle, P., Muller, K., Wiegand, T.: 3D video: Acquisition, coding, and display. In: Proc. IEEE ICCE, pp. 127–128 (2010)

[8] Lim, J.S., Reeve, H.C.: Reduction of blocking effect in image coding. Opt. Eng. 23, 34–37 (1984)

[9] Ramamurthi, B., Gersho, A.: Nonlinear space-variant postprocessing of block coded images. IEEE Trans. Acoust., Speech, Signal Process. 34, 1258–1268 (1986)

[10] List, P., Joch, A., Lainema, J., Bjntegaard, G., Karczewicz, M.: Adaptive deblocking filter. IEEE Trans. Circuits Syst. Video Technol. 13, 614–619 (2003)

[11] Xiong, Z., Orchard, M., Zhang, Y.Q.: A deblocking algorithm for JPEG compressed images using overcomplete wavelet representations. IEEE Trans. Circuits Syst. Video Technol. 7, 433–443 (1997)

[12] Liew, A.W.C., Yan, H.: Blocking artifacts suppression in block-coded images using overcomplete wavelet representation. IEEE Trans. Circuits Syst. Video Technol. 14, 450–461 (2004)

[13] Rosenholtz, R.E., Zakhor, A.: Iterative procedures for reduction of blocking effects in transform image coding. In: Proc. SPIE Image Processing Algorithms and Techniques II, pp. 116–126 (1991)

[14] Kim, Y., Park, C.S., Ko, S.J.: Fast POCS based post-processing technique for HDTV. IEEE Trans. Consumer Electronics, 1438–1447 (2003)

[15] Alter, F., Durand, S.Y., Froment, J.: Deblocking DCT-based compressed images with weighted total variation. In: Proc. IEEE ICASSP, pp. 17–21 (2004)

[16] Do, Q.B., Beghdadi, A., Luong, M.: A new adaptive image post-treatment for deblocking and deringing based on total variation method. In: Proc. ISSPA, pp. 10–13 (2010)

[17] Sun, D., Cham, W.K.: Postprocessing of low bit-rate block DCT coded images based on a fields of experts prior. IEEE Trans. Image Process. 16, 2743–2751 (2007)

[18] Meier, T., Ngan, T.K., Crebbin, G.: Reduction of blocking artifacts in image and video coding. IEEE Trans. Circuits Syst. Video Technol. 9, 490–500 (1999)

[19] Elad, M., Aharon, M.: Image denoising via sparse and redundant representations over learned dictionaries. IEEE Trans. Image Process. 15, 3736–3745 (2006)

[20] Carvajalino, J.M.D., Sapiro, G.: Learning to sense sparse signals: Simultaneous sensing matrix and sparsifying dictionary optimization. IEEE Trans. Image Process. 18, 1395–1408 (2009)

[21] Yang, J., Wright, J., Huang, T., Ma, Y.: Image super-resolution via sparse representation. IEEE Trans. Image Process. 19, 2861–2873 (2010)

[22] Jung, C., Jiao, L.C., Liu, B., Qi, H., Sun, T.: Toward high-quality image communications: inverse problems in image processing. Optical Engineering 51, 100901 (2012)

[23] Jung, C., Jiao, L.C., Qi, H., Sun, T.: Image deblocking via sparse representation. Signal Processing: Image Communication 27, 663–677 (2012)

[24] Foi, A., Katkovnik, V., Egiazarian, K.: Pointwise shape adaptive DCT for high quality denoising and deblocking of grayscale and color images. IEEE Trans. Image Process. 16, 1395–1411 (2007)

Efficient HEVC to H.264/AVC Transcoding with Fast Intra Mode Decision

Jun Zhang[1,2], Feng Dai[1], Yongdong Zhang[1], and Chenggang Yan[1,2]

[1] Advanced Computing Research Laboratory, Beijing Key Laboratory of Mobile Computing
and Pervasive Device, Institute of Computing Technology,
Chinese Academy of Sciences, Beijing, China
[2] University of Chinese Academy of Sciences, Beijing, China
{zhangjun01,fdai,zhyd,yanchenggang}@ict.ac.cn

Abstract. High Efficiency Video Coding (HEVC) standard will soon reach its final draft. To provide the widely deployed H.264/AVC devices with HEVC video contents, transcoding pre-encoded HEVC video into H.264/AVC format is highly necessary. Computational complexity of H.264 hinders real-time transcoding. In this paper, we propose an efficient HEVC to H.264 intra frame transcoder to accelerate the time-consuming H.264 intra mode decision while ensure rate distortion (RD) performance. The proposed transcoder incorporates a support vector machine (SVM) based macroblock (MB) partition mode decision and a fast prediction mode decision. Compared with the reference transcoder which employs exhaustive search mode decision, our proposed transcoder can save 68.83% of transcoding time with negligible 2.32% bit-rate increase on average.

Keywords: Video Transcoding, HEVC, H.264/AVC, Fast Mode Decision.

1 Introduction

In order to satisfy the increasing demand for better visual quality, VCEG and ISO/IEC have formed a group named Joint Collaborative Team on Video Coding (JCT-VC) to develop a new video coding standard HEVC [1] with the target of outperforming all existing standards. HEVC will soon reach its Final Draft as an International Standard, so more and more videos are to be encoded in this outstanding next generation standard. However, as H.264/AVC [2] is now being widely adopted, there will be inevitably a long term co-existence of them and the coming HEVC videos need to be converted into H.264 format to play on legacy H.264 compatible devices. For example, a high definition HEVC digital TV program cannot be displayed on a mobile phone that supports only H.264 unless transcoded. Thus, transcoding pre-encoded HEVC video to H.264 format is definitely needed.

Due to complex coding tools, H.264/AVC encoding is very time-consuming and many researches on accelerating it can be seen in [3-8]. By the same token, the encoder side of an HEVC to H.264 transcoder also needs to be accelerated for real-time applications.

S. Li et al. (Eds.): MMM 2013, Part I, LNCS 7732, pp. 295–306, 2013.
© Springer-Verlag Berlin Heidelberg 2013

In H.264 encoding, intra mode decision plays an important role and it's computationally complex because of so many coding modes to choose from. In order to speed up this process, many efficient algorithms have been proposed in the past several years which can be classified into two categories. The first one generally employs a simplified computation of RD cost. In [5], the author estimates rate using ρ-domain model based on the number of zero quantized transform coefficients. In [6] and [7] SATD (sum of absolute transformed differences) based methods are used in which the standard deviation of SATD coefficients are used to estimate rate and show better efficiency than [5]. The second category picks out unnecessary modes and eliminates them. In [8], the variance of a MB is used to decide the partition mode (16×16, 8×8, 4×4) and filter based approach is adopted to reduce the candidate prediction modes. A MPEG-2 to H.264/AVC transcoder proposed in [9] fully utilizes the DCT coefficients contained in the incoming MPEG-2 stream to do fast mode decision. Learning based approaches [10-12] can get further acceleration. To the best of our knowledge, no works about HEVC/H.264 transcoding have been published yet. Methods listed above can't be directly employed to HEVC/H.264 transcoder because differences between these two standards are relatively significant.

This paper proposes an efficient HEVC to H.264/AVC intra frame transcoder with fast MB partition mode decision and fast prediction mode decision. Experimental results show that our proposed transcoder can get good performances with negligible quality loss but considerable time saving, as compared with the method of exhaustive search.

The remainder of this paper is organized as follows. In section 2 we briefly review intra prediction and mode decision. Our proposed transcoder is fully expounded in section 3. Then in section 4 we show our experimental results. Finally this paper is briefly concluded.

2   Intra Prediction

2.1   Intra Prediction in HEVC

The concept of MB in previous video coding standards is not directly inherited by HEVC. Instead, a more flexible block splitting manner is adopted. A picture is firstly divided uniformly into unlapped square units named Largest Coding Unit (LCU) which is the top-most level unit in the splitting hierarchy. LCUs are split recursively into four equally sized units in a quad-tree manner and a leaf node of the resulting recursive quad-tree is called a Coding Unit (CU) which also has square size. An intra CU can be further partitioned into four Prediction Units (PUs) which are the basic units carrying prediction information in intra slices. Additionally, there is a concept of Transform Unit (TU) which is the basic unit for intra prediction and transform and it's limited to a CU. It should be noticed that the PU and TU shape may not be square in inter slices but they are always square in intra slices.

In order to represent complex textures or image contents with different directions more efficiently than H.264/AVC, HEVC adopts a more flexible direction representation method which can provide up to 33 directional intra modes and improved accuracy as shown in Figure 1. In addition, two non-directional modes named DC and planar are available. The number 35 mode which means predicting from luma component belongs only to chroma component.

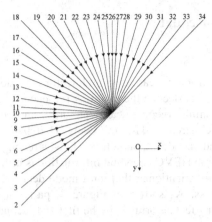

Fig. 1. The 33 directional intra prediction modes in HEVC for luma component

2.2   Intra Prediction in H.264/AVC

In H.264/AVC, a MB can be partitioned into blocks of size 16×16, 8×8 or 4×4 for intra prediction of its luma component, while both chroma components are predicted as a single block. For a 4×4 luma block, up to nine prediction modes can be used, including the DC mode and 8 directional modes. The prediction modes for 8×8 luma blocks are exactly the same as those for 4×4 blocks expect that the block to be predicted is 8×8. For a 16×16 luma block, there are four modes namely vertical, horizontal, DC and plane respectively can be use. The prediction modes for chroma component are very similar to those of 16×16 luma block except for different numbering. Because intra prediction of chroma component is much computationally easier than luma, we don't consider it in this paper.

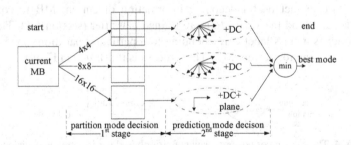

Fig. 2. H.264 intra mode decision for luma component

H.264 intra mode decision tries to find the best partition mode and prediction mode combination which minimize the RD cost for current MB. This process can be modeled as a two-stage decision process as shown in Figure 2 from which we can see that pursuing the best coding mode means searching for a two-stage path from the start to end with the minimum RD cost. The best partition mode is the choice in the first stage and the best prediction mode is the choice in the second stage.

3   Proposed Transcoder

3.1   Architecture

The conceptually straightforward cascaded architecture is adopted in this paper as shown in Figure 3. We firstly decode the inputting pre-encoded HEVC stream using a HEVC decoder, generating reconstructed pictures and decoding information contained in the HEVC stream. The following H.264/AVC encoder takes the reconstructed pictures and encodes it, outputting H.264 stream.

During the encoding, the HEVC decoding information is utilized to accelerate intra mode decision. It has been mentioned that intra mode decision can be modeled as a two-stage decision process. As shown in Figure 3, partition mode and prediction mode are determined using fast algorithms in the first and second stage respectively.

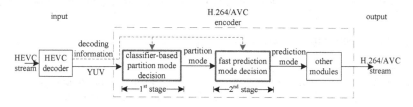

Fig. 3. Outline of our proposed transcoder

3.2   Classifier-Based Partition Mode Decision

In H.264 full search mode decision, all three partition modes 16×16, 8×8 and 4×4 are computed to find the best, so it's very time-consuming. To ease the computational overhead, we model the partition mode decision as a classification problem. Specifically, we extract useful decoding information of current MB to compose a feature vector and send the vector to a pre-trained classifier (section 4.1). The output of the classifier is a class label corresponding to a particular partition mode which is probably the best one for current MB. This process is illustrated in Figure 4.

Fig. 4. The pipeline of our proposed classifier based partition mode decision

To make a compromise between coding quality and computational load, instead of the direct 3-class classifier, we adopt a 2-class classifier. As [8] does, 8×8 mode is always computed; the classifier outputs class label 0 or 1 indicating whether 16×16 or 4×4 mode should be additionally computed and then compared with 8×8 mode to find the best. Thus, totally two modes are computed rather than full search.

There is a lot of information contained in the inputting HEVC stream, so which information to extract to compose the feature vector may significantly influence the classification accuracy and eventually the coding quality. An ideal feature vector should be able to reflect the complexity extent of a MB. After a lot of experiments, we decide to use the information listed below as features to compose the vector.

Prediction Bits. During the HEVC decoding process, we count for each MB area (a 16×16 area) the bits that are used for coding HEVC syntax elements related to intra prediction including CU splitting flag, PU partition flag and intra prediction mode signaling. We call this feature *PB* in this paper. *PB* is the first dimension of the feature vector.

As the concept of MB does not exist in HEVC but a flexible block splitting, during the counting, we convert CUs to MB areas as the example shown in Figure 5 in which the highlighted block represents current MB area. Figure 5 (a) shows the CU splitting. If current MB lies precisely on a CU of size 16×16 like Figure 5 (b), the *PB* of this CU is directly assigned to this MB; if current MB area covers more than one CUs as (c) shows, the *PB* of this MB area is set to the sum of *PB*s of the covered CUs; otherwise, if the size of a CU is larger than 16×16 as Figure5 (d) shows, then *PB* of each contained MB is set to the complete *PB* of this CU divided by the number of MBs in the CU.

PB is able to reflect the coding complexity of current MB area to some extent because a complicated MB area usually needs to be more accurately predicted in order to minimize the residual signal energy and get better RD performance, which leads to large *PB*.

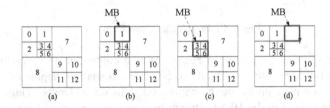

Fig. 5. The relationship between HEVC CUs and H.264/AVC MB areas

Residual Bits. It's not convincing enough to measure the complexity of a MB area only by *PB* because it is the residual signal bits that occupies most part in the generated bits stream. Like *PB*, for each MB area, we count the bits that are used for coding HEVC syntax elements related to residual signal, including TU splitting flag, coded block flag (Cbf) and transform coefficients. This feature is called *RB* in this paper. The conversion from CUs to MBs is performed exactly the same way as the *PB* computing does as shown in Figure 5.

The average quantization parameter (AQP) in each MB area is also computed from the HEVC stream using which the residual complexity (*RC*) can be formulated as (1). For simplicity but without loss of generality, in our implementation the QP for every frame and block of the inputting HEVC stream is constant so AQP is same for every MB area. Thus, only *RB* is needed to reflect the complexity of a MB area and we put it in the second dimension of the feature vector.

$$RC = AQP \cdot RB \ . \tag{1}$$

Splitting Depth. As mentioned, HEVC adopts a highly flexible block splitting method which recursively split LCU into four equally sized units. Similarly, a CU can further be split recursively. Both these two processes are performed in a quad-tree manner with the first tree called coding tree ($CTree$) whose leaf nodes are CUs and the second one transform tree ($TTree$) whose leaf nodes are TUs. A $TTree$ is rooted by a CU. TUs are the basic units for intra prediction and transform. An example of unit partition and its quad-tree representation are shown in Figure 6 (a) and (b) respectively. Thin lines (black) represent CU splitting and coding tree while bold (red) lines represent TU splitting and transform tree. In Figure 6 (b), hollow circles mean CUs and solid squares (red) represent TUs. Leaf nodes attached with both shapes mean that these CUs are predicted and transformed as a whole without further splitting, i.e. they contain only one TU.

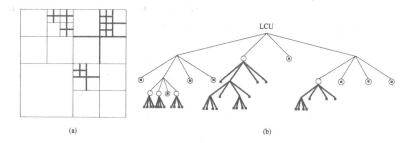

(a)   (b)

Fig. 6. Coding tree and transform tree

These two splitting processes in HEVC offer a content-adaptive coding method during which flat areas are most likely to be coded in shallow depths of $CTree$ and $TTree$ while complex areas are probably coded in deep depths. Thus, the depth information contained in the HEVC stream may play an important role in estimating spatial complexity. For a MB area, we calculate the average depth of $CTree$ and average depth of $TTree$ within it by simple mean computations utilizing the decoding information. These two features are named D_CTree and D_TTree respectively and act as the next two dimensions of the feature vector.

$$FV = (PB, RB, D\_CTree, D\_TTree) \ . \tag{2}$$

By now, the feature vector can be formulated as (2). The classifier is trained as follows. Pre-encoded HEVC files are decoded to generate reconstructed YUV files and feature vectors of each MB area. We then encode the reconstructed YUVs using a full search H.264 encoder. During the encoding, if the optimal partition mode for one MB decided by the encoder is 16×16 or 4×4, it's taken as *ground truth* and recorded along with the corresponding feature vector of this MB, which are jointly called a training instance. Extensive instances are inputted into a classifier to train it.

During transcoding, the feature vector of each MB area is generated and sent to the pre-trained classifier to directly decide the best partition mode.

3.3   Fast Prediction Mode Decision

A number of prediction modes can be used in H.264 intra coding, especially for 4×4 and 8×8 blocks. It's time-consuming to do full search. In this section we propose a fast intra prediction mode decision for HEVC to H.264 transcoding. We select a subset of the H.264 prediction modes to do mode decision, denoted as M. Only the modes in the subset M are computed and others are omitted. The modes in M are called candidate modes. In our proposed algorithm, DC and the most probable mode (MPM, 8×8 and 4×4 blocks only) are always computed due to their high probability of being the best [8,9], i.e. M is initialized to {DC} or {DC, MPM}. By referring corresponding HEVC modes, the other modes are selectively added into M as follows.

In HEVC, there are 35 intra prediction modes for luma component including two non-directional modes planar (0) and DC (1) and 33 directional ones (2 to 34) as shown in Figure 1. The minimal coding block in HEVC is 4×4, so the prediction modes of a certain area can be expressed, stored and accessed in 4×4 block unit as HM [14] does. We assume that the corresponding HEVC prediction modes for a H.264 block area are $m_0, m_1, ..., m_{n-1}$ where m_k ($k = 0,1, ... n - 1$, $n=1$ for 4×4 block, n=4 for 8×8 block and $n=16$ for 16×16 block) means the kth HEVC prediction mode in Z-scan order within this block as shown in Figure 7.

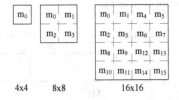

Fig. 7. HEVC prediction modes in a H.264 block. A square represents a 4×4 block area

1) All Non-directional. We observe by experiments that if all HEVC prediction modes within one block are planar or DC, i.e. $m_k \leq 1$ ($k = 0,1, ..., n - 1$), then this block will be coded in H.264 vertical or horizontal mode with a high probability. So if current block meets this condition, $M=M \cup \{0,1\}$.

2) Non-directional and Directional. However, if both non-directional and directional HEVC modes exist in a block, then $M=\{0,1,2,3\}$ for 16×16 block and $M=\{0,1,2,3,4,5,6,7,8\}$ for 8×8 or 4×4 block. This means all modes will be computed because the dominant texture direction in this block is not obvious.

3) All Directional If HEVC prediction modes within current block are all directional ones which means $m_k \geq 2$ ($k = 0,1, ..., n - 1$), the following procedures apply.

Some Definitions. As shown in Figure 1, we denote the angles of the 33 directional modes relative to x-axis as α_i ($i = 2,3, ..., 34, -\pi < \alpha_i \leq \pi$). Similarly, we denote the angles of eight H.264/AVC directional intra modes as β_i ($0 \leq i \leq 8, i \neq 2, -\pi < \beta_i \leq \pi$).

We further define the angle distance $D_{\gamma,\eta}$ between two angles γ and η $(-\pi < \gamma, \eta \le \pi)$ as (3) where *abs* means absolute value and *min* returns the smaller of its two operands.

$$D_{\gamma,\eta} = min\{abs(\gamma - \eta), \pi - abs(\gamma - \eta)\} . \qquad (3)$$

Based on definitions above, we define the mean mode m^* of a series of HEVC directional modes $m_0, m_1, \ldots, m_{n-1}$ (between 2 and 34) as (4).

$$m^* = arg \min_i \left(\sum_{k=0}^{n-1} D_{\alpha_i, \alpha_{m_k}} \right) . \qquad (4)$$

Candidates Selection. We firstly compute the mean mode m^* of all the HEVC modes $m_0, m_1, \ldots, m_{n-1}$ contained in this block as formula (4). Then,

a) For 8×8 or 4×4 block, we find that the probability of mode 0 or 1 being the best is relatively high, so they are always selected, $M=M\cup\{0,1\}$. We then compute the angle distances between m^* and the rest six H.264 directional modes, and denoted them as $D_{\alpha_{m^*}, \beta_j}$ with $3 \le j \le 8$. We select the three smallest angle distances and the corresponding mode j is added into M, $M=M\cup\{j\}$.
b) For 16×16 block, if $6 \le m^* \le 14$, $M=M\cup\{1\}$; if $22 \le m^* \le 30$, $M=M\cup\{0\}$; otherwise, $M=M\cup\{0,1,3\}$.

Summary of the proposed algorithm are given in Table 1.

Table 1. Candidate modes selection in the proposed fast prediction mode decision

block size / HEVC mode	4×4	8×8	16×16
all $m_k \le 1$	M={0, 1, 2, MPM}		M={0, 1, 2}
not all $m_k \le 1$	—	M={0,1,…,8}	M={0,1,2,3}
all $m_k \ge 2$	calculate m^* as formula (4)		
	M={three modes between 3 and 8}∪{0,1,2,MPM}.		if 6≤m^*≤14, M={1,2}; if 22≤m^*≤30, M={0,2}; else M={0,1,2,3}.

4   Experimental Results

Using H.264/AVC reference software JM18.3 [13], HEVC reference software HM6.1 [14] and a well-known library libsvm3.12 [15] for support vector machine (SVM), all of which are the latest version at the start of our work, we evaluate the performance of our proposed heterogeneous HEVC to H.264/AVC transcoder. The experiment environments and configurations are as follows.

a) Implemented on a PC with Intel 3.10GHz i5-2400 processor and 4GB RAM.
b) Nine standard test sequences are used, including 1080p *Kimono, ParkSene, Cactus, BasketballDrive, BQTerrace, Tennis* and 720p *Vidio1, Vidio3 Vidio4*. The first 200 frames of each sequence are used.

c) Original raw YUV videos are firstly encoded by HM encoder to get pre-encoded HEVC stream files using the default high efficiency-all intra configuration file. Specifically, all frames are encoded into intra pictures, LCU size is set to 64x64, QP to 32.

d) For the H.264/AVC encoder, High profile is adopted. All incoming frames are encoded as intra; QP is set to 24, 28, 32, and 36 respectively to get the experimental data; RDO is set on; PCM is disabled.

In the next sub-sections, we evaluate the performance of our proposed algorithms separately and jointly in the transcoding context. The experimental results of the corresponding transcoders are compared in terms of BD-rate [16] and time reduction to those of the reference cascaded transcoder in which the exhaustive search mode decision is employed. Average PSNR of all three components defined as $(6 * YPSNR + UPSNR + VPSNR)/8$ is used to calculate average BD-rate.

Table 2. Performance comparison between the reference transcoder and our transcoder

Class	Sequence	1st stage		2nd stage		Two-Stage	
		BD-rate (%)	Time (%)	BD-rate (%)	Time (%)	BD-rate (%)	Time (%)
1080p	Kimono	+0.00	-51.84	+3.53	-52.55	+3.54	-75.16
	ParkScene	+0.02	-32.66	+2.75	-48.64	+2.75	-64.21
	Cactus	+0.05	-37.74	+2.24	-45.44	+2.29	-63.24
	BasketballDrive	+0.38	-47.96	+1.33	-41.75	+1.75	-67.16
	BQTerrace	+0.13	-32.04	+1.60	-40.21	+1.75	-57.77
	Tennis	+0.05	-54.47	+2.73	-42.93	+2.77	-72.62
720p	Vidio1	+0.10	-51.38	+2.20	-45.57	+2.28	-71.00
	Vidio3	+0.20	-46.37	+1.96	-45.27	+2.12	-68.27
	Vidio4	+0.13	-51.43	+1.98	-47.75	+2.11	-72.38
	average	+0.13	-45.10	+2.21	-45.57	+2.32	-68.63

4.1   Evaluation of the SVM-Based Partition Mode Decision

Training the Classifier We choose SVM due to its excellent classification accuracy. Four 1080p video materials including smooth and textured ones in spatial variation namely *BQTerrace*, *BasketballDrive*, *Cactus* and *Kimono* are firstly encoded by HM encoder to generate pre-encoded HEVC files. These files are then decoded by HM decoder to produce reconstructed YUV files and corresponding feature vectors of each MB area as specified in section 3.2. The reconstructed YUVs are then encoded by JM18.3. During the encoding process, if the best partition mode for one MB decided by JM is 16x16 or 4x4, it's taken as the ground truth. The ground truth and corresponding feature vector of a MB are jointly called a training instance. We randomly select 10,000 instances which are inputted into a SVM to train it. Cross-fold validation is performed. The resultant 2-class classifier achieves a classification accuracy of about 98% on the whole training set.

The training process is performed offline and the final output is a SVM mode file. The transcoder loads the model file into memory on its start-up and it classifies the incoming feature vector of a MB according to the model file.

Transcoding We replace the full search approach in JM18.3 encoder with SVM-based partition mode decision while the prediction mode decision remains unchanged, i.e. try all available prediction modes. The performance comparison to the reference transcoder is shown in *1st stage* column of Table 2. We can see time reduction is relatively low for textured videos such as *BQTerrace* and *Cactus*, and relatively high for smooth ones such as *Kimono*. The reason is that 4×4 mode is much more computationally intensive than 16×16 but may be skipped without computing for less complex videos [8]. On the whole, the performance of the proposed algorithm is satisfactory with almost half time saving but tiny bit-rate increase.

4.2   Evaluation of the Fast Prediction Mode Decision

In order to verify the proposed prediction mode decision algorithm separately, we implement it based on JM18.3 encoder with all three partition modes (16×16, 8×8 and 4×4) computed. The results are shown in *2nd stage* column of Table 2 from which we can see that the average time reduction is nearly the same as *1st stage* column with a little BD-rate increase.

4.3   Evaluation of the Two-Stage Fast Mode Decision

We incorporate both these two fast mode decision algorithms into JM18.3 to get the final proposed transcoder. The RD performance and transcoding speed relative to the reference transcoder are shown in *Two-Stage* column of Table 2.

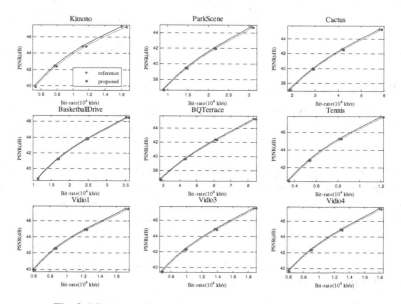

Fig. 8. RD curves of the reference and our proposed transcoder

As shown in Table 2, 68.63% transcoding time can be saved with 2.32% rate increase on average. Obviously the RD performance loss is negligible in view of the considerable time reduction. The RD plots of transcoding are depicted in Figure 8. The negligible differences between these two curves and the transcoding acceleration verify the efficiency of our proposed algorithms.

5   Conclusion

In this paper, an efficient HEVC to H.264 intra frame transcoder is proposed. The transcoder incorporates a SVM-based partition mode decision algorithm and a fast prediction mode decision algorithm. Experimental results obtained by implementing these two algorithms separately and jointly in transcoding context demonstrate their effectiveness. Using our proposed transcoder, considerable time saving can be obtained with negligible bit-rate increase.

Acknowledgments. This work is supported by National Nature Science Foundation of China (61102101, 61272323), National Key Technology Research and Development Program of China (2012BAH06B01), Co-building Program of Beijing Municipal Education Commission.

References

1. Bross, B., Han, W.-J., Ohm, J.-R., Sullivan, G.J., Wiegand, T.: High Efficiency Video Coding (HEVC) Text Specification Draft 6, JCTVC-H1003, San José, CA, USA, pp. 1–10 (February 2012)
2. Draft ITU-T Recommendation and Final Draft International Standard of Joint Video Specification, document JVT-G050.doc, ITU-I Rec. H.264 and ISO/IEC 14496-10 AVC (2003)
3. Zhang, Y., Yan, C., Dai, F., Ma, Y.: Efficient parallel framework for H.264/AVC deblocking filter on many-core platform. IEEE TMM 14(1), 510–524 (2012)
4. Yan, C., Dai, F., Zhang, Y., Ma, Y., Chen, L., Fan, L., Zheng, Y.: Parallel deblocking filter for H.264/AVC implemented on Tile64 platform. In: ICME 2011, Barcelona, Spain, pp. 1–6 (2011)
5. Kim, H., Altunbasak, Y.: Low-complexity macroblock mode selection for H.264/AVC encoder. In: ICIP 2004, October 24-27, pp. 765–768 (2004)
6. Lee, Y.-M., Sun, Y.-T., et al.: SATD-Based Intra Mode Decision for H.264/AVC Video Coding. IEEE TCSVT 20(3), 463–469 (2010)
7. Lin, Y., Lee, Y.-M., et al.: Efficient Algorithm for H.264/AVC Intra frame Video Coding. IEEE TCSVT 20(10), 1367–1372 (2010)
8. Yi-Hsin, H., Ou, T.-Sh., et al.: Fast Decision of Block Size, Prediction Mode, and Intra Block for H.264 Intra Prediction. IEEE TCSVT 20(8), 1122–1132 (2010)
9. Xingang, L., Yoo, K.-Y., et al.: Low Complexity Intra Prediction Algorithm for MPEG-2 to H.264/AVC Transcoder. IEEE Trans. Cons. Electro. 56(2), 987–994 (2010)
10. Chiang, C.-K., Pan, W.-H., et al.: Fast H.264 Encoding Based on Statistical Learning. IEEE TCSVT 21(9), 1304–1315 (2011)

11. Lu, Z.-Y., Jia, K.-B., Siu, W.-C.: Low-complexity Intra Prediction Algorithm for Video Down-Sizing Transcoder. In: VCIP 2011, vol. 6-9, pp. 1–4 (2011)
12. Fern, G., Kalva, H., et al.: An MPEG-2 to H.264 Video Transcoder in the Baseline Profile. IEEE TCSVT 20(5), 763–768 (2010)
13. JM Reference Software Version 18.3, http://iphome.hhi.de/suehring/tml/
14. HM Reference Software Version 6.1, https://hevc.hhi.fraunhofer.de/svn-/svn_HEVCSoftware/
15. Libsvm Version 3.12, http://www.csie.ntu.edu.tw/~cjlin/libsvm/
16. Bjontegaard, G.: Calculation of average PSNR differences between RD-curves. In: Document VCEG-M33, 13th Meeting VCEG, Austin, Texas, USA, pp. 2–4 (April 2001)

SSIM-Based End-to-End Distortion Model for Error Resilient Video Coding over Packet-Switched Networks

Lei Zhang, Qiang Peng, and Xiao Wu

Southwest Jiaotong University, Chengdu, China
swjtu_zl@yahoo.cn, {qpeng,wuxiao}@home.swjtu.edu.cn

Abstract. Conventional end-to-end distortion models measure the overall distortion based on independent estimation of the source distortion and channel distortion. However, they are not correlating well with perceptual characteristics in which a strong dependency exists among the source distortion, channel distortion and video content. As most compressed videos are represented to human users, perception-based end-to-end distortion model should be developed for error resilient video coding. In this paper, we propose a SSIM-based end-to-end distortion model to optimally estimate the overall perceptual distortion due to quantization, error concealment and error propagation. Experiments show that the proposed end-to-end distortion model can bring significant visual quality improvement for H.264/AVC video coding over packet-switched networks.

Keywords: Error resilience, end-to-end distortion model, structural similarity.

1   Introduction

It is known that most video coding standards use the transform coding and motion compensated prediction to achieve high compression, which creates strong spatial-temporal dependency in compressed videos. Thus, transmitting the highly compressed video streams over packet-switched networks may suffer from spatial-temporal error propagation and lead to severe quality degradation at the decoder side. To protect compressed videos from packet loss, error resilient video coding becomes a crucial requirement. For the given transmission conditions, such as bit-rate and packet-loss ratio, the target of error resilient video coding is to minimize the distortion at the receiver [1]:

$$\min\{D\} \quad s.t. \ R \le R_T \quad and \quad p \tag{1}$$

where D and R denote the distortion at the receiver and bit-rate, respectively. R_T is the target bit-rate and p is the packet-loss ratio.

To solve above minimization problem, the accurate estimation of the overall distortion plays an important role in error resilient video coding. Several end-to-end distortion models (also known as joint source-channel distortion model) were proposed. For instance, the work in [1] developed a ρ-domain end-to-end distortion

S. Li et al. (Eds.): MMM 2013, Part I, LNCS 7732, pp. 307–317, 2013.

model to estimate the distortion due to bit errors. The well-known recursive optimal pixel estimation (ROPE) model [2] and its extension [3] were presented to estimate the distortion due to packet loss. Recently, a novel channel distortion model was developed for generic multi-view video transmission over packet-switched networks [4]. Since these models are all derived in terms of mean squared error (MSE), the end-to-end distortion can be individually and independently estimated by the source distortion and channel distortion. The general formulation of the MSE-based end-to-end distortion can be shown as

$$D = (1-p) \cdot D_Q + p \cdot D_C + (1-p) \cdot D_{P\_f} + p \cdot D_{P\_c} \qquad (2)$$

where D_Q denotes source distortion due to the quantization. D_C, D_{P_f} and D_{P_C} represent the channel distortion due to the error concealment, the error propagation from the reference frames, and the error propagation from the concealment frames, respectively. Such simple model in Equation (2) is appealing because it is easy to calculate and has clear physical meanings. However, due to the perceptual distortion is dependent on the video content, identical MSE values may amount to very different levels of perceptual distortion. Thus, the conventional MSE-based end-to-end distortion models are not correlating well with human perceptual characteristics.

As most compressed videos are represented to human users, minimizing the perceptual distortion of compressed videos from transmission errors becomes an important task. The goal of this paper is to propose a perception-based end-to-end distortion model that can provide more accurate estimation of the overall perceptual distortion. It refers to two major contributions: 1) a perceptual-based quantization distortion model; 2) a perceptual-based error propagation model, which are useful to estimate the content-adaptive perceptual distortion. Our extensive experimental results demonstrate that proposed end-to-end distortion model can bring significant visual quality improvement for H.264/AVC video coding over packet-switched networks.

2   SSIM-Based End-to-End Distortion Model

To estimate the overall perceptual distortion of decoded videos, we adopt the structural similarity (SSIM) index [5] as the perceptual distortion metric due to its best trade off among simplicity and efficiency. Three important perceptual components, luminance, contrast and structure, are combined as an overall similarity measure. For two images x and y, the SSIM index is defined as follows:

$$SSIM(x, y) = l(x, y) \cdot c(x, y) \cdot s(x, y)$$
$$= \frac{2\mu_x\mu_y + c_1}{\mu_x^2 + \mu_y^2 + c_1} \cdot \frac{2\sigma_{xy} + c_2}{\sigma_x^2 + \sigma_y^2 + c_2} \qquad (3)$$

where $l(x, y)$, $c(x, y)$ and $s(x, y)$ represent the luminance, contrast and structure perceptual components, respectively. μ, σ^2 and σ_{xy} are the mean, variance and cross covariance. c_1 and c_2 are used to avoid the instability when means or variances are very close to zero.

Based on the perceptual distortion metric, we develop a novel end-to-end distortion model as follows. Let b denote the original block and $\tilde{b}$ is the corresponding reconstruction block at the decoder. $\hat{r}$ and $\tilde{r}$ represent the prediction block of b at the encoder and at the decoder, respectively. e denotes the prediction residual and its reconstruction value is $\hat{e}$. If the block is received correctly, $\tilde{b} = \tilde{r} + \hat{e}$. When the block is lost, an error concealment technique is used to estimate the missing content. Let $\hat{c}$ and $\tilde{c}$ represent the concealment block of b at the encoder and at the decoder, respectively. In this case, $\tilde{b} = \tilde{c}$. We assume that the packet loss ratio p is available at the encoder throughout this paper. This can be either specified as part of the initial negotiations, or adaptively calculated from information provided by the transmission protocol [2]. Thus, the general SSIM-based end-to-end distortion can be expressed as

$$
\begin{aligned}
D_{ssim}(b, \tilde{b}) &= (1-p) \cdot (1 - E\{SSIM(b, \tilde{r}+\hat{e})\}) + p \cdot (1 - E\{SSIM(b, \tilde{c})\}) \\
&= (1-p) \cdot \Phi_r[1 - SSIM(b, \hat{r}+\hat{e}), E\{1 - SSIM(r, \tilde{r})\}] \\
&\quad + p \cdot \Phi_c[1 - SSIM(b, \hat{c}), E\{1 - SSIM(c, \tilde{c})\}] \\
&= (1-p) \cdot \Phi_r[D_{SSIM}(b, \hat{r}+\hat{e}), D_{SSIM}(r, \tilde{r})] \\
&\quad + p \cdot \Phi_c[D_{SSIM}(b, \hat{c}), D_{SSIM}(c, \tilde{c})]
\end{aligned}
\tag{4}
$$

where $E\{\cdot\}$ is the expectation operator and $\Phi[\cdot]$ is the error propagation function. $D_{SSIM}(b, \hat{c})$ denotes the concealment distortion, which can be calculated at the encoder. $D_{SSIM}(r, \tilde{r})$ and $D_{SSIM}(c, \tilde{c})$ represent the end-to-end distortion of the prediction block and concealment block, which can be recursively computed by Equation (4). With such a formulation, we need to estimate the quantization distortion $D_{SSIM}(b, \hat{r}+\hat{e})$ and the error propagation function $\Phi[\cdot]$. In the following section, we will make a development of the two terms based on content dependency.

3   Development of Quantization Distortion Model

In this section, we aim to estimate the quantization distortion $D_{SSIM}(b, \hat{r}+\hat{e})$ at the block-level (the 4×4 transform and quantization unit is used throughout this paper). Since the quantization distortion only can be obtained after the reconstruction at the encoder, the proposed distortion estimation can avoid the two-pass coding and reduce the computational. According to SSIM index, the quantization distortion can be derived as

$$
D_{SSIM}(b, \hat{r}+\hat{e}) = 1 - SSIM(b, \hat{r}+\hat{e}) = 1 - l(b, \hat{r}+\hat{e}) \cdot c(b, \hat{r}+\hat{e}) \cdot s(b, \hat{r}+\hat{e})
\tag{5}
$$

with

$$l(b, \hat{r} + \hat{e}) = \frac{2\mu_b \mu_{\hat{r}} + 2\mu_b \mu_{\hat{e}} + c_1}{\mu_b^2 + \mu_{\hat{r}}^2 + \mu_{\hat{e}}^2 + 2\mu_{\hat{r}}\mu_{\hat{e}} + c_1} \tag{6}$$

$$c(b, \hat{r} + \hat{e}) \cdot s(b, \hat{r} + \hat{e}) = \frac{2\sigma_{b\hat{r}} + 2\sigma_{b\hat{e}} + c_2}{\sigma_b^2 + \sigma_{\hat{r}}^2 + \sigma_{\hat{e}}^2 + 2\sigma_{\hat{r}\hat{e}} + c_2} \tag{7}$$

We assume the prediction residual follows a zero-mean Laplacian distribution. Now the question is how to obtain the following three terms: 1) the variance of reconstructed prediction residual; 2) the cross covariance between the reconstructed prediction residual and current block; 3) the cross covariance between the reconstructed prediction residual and prediction block.

In [6], it is reported that the DCT coefficients of prediction residual closely follows a zero-mean Laplacian distribution. Based on this phenomenon, the work in [7] proved that the reconstruction distortion from the prediction residual can be estimated by the Laplacian parameter and the quantization step. Extending the derivation in [7] into pixel-domain, we establish the following two estimation models for above terms where α and β denote the Laplacian parameter, x represents the current block or the prediction block. QP is the quantization parameter in H.264/AVC. With extensive simulations, the scaling map M_{var} and M_{cov} shown in Fig.1 can be constructed as look-up tables.

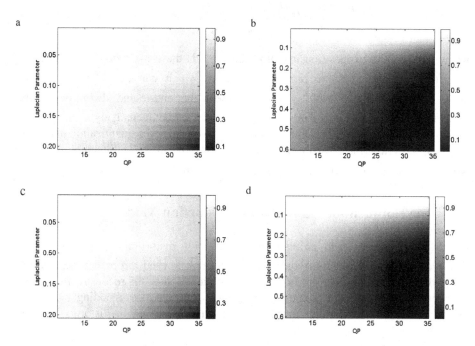

Fig. 1. The scaling maps: (a) for residual variance by intra-coding, (b) for residual variance by inter-coding, (c) for residual cross covariance by intra-coding, (b) for residual cross covariance by inter-coding

$$\sigma_{\hat{e}}^2 = \sigma_e^2 \cdot M_{\text{var}}(\alpha, QP), \quad \alpha = \sqrt{2/\sigma_e^2} \tag{8}$$

$$\sigma_{x\hat{e}} = \sigma_{xe} \cdot M_{\text{cov}}(\beta, QP), \quad \beta = \sqrt{2/\sigma_{xe}} \tag{9}$$

To demonstrate the estimation results of quantization distortion, four HD sequences (640×360) [8] are tested: 'Crow_run', 'In_to_tree', 'Ducks_take_off' and 'Old_town_cross'. All frames of each sequence are coded as intra-frame (I frame) and inter-frames (P frames) with constant quantization parameters [20, 25, 30, 35], respectively. Table 1 shows the mean absolute deviation (MAD) between the actual and estimated quantization distortion. It is evident that the proposed models can provide an accurate estimation.

Table 1. MAD between actual and estimated quantization distortion

Test	$\sigma_{\hat{e}}^2$	$\sigma_{x\hat{e}}$	l	$c \cdot s$	D_{SSIM}
MAD (Intra-coding)	5.8118	3.4852	0.0001	0.0067	0.0067
MAD (Inter-coding)	2.4700	3.0642	0.00004	0.0056	0.0056

4   Development of Error Propagation Function

The error propagation distortion is the key component of end-to-end distortion model. Different from the independent estimation in conventional MSE-based end-to-end distortion model, the perceptual error propagation distortion is dependent on the source distortion or the concealment distortion. In this section, our primary goal is to develop the error propagation functions to estimate the overall perceptual distortion for a given quantization distortion or concealment distortion.

The error propagation functions are motivated by three observations. The first observation is related to the impact of error propagation on the three components of SSIM. Let Q_{att} denotes the quality attenuation due to the error propagation

$$Q_{att}(x) = SSIM(x, \tilde{x}) / SSIM(x, \hat{x}) \tag{10}$$

To illustrate the fact, Q_{att} is measured by three different similarity metrics: 1) luminance component of SSIM; 2) the contrast and structure components of SSIM; 3) SSIM index. For instance as shown in Fig. 2, it can be seen that the contrast and structure components have the similar changes with SSIM. On the other hand, the impact of error propagation on the luminance component is limited.

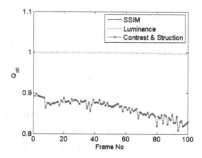

Fig. 2. The quality attenuation due to the transmission distortion

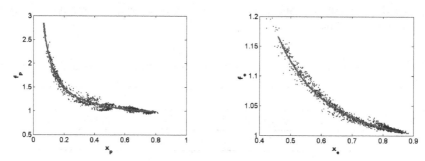

Fig. 3. The relationship between f_p and x_p

Fig. 4. The relationship between f_e and x_e

The second observation is made on the relationship f_p between the quality attenuation of current block b and the quality attenuation of concealment block (prediction block) p.

$$f_p = Q_{att}(b_p)/Q_{att}(p) = \frac{\mathrm{SSIM}(b, \tilde{p})}{\mathrm{SSIM}(b, \hat{p})} / \frac{\mathrm{SSIM}(p, \tilde{p})}{\mathrm{SSIM}(p, \hat{p})} \qquad (11)$$

$$x_p = Q_{att}(p) \cdot \mathrm{SSIM}(b, \hat{p}) \qquad (12)$$

Fig. 3 shows the simulation results that the quality attenuation of current block, in terms of SSIM, does not turn out be same as the quality attenuation of concealment block (prediction block). Moreover, it denotes that less structural similarity between current block and concealment block (prediction block) will lead to less quality attenuation.

The third observation is related to the impact of prediction residual on the decoded quality. Let Q_{enh} denote the quality enhancement due to the prediction residual, f_e denote the relationship between the quality attenuation of current block b with and without prediction residual.

$$Q_{enh}(x, x_p, x_e) = SSIM(x, x_p + x_e) / SSIM(x, x_p) \quad (13)$$

$$f_e = Q_{att}(b_{p+e}) / Q_{att}(b_p) = \frac{SSIM(b, \tilde{p}+\hat{e})}{SSIM(b, \hat{p}+\hat{e})} / \frac{SSIM(b, \tilde{p})}{SSIM(b, \hat{p})} \quad (14)$$

$$x_e = Q_{att}(b_p) / Q_{enh}(b, \hat{p}, \hat{e}) \quad (15)$$

Fig. 4 shows the simulation results that a larger prediction residual means a better decoded quality of current block, and smaller influence will be induced with error propagation from prediction block.

Based on these observations, an effective approximation of error propagation are developed as

$$\Phi_c = 1 - x_p \cdot f_p(x_p) \quad (16)$$

$$\Phi_r = 1 - SSIM(b, \hat{p}+\hat{e}) \cdot Q_{att}(b_p) \cdot f_e(x_e) \quad (17)$$

To demonstrate the estimation results of error propagation functions, same sequences are tested as the quantization distortion. All frames of each sequence are coded with error propagation form concealment blocks and error propagation form prediction blocks, respectively. Table 2 shows the MAD between the actual and estimated overall end-to-end distortion. Since the perceptual distortion is not easy to model accurately, the results show that the simplified models can achieve a good trade-off between the complexity and accuracy.

Table 2. MAD between actual and estimated overall end-to-end distortion

Test	Error propagation from concealment block	Error propagation from prediction block
MAD	0.0151	0.0172

5   Experimental Results

It is widely recognized that intra-update is an effective approach for error resilient video coding, because the decoding of an intra-coding block does not need the information from its previous frames. Thus, we incorporate the proposed SSIM-based end-to-end distortion model and MSE-based end-to-end distortion model into the mode selection to improve the RD performance over packet-switched networks, respectively. Then the optimization problem in Equation (1) can be converted to the problem of model selection between intra-coding and inter-coding as follows

$$\min\{J(mode)\} = \min\{D(mode \mid p, QP) + \lambda \cdot R(mode \mid QP)\} \qquad (18)$$

where D and R denote the end-to-end distortion and bit-rate of current coding block. *mode* denotes the coding mode. QP is the quantization parameter, which is determined by the target bit-rate. According to [2, 9,10], the Lagrange multiplier λ is determined as follows

$$\lambda = \begin{cases} 0.85 \cdot 2^{(QP-12)/3} & \text{for MSE-based end-to-end distortion} \\ \omega \cdot (-\dfrac{\beta}{R_T}) \cdot \overline{D'} & \text{for SSIM-based end-to-end distortion} \end{cases} \qquad (19)$$

where $\overline{D'}$ denotes the average distortion of previous coding units without the transmission error. R_T is the target bit-rate and β is a negative constant. ω is used to adjust the percentage of intra-coding macroblocks

$$\omega = 2^{(QP - \overline{QP})/3} \qquad (20)$$

where $\overline{QP}$ is the average quantization parameter of previous coding units.

5.1   Evaluation of RD Performance

To verify the performance of the proposed end-to-end distortion model, four CIF (352×288) sequences: 'Flower', 'Football', 'Mobile', 'Stefan' and four HD (640×360) sequences: 'Crow_run', 'In_to_tree', 'Ducks_take_off', 'Old_town_cross' are tested in the experiments. Each sequence has been coded with IPPP mode (all P frames except the first frame). P frames is subject to random packet loss. According to the Internet error pattern tested by VCEG [11], the test sequences are encoded in 10% and 20% packet-loss ratio with a random pattern and repeated 30 times, respectively. All the distortion results in the following are the averaged values.

The results with different bit-rates and packet-loss ratios for all sequences are listed in Table 3. Fig. 5 (a) and (b) demonstrate the Rate-SSIM curves of 'In_to_tree' and 'Football' sequences, respectively. The initial QP is equal to 28 and 36. The "SSIM-ER" shows the error resilient coding results based on the proposed end-to-end distortion model and the "MSE-ER" denotes the error resilient coding results based on MSE-based end-to-end distortion model. The experiment results support the claim that the proposed scheme yields consistent and significant gains over 'MSE-ER' for all sequences. The average gain of SSIM is 0.024 for the same bit-rate, which is approximately equal to average 20% bit-rate reduction for the same SSIM.

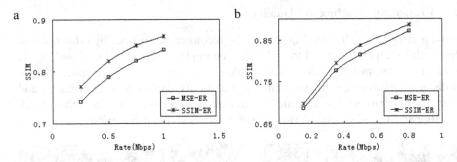

Fig. 5. Overall Rate-SSIM performance: (a) 'In_to_tree' with 20% packet-loss; (b) 'Football' with 10% packet-loss

Table 3. Simulation results with different packet-loss ratios and bit-rates

Test Sequences	Bit-rate (Mbps)	Packet-loss ratio: 20%			Packet-loss ratio: 10%		
		MSE-ER	SSIM-ER	ΔSSIM	MSE-ER	SSIM-ER	ΔSSIM
Crow_run	2.00	0.7945	0.8176	+0.023	0.8078	0.8275	+0.020
	1.00	0.7023	0.7303	+0.028	0.7064	0.7323	+0.026
Ducks_take_off	1.50	0.7989	0.8150	+0.016	0.8107	0.8232	+0.013
	0.40	0.5674	0.6159	+0.049	0.5685	0.6205	+0.052
In_to_tree	1.00	0.8417	0.8677	+0.026	0.8599	0.8834	+0.024
	0.50	0.7894	0.8196	+0.030	0.8057	0.8345	+0.029
old_town_cross	0.75	0.8801	0.8873	+0.007	0.9039	0.9082	+0.004
	0.15	0.7088	0.7290	+0.020	0.7117	0.7303	+0.019
Flower	1.00	0.8548	0.8757	+0.021	0.8854	0.9011	+0.016
	0.40	0.7223	0.7540	+0.032	0.7314	0.7634	+0.032
Football	0.80	0.8653	0.8831	+0.018	0.8712	0.8867	+0.016
	0.35	0.7738	0.7938	+0.020	0.7766	0.7943	+0.018
Mobile	1.20	0.8528	0.8678	+0.015	0.8851	0.8962	+0.011
	0.40	0.6248	0.6536	+0.029	0.6423	0.6985	+0.056
Stefan	0.80	0.8574	0.8728	+0.015	0.8814	0.8993	+0.018
	0.30	0.7111	0.7454	+0.034	0.7395	0.7713	+0.032
Average	-	0.7715	0.7955	+0.0239	0.7867	0.8106	+0.0241

5.2   Evaluation of Subjective Quality

Finally, a comparison on visual quality of the reconstructed images by error resilient video coding with different end-to-end distortion model is shown in Fig. 6 and Fig. 7. For the similar bit-rate, the error resilient video coding based on the proposed end-to-end distortion model can provide better visual quality due to more information and details have been protected from the transmission errors. On the other hand, conventional 'MSE-ER' scheme suffers with larger perceptual distortion.

Fig. 6.61st frame of 'football' encoded at 0.5Mbps with 10% packet-loss: (a) Original frame, (b) MSE-ER (SSIM: 0.844), (c) SSIM-ER (SSIM: 0.864)

Fig. 7. 33rd frame of 'mobile' encoded at 0.2Mbps with 20% packet-loss: (a) Original frame, (b) MSE-ER (SSIM: 0.483, (c) SSIM-ER (SSIM: 0.525)

6   Discussion

In this paper, we propose a SSIM-based end-to-end distortion model for H.264/AVC video coding over packet-switched networks. This model is useful to estimate the overall perceptual distortion of quantization, error concealment and error propagation. We integrate the proposed end-to-end distortion model into the error resilient video coding framework to optimally select the coding mode. Simulation results show that the proposed scheme consistently outperforms MSE-based error resilient video coding scheme, at the cost of modest additional complexity. More perceptual features will be considered to optimize the end-to-end distortion model in the future work.

Acknowledgments. The work described in this paper was supported by the National Natural Science Foundation of China (No. 61071184, 60972111, 61036008), Research Funds for the Doctoral Program of Higher Education of China (No. 20100184120009, 20120184110001), Program for Sichuan Provincial Science Fund for Distinguished Young Scholars (No. 2012JQ0029), the Fundamental Research Funds for the Central Universities (Project no. SWJTU09CX032, SWJTU10CX08, SWJTU11ZT08), and Open Project Program of the National Laboratory of Pattern Recognition (NLPR).

References

1. He, Z.H., Cai, J.F., Chen, C.W.: Joint source channel rate-distortion analysis for adaptive mode selection and rate control in wireless video coding. IEEE Transaction Circuits System Video Technology 12, 511–523 (2002)
2. Zhang, R., Regunathan, S.L., Rose, K.: Video coding with optimal inter/intra-mode switching for packet loss resilience. IEEE Journal on Selected Areas in Communications 18, 966–976 (2000)
3. Wang, Y., Wu, Z.Y., Boyce, J.M.: Modeling of transmission-loss-induced distortion in decoded video. IEEE Transaction Circuits System Video Technology 16, 716–732 (2006)
4. Zhou, Y., Hou, C.P., Xiang, W., Wu, F.: Channel distortion modeling for multi-view video transmission over packet-switched networks. IEEE Transaction Circuits System Video Technology 21, 1679–1692 (2011)
5. Wang, Z., Bovik, A.C., Sheikh, H.R., Simoncelli, E.P.: Image quality assessment: From error visibility to structural similarity. IEEE Transaction on Image Processing 13, 600–612 (2004)
6. Lam, E., Goodman, J.: A mathematic analysis of the DCT coefficient distribution for images. IEEE Transaction on Image Processing 9, 1661–1666 (2000)
7. Yang, T.W., Zhu, C., Fan, X.J., Peng, Q.: Source distortion temporal propagation model for motion compensated video coding optimization. In: IEEE International Conference on Multimedia and Expo (ICME), pp. 85–90. IEEE Press, New York (2012)
8. Xiph.org Video Test Media, http://media.xiph.org/video/derf/
9. Ou, T.S., Huang, Y.H., Chen, H.H.: SSIM-based perceptual rate control for video coding. IEEE Transaction Circuits System Video Technology 21, 682–691 (2011)
10. Wiegand, T., Girod, B.: Lagrange multiplier selection in hybrid video coder control. In: IEEE International Conference on Image Processing (ICIP), pp. 542–545. IEEE Press, New York (2001)
11. Wenger, S.: Error Patterns for Internet Experiments, ftp://ftp.imtc-files.org/jvt-experts/9910_Red/Q15-I16r1.zip

A Novel and Robust System for Time Recognition of the Digital Video Clock Using the Domain Knowledge

Xinguo Yu[1,2], Tie Rong[1], Lin Li[1], and Hon Wai Leong[3]

[1] National Engineering Research Center for E-Learning, Central China Normal University,
Wuhan, China 430079
xgyu@mail.ccnu.edu.cn
[2] Institute for Infocomm Research, 1 Fusionopolis Way, #21-01 Connexis, Singapore 138632
xinguo@i2r.a-star.edu.sg
[3] Dept of Computer Science, National University of Singapore, 3 Science Drive 2,
Singapore 117543
leonghw@comp.nus.edu.sg

Abstract. This paper presents a novel and robust system for recognizing the time of the digital video clock by using the domain knowledge. This system comprises a set of the functions for the time recognition so that the user can conveniently use these functions to recognize the time of digital video clocks or to execute some steps of recognizing time. These functions are novel and robust because they use the novel methods derived from the domain knowledge of the digital video clock. These methods are region second periodicity, global maximum model, digit location model, digit-sequence recognition, and on-the-fly SVM. Experimental results show that both the functions and the system can achieve very high recognition accuracy.

Keywords: Digital Video Clock, Time Recognition, Second Periodicity, Mixture Gaussian, On-the-fly SVM.

1   Introduction

Time recognition of the digital video clock is a crucial problem in some applications of video analysis and video annotation and as such it has been attracting a good attention from researchers [1-8]. On the other hand, time recognition is a challenge problem because the clock digits expressing time are in low resolution and the video clocks varies in color, size, layout, font type, etc., as Fig 1 shows. Time recognition of the digital video clock is a very interesting and special character (*or text*) recognition problem in video analysis and annotation. The first thought to solve this problem is to use OCR (Optical Character Recognition) algorithms to recognize the clock digits. Unluckily, the experiments showed that the OCR algorithms cannot robustly recognize the clock digits due to these digits are in very low resolution and blur [3]. This failure drives the researchers to seek the methods that use the domain knowledge of the digital video clock. Li *et al* [1-2] first proposed to detect the transit frame of both clock-second and ten-second by using correlation periodicity analysis (*a clock-second transit frame is the frame that clock-second place transits its digit*).

S. Li et al. (Eds.): MMM 2013, Part I, LNCS 7732, pp. 318–326, 2013.

Papers [1-2] called correlation periodicity analysis as the temporal neighboring pattern similarity (TNPS); they used TNPS to detect the transit frame of both clock-second and ten-second places [1-2]. And then they used the ten-second transit frame to infer the digits on the clock-second place. Once they know the second-digit of each frame they create the templates of digit 0 to 9. Then they use the created templates to recognize the digits on ten-second, minute, and ten-minute places. However, this first time recognition algorithm of the digital video clock has two demerits. The first demerit is that they modeled the transit detection as local minima detection problem. The other is that that algorithm has a long lag due to that they used the ten-second transit frame to infer the digit on the second place. To overcome these two demerits Yu *et al* [3] proposed two new techniques: global maxima model for clock-second transit detection and digit-sequence recognition for recognizing digits on clock-second and ten-second places. The global maxima model proposed in Yu's paper [3] makes the clock-second transit detection very robust. After we know the clock-second transit frames the *periodic increasing sequence* of 10-digit can be obtained from a 10-second video clip. Recognizing them as a sequence is very robust, i.e. it solved the problem that recognizing these digits one by one is not robust. This digit-sequence recognition technique also can recognize the digit on ten-second place by analyzing a 22-second clip. On the contrary to Li's papers [1-2], Yu's paper [3] used the frame that the clock-second place appears the first "0" to infer the ten-second transit frame. Though Yu's paper made the big improvement on the robustness of time recognition algorithm, it still has two demerits. The first one is that it assumes that both the digit color and the digit locations are known. The other is that the procedure recognizing minute and ten-minutes by template matching is not robust.

In this paper we present a novel and robust system for recognizing the time of the various types of the digital video clocks. This system achieves its robustness by adopting and creating the novel methods deriving from the domain knowledge of the digital clock. Particularly, it combines the second-region localization method developed in papers [1-2] and the mature image processing methods to form a procedure to identify the second region. This paper develops a knowledge-based procedure for the system to robustly acquire the digit and background colors, which solves the problem of the color variation of clocks. This paper also addresses the problem

Fig. 1. The samples of various digit video clock boards from soccer vides

of recognizing the digits on minute places. We propose an on-the-fly SVM (Support Vector Machine) method to solve this problem. The novelty of this method is that it does not train the SVM manually but that it uses the digits on second place to train SVM during the algorithm running. The advantage of this method is that both the instances to be recognized and the instances for training come from the same video.

The rest of the paper is organized as follows. Section 2 describes the structure of the time recognition system. Section 3 presents the procedure of digits localization of the digital video clock. Section 4 presents the procedure of s-digit and ts-digit recognition. Section 5 presents the procedure for minute recognition using the on-the-fly SVM method. Section 6 gives the experimental results. We draw the conclusion of the paper in Section 7.

2   System Structure of Time Recognition

This section first depicts the system structure of time recognition and then briefly explains the functions of the system. Normally the digital video clock has four clock-digits representing second, ten second, minute, and ten minute, denoted as s-digit, ts-digit, m-digit, and tm-digit in the rest of the paper, respectively.

As Fig 2 shows the system has a hierarchical structure and it comprises four main components: digits localization, s-digit recognition, ts-digit recognition, and minute recognition. *Digits localization* is the first component and the most tedious one in this system. It includes four steps. The first step called static region detection aims to detect the static regions of the given video. For static region detection we adopt the mature method in the literature. Characters localization is the second step. This step is to identify the character regions by using color segmentation and CCA (connected component analysis) techniques. The third step called s-digit localization localizes s-digit by using region second periodicity method. Region second periodicity method bases on the fact that s-digit changes its digit every second. This method was proposed in paper [1-2] and was improved in paper [3]. Remaining digits localization is the fourth step. This step aims to find the bounding boxes of ts-digit, m-digit, and tm-digit by using digit layout model methods. After we have known the locations of the four digits we can recognize them. We first recognize s-digit by using color conversion and digit-sequence recognition. Color conversion converts the digit and background pixels into white and black respectively, whereas digit-sequence recognition recognizes the digits on s-digit regions.

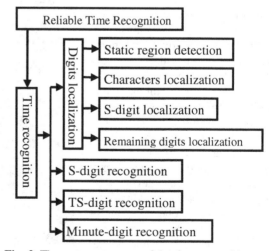

Fig. 2. The system structure of the time recognition of the digital video clock

Similarly, we use digit-sequence method to recognize the digits on ts-digit place. We develop an on-the-fly SVM procedure to recognize both m-digit and tm-digit, i.e. to find the minutes of the time. This on-the-fly SVM method can robustly recognize these digits because the instances in both training and working phases come from the same source (*same video*).

The time recognition function using the above-described four components is very robust, but it sometimes still fails to get the time due to the various reasons in long run. To counteract this issue we propose a reliable time recognition method. This method recognizes time two times and checks the consistence of them according to the following consistence principle.

The Consistent Principle of Time: The output of the time recognition algorithm is a pair (T, F), where T is time in second and F is the frame number on which the time is presented. Let (T1, F1) and (T2, F2) be the outputs of two calls of the time recognition algorithm and T2 > T1. We say that these two outputs are consistent if and only if they meet the following conditions.

$$|R*(T2 - T1) - (F2-F1)| < 2. \qquad (1)$$

where R is the frame rate of the video and "<2" is for tolerating the 1 frame shifting of the transit frame.

3   Digits Localization

Digits localization has four steps and the first three steps adopt the methods presented in papers [1-3]. This section only presents the details of step 4, which has two procedures: digits extraction and remaining digits localization.

3.1   Digits Extraction

Here we describe the method for acquiring the digit and background colors for the digit instances of the given video. In our algorithm, we collect all the instances of

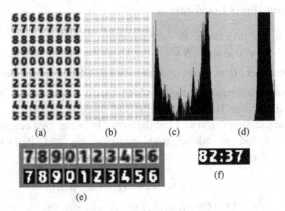

Fig. 3. The procedure illustration and samples of the digit extraction. The digit and background samples of s-digit are shown in (a) and (b) respectively; the histograms of (a) and (b) are given in (c) and (d); 10 s-digits and their converted version are given in (e) and a color converted clock is given in (f).

the digits on s-digit from a 10-second long clip as Fig 3(a) shows. Fig 3(c) is the histogram of these digit instances, called the *instance histogram*. Fig 3(d) is the histogram of the pixels surrounding the 4-digit area, called background histogram. The *background histogram* in Fig 3(d) indicates the color distribution of the digit background. Hence we can obtain the *digit histogram* from the instance histogram subtracting the background histogram. Then we can get the digit color Gaussian from the digit histogram. Next we use the acquired digit color Gaussian to identify the pixels on the digit on the video clock board. At the same time we change the value of each digit pixel into 255 (white) and the one of each digit background pixel into 0 (black) as Fig 3(e-f) shows. Based on the above discussion we form the procedure given in Fig 4 to extract digits and to convert digit instances for the digital video clock.

```
Step 1. Collect all the digit instances for 0 to
        9 and the background instances from a 10-
        second clip;
Step 2. Create the histograms of all the digit
        instances and the background histogram. Then
        identify the digit color portion. Then
        obtain the Gaussian that fit the digit color
        portion in the instance histogram.
Step 3. Use this acquired digit color Gaussian to
        identify the digit pixels.
Step 4. Set digit and background pixels into
        white and black respectively.
```

Fig. 4. The procedure of the digit extraction and conversion

3.2   Remaining Digits Localization

Here we present the procedure to determine the bounding boxes of the remaining three clock-digits. Assume that $B = (r,c,w,h)$ is the box of s-digit determined in the preceding section. Let $B_i = (r_i,c_i,w_i,h_i)$ for $i=1,2,3,4$ denote the boxes of s-digit, ts-digit, m-digit, and tm-digit in a digital video clock, respectively. For convenience, we will alternatively use B or B_1 to represent the box of s-digit. Due to the four boxes have the same dimension and the distances between two neighbor boxes are the same except a colon is added into between B_2 and B_3 we can represent $B_i = (r_i,c_i,w_i,h_i)$ for $i=1,2,3,4$ as follows.

$$\Theta : \begin{cases} (r_2,c_2,w_2,h_2) = (r,c-d_1,w,h), \\ (r_3,c_3,w_3,h_3) = (r,c-d_1-d_2,w,h), \\ (r_4,c_4,w_4,h_4) = (r,c-2d_1-d_2,w,h). \end{cases} \quad (2)$$

We use a Hough-like procedure to determine d_1 and d_2. And hence the locations of the four clock-digits are determined. The Hough space is:

$$H = \{(d_1, d_2): \ \alpha_1^1 \le d_1 \le \alpha_2^1 \ \& \ \alpha_1^2 \le d_1 \le \alpha_2^2\} \qquad (3)$$

where $\alpha_1^1, \alpha_2^1, \alpha_1^2, \alpha_2^2$ are integer and constant.

We define $E(p) = w - d(p, l_i)$ where l_i is the middle line of B_i and $d(p, l_i)$ is the distance from p to l_i. Then the measure function of each cell of H can be defined as

$$M(d_1, d_2) = \sum_{i=2}^{4} \sum_{p \in B_i} E(p) \qquad (4)$$

The pair (d_1, d_2) corresponding to the maximum of bi-variables function $M(d_1, d_2)$ in H is the wanted answer.

4   Recognition of Two Second-Digits

Now we collect all image instances of "0" to "9" in 11 seconds at s-digit box without knowing what digit is on which instance yet. The s-digits in the frames from k*W+1 to (k+1)*W are the same if frame 1 is s-digit transition frame. Without loss of generality, we assume that the s-digit in the frames from 1 to W is "0". Thus, the s-digit in the frames k*W+1 to (k+1)*W is number k. In other words, the s-digits in the frames from 1 to 11*W form a digit sequence. We select frame (k+0.5)*W to represent the frame from k*W+1 to (k+1)*W, denoted as F^k. Let D(j) be the standard template of digit "j", j=0, 1,2,...,9. Then N(x), the measurement of the sequence starting with x, is defined as follows.

$$N(x) = \sum_{r=0}^{9} M\ (D((x+r)\%10\ ), ROI((r+0.5)*W) \qquad (5)$$

where M(ROI1, ROI2) is the measure function of the two region similarity and N(x) is defined on {0, 1, 2,..., 9} and % means mode. Then we identify the maximum point of N(x), which tells us the s-digits on any frame. For instance, assume that x=0 is the maximum point of N(x). Then the s-digit on frames k*W+1 to (k+1)*W is digit "k" for k=0 to 9.

Thus we finish the s-digit recognition and the ts-digit recognition can be done in a similar way.

5   Minute Recognition Using On-the-fly SVM

This section presents an on-the-fly SVM (Support Vector Machine) for recognizing the digits on both m-digit and tm-digit places. Here on-the-fly SVM means that the SVM is trained in the running phase and it has no offline training. This method uses the digit and the blank instances collected from s-digit place to train a SVM, and then use the trained SVM to recognize the digits and the blank on minute places. Without the loss of generality we assume that s-digit of from k*R+1 to (k+1)*R frames is digit k for k=0 to 9. Then we collect 2R-10 instances of k from frame k*R+2 to (k+1)*R-3 and (k+10)*R+2 to (k+11)*R-3. We use the area surrounding the digits to create a blank instance for each frame as Fig 3(b) shows. We use these instances to train the SVM.

We acquire the instances of m-digit and tm-digit from frame k*R+2 to (k+1)*R-3. We put this digit or blank instances into the trained SVM. Then these SVM will tell us what digits are on them. To achieve the robustness of this recognition the trained SVM recognizes a set of the instances that show the same digit. The most occurrence digit of the multiple SVM outputs is considered as its digit of this place. This on-the-fly SVM procedure is given in Fig 5.

```
Step 1: Collect the instances from s-digit
   place;
Step 2: Train a SVM using the obtained
   instances;
Step 3: Collect the instance of m-digit and tm-
   digit;
Step 4: Recognize the instances and identify the
   digit for both m-digit and tm-digit.
```

Fig. 5. The on-the-fly SVM procedure for recognizing m-digit and tm-digit

6   Minute Recognition Using On-the-fly SVM

We implemented our time recognition system in Visual C++ and tested the system and its functions on 220 mepg1 and mepg2 clips. To show the good quality of the system we test the on-the-fly SVM and time recognition functions and compare them with the functions in the literature and the usual SVM.

6.1   Performance of the On-the-fly SVM

Here we compare the performances of the three methods for recognizing minute and ten-minute of the digital video clock. They are the template matching method used in papers [1-3], the offline SVM method, and the on-the-fly SVM method and they are called as template, offline, and on-the-fly, respectively for short. The offline method trains a uniform SVM to recognize m-digit and tm-digit for all the videos, whereas the on-the-fly method trains a SVM for a video in an on-the-fly way. Compared with the offline SVM the on-the-fly SVM has two merits. The first merit is that it can recognize both m-digits and tm-digits in a higher accuracy because the training samples and work samples are from the same video. The other is that it does not need the instance normalization. Due to that the localized digit boxes may have half a pixel discrepancy so we takes their instances by using multiple boxes produced through moving 1 pixel left or right. Recognize m-digit and tm-digit by using the instances taken from the localized box and by multiple boxes are labeled as "1-box" and "m-box" in Table 1 respectively. Table 1 shows that the method developed in this paper has the best performance, whose accuracy is higher than 92%.

Table 1. Comparison on the minute recognition accuracy of the different methods

template		offline		on-the-fly	
1-box	M-box	1-box	M-box	1-box	M-box
46.4%	63.41%	68.2%	86.8%	97.3%	98.6%
102/220	140/220	150/220	191/220	214/220	217/220

6.2   Comparison on Different Time Recognition Functions

Here we compare the time recognition function denoted as RTF and the reliable time recognition function denoted as RRTF developed in this paper with the time recognition function developed in [3]. The function presented in [3] is denoted as RTF1; RTF1 strengthened by the digit color conversion presented in this paper is denoted as RTF2. The results of these four functions are given in Table 2. In Table 2, column "s-digit", "ts-digit", "m-digit", and "tm-digit" give the number of clips that the corresponding digit place are correctly recognized and their percentages by observing video 22 seconds except RRTF. Notice that RRTF needs to observe a 24-second clip. Column "time" gives the number of clips that the algorithm can recognize all 4 four digits correctly. Table 2 shows that RTF and RRTF can achieve a higher accuracy, compared with RTF1 and RTF2 for each digit of the digital video clock.

Table 2. Comparison on the time recognition accuracy of RRTF, RTF, RTF1, and RTF2 on 220 soccer video clips

Algorithms	s-digit	ts-digit	m-digit	tm-digit	time
	180/220	180/180	178/180	177/180	176/220
RTF1	81.8%	100.0%	98.9%	98.3%	80.0%
	220/220	220/220	217/220	216/220	214/220
RTF 2	100.0%	100.0%	98.6%	98.2%	97.3%
	220/220	220/220	219/220	218/220	217/220
RTF	100.0%	100.0%	99.5%	99.1%	98.6%
	220/220	220/220	220/220	220/220	220/220
RRTF	100.0%	100.0%	100.0%	100.0%	100.0%

7   Conclusions and Future Work

We have presented a system for recognizing the time of the digital video clock, which comprises all the functions for time recognition from static region detection to digits localization and then to time recognition. This system can reliably recognize the time of various types of video clocks due to that it employed a series of effective and robust methods derived from the domain knowledge of the digital video clock. Three methods are proposed in this paper and the others are inherited from the literature. In the technology development this paper has the following four contributions. First, it developed a very novel and robust time recognition system. Second, it developed a robust digit extraction method, which can accurately identify the font color and digit pixels of the given video. Third, it formed an s-digit localization procedure by fusing the image processing techniques and the global maximum model of region second periodicity method. Fourth, it developed an on-the-fly SVM method. Compared with the offline SVM, i.e. usual SVM, the on-the-fly SVM has a better performance.

In the near future we plan to develop a method to quickly localize the s-digit because the current method is too tedious. Then we will make a time recognition dynamic link library for public use. We will apply the time recognition library to recognize the time of the digital video clock of other types of videos such as basketball videos.

References

1. Li, Y., Wan, K., Yan, X., Yu, X., Xu, C.: Video clock time recognition based on temporal periodic pattern. In: ICASSP 2006, vol. II, pp. 653–656 (2006)
2. Li, Y., Xu, C., Wan, K., Yan, X., Yu, X.: Reliable video clock time recognition. In: ICPR 2006, vol. 4, pp. 128–131 (2006)
3. Yu, X., Li, Y., Lee, W.: Robust time recognition of video clock based on digit transit detection and digit-sequence recognition. In: ICPR, pp. 1–4 (2008)
4. Bu, F., Sun, L.-F., Ding, X.-F., Miao, Y.-J., Yang, S.-Q.: Detect and Recognize Clock Time in Sports Video. In: Huang, Y.-M.R., Xu, C., Cheng, K.-S., Yang, J.-F.K., Swamy, M.N.S., Li, S., Ding, J.-W. (eds.) PCM 2008. LNCS, vol. 5353, pp. 306–316. Springer, Heidelberg (2008)
5. Jung, K., Kim, K.I., Jain, A.K.: Text information extraction in images and video: a survey. Pattern Recognition 37(5), 977–997 (2004)
6. Wang, M., Hong, R., Li, G., Zha, Z., Yan, S., Chua, T.-S.: Event driven web video summarization by tag localization and key-shot identification. IEEE Transactions on Multimedia 14(4), 975–985 (2012)
7. Xu, C., Zhang, Y., Zhu, G., Rui, Y., Lu, H., Huang, Q.: Using webcast text for semantic event detection in broadcast sports video. IEEE Transactions on Multimedia 10(7), 1342–1355 (2008)
8. Yu, X., Li, L., Leong, H.W.: Interactive Broadcast Services for Live Soccer Video Based on Instant Semantics Acquisition. J. of Visual Communication and Image Representation 20(2), 117–130 (2009)

On Modeling 3-D Video Traffic

M.E. Sousa-Vieira

Department of Telematics Engineering
University of Vigo, Spain

Abstract. Today's 3-D video has become feasible on consumer electronics platforms through advances in display technology, signal processing,
transmission technology, circuit design and computer power. In order to
develop research studies on network transport of 3-D video, adequate
traffic models are necessary. Particularly, the efficient and on-line generation of synthetic paths is fundamental for simulation studies. In this
work we check the suitability of the $M/G/\infty$ process for modeling the
correlation structure of 3-D video.

Keywords: 3-D video, traffic modeling, $M/G/\infty$ process, Whittle
estimator.

1 Introduction

With the increasing popularity of multimedia applications, video data represents
a large portion of the traffic in modern networks.

H.264/AVC and the scalability extensions [1] have recently become some of the
most widely accepted video coding standards, because they have demonstrated
significantly improved coding efficiency, substantially enhanced error robustness
and increased flexibility and scope of applicability relative to prior video coding
standards, such as H.263 and MPEG2.

Since developing these standards, the Joint Video Team of the ITU-T Video
Coding Experts Group (MPEG) has also standardized an extension of that technology that is referred to as multiview video coding (MVC) [2]. Multiview video
provides several views taken from different perspectives of a given scene, that
give viewers the perception of depth and is therefore commonly referred to as
three-dimensional (3-D) video.

Today's 3-D video has become feasible on consumer electronics platforms
through advances in display technology, signal processing, transmission technology, circuit design and computer power.

Related to multiview video compression, multiview video coding exploits the
redundancies across different views of the same scene in addition to the temporal
and intra-view spatial redundancies exploited in single-view encoding. As an
example a prediction structure is shown in Figure 1.

The main approaches for representing and encoding 3-D video are the multiview video (MV) representation, which exploits the redundancies between the
different views, the frame sequential representation (FS), which merges the views

S. Li et al. (Eds.): MMM 2013, Part I, LNCS 7732, pp. 327–335, 2013.

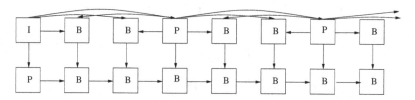

Fig. 1. Inter-view prediction in multiview video compression

to form a single sequence and applies conventional single-view encoding and the side-by-side (SBS) representation, which reduces the horizontal resolution of the views and combines them to form a single frame sequence for single-view encoding.

In order to develop research studies on network transport of 3-D video, adequate traffic models are necessary. Particularly, the efficient and on-line generation of synthetic paths is fundamental for simulation studies.

Several traffic studies have convincingly show the existence of persistent correlations in several kinds of traffic [3–6], as VBR video, and that the impact of the correlation on the performance metrics may be drastic, and several works have been conducted in modeling VBR video traffic based on different stochastic processes [7–13] that display different forms of correlation.

We focus on the $M/G/\infty$-type processes [14, 8, 15] for its theoretical simplicity, its flexibility to exhibit both Short-Range Dependence (SRD) and Long-Range Dependence (LRD) in a parsimonious way and its advantages for simulation studies, such as the possibility of on-line generation and the lower computational cost [16].

In order to apply a model to the synthetic generation of traces with a correlation structure similar to that of real sequences, a fundamental problem is the estimation of the parameters of the model. Between the methods proposed in the literature to deal with persistent correlations [17–20], those based on the Whittle estimator are especially interesting because they permit to fit the whole spectral density and to obtain confidence intervals of the estimated parameters.

In this work we check the suitability of the $M/G/\infty$ process for modeling the correlation structure of 3-D video.

The remainder of the paper is organized as follows. We begin reviewing the main concepts related to the $M/G/\infty$ process in Section 2 and those related to the Whittle estimator in Section 3. In Section 4 we explain the $M/G/\infty$-based model that we consider in this work and in Section 5 we apply it to the modeling the correlation structure of 3-D video traffic at the GoP level. Finally, concluding remarks are summarized in Section 6.

2 $M/G/\infty$ Process

The $M/G/\infty$ process [21] is a stationary version of the occupancy process of an $M/G/\infty$ queueing system. In this queueing system, customers arrive according to a Poisson process, occupy a server for a random time with a generic distribution X with finite mean, and leave the system.

Though the system operates in continuous time, it is easier to simulate it in discrete-time, so this will be the convention henceforth [15]. The number of busy servers at time $t \in \mathbb{Z}^+$ is $Y_t = \sum_{i=1}^{\infty} A_{t,i}$, where $A_{t,i}$ is the number of arrivals at time $t - i$ which remain active at time t, i.e., the number of active customers with age i. For any fixed t, $\{A_{t,i}, i = 1, \ldots \}$ are a sequence of independent and identically distributed (iid) Poisson variables with parameter $\lambda \mathbb{P}(X \geq i)$, where λ is the rate of the arrival process. The expectation and variance of the number of servers occupied at time t is

$$\mathbb{E}(Y_t) = \mathrm{Var}(Y_t) = \lambda \sum_{i=1}^{\infty} \mathbb{P}(X \geq i) = \lambda \mathbb{E}(X).$$

The discrete-time process $Y_t, t = 0, 1, \ldots$ is time-reversible and wide-sense stationary, with autocovariance function

$$\gamma(h) = \mathrm{Cov}(Y_{t+h}, Y_t) = \lambda \sum_{i=h+1}^{\infty} \mathbb{P}(X \geq i), \quad h = 0, 1, \ldots$$

The function $\gamma(h)$ determines completely the expected service time

$$\mathbb{E}(X) = \frac{\gamma(0)}{\lambda}$$

and the distribution of X, the service time, because

$$\mathbb{P}(X = i) = \frac{\gamma(i-1) - 2\gamma(i) + \gamma(i+1)}{\lambda}, \quad i = 1, 2, \ldots . \tag{1}$$

By (1), the autocovariance is a non-negative convex function. Alternatively, any real-valued sequence $\gamma(h)$ can be the autocovariance function of a discrete-time $M/G/\infty$ occupancy process if and only if it is decreasing, non-negative and integer-convex [8]. In such a case, $\lim_{h \to \infty} \gamma(h) = 0$ and the probability mass function of X is given by (1).

If $A_{0,0}$ (i.e., the initial number of customers in the system) follows a Poisson distribution with mean $\lambda \mathbb{E}(X)$, and their service times have the same distribution as the residual life $\widehat{X}$ of the random variable X

$$\mathbb{P}(\widehat{X} = i) = \frac{\mathbb{P}(X \geq i)}{\mathbb{E}(X)},$$

then $\{Y_t, t = 0, 1, \ldots \}$ is strict-sense stationary, ergodic, and enjoys the following properties:

1. The marginal distribution of Y_t is Poissonian for all t, with mean value $\mu = \mathbb{E}(Y_t) = \lambda \mathbb{E}(X)$.
2. The autocovariance function is $\gamma(h) = \gamma(0)\mathbb{P}(\widehat{X} > h) \quad \forall h \geq 0$.

If the autocovariance function is summable the process exhibits SRD. Conversely, if the autocovariance function is not summable, the process exhibits LRD. In particular, the M/G/∞ process exhibits LRD when X has infinite variance, as in the case of some heavy-tailed distributions. The latter are the discrete probability distribution functions satisfying $\mathbb{P}(X > k) \sim k^{-\alpha}$ asymptotically as $k \to \infty$.

3 Whittle Estimator

Let $f_\theta(\lambda)$ be the spectral density function of a zero-mean stationary Gaussian stochastic process, X, where $\theta = (\theta_1, \ldots, \theta_M)$ is a vector of unknown parameters that is to be estimated from observations. Let

$$I_{X^N} = \frac{1}{2\pi N} \| \sum_{i=0}^{N-1} X_i e^{-j\lambda i} \|^2$$

be the periodogram of a sample of size N of the process X. The approximate Whittle estimator [17] is the vector $\widehat{\theta} = (\hat{\theta}_1, \ldots, \hat{\theta}_M)$ minimizing, for a given sample X^N of size N of X, the statistic

$$Q_{X^N}(\theta) \triangleq \frac{1}{2\pi} \left(\int_{-\pi}^{\pi} \frac{I_{X^N}(\lambda)}{f_\theta(\lambda)} d\lambda + \int_{-\pi}^{\pi} \log f_\theta(\lambda) d\lambda \right). \tag{2}$$

If θ° is the true value of θ, then

$$\lim_{N \to \infty} \Pr\left(\|\hat{\theta} - \theta^\circ\| < \epsilon \right) = 1,$$

for any $\epsilon > 0$, namely, $\hat{\theta}$ converges in probability to θ° (is a weakly consistent estimator). It is also asymptotically normal, since $\sqrt{N}(\hat{\theta} - \theta^\circ)$ converges in distribution to ζ, as $N \to \infty$, where ζ is a zero-mean Gaussian vector with matrix of covariances known. Thus, from this asymptotic normality, confidence intervals of the estimated values can be computed.

4 M/G/∞-Based Model

In this work we consider as distribution for the service time the discrete-time distribution S, proposed in [15]. Its main characteristic is that of being a heavy-tailed distribution with two parameters, α and m, a feature that allows to model simultaneously the short-term correlation behavior (by means of the one-lag autocorrelation coefficient r(1)) and the long-term correlation behavior (by means of the H [22] parameter) of the occupancy process. Specifically, the autocorrelation function of the resulting M/S/∞ process is

$$r(k) = \begin{cases} 1 - \dfrac{\alpha - 1}{m\alpha} k & \forall k \in (0, \text{m}] \\ \dfrac{1}{\alpha} \left(\dfrac{\text{m}}{k} \right)^{\alpha - 1} & \forall k \geq \text{m}, \end{cases}$$

with

$$\alpha = 3 - 2H,$$

$$m = \begin{cases} (\alpha r(1))^{\frac{1}{\alpha-1}} & \forall r(1) \in (0, \frac{1}{\alpha}] \\ \dfrac{\alpha - 1}{\alpha - \alpha r(1)} & \forall r(1) \in [\frac{1}{\alpha}, 1). \end{cases}$$

If $\alpha \in (1, 2)$ then $H \in (0.5, 1)$, and $\sum_{k=0}^{\infty} r(k) = \infty$. Hence, in this case this correlation structure gives rise to an LRD process.

The spectral density, needed to use the Whittle estimator, is given by [20]

$$f_X(\lambda) = f_{\{m,\alpha\}}(\lambda) \stackrel{\Delta}{=} \text{Var}\,[X] \left\{ \frac{m^{\alpha-1}}{\alpha \cos(\lambda) - \alpha} f_h(\lambda) + \frac{1}{2\pi} \right.$$

$$\left. + \frac{1}{\pi} \sum_{k=1}^{\lfloor m \rfloor} \cos(\lambda k) \left[\frac{\alpha(m-k)+k}{m\alpha} - \frac{1}{\alpha} \left(\frac{m}{k}\right)^{\alpha-1} \right] \right\} \quad \forall \lambda \in [-\pi, \pi],$$

where f_h is the spectral density of a FGN [23] process with $h = \frac{(1-\alpha)}{2}$, scaled by the variance.

In order to improve the adjustment of the short-term correlation of the previous process, in [24] has been proposed to add an autoregressive filter.

Specifically, we focus on the particular case of an AR(1) filter. If Y is the M/S/∞ original process, the new one is obtained as $X_n = \alpha_1 X_{n-1} + Y_n$.

The mean values and covariances are related by

$$E\,[X] = \frac{E\,[Y]}{1 - \alpha_1},$$

$$\gamma^X(k) = \frac{\gamma^Y(k) + \sum_{i=1}^{\infty} \gamma^Y(k+i)\alpha_1^i + \sum_{i=1}^{\infty} \gamma^Y(k-i)\alpha_1^i}{1 - \alpha_1^2}.$$

The spectral density results

$$f_X(\lambda) = \frac{f_Y(\lambda)}{|1 - \alpha_1 e^{j\lambda}|^2} \quad \forall \lambda \in [-\pi, \pi].$$

We denote the resulting process as M/S/∞-AR.

5 Modeling the Correlation Structure of 3-D Video Traffic at the GoP Level

We consider, as an example, the following empirical video traces of the Group of Pictures (GoP) sizes available at [25].

- 'Alice in Wonderland' (a fantasy movie):
  - T-1-left: G16B1 GoP-format and left view.
  - T-1-right: G16B1 GoP-format and right view.
  - T-2-left: G16B7 GoP-format and left view.
  - T-2-right: G16B7 GoP-format and right view.

- 'Monsters vs Aliens' (a computer-animated fiction movie):
  - T-3-left: G16B1 GoP-format and left view.
  - T-3-right: G16B1 GoP-format and right view.
  - T-4-left: G16B7 GoP-format and left view.
  - T-4-right: G16B7 GoP-format and right view.

G16B1 and G16B7 refer to the IBBBBBBBBBBBBBBB and IBBBBBBBPBBBBBBB GoP formats, respectively.

The estimations of the parameters of the M/S/∞-AR process, computed via the Whittle estimator, are as follows:

- T-1-left: $\widehat{\alpha_1} = -0.03$, $\widehat{r(1)} = 0.94$, $\widehat{H} = 0.89$.
- T-1-right: $\widehat{\alpha_1} = 0.26$, $\widehat{r(1)} = 0.79$, $\widehat{H} = 0.75$.
- T-2-left: $\widehat{\alpha_1} = 0.02$, $\widehat{r(1)} = 0.93$, $\widehat{H} = 0.92$.
- T-2-right: $\widehat{\alpha_1} = 0.28$, $\widehat{r(1)} = 0.78$, $\widehat{H} = 0.74$.
- T-3-left: $\widehat{\alpha_1} = 0.47$, $\widehat{r(1)} = 0.51$, $\widehat{H} = 0.81$.
- T-3-right: $\widehat{\alpha_1} = 0.41$, $\widehat{r(1)} = 0.71$, $\widehat{H} = 0.81$.
- T-4-left: $\widehat{\alpha_1} = 0.43$, $\widehat{r(1)} = 0.56$, $\widehat{H} = 0.81$.
- T-4-right: $\widehat{\alpha_1} = 0.43$, $\widehat{r(1)} = 0.59$, $\widehat{H} = 0.81$.

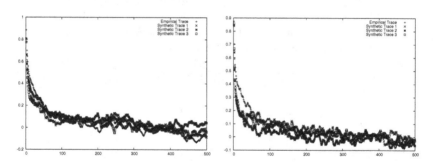

Fig. 2. 'Alice in Wonderland' (G16B1 GoP format). Adjustment of the autocorrelation - left view (left) and right view (right).

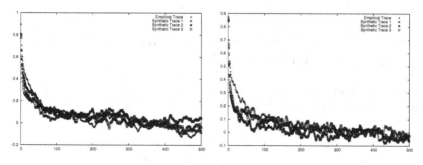

Fig. 3. 'Alice in Wonderland' (G16B7 GoP format). Adjustment of the autocorrelation - left view (left) and right view (right).

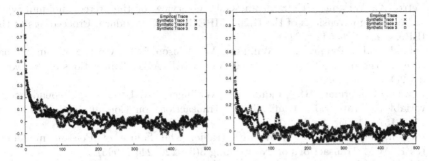

Fig. 4. 'Monsters vs Aliens' (G16B1 GoP format). Adjustment of the autocorrelation - left view (left) and right view (right).

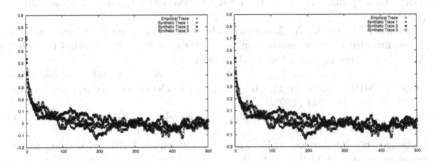

Fig. 5. 'Monsters vs Aliens' (G16B7 GoP format). Adjustment of the autocorrelation - left view (left) and right view (right).

In Figs. 2, 3, 4, and 5 we represent the autocorrelation function of synthetic traces of the M/S/∞-AR process. We can observe a good math with the empirical traces in all cases.

6   Conclusions

In this paper, we have checked the suitability of LRD models based on the M/G/∞ process for modeling 3-D video traffic. Due to the high values of the estimated Hurst parameter we can conclude that LRD is a good hypothesis for these video sequences.

The proposed model enjoy several interesting features: highly efficient, on-line generation and the possibility of capturing the whole correlation structure in a parsimonious way.

References

1. Schwarz, H., Marpe, D., Wiegand, T.: Overview of the scalable video coding extension of the H.264/AVC standard. IEEE Transactions on Circuits and Systems for Video Technology 17(9), 1103–1120 (2007)

2. Vetro, A., Wiegand, T., Sullivan, G.J.: Overview of the stereo and multiview video coding extensions of the H.264/MPEG-4 AVC standard. Proceedings of the IEEE 4(99), 626–642 (2011)
3. Leland, W.E., Taqqu, M.S., Willinger, W., Wilson, D.V.: On the self-similar nature of Ethernet traffic (extended version). IEEE/ACM Transactions on Networking 2(1), 1–15 (1994)
4. Beran, J., Sherman, R., Taqqu, M.S., Willinger, W.: Long-range dependence in variable bit rate video traffic. IEEE Transactions on Communications 43(2-4), 1566–1579 (1995)
5. Paxson, V., Floyd, S.: Wide-area traffic: The failure of Poisson modeling. IEEE/ACM Transactions on Networking 3(3), 226–244 (1995)
6. Crovella, M.E., Bestavros, A.: Self-similarity in world wide web traffic: Evidence and possible causes. IEEE/ACM Transactions on Networking 5(6), 835–846 (1997)
7. Garrett, M.W., Willinger, W.: Analysis, modeling and generation of self-similar VBR video traffic. In: Proc. ACM SIGCOMM 1994, London, UK, pp. 269–280 (1994)
8. Krunz, M., Makowski, A.: Modeling video traffic using $M/G/\infty$ input processes: A compromise between markovian and LRD models. IEEE Journal on Selected Areas in Communications 16(5), 733–748 (1998)
9. Lombardo, A., Morabito, G., Schembra, G.: An accurate and treatable Markov model of MPEG video traffic. In: Proc. IEEE INFOCOM 1998, San Francisco, CA, USA, pp. 217–224 (1998)
10. Ma, S., Ji, C.: Modeling video traffic using wavelets. IEEE Communications Letters 2(4), 100–103 (1998)
11. Melamed, B., Pendarakis, D.E.: Modeling full-length VBR video using Markov-renewal-modulated TES models. IEEE Journal on Selected Areas in Communications 16(5), 600–611 (1998)
12. Frey, M., Nguyen-Quang, S.: A Gamma-based framework for modeling variable bit rate MPEG video sources: The GoP GBAR model. IEEE/ACM Transactions on Networking 8(6), 710–719 (2000)
13. Sarkar, U.K., Ramakrishnan, S., Sarkar, D.: Modeling full-length video using Markov-modulated Gamma-based framework. IEEE/ACM Transactions on Networking 11(4), 638–649 (2003)
14. Cox, D.R.: Long-range dependence: A review. In: David, H.A., David, H.T. (eds.) Statistics: An Appraisal, pp. 55–74. Iowa State University Press (1984)
15. Suárez, A., López, J.C., López, C., Fernández, M., Rodríguez, R., Sousa, M.E.: A new heavy-tailed discrete distribution for LRD $M/G/\infty$ sample generation. Performance Evaluation 47(2/3), 197–219 (2002)
16. Sousa, M.E., Suárez, A., Fernández, M., López, C., Rodríguez, R.F.: A highly efficient $M/G/\infty$ generator of self-similar traces. In: 2006 Winter Simulation Conference, Monterey, CA, USA, pp. 2146–2153 (2006)
17. Whittle, P.: Estimation and information in stationary time series. Arkiv Matematick 2(23), 423–434 (1953)
18. Abry, P., Veitch, D.: Wavelet analysis of long-range dependent traffic. IEEE Transactions on Information Theory 44(1), 2–15 (1998)
19. Taqqu, M.S., Teverovsky, V.: On estimating the intensity of long-range dependence in finite and infinite variance time series. In: Adler, R., Feldman, R., Taqqu, M.S. (eds.) A Practical Guide to Heavy Tails, pp. 177–217. Birkhäuser (1998)
20. Sousa, M.E., Suárez, A., López, J.C., López, C., Fernández, M., Rodríguez, R.F.: Application of the Whittle estimator to the modeling of traffic based on the $M/G/\infty$ process. IEEE Communications Letters 11(10), 817–819 (2007)

21. Cox, D.R., Isham, V.: Point Processes. Chapman and Hall (1980)
22. Hurst, H.E.: Long-term storage capacity of reservoirs. Transactions of the American Society of Civil Engineers 116, 770–799 (1951)
23. Beran, J.: Statistics for Long-Memory Processes. Chapman and Hall (1994)
24. Sousa, M.E., Suárez, A., Rodríguez, R.F., López, C.: Flexible adjustment of the short-term correlation of LRD M/G/∞-based processes. Electronic Notes in Theoretical Computer Science (2010)
25. Online, http://trace.eas.asu.edu

A Low-Complexity Quantization-Domain H.264/SVC to H.264/AVC Transcoder with Medium-Grain Quality Scalability

Lei Sun[1], Zhenyu Liu[2], and Takeshi Ikenaga[1]

[1] Graduate School of Information, Production and Systems,
Waseda University, Kitakyushu, 808-0135 Japan
[2] Tsinghua National Laboratory for Information Science and Technology,
Tsinghua University, Beijing, 100084 China
sunlei@ruri.waseda.jp, liuzhenyu73@mail.tsinghua.edu.cn,
ikenaga@waseda.jp

Abstract. Scalable Video Coding (SVC) aiming to provide the ability to adapt to heterogeneous requirements. It offers great flexibility for bitstream adaptation in multi-point applications. However, transcoding between SVC and AVC is necessary due to the existence of legacy AVC-based systems. This paper proposes a fast SVC-to-AVC MGS (Medium-Grain quality Scalability) transcoder. A quantization-domain transcoding architecture is proposed for transcoding non-KEY pictures in MGS. KEY pictures are transcoded by drift-free architecture so that error propagation is constrained. Simulation results show that proposed transcoder achieves averagely 37 times speed-up compared with the re-encoding method with acceptable coding efficiency loss.

Keywords: low complexity, quantization domain, SVC/AVC transcoding, MGS.

1 Introduction

With the intention of providing scalability to diverse environments in multi-point applications, SVC enables transmission of a single bitstream containing multiple subset bitstreams. The subset bitstreams are organized in layered structure efficiently and can be extracted according to different requirements [1, 2]. Three main scalabilities are provided, i.e., spatial, temporal and quality. Performance of SVC and key technologies are described in literatures [3–5].

For multi-point applications involving terminals with different network bandwidth, display resolution and processing ability, SVC is expected to be a promising solution. Unlike the traditional transcoding-based systems, SVC-based system only requires lightweight operations for bitstream adaptation. However, as an undeniable fact, the current applications are mostly adopting legacy AVC-based systems. To achieve interoperability, transcoding between SVC format and AVC format is needed. In this paper, we focus on the ultra-low-delay transcoding from SVC to AVC with MGS Scalability (as will be explained in Section 2).

S. Li et al. (Eds.): MMM 2013, Part I, LNCS 7732, pp. 336–346, 2013.

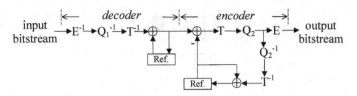

Fig. 1. Re-encoding transcoding architecture. (E: entropy coding, $Q_i(i = 1, 2)$: quantization, T: DCT transform, $Ref.$: reference picture buffer, superscript "-1": inverse process.)

SVC standard provides a special encoder-side configuration for SVC-to-AVC rewriting [6], which requires the encoding process modification at the sender side. It is actually a particular functionality during bitstream generation, rather than a real transcoder. When the sender side uses normal SVC configuration (i.e., without the rewriting functionality), this method does not work.

Transcoding has been an important task for bitstream adaptation or format conversion, due to the continuous progress in video coding standards [7–9]. A straightforward solution is the "re-encoding" method (Figure 1). It fully decodes the input bitstream and then fully re-encodes the decoded pictures, consuming intensive computations. For single layer transcoding, many works have been done. Literatures [10–17] are based on the motion reuse (MR) transcoding architecture. The modes and motion vectors (MVs) of input bitstream are utilized and refined to accelerate the motion estimation (ME) process of encoder. MR based works only accelerates the ME part of the re-encoding method, and the speed-up is restricted by existence of other components such as DCT transforms. In [18] the authors merge the decoder and encoder MCP (motion-compensated prediction) loops under the assumption that motion data are the same for decoder and encoder, referred as Single-Loop (SL) transcoding architecture (Figure 2). This architecture is free of drift (error propagation), and one inverse transform and one picture buffer are reduced. A further accelerated transcoding architecture is proposed by [19], in which transforms are totally removed and motion compensation is directly performed on DCT transform coefficients (denoted as MC-DCT). MC-DCT needs floating-point matrix multiplication which is quite costly and diminishes the speed gain. Another common known architecture is the open-loop (OL) transcoding architecture (Figure 3), for which the drift problem is quite severe. The SL and OL architectures are often referred as

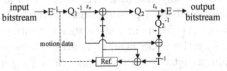

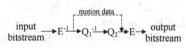

Fig. 2. Single-loop (SL) transcoding architecture

Fig. 3. Open-loop (OL) transcoding architecture

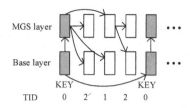

MGS layer

Base layer

KEY   KEY

TID   0   2'   1   2   0

Fig. 4. Hierarchical-P with KEY pictures. (GOPSize = 4)

frequency-domain transcoding methods since there is no transform operations on the main route. Relatively, the re-encoding and MR architectures are usually mentioned as pixel-domain transcoding methods.

In this paper, a fast SVC-to-AVC MGS transcoding architecture is proposed. Significant speed-up is achieved by proposed quantization-domain transcoding for non-KEY pictures. KEY pictures are transcoded with drift-free SL architecture such that drift will be restricted between KEY pictures.

The rest of this paper is organized as follows. Section 2 describes SVC features. Section 3 shows proposed quantization-domain transcoding architecture for non-KEY pictures, and Section 4 explains KEY pictures transcoding. Simulation results are given in Section 5 and conclusions are drawn in Section 6.

2   Scalable Video Coding

2.1   Medium-Grain Quality Scalability

MGS includes two main features, i.e., coefficients partitioning and KEY picture concept. Coefficients partitioning allows to distribute the transform coefficients among several NALUs (Network Abstraction Layer Units). The coefficients are divided into groups, and each group corresponds to one NALU. Up to 16 NALUs are possible, and by discarding several of them flexible packet-based quality scalability is provided. In transcoding, it is very easy to parse the input NALUs and reassemble the transform coefficients.

Another feature is the KEY picture concept, which is based on hierarchical prediction structure. Figure 4 shows a hierarchical-P prediction structure for MGS encoding [20] (we target for low-delay applications where highly delayed B pictures are rarely used). TID represents the temporal layer ID. Grey-colored pictures are KEY pictures ($TID = 0$), which only use other KEY pictures for prediction. Base layer KEY picture is predicted from previous base layer KEY picture and MGS layer KEY picture is predicted from current base layer picture. Non-KEY picture ($TID > 0$) is predicted by the MGS layer of nearest previous picture with smaller TID. Such prediction structure can constrain the drift due to discarded packets within a GOP (Group Of Pictures).

2.2 Coding Modes in SVC

SVC introduces inter-layer prediction (ILP) schemes while inheriting the AVC coding modes (INTER and INTRA). Three kinds of ILPs are introduced to explore the correlation between base layer (BL) and enhancement layer (EL). They are inter-layer residual, inter-layer intra and inter-layer motion predictions (denoted as IL_Residual, IL_Intra and IL_Motion predictions hereafter).

IL_Intra prediction predicts the original enhancement layer input picture using the upsampled base layer reconstructed picture. IL_Residual prediction tries to predict the residual data generated by INTER prediction. The residual generated by normal INTER prediction is predicted by the upsampled base layer reconstructed residual signal. IL_Motion prediction tries to reduce the size of motion data for INTER coded MBs. The upsampled base layer mode and MV information is utilized to predict the enhancement layer motion data. More descriptions about ILPs can be found in [1].

The IL_Intra prediction is totally independent from the AVC INTRA or INTER modes, while the IL_Residual and IL_Motion predictions are additional refinements based on AVC INTER mode. Thus the coding modes in SVC are shown in Table 1. It is also possible that IL_Residual and IL_Motion both exist for an INTER MB. In such case, it is considered as IL_Residual. For short, "INTER with IL_Residual" and "INTER with IL_Motion" will be denoted as IL_Residual and IL_Motion hereafter.

Table 1. Coding modes in SVC

Inherited modes	Newly introduced modes
INTRA INTER without ILP	IL_Intra INTER with IL_Residual INTER with IL_Motion

3 Non-KEY Pictures Transcoding

3.1 Quantization-Domain Single-Loop Transcoding for IL_Residual MBs

In this subsection, a special frequency-domain transcoding architecture is derived for IL_Residual transcoding, namely quantization-domain single-loop (QDSL) transcoding. Let's start from the drift-free SL architecture as shown in Figure 2. Two signals s_n and t_n are shown and the relation between them is shown in (1).

$$t_n = Q_2(s_n + T(MC(T^{-1}(s_{n-1} - Q_2^{-1}(t_{n-1}))))) \tag{1}$$

Here the $MC(.)$ represents the motion compensation operation corresponding to the bottom addition symbol in Figure 3. By assuming a distributive property for quantization (which is actually not true; same implication for following "assuming"), Equation (1) is modified to Equation (2).

$$t_n = Q_2(s_n) + Q_2(T(MC(T^{-1}(s_{n-1} - Q_2^{-1}(t_{n-1}))))) \tag{2}$$

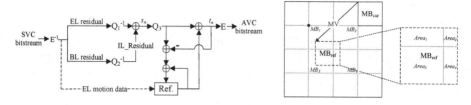

Fig. 5. Quantization-domain single-loop (QDSL) transcoding architecture

Fig. 6. MC in quantization domain

Then by assuming a commutative property between DCT transform and motion compensation, Equation (2) is further modified to Equation (3), also based on the fact that DCT transform is a lossless operation.

$$t_n = Q_2(s_n) + Q_2(MC(s_{n-1} - Q_2^{-1}(t_{n-1}))) \tag{3}$$

By assuming the commutative property between quantization and motion compensation, Equation (3) is changed to Equation (4).

$$t_n = Q_2(s_n) + MC(Q_2(s_{n-1} - Q_2^{-1}(t_{n-1}))) \tag{4}$$

Finally, by applying the previously assumed distributive property of quantization operation, Equation (5) is obtained.

$$t_n = Q_2(s_n) + MC(Q_2(s_{n-1}) - t_{n-1}) \tag{5}$$

The second term of Equation (5) implies a motion compensation on quantized transform coefficients. The first term is easily obtained by quantizing the input signal s_n. A corresponding architecture based on Equation (5) for IL_Residual is shown in Figure 5. Proposed QDSL architecture eliminates DCT transforms and further reduces one inverse quantization component comparing with reference [19].

Figure 6 illustrates the motion compensation method in quantization domain. MB_{cur} is the current MB to be coded and MB_{ref} is the reference MB. Dotted lines are aligned MB boundaries. If the motion vector points to an intersection of dotted lines, e.g. the intersection near the word "MB_1", then the prediction signal can be easily decided as the quantized transform coefficients of MB_1. When the MV does not point to an intersection, the prediction is composed by weighted sum of several related MBs. $MB_i(i = 1..4)$ are MBs overrode by MB_{ref}. The right sub-figure in Figure 10 enlarges the overrode area consisting of 4 regions. The areas of these regions are denoted by $Area_i(i = 1..4)$. Let $Coeff(MB_i)$ be the quantized coefficients matrix of MB_i. The prediction signal is generated by Equation (6). Partition or sub-partition motion compensation is done in a similar way.

$$PRED = \frac{\sum_{i=1}^{4}[Area_i \times Coeff(MB_i)]}{Area_1 + Area_2 + Area_3 + Area_4} \tag{6}$$

3.2   Quantization-Domain Intra Prediction for IL_Intra MBs

Similar to the derivation in previous subsection, Equation (7) can be obtained in the context of intra prediction. Here the $I\_PRED(.)$ is the intra prediction operation. The second term implies quantization-domain intra prediction (QDIP).

$$t_n = Q_2(s_n) + I\_PRED(Q_2(s_{n-1}) - t_{n-1}) \tag{7}$$

In the pixel domain, neighboring pixels are used to form an intra prediction. But in quantization-domain where coefficients are concentrated in upper-left corner, extracting corresponding coefficients for those neighboring pixels is difficult. In proposed transcoder, QDIP is accomplished by approximating the prediction signal using neighboring 4x4 blocks. Figure 7 shows the intra 16x16 prediction in quantization domain. In the leftmost sub-figure, $B_i(i = 1..8)$ are the neighboring 4x4 blocks containing quantized coefficients. When intra 16x16 prediction mode is vertical or horizontal, the prediction is formed by extending neighboring blocks along the prediction direction. For other modes (DC/plane), the prediction is formed by averaging the vertical and horizontal predictions. $B_{ij}(i = 1..4, j = 5..8)$ is the average of B_i and B_j.

Intra 4x4 predictions are processed as Figure 8. B_{cur} is the current 4x4 block to be predicted and $B_i(i = 1..4)$ are neighboring blocks. $(X, A..L)$ are positions used for intra prediction in pixel domain. For mode 0, 1 or 8, the pixels used for intra 4x4 prediction belong to one particular neighboring block. This block is selected to approximate the prediction signal. For mode 2 which uses the mean value of $(A..B, I..L)$, the average of B_2 and B_4 is selected as the prediction. For the rest modes, a weighted average of neighboring blocks is formed as the prediction. The weight depends on how much one block contributes to the prediction signal. For example, Figure 9 shows the mode 3 prediction (each square corresponds to one pixel position), where there are totally 7 predictor values. Table 2 shows the corresponding positions using these predictors. Function $Con(.)$ represents the contribution of block B_i to each predictor. It equals the sum of weights of B_i pixels used for the predictor, multiplied by the number of blocks using this predictor. For example, in mode 3 the predictor $(C + 2D + E)/4$ uses C, D in B_2 and their weights are 1/4 & 2/4. Number of blocks using this predictor is 3 and thus the contribution of B_2 for this predictor is $(1/4 + 2/4) \times 3 = 9/4$. The total contributions for B_2 and B_3 are 25/4 & 39/4, and thus the prediction is

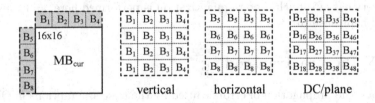

vertical   horizontal   DC/plane

Fig. 7. Intra 16x16 prediction

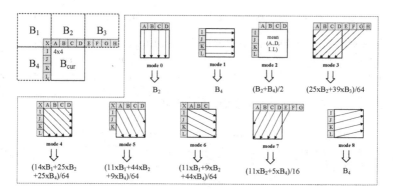

Fig. 8. Intra 4x4 prediction

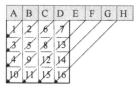

Fig. 9. Intra 4x4 mode 3 prediction

Table 2. Mode 3 prediction

Pixel no.	Predictor	Con(B_2)	Con(B_3)
1	(A+2B+C)/4	1	0
2,3	(B+2C+D)/4	2	0
4,5,6	(C+2D+E)/4	9/4	3/4
7,8,9,10	(D+2E+F)/4	1	3
11,12,13	(E+2F+G)/4	0	3
14,15	(F+2G+H)/4	0	2
16	(G+2H+H)/4	0	1

formed by $(25/4 \times B_2 + 39/4 \times B_3)/(25/4 + 39/4) = (25 \times B_2 + 39 \times B_3)/64$. The predictions for other modes are calculated similarly, and the final results are shown in Figure 13.

3.3   Quantization-Domain Copy for Other MBs

For MBs with other coding modes (INTRA/INTER without ILP/IL_Motion), a quantization-domain copy (QDC) method is applied. Different from IL_Residual or IL_Intra, for these modes the residual generation process in SVC is identical to AVC encoding. The input MB is entropy decoded only, and the quantized residual coefficients along with the motion data are copied into AVC bitstream directly. In case of IL_Motion, the motion data need to be reconstructed first, which is very easy. Note that no re-quantization is performed here, and thus the residual is kept accurate.

4   KEY Picture Transcoding

To constrain the propagation of errors caused by the false assumptions in QDSL deduction and MC/intra prediction approximations , KEY pictures are transcoded based on drift-free single-loop architecture. In hierarchical-P prediction structure, MGS layer KEY pictures are predicted by base layer KEY

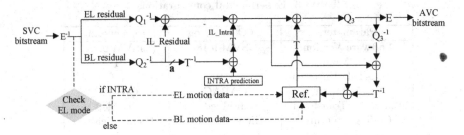

Fig. 10. KEY picture transcoding architecture

pictures. However, after transcoding the base layer pictures will be all discarded. Thus motion re-estimation is necessary for these KEY pictures since a new reference picture must be selected. In proposed transcoder, the previous MGS layer KEY picture is selected as the reference picture. Two merits can be obtained by such selection. Firstly, the prediction structure will remain as a single-layer hierarchical-P structure, by which the temporal scalability is kept. Secondly, due to the high correlation between MGS layer and base layer, the motion data from base layer prediction can be reused for MGS layer.

The above discussion solves INTER MBs in MGS layer (INTER, IL_Residual, IL_Motion). For IL_Intra MB, the base layer prediction mode of current frame is reused since there is no mode information transmitted in MGS layer. For INTRA MB, the MGS layer mode is directly reused. The proposed single-loop based BL-copy architecture is shown in Figure 10. The EL mode is checked to decide which motion data to be utilized. Note that EL does not need inverse transform or motion compensation. Partial decoding is also performed for BL decoding. Only MBs used for IL_Residual and IL_Intra predictions will be decoded. Besides, decoding of MBs used for IL_Residual prediction stops at position **a** since BL reconstruction signal is not needed.

5 Simulation Results

In this section, the proposed transcoder is applied to several publicly available sequences and the results are shown. 6 sequences are encoded with 3-layer MGS scalability, and then the encoded bitstreams are transcoded into AVC format with highest MGS layer quality. *Akiyo* and *bus* are CIF (352x288) sequences. *cheer_leaders* and *flower_garden* are VGA (640x480) sequences. *vidyo1* and *parkrun* are 720p (1280x720) sequences. The main configuration parameters are shown in Table 3. All experiments are performed on an Intel Core 2 (2.67GHz) computer with 2.0GB RAM.

Besides the proposed method, 3 methods are used for comparison - re-encoding, single loop (SL) and open loop (OL). The implementation of re-encoding and OL methods are straightforward. The SL method is implemented based on reference [18]. Table 4 shows the computational time comparisons. Three criteria are shown, i.e., total transcoding time (C1), time saving (C2) and time per frame

Table 3. Experimental configurations

Parameters	SVC encoding	AVC encoding
Software Version	JSVM 9.18	JSVM 9.18
AVCMode	0	1
FramesToBeEncoded	150	150
SymbolMode	CABAC	CABAC
Enable8x8Transform	disabled	disabled
CodingStructure	Hierarchical-P	Hierarchical-P
NumRefFrames	5	5
SearchMode	4 (FastSearch)	4 (FastSearch)
SearchRange	16 for CIF/VGA, 32 for 720p	16 for CIF/VGA 32 for 720p
Quantization Parameter	28 for BL, 24 for EL	20/24/28/32
Loop Filter	enabled	enabled
DisableBSlices	1 (B-slice disabled)	1 (B-slice disabled)
GOPSize	4	4
MGSVectorX(X=0,1,2)	3,3,10	-
InterLayerPred	2 (adaptive)	-
AVCRewriteFlag	0 (disabled)	-

Table 4. Computational time comparisons

Sequence	Re-encoding	Single Loop			Open Loop			Proposal		
	C1	C1	C2	C3	C1	C2	C3	C1	C2	C3
akiyo	194.1	23.5	87.9	157	1.5	99.2	10	3.1	**98.4**	**21**
bus	230.3	26.2	88.6	175	1.9	99.1	13	6.9	**97.0**	**46**
cheer_leaders	738.0	97.6	86.8	651	6.9	99.1	46	25.2	**96.6**	**168**
flower_garden	668.8	80.3	88.0	535	5.7	99.1	38	20.3	**97.0**	**135**
vidyo1	1942.4	242.3	87.5	1615	14.2	99.3	94	3.9	**97.9**	**266**
parkrun	2209.3	281.2	87.3	1875	18.3	99.2	122	73.5	**96.7**	**490**
average	-	-	87.7	-	-	99.2	-	-	**97.3**	-

Criteria: C1: time(s), C2: time saving(%), C3: time/frame(ms)

(C3). The re-encoding method is selected as the comparison base, and the time saving for other methods is calculated by comparing with re-encoding method. The bolded figures in Table 4 show the time saving for our proposal relative to the re-encoding method, as well as the processing time per frame. Time saving ranges from 96.6% up to 98.4%, and the average time saving is 97.3% corresponding to a 37 times speed-up. Processing time per frame ranges from 490 ms down to 21 ms. Comparing with SL methods, proposed transcoder achieves averagely 4.6 times speed-up. OL method is about 3.4 times faster than proposed method.

To give intuitive coding efficiency comparisons, Figure 11 is provided which shows the R-D (rate-distortion) curves for tested sequences. It is obvious that for all sequences the re-encoding method performs best (topmost curve), following by SL and proposal curves sequentially with similar small gaps. OL method is much worse than the other 4 methods, mostly 3 to 4 dB lower.

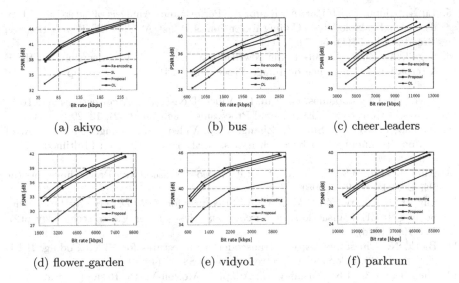

(a) akiyo   (b) bus   (c) cheer_leaders

(d) flower_garden   (e) vidyo1   (f) parkrun

Fig. 11. R-D curves comparison

6   Conclusions

This paper proposes a fast SVC-to-AVC MGS transcoder. Non-KEY frames of input MGS bitstream are transcoded using a quantization-domain transcoding architecture. KEY frames are transcoded using a drift-free transcoding method to constrain the error propagation. Simulation results show that proposed method gains 37 times speed-up comparing with the re-encoding method with acceptable coding efficiency loss.

Acknowledgement. This work was supported by KAKENHI (23300018) and KDDI foundation, Japan.

References

1. Schwarz, H., Marpe, D., Wiegand, T.: Overview of the Scalable Video Coding Extension of the H.264/AVC Standard. IEEE Transactions on Circuits and Systems for Video Technology 17(9), 1103–1120 (2007)
2. Schwarz, H., Wien, M.: The Scalable Video Coding Extension of the H.264/AVC Standard [Standards in a Nutshell]. IEEE Signal Processing Magazine 25(2), 135–141 (2008)
3. Oelbaum, T., Schwarz, H., Wien, M., Wiegand, T.: Subjective Performance Evaluation of the SVC Extension of H.264/AVC. In: IEEE International Conference on Image Processing (ICIP), San Diego, pp. 2772–2775 (2008)
4. Jang, E.D., Kim, J.G., Thang, T.C., Kang, J.W.: Adaptation of Scalable Video Coding to packet loss and its performance analysis. In: International Conference on Advanced Communication Technology (ICACT), Phoenix Park, vol. 1, pp. 696–700 (2010)

5. Li, X., Amon, P., Hutter, A., Kaup, A.: Performance Analysis of Inter-Layer Prediction in Scalable Video Coding Extension of H.264/AVC. IEEE Transactions on Broadcasting 57(1), 66–74 (2011)
6. Segall, A., Zhao, J.: Bit-stream rewriting for SVC-to-AVC conversion. In: IEEE International Conference on Image Processing (ICIP), San Diego, pp. 2776–2779 (2008)
7. Vetro, A., Christopoulos, C., Sun, H.: Video Transcoding Architectures and Techniques: An Overview. IEEE Signal Processing Magazine 20(2), 18–29 (2003)
8. Ahmad, I., Wei, X., Sun, Y., Zhang, Y.Q.: Video Transcoding: An Overview of Various Techniques and Research Issues. IEEE Transactions on Multimedia 7(5), 793–804 (2005)
9. Xin, J., Lin, C., Sun, M.: Digital Video Transcoding. Proceedings of the IEEE 93(1), 84–97 (2005)
10. Sun, H., Kwok, W., Zdepski, J.W.: Architectures for MPEG Compressed Bitstream Scaling. IEEE Transactions on Circuits and Systems for Video Technology 6(2), 191–199 (1996)
11. Bjork, N., Christopoulos, C.: Transcoder Architectures for Video Coding. IEEE Transactions on Consumer Electronics 44(1), 88–98 (1998)
12. Shen, B., Sethi, I.K., Vasudev, B.: Adaptive Motion-Vector Resampling for Compressed Video Downscaling. IEEE Transactions on Circuits and Systems for Video Technology 9(6), 929–936 (1999)
13. Zhang, P., Liu, Y., Huang, Q., Gao, W.: Mode Mapping Method for H.264/AVC Spatial Downscaling Transcoding. In: IEEE International Conference on Image Processing (ICIP), Singapore, vol. 4, pp. 2781–2784 (2004)
14. Youn, J., Sun, M., Lin, C.: Motion Vector Refinement for High-Performance Transcoding. IEEE Transactions on Multimedia 1(1), 30–40 (1999)
15. Shanableh, T., Ghanbari, M.: Heterogenesous Video Transcoding to Lower Spatio-Temporal Rsolutions and Different Encoding Formats. IEEE Transactions on Multimedia 2(2), 101–110 (2000)
16. Escribano, G., Kalva, H., Cuenca, P., Barbosa, L., Garrido, A.: A Fast MB Mode Decision Algorithm for MPEG-2 to H.264 P-Frame Transcoding. IEEE Transactions on Circuits and Systems for Video Technology 18(2), 172–185 (2008)
17. Tang, Q., Nasiopoulos, P.: Efficient Motion Re-Estimation with Rate-Distortion Optimization for MPEG-2 to H.264/AVc Transcoding. IEEE Transactions on Circuits and Systems for Video Technology 20(2), 172–185 (2010)
18. Assuncao, P., Ghanbari, M.: Post-Processing of MPEG2 Coded Video for Transmission at Lower Bit Rates. In: IEEE International Conference on Acoustics, Speech, and Signal Processing (ICASSP), vol. 4, pp. 1998–2001 (1996)
19. Assuncao, P., Ghanbari, M.: A Frequency-Domain Video Transcoder for Dynamic Bit-Rate Reduction of MPEG-2 Bit Streams. IEEE Transactions on Circuits and Systems for Video Technology 8(8), 953–967 (1998)
20. Hong, D., Horowitz, M., Eleftheriadis, A., Wiegand, T.: H.264 Hierarchical P Coding in the Context of Ultra-Low Delay, Low Complexity Applications. In: Picture Coding Symposium (PCS), Nagoya, pp. 146–149 (2010)

Evaluation of Product Quantization for Image Search

Wei-Ta Chu[1], Chun-Chang Huang[1], and Jen-Yu Yu[2]

[1] Department of Computer Science and Information Engineering
National Chung Cheng University, Taiwan
wtchu@cs.ccu.edu.tw, icecandyandscheme@gmail.com
[2] Information and Communication Research Lab,
Industrial Technology Research Institute, Hsinchu, Taiwan
KevinYu@itri.org.tw

Abstract. Product quantization is an effective quantization scheme, with that a high-dimensional space is decomposed into a Cartesian product of low-dimensional subspaces, and quantization in different subspaces is conducted separately. We briefly discuss the factors for designing a product quantizer, and then design experiments to comprehensively investigate how these factors influence performance of image search. By this evaluation we reveal design principles that have not been well investigated before.

Keywords: Product quantization, image search.

1 Introduction

Measuring distances, e.g., Euclidean distances, between data points is a fundamental step for data clustering, classification, and retrieval. In many applications, such as video annotation [10] and image retrieval [11], with such distances the nearest neighbors to a query data point are thus determined. However, in many multimedia applications, we have to process tremendous amount of high-dimensional data points. Exact and exhaustive distance measurement thus suffers from efficiency issues and curse of dimensionality.

Rich methods have been proposed for efficient multidimensional indexing. For example, the KD-tree [1] describes data points by dividing the data space, and search time can be effectively reduced. However, for high-dimensional data, the method is not much more efficient than the brute-force search. Approximate nearest neighbor (ANN) search is thus proposed. Instead of finding the exact nearest neighbors, ANN approaches find N data points that are nearest to a query point with high probability. The Euclidean Locality-Sensitive Hashing (E2LSH) [2], for example, is a widely adopted ANN approach that hashes similar high-dimensional data points into the same bucket with high probability. However, memory needed to store the indexing structure of E2LSH is critical. For applications such as large-scale scene recognition or content-based image retrieval, the trade-off of memory usage of indexing structure and search efficiency becomes severe.

S. Li et al. (Eds.): MMM 2013, Part I, LNCS 7732, pp. 347–356, 2013.

Jegou et al. [3] propose a scheme called product quantization (PQ) to construct a large codebook with little cost. The concept is to "decompose a high-dimensional space into a Cartesian product of low-dimensional spaces and to quantize each subspace separately." A high-dimensional vector $x = (x_1, x_2, ..., x_D)$ is first divided into lower-dimensional subvectors $\{u_1, u_2, ..., u_p\}$, where $u_i = (x_{(i-1) \times d+1}, x_{(i-1) \times d+2}, ..., x_{i \times d})$ is the ith d-dimensional subvector, $d \ll D$. For the training vectors $\{x_k\}_{k=1}^N$, their ith corresponding subvectors $\{u_i\}$ are collected and clustered by the K-means algorithm to form a small codebook. The codebooks constructed based on each set of subvectors are then combined by the Cartesian product to form a large codebook. This scheme effectively constructs a large codebook for multidimensional indexing. PQ has been demonstrated to be scalable to large amount of data [3] and suitable to mobile visual search [9].

Overall, PQ significantly reduces the time required to construct a large codebook, reduces the influence of the curse of dimensionality, and reduces the memory space for storing the codebook. As stated in [3], performance of PQ can be analyzed from the following three aspects, but the authors did not conduct deep investigation in their paper.

- The way to splitting vectors
- The dimension of subvectors
- The number of quantization levels for each subquantizer

Details of these factors will be described in the next section. In this paper, we describe images by bag of visual words based on various codebooks, which are constructed by product quantizers based on different settings. We investigate how different codebooks affect performance of content-based image retrieval. From experimental results, we make a few suggestions for codebook construction.

In the following text, we briefly describe product quantization and the issues to be addressed in Section 2. Section 3 describes evaluation settings, including feature extraction and construction of product quantizers. Various experimental results and discussion are provided in Section 4, followed by the conclusion in Section 5.

2   Product Quantization

Let $x \in R^d$ be a query vector and $\mathcal{Y} = \{y_1, ..., y_n\}$ be the set of vectors in a database. We wish to find the nearest neighbor of x from the database. By the definition of asymmetric distance [3], the approximate distance between x and a database vector y_i is computed as

$$d_c(x, y_i)^2 = \|x - q_c(y_i)\|^2, \tag{1}$$

where $q_c(\cdot)$ denotes a quantizer with K centroids. The vector $c_i = q_c(y_i) \in R^d$ is encoded by $\log_2 K$ bits if K is a power of 2.

The approximate nearest neighbor of x is obtained by minimizing the distance

$$NN_a(x) = \arg\min_i \|x - q_c(y_i)\|^2. \tag{2}$$

Note that the "asymmetric" distance means the query vector x is not quantized, but the database vector y_i is. The results reported in [3] show that the asymmetric version significantly outperforms the symmetric version (quantizing both the query vector and database vectors) in terms of memory usage and search accuracy.

To obtain good approximation, the number of quantization levels K should be large. For example, $K = 2^{64}$ if a 64-bit code is desired. However, learning such a big codebook needs tremendous computation, and storing 2^{64} centroids in floating points is not feasible. The work in [3] addresses this issue by decomposing a high-dimensional space into a Cartesian product of low-dimensional subspaces, and quantizes each subspace separately. A database vector $y \in R^d$ is first equally split into m subvectors $y^1, ..., y^m \in R^{d/m}$. A product quantizer is defined as a function

$$q_c(y) = (q^1(y^1), ..., q^m(y^m)), \tag{3}$$

which maps the vector y to a tuple of indices. Each quantizer $q^j(\cdot)$ has K_s quantization levels, and is learned by the K-means algorithm based on the set of the jth subvectors $\{y^j\}$ in the database.

With the product quantization scheme, the combination of q^j, $j = 1, ..., m$, forms a large codebook consisting of $K = (K_s)^m$ centroids. The square distance in eq. (1) is calculated using the decomposition

$$d_c(x, y_i)^2 = \|x - q_c(y_i)\|^2 = \sum_{j=1}^m \|x^j - q^j(y^j)\|^2, \tag{4}$$

where x^j is the jth subvector of x. Before calculating the distance described in eq. (4), m tables are constructed for a given query. For the ith entry of the jth table, the distance between the subvector x^j and the ith centroid of q^j, $i = 1, ..., K_s$, is stored. The complexity of table generation is $O(\frac{d}{m} \times m \times K_s) = O(d \times K_s)$. When $K_s \ll n$, this complexity can be ignored compared to the summation cost of $O(d \times n)$ in eq. (1).

To briefly describe the idea of product quantization, both [3] and the aforementioned example assume that a high-dimensional vector is equally split into m subvectors. However, different splitting schemes may have significantly different performance. We can investigate a product quantizer from three aspects:

- The way to splitting vectors: It would be better that different subvectors are uncorrelated. On the contrary, consecutive components are usually correlated, and they are better quantized using the same subquantizer.
- The dimension of subvectors: Dimensions of subvectors should be appropriately chosen to avoid the curse of dimensionality.

- The number of quantization levels for each subquantizer: Numbers of quantization levels for different subvectors may be different, depending on the "importance" of corresponding subvectors for a given task.

It is obvious that these three issues are not independent. In [3], by equally splitting SIFT descriptors [4] and GIST descriptors [5] into four or eight subvectors, the authors verify that different component grouping schemes give significantly different performance. They just briefly discuss the first issue. In this paper, we would deeply investigate search accuracy and memory usage from all these three aspects.

3 Evaluation Settings

3.1 Dataset and Features

Similar to [3], we extract feature vectors from three datasets: learning, database, and query. The learning set comes from the first 100k images of the MIRFLICKR image collection [6]. Based on the learning set, various product quantizers are constructed. Both the database set and the query set are from the Holidays image collection [7]. Given a query image, we would like to find its nearest images, based on the representation with product quantization. Figure 1 shows some sample images from the Holidays image collection. They were all captured by amateur photographers with low-end cameras. In retrieving nearest images, some of them are relatively simpler (the left part of Figure 1), and some of them are extremely challenging (the right part).

In contrast to [3], we extract pyramid of histogram of oriented gradients (PHOG) [8] to describe images. For each image, oriented gradients are first computed using a 3×3 Sobel mask with Gaussian smoothing. Orientations of pixels are quantized into B bins. For a given image region, a B-dimensional histogram of oriented gradients (HOG) can be constructed. To form the level ℓ pyramid, the image is divided into 2^ℓ cells horizontally and vertically. The final level ℓ PHOG descriptor is a concatenation of all HOG descriptors extracted from cells in levels $0, 1, ..., \ell$. Therefore, for example, a level 1 PHOG descriptor has dimension of $(2^0 \times 2^0 + 2^1 \times 2^1) \times B = 5B$, and a level 2 PHOG descriptor has dimension of $(1 + 4 + 16) \times B = 21B$.

Figure 2 illustrates examples of level 1 PHOG and level 2 PHOG if $B = 16$. The first 16 components (denoted by h_0) in level 1 PHOG correspond to the HOG of the cell in level 0, and the following 64 components (denoted by h_1) correspond to the HOG of the 4 cells in level 1. Similarly, the last 256 components (denoted by h_2) in level 2 PHOG correspond to HOG of the 16 cells in level 2.

The reasons to use PHOG for evaluation are that different scales of information are concatenated as a high-dimensional vector, and components in different dimensions are often not independent. Figure 3 shows two sample correlation matrices between components in different dimensions of level 1 PHOG and level 2 PHOG, respectively. We clearly see that some dimensions are correlated, and from the viewpoint of subvector splitting, they are better categorized into the same subvector. In contrast to

SIFT that just encodes gradient information of a local patch and GIST that encodes the global shape distribution, we can design much more variations of splitting schemes for PHOG, and investigate how quantization within levels and across levels affect the performance.

Fig. 1. Sample images from the Holidays image collection

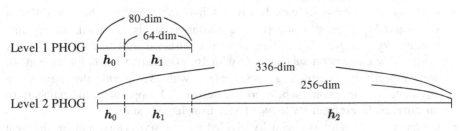

Fig. 2. Illustration of level 1 PHOG and level 2 PHOG

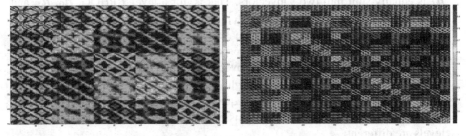

Fig. 3. Two sample correlation matrices for the level 1 PHOG (left) and level 2 PHOG (right). Gradients in both PHOGs are quantized into 32 bins.

3.2   Experiment Settings

We design experiments to investigate how different product quantization schemes influence the performance of nearest image search. The search quality is measured by recall@100, i.e., the proportion of query vectors for which the nearest neighbor is ranked in the first 100 positions [3]. To construct different PHOGs, the orientation of gradients is quantized into 16 or 32 bins ($B = 16$ or 32). Two levels of PHOGs, i.e., level 1 and level 2, are extracted for evaluation. Table 1 shows the overview of our experiment settings.

Two main factors determine how we implement product quantization: how to split a high-dimensional vector into subvectors, and the length of code using to represent a subvector (the number of quantization level of each subquantizer determines the code length). By varying the two factors, we divide our experiments into three categories:

- Category 1: A vector is equally divided into m subvectors, and for different subquantizers the same number of quantization levels is used. As shown in Table 1, a level 1 PHOG may be equally divided into 5, 10, or 20 subvectors. Note that the first two of the 10 subvectors, for example, correspond to the HOG of the cell at level 0. That is, the h_0 in Figure 2 is split into two subvectors v_0 and v_1. Components in the same subvector v_i, $i = 0, 1$, belong to the same level of HOG. However, the components v_0 convey information of orientations different from that conveyed by the components in v_1. A level 2 PHOG may be equally divided into 21, 42, or 84 subvectors.

- Category 2: A vector is equally divided into m subvectors, but for different subquantizers different numbers of quantization levels are used. In the first subcategory Category 2-1, taking the same instance as Category 1, we may respectively construct a subquantizer with 16 quantization levels for v_0 and v_1, and respectively construct a subquantizer with 32 quantization levels for $v_2, ..., v_9$. In the second subcategory Category 2-2, oppositely, the numbers of quantization levels from v_0 to v_9 change from large to small.

- Category 3: A vector is unequally divided into m subvectors, and for different subquantizers the same number of quantization levels is used. The level ℓ HOG h_ℓ is equally split into two subvectors. That is, for a level 2 PHOG, h_0, h_1, and h_2 are split into six subvectors, $\{v_0, v_1\}$, $\{v_2, v_3\}$, and $\{v_4, v_5\}$, respectively. If $B = 16$, the dimensions of the three sets of subvectors are 8, 32, and 128, respectively. Note that v_0 and v_1 respectively represents half of orientation information of level 0 HOG. The subvectors v_2 and v_3 respectively represents the level 1 HOG in the left half and right half of an image (please refer to [8] for details of PHOG). We can further equally split each subvectors into shorter subvectors. In this category, dimensions of subvectors corresponding to different levels are different.

Table 1. Overview of experiment settings. The bold-faced numbers correspond to the examples we give in the text.

	#bins of HOG	PHOG levels	#subvectors (m)	#quantization levels
Cat 1	$\{16, 32\}$	$\{1, 2\}$	$\{5, 10, 20\}$ $\{21, 42, 84\}$	$\{8, 16, 32, 64, 128\}$
Cat 2-1	$\{16, 32\}$	$\{1, 2\}$	$\{5, 10, 20\}$ $\{21, 42, 84\}$	$\{(8, 16), (16, 32), (32, 64)\}$ $\{(8, 16, 32) (16, 32, 64), (32, 64, 128)\}$
Cat 2-2	$\{16, 32\}$	$\{1, 2\}$	$\{5, 10, 20\}$ $\{21, 42, 84\}$	$\{(64, 32), (32, 16), (16, 8)\}$ $\{(128, 64, 32) (64, 32, 16), (32, 16, 8)\}$

4   Experiments

4.1   Experiment Settings

We first describe performance variations of different splitting schemes, i.e., Category 1 vs. Category 3. In Category 1, subvectors corresponding to different levels have the same dimension, while in Category 3, subvectors corresponding to level 0 have lower dimension than that corresponding to level 1, and so on.

Figure 4 shows performance variations of different splitting schemes, based on level 1 and level 2 PHOGs. From both subplots, we clear see that with non-uniform splitting (Category 3), we achieve the same recall@100 with shorter code lengths than that based on uniform splitting (Category 1). This trend conforms to studies in many multimedia applications. Generally, slightly better performance is obtained based on PHOGs with gradients quantizing into 32 bins. We also observe that level 2 PHOGs provide slightly better performance. Therefore, in the following experiments, we mainly do performance comparison based on level 2 PHOGs with gradients quantizing into 32 bins.

4.2   Number of Quantization Intervals

Experiments in Category 2 are designed to investigate how numbers of quantization intervals for different levels of HOG influence the nearest image search. Figure 5 shows performance variations of experiments in Category 2-1 and Category 2-2. As we can see, Category 2-2 obviously provides better performance because the same search accuracy can be achieved based on codes of shorter lengths. In Category 2-2, coarse representations of HOGs (levels 0/1) are more finely quantized. This figure reveals a fact about product quantization that has never been verified before: we are able to derive a more effective descriptor if more important information (in this case, lower level HOGs) is quantized more finely.

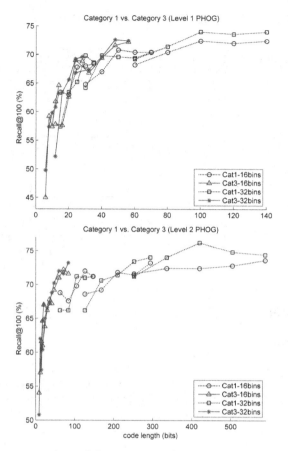

Fig. 4. Performance comparison of Category 1 vs. Category 3 experiments, based on level 1 (top) and level 2 (bottom) PHOGs

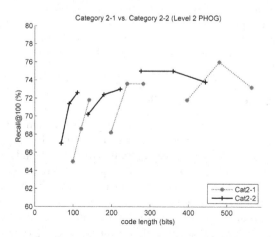

Fig. 5. Performance comparison of two experiments in Category 2, based on level 2 PHOGs

4.3   The Influence on Retrieval Rank

All experiments described above and the ones reported in [3] show retrieval performance in terms of recall@100, but not reflect how different product quantization schemes influence the rank of retrieval results. Based on the Category 2 settings, we evaluate the rank of the first correctly retrieved image for each query. Figure 6 shows the average rank versus code lengths. The curves again show that Category 2-2 provides better retrieval results, because lower rank values mean this representation more accurately captures the characteristics of images. Combining the results in Figures 5 and 6, we know that if more important information is quantized more finely, both recall (represented by recall@100 in Figure 5) and precision (represented by rank values in Figure 6) can be improved.

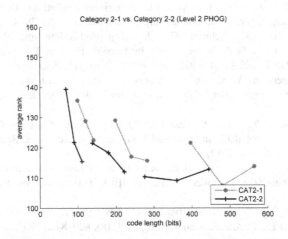

Fig. 6. Average retrieval rank of two experiments in Category 2, based on level 2 PHOGs

5   Conclusion

In this paper we design experiments to deeply investigate how different product quantization schemes influence the performance of nearest image retrieval. Based on the PHOG descriptors, three categories of experiments are conducted for evaluating the factors of how subvectors are split and how subvectors are quantized. The experimental results reveal some information that has not been verified before: 1) appropriate nonuniform splitting yields better performance; 2) if more important information is quantized more finely, both recall and precision can be improved. In the future, product quantization can be applied to applications with large-scale nearest neighbor search, with appropriate designs of splitting and subvector quantization.

Acknowledgement. The work was partially supported by the National Science Council of Taiwan, Republic of China under research contract NSC 101-2221-E-194-055-MY2.

References

1. Friedman, J.H., Bentley, J.L., Finkel, R.A.: An algorithm for finding best matches in logarithmic expected time. ACM Trans. on Mathematical Software 3(3), 209–226 (1977)
2. Datar, M., Immorlica, N., Indyk, P., Mirrokni, V.: Locality-sensitive hashing scheme based on p-stable distributions. In: Proc. of Annual Symposium on Computational Geometry, pp. 253–262 (2004)
3. Jegou, H., Douze, M., Schmid, C.: Product quantization for nearest neighbor search. IEEE Trans. Pattern Analysis and Machine Intelligence 33(1), 117–128 (2011)
4. Lowe, D.: Distinctive image features from scale invariant keypoints. International Journal of Computer Vision 60(2), 91–110 (2004)
5. Oliva, A., Torralba, A.: Modeling the shape of the scene: a holistic representation of the spatial envelope. International Journal of Computer Vision 42(3), 145–175 (2001)
6. Huiskes, M.J., Lew, M.S.: The MIR Flickr retrieval evaluation. In: Proc. of ACM International Conference on Multimedia Information Retrieval, pp. 39–43 (2008)
7. Jegou, H., Douze, M., Schmid, C.: Hamming Embedding and Weak Geometric Consistency for Large Scale Image Search. In: Forsyth, D., Torr, P., Zisserman, A. (eds.) ECCV 2008, Part I. LNCS, vol. 5302, pp. 304–317. Springer, Heidelberg (2008)
8. Bosch, A., Zisserman, A., Munoz, X.: Representing shape with a spatial pyramid kernel. In: Proc. of ACM International Conference on Image and Video Retrieval, pp. 401–408 (2007)
9. Wang, C., Duan, L.-Y., Wang, Y., Gao, W.: PQ-WGLOH: A bit-rate scalable local feature descriptor. In: Proc. of IEEE International Conference on Acoustics, Speech, and Signal Processing, pp. 941–944 (2012)
10. Wang, M., Hua, X.-S., Tang, J., Hong, R.: Beyond distance measurement: constructing neighborhood similarity for video annotation. IEEE Transactions on Multimedia 11(3), 465–476 (2009)
11. Wang, M., Yang, K., Huan, X.-S., Zhang, H.-J.: Towards a relevant and diverse search of social images. IEEE Transactions on Multimedia 12(8), 829–842 (2010)

Rate-Quantization and Distortion-Quantization Models of Dead-Zone Plus Uniform Threshold Scalar Quantizers for Generalized Gaussian Random Variables

Yizhou Duan, Jun Sun[*], and Zongming Guo

Institute of Computer Science and Technology, Peking University, Beijing, China
{duanyizhou,sunjun,guozongming}@pku.edu.cn

Abstract. This paper presents the rate-distortion modeling of the dead-zone plus uniform threshold scalar quantizers with nearly-uniform reconstruction quantizers (DZ+UTSQ/NURQ) for generalized Gaussian distribution (GGD). First, we rigorously deduce the analytical rate-quantization (R-Q) and distortion-quantization (D-Q) functions of DZ+UTSQ/NURQ for Laplacian distribution (an important special case of GGD). Then we heuristically extend these results and obtain the R-Q and D-Q models for GGD under DZ+UTSQ/NURQ. The effectiveness of the proposed GGD R-Q and D-Q models is well confirmed from low to high bit rate via extensive simulation experiments, promising the efficiency and accuracy to guide various video applications in practice.

Keywords: Rate-distortion modeling, generalized Gaussian distribution, dead-zone plus uniform threshold quantization, video coding.

1 Introduction

As one of the key issues in digital video coding realm, rate-distortion (R-D) modeling aims to reveal the relationship between entropy rate and reconstruction distortion of any given compression method. In practical video coding systems where transform-based quantization scheme is widely implemented [1], both rate and distortion are generally controlled by the quantization step. Therefore, the R-D relationship is typically characterized by the rate-quantization (R-Q) model and the distortion-quantization (D-Q) model, which actually represent the target of R-D modeling.

Since the R-D relationship is the foundation of essential video coding technologies like rate control and R-D optimization, the deduction of accurate R-Q and D-Q models has received increasing research attentions in recent years. By using analytical or heuristic method, several representative R-Q and D-Q models have been developed [1], [2], [3], [4], providing useful references for video coding. However, the existing work of R-D modeling still has limitations in the following two aspects. First, for the heuristic models [3], [4], the control parameters are determined according to the actual coding data or rate control performance. Without rigorous analysis of video

---

[*] Corresponding author.

S. Li et al. (Eds.): MMM 2013, Part I, LNCS 7732, pp. 357–367, 2013.
© Springer-Verlag Berlin Heidelberg 2013

source, these models can be efficient in some situations but may not effectively contribute to different video applications. Second, the analytical models [1], [2] are deduced base on the R-D relationship that is applicable only in very high bit rate, while the source distribution and quantizers are selected mainly for the concern of simplicity rather than accuracy [5]. As a result, these models may deviate from the practical systems in the actual video coding environment.

To cope with above limitations, we focus on the R-D modeling of the dead-zone plus uniform threshold scalar quantizers (DZ+UTSQ) with nearly-uniform reconstruction quantizers (NURQ) for generalized Gaussian distribution (GGD). DZ+UTSQ/NURQ scheme is now adopted by all the major international video coding standards as recommendation [6], and GGD has been proposed as an excellent choice [2] to well represent the actual DCT coefficients. There are two main points we have contributed in this paper. First, we rigorously deduce the closed-form R-Q and D-Q functions of DZ+UTSQ/NURQ for Laplacian distribution (an important special case of GGD). As the theoretical foundation, these functions for the first time precisely describe the R-D relationship of Laplacian sources under the most universal quantizers in the practical video coding systems. Second, we further extend the Laplacian results and obtain the R-Q and D-Q models of DZ+UTSQ/NURQ for GGD, which can flexibly adapt to various real video sources. Extensive simulation experiments from low to high bit rate demonstrate superior accuracy of the proposed GGD models compared with other representative models.

The rest of this paper is organized as follows. The background knowledge of GGD and DZ+UTSQ/NURQ is briefly introduced in Section 2. The deduction of the novel GGD R-Q and D-Q models are provided in Section 3 and Section 4, followed by the rigorous simulation experiments of the proposed models in Section 5. Finally, concluding remarks are given in Section 6.

2   Background Knowledge

2.1   Zero-Mean GGD

For simplicity, we use GGD to denote zero-mean GGD in this paper. The probability density function (PDF) of GGD can be expressed as follows:

$$p(x) = \frac{g_1(\alpha)}{\beta} \exp\left(-\left[g_2(\alpha)\frac{|x|}{\beta}\right]^\alpha\right) , \tag{1}$$

with

$$g_1(\alpha) = \frac{\alpha[\Gamma(3/\alpha)]^{1/2}}{2[\Gamma(1/\alpha)]^{3/2}} \quad \text{and} \quad g_2(\alpha) = \left[\frac{\Gamma(3/\alpha)}{\Gamma(1/\alpha)}\right]^{1/2} , \tag{2}$$

where α is the shape parameter and β is the standard deviation. Specifically, for GGD shapes $\alpha = 1$, $\alpha = 2$ and $\alpha \to \infty$, GGD becomes Laplacian, Gaussian and uniform distribution, as is shown in Fig. 1. By adjusting α and β, GGD can well adapt to the DCT coefficients of various source signals.

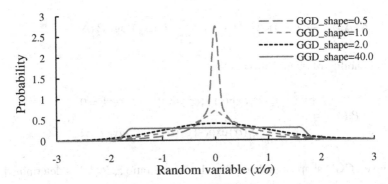

Fig. 1. Zero-mean generalized Gaussian distribution with different shape parameters

2.2 DZ+UTSQ/NURQ Scheme

The complete DZ+UTSQ/NURQ scheme is illustrated in Fig. 2, which consists of the DZ+UTSQ *classification* rule and NURQ *reconstruction* rule, both are symmetric about zero. In classification, each input signal is mapped to the integer-valued quantization index according to *quantization threshold T*. In reconstruction, a real-valued *reconstruction level R* is produced for each quantization index. For the index $k \geq 1$, the decision interval is $[T_k, T_{k+1})$ and for $k \leq -1$, the interval is $(T_{-k-1}, T_{-k}]$. For the index 0, the interval is the dead-zone region given by (T_{-1}, T_1). The quantization threshold T_k and the reconstruction level R_k are expressed as:

$$T_k = \begin{cases} (k-1+z)\Delta, & k \geq 1 \\ T_{-k}, & k \leq -1 \end{cases} \quad \text{and} \quad R_k = \begin{cases} 0, & k = 0 \\ (k+p)\Delta, & k \geq 1 \\ R_{-k}, & k \leq -1 \end{cases}, \quad (3)$$

where z is the dead-zone ratio which is half size of the dead-zone region, and p is the reconstruction offset from the corresponding index. Both z and p are normalized by the quantization step Δ.

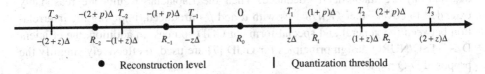

Fig. 2. Illustration of the DZ+UTSQ/NURQ scheme

2.3   R-Q and D-Q Formulation

For GGD, with an arbitrary variable $\delta > 0$, we rewrite the quantization step Δ as $\Delta_\beta = \beta\delta$. In the symmetric DZ+UTSQ rule, the entropy rate is calculated by the R-Q function:

$$H(\Delta_\beta) = -P_0(\Delta_\beta)\log_2 P_0(\Delta_\beta) + 2\sum_{k=1}^{+\infty} -P_k(\Delta_\beta)\log_2 P_k(\Delta_\beta) , \qquad (4)$$

with the probability of quantization index k:

$$P_k(\Delta_\beta) = \begin{cases} \int_{-z\delta}^{z\delta} g_1(\alpha)\exp(-[g_2(\alpha)\,|\,x\,|]^\alpha)dx, & k = 0 \\ \int_{(z+k-1)\delta}^{(z+k)\delta} g_1(\alpha)\exp(-[g_2(\alpha)\,|\,x\,|]^\alpha)dx, & k \geq 1 \end{cases}. \qquad (5)$$

For any given GGD shape α and constant dead-zone ratio z, $P_k(\Delta_\beta)$ is determined only by the variable δ. Therefore, the entropy rate is just a function of δ.

Likewise, in the symmetric NURQ rule, the MSE reconstruction distortion is denoted by the D-Q function:

$$D(\Delta_\beta) = 2\sum_{k=0}^{+\infty} D_k(\Delta_\beta) , \qquad (6)$$

with the distortion of each reconstruction level:

$$D_k(\Delta_\beta) = \begin{cases} \beta^2 \int_0^{z\delta} x^2 g_1(\alpha)\exp(-[g_2(\alpha)\,|\,x\,|]^\alpha)dx, & k = 0 \\ \beta^2 \int_{(z+k-1)\delta}^{(z+k)\delta} [x-(p+k)\delta]^2 g_1(\alpha)\exp(-[g_2(\alpha)\,|\,x\,|]^\alpha)dx, & k \geq 1 \end{cases}. \qquad (7)$$

From (6) and (7), it is easy to know that for any given α, z and offset p, the distortion can also be regarded as a function of variable δ, while β only serves as a scaling factor.

However, due to the exponential PDF of GGD, it is incapable to directly obtain the closed-form R-Q and D-Q functions from (4)–(7). In the following two sections, we composite the analytical and heuristic approaches to give a precise approximation of the GGD R-Q and D-Q functions under the general form of DZ+UTSQ/NURQ. First, the R-Q and D-Q functions of DZ+UTSQ/NURQ for GGD shape $\alpha = 1$ (Laplacian distribution) are theoretically deduced. Then the Laplacian results are reasonably extended to cover other GGD cases with α and β introduced, implying the interrelation between the general and special forms of GGD sources. In addition, the efficient DZ+UTSQ/NURQ design principles for GGD [7] are used to effectively simplify the proposed D-Q model.

3   R-Q Model of DZ+UTSQ/NURQ for GGD

The PDF of GGD shape $\alpha = 1$ (Laplacian distribution) can be expressed as:

$$p(x) = \frac{\mu}{2} e^{-\mu|x|} , \tag{8}$$

with the standard deviation $\beta = \sqrt{2}/\mu$. For Laplacian distribution, from (5) the probability of quantization index k under DZ+UTSQ classification rule is obtained as:

$$P_k(\Delta) = \begin{cases} \int_{-z\Delta}^{z\Delta} p(x)dx = 1 - e^{-z\mu\Delta}, & k = 0 \\ \int_{(z+k-1)\Delta}^{(z+k)\Delta} p(x)dx = \frac{1}{2} e^{-(z+k-1)\mu\Delta}(1 - e^{-\mu\Delta}), & k \geq 1 \end{cases} . \tag{9}$$

Based on equation (4), the entropy rate is calculated:

$$\begin{aligned} H(\Delta) &= -P_0(\Delta)\log_2 P_0(\Delta) - (1 - e^{-\mu\Delta})\log_2\left[(1 - e^{-\mu\Delta})/2\right]\sum_{k=1}^{+\infty} e^{-(z+k-1)\mu\Delta} \\ &\quad + (1 - e^{-\mu\Delta})\log_2 e \sum_{k=1}^{+\infty}\left[(z+k-1)\mu\Delta e^{-(z+k-1)\mu\Delta}\right] \\ &= -P_0(\Delta)\log_2 P_0(\Delta) - \left[(1 - e^{-\mu\Delta})e^{-(z-1)\mu\Delta}\log_2\left[(1 - e^{-\mu\Delta})/2\right]\right]\sum_{k=1}^{+\infty} e^{-k\mu\Delta} \\ &\quad + \left[\mu(z-1)\Delta(1 - e^{-\mu\Delta})e^{-(z-1)\mu\Delta}\log_2 e\right]\sum_{k=1}^{+\infty} e^{-k\mu\Delta} \\ &\quad + \left[(1 - e^{-\mu\Delta})e^{-(z-1)\mu\Delta}\log_2 e\right]\sum_{k=1}^{+\infty} k\mu\Delta e^{-k\mu\Delta} \\ &= -(1 - e^{-\mu\Delta})\log_2(1 - e^{-\mu\Delta}) - \left[(1 - e^{-\mu\Delta})e^{-(z-1)\mu\Delta}\log_2\left[(1 - e^{-\mu\Delta})/2\right]\right]\sum_{k=1}^{+\infty} e^{-k\mu\Delta} \\ &\quad + \frac{1}{\ln 2}\left[(z-1)\mu\Delta(1 - e^{-\mu\Delta})e^{-(z-1)\mu\Delta}\right]\sum_{k=1}^{+\infty} e^{-k\mu\Delta} \\ &\quad + \frac{1}{\ln 2}\left[(1 - e^{-\mu\Delta})e^{-(z-1)\mu\Delta}\right]\sum_{k=1}^{+\infty} k\mu\Delta e^{-k\mu\Delta} . \end{aligned} \tag{10}$$

By simple accumulation of the infinite series, we have:

$$\sum_{k=1}^{+\infty} e^{-k\mu\Delta} = \frac{e^{-\mu\Delta}}{1 - e^{-\mu\Delta}} . \tag{11}$$

And in Appendix, we also prove that

$$\sum_{k=1}^{+\infty} k\mu\Delta e^{-k\mu\Delta} = \frac{\mu\Delta e^{-\mu\Delta}}{(1 - e^{-\mu\Delta})^2} . \tag{12}$$

By using (10)–(12), the closed-form R-Q function of DZ+UTSQ/NURQ for Laplacian distribution is obtained:

$$
\begin{aligned}
H(\Delta) &= e^{-z\mu\Delta} + \left[(z-1)\mu\Delta e^{-z\mu\Delta} - \ln(1-e^{-z\mu\Delta}) \right] / \ln 2 + \mu\Delta e^{-z\mu\Delta} / \left[(1-e^{-\mu\Delta})\ln 2 \right] \\
&= e^{-z\mu\Delta} - \ln(1-e^{-z\mu\Delta}) / \ln 2 + \mu\Delta e^{-z\mu\Delta}(z+e^{-\mu\Delta}-ze^{-\mu\Delta}) / \left[(1-e^{-\mu\Delta})\ln 2 \right]
\end{aligned} \tag{13}
$$

In particular, when $z = 1$, (13) is simplified as:

$$
H(\Delta) = e^{-\mu\Delta} - \ln(1-e^{-\mu\Delta}) / \ln 2 + \mu\Delta e^{-\mu\Delta} / \left[(1-e^{-\mu\Delta})\ln 2 \right] , \tag{14}
$$

which is the R-Q function obtained by Sun *et al.* [8] for Laplacian distribution in MPEG-4 FGS coding (a special case of DZ+UTSQ/NURQ with $z = 1$).

Laplacian distribution is an important special case of GGD with shape $\alpha = 1$ and standard deviation $\beta = \sqrt{2}/\mu$. It is observed from the GGD PDF that apart from in the gamma function $\Gamma(\cdot)$, α only appears as the exponential part of the input variable. Heuristically, by introducing α and β to extend the Laplacian R-Q function to various GGD sources, we can rewrite $\mu\Delta$ of (13) into $(\sqrt{2}\Delta/\beta)^{\alpha}$ or $\sqrt{2}\Delta^{\alpha}/\beta$. Note that with an arbitrary variable $\delta > 0$ satisfying $\delta = \Delta/\beta$, the R-Q function of GGD is only a function of δ and is not related to β (see Section 2.3). Thus, it is reasonable to use $(\sqrt{2}\Delta/\beta)^{\alpha}$ rather than $\sqrt{2}\Delta^{\alpha}/\beta$ to replace $\mu\Delta$ so that the GGD R-Q model is consistent with our deduction. Hence, (13) is changed into

$$
\begin{aligned}
H(\Delta) &= -\ln(1-e^{-z(\sqrt{2}\Delta/\beta)^{\alpha}}) / \ln 2 + e^{-z(\sqrt{2}\Delta/\beta)^{\alpha}} \\
&+ (\sqrt{2}\Delta/\beta)^{\alpha} e^{-z(\sqrt{2}\Delta/\beta)^{\alpha}} (z+e^{-(\sqrt{2}\Delta/\beta)^{\alpha}} - ze^{-(\sqrt{2}\Delta/\beta)^{\alpha}}) / \left[(1-e^{-(\sqrt{2}\Delta/\beta)^{\alpha}})\ln 2 \right]
\end{aligned} \tag{15}
$$

Equation (15) is the proposed R-Q model for GGD under DZ+UTSQ/NURQ, which provides different GGD parameters and dead-zone ratios to adapt to various source signals and quantizer implementations.

4   D-Q Model of DZ+UTSQ/NURQ For GGD

Similar to the process of obtaining the R-Q model, we first deduce the Laplacian D-Q function of DZ+UTSQ/NURQ. For Laplacian distribution, from (7) the MSE of each reconstruction level under NURQ reconstruction rule is obtained as:

$$
D_k(\Delta) = \begin{cases} \int_0^{z\Delta} x^2 \dfrac{\mu}{2} e^{-\mu x} dx, & k = 0 \\[2ex] \int_{(z+k-1)\Delta}^{(z+k)\Delta} [x-(p+k)\Delta]^2 \dfrac{\mu}{2} e^{-\mu x} dx, & k \geq 1 \end{cases} \tag{16}
$$

Using integration by part, (16) can be evaluated separately. When $k \geq 1$ we have

$$
\begin{aligned}
D_k(\Delta) =& -\frac{1}{2}\big[(z+k)\Delta\big]^2 e^{-(z+k)\mu\Delta} + \frac{1}{2}\big[(z+k-1)\Delta\big]^2 e^{-(z+k-1)\mu\Delta} \\
& -\frac{1}{\mu}\Big[(z+k)\Delta e^{-(z+k)\mu\Delta} - (z+k-1)\Delta e^{-(z+k-1)\mu\Delta}\Big] - \frac{1}{\mu^2}\Big[e^{-(z+k)\mu\Delta} - e^{-(z+k-1)\mu\Delta}\Big] \\
& +(p+k)\Delta^2\Big[(z+k)e^{-(z+k)\mu\Delta} - (z+k-1)e^{-(z+k-1)\mu\Delta}\Big] \\
& +\frac{(p+k)\Delta^2}{\mu}\Big[e^{-(z+k)\mu\Delta} - e^{-(z+k-1)\mu\Delta}\Big] - \frac{(p+k)^2\Delta^2}{2}\Big[e^{-(z+k)\mu\Delta} - e^{-(z+k-1)\mu\Delta}\Big]
\end{aligned}
\tag{17}
$$

And when $k = 0$, we have

$$
D_0(\Delta) = \frac{\mu^{-2}}{2}\Big[2 - e^{-z\mu\Delta}\big[(z\mu\Delta)^2 + 2z\mu\Delta + 2\big]\Big] .
\tag{18}
$$

Although (17) is already a closed-form expression to denote the distortion of the reconstruction level $(p+k)\Delta$, it still needs to be simplified for accumulation. Under the classifier of DZ+UTSQ, the uniform reconstruction quantizer (URQ) has been proved optimal for Laplacian source and near-optimal for other GGD sources [9], being an effective sub-optimal case of NURQ [6]. Thus, by using URQ to efficiently simplify NURQ [7], we set p to 0 in (17) and obtain a precise approximation:

$$
\begin{aligned}
D_k(\Delta) =& \frac{\mu^{-2}}{2}\Big[\big[(z-1)\mu\Delta\big]^2 + 2(z-1)\mu\Delta + 2\Big]e^{-(z+k-1)\mu\Delta} \\
& -\frac{\mu^{-2}}{2}\Big[(z\mu\Delta)^2 + 2z\mu\Delta + 2\Big]e^{-(z+k)\mu\Delta}
\end{aligned}
\tag{19}
$$

Using (11), (18) and (19), according to equation (6), the D-Q function of DZ+UTSQ/NURQ for Laplacian distribution is calculated as follows:

$$
\begin{aligned}
D(\Delta) =& \mu^{-2}\Big[2 - e^{-z\mu\Delta}\big[(z\mu\Delta)^2 + 2z\mu\Delta + 2\big]\Big] \\
& +\mu^{-2}\Big[\big[(z-1)\mu\Delta\big]^2 + 2(z-1)\mu\Delta + 2\Big]e^{-(z-1)\mu\Delta}\sum_{k=1}^{+\infty} e^{-k\mu\Delta} \\
& -\mu^{-2}\Big[(z\mu\Delta)^2 + 2z\mu\Delta + 2\Big]e^{-z\mu\Delta}\sum_{k=1}^{+\infty} e^{-k\mu\Delta} \\
=& \frac{1}{\mu^2}\Big[2 - e^{-z\mu\Delta}\big[(z\mu\Delta)^2 + 2z\mu\Delta + 2\big]\Big] \\
& +\frac{e^{-z\mu\Delta}}{\mu^2(1-e^{-\mu\Delta})}\Big[\big[(z-1)\mu\Delta\big]^2 + 2(z-1)\mu\Delta + 2\Big] \\
& -\frac{e^{-(z+1)\mu\Delta}}{\mu^2(1-e^{-\mu\Delta})}\Big[(z\mu\Delta)^2 + 2z\mu\Delta + 2\Big] \\
=& \frac{2}{\mu^2} + \frac{e^{-z\mu\Delta}}{\mu^2(1-e^{-\mu\Delta})}\Big[\mu^2\Delta^2(1-2z) - 2\mu\Delta\Big]
\end{aligned}
\tag{20}
$$

In particular, when $z = 1$, (20) is further simplified as:

$$D(\Delta) = \frac{e^{-\mu\Delta}}{\mu^2}\left[2e^{\mu\Delta} - \left[(\mu\Delta)^2 + 2\mu\Delta + 2\right]\right]$$
$$+ \frac{e^{-2\mu\Delta}}{\mu^2(1 - e^{-\mu\Delta})}\left[2e^{\mu\Delta} - \left[(\mu\Delta)^2 + 2\mu\Delta + 2\right]\right] \quad , \qquad (21)$$
$$= \left[2e^{\mu\Delta} - \left[(\mu\Delta)^2 + 2\mu\Delta + 2\right]\right] / \left[\mu^2(1 - e^{-\mu\Delta})\right]$$

which is the D-Q function obtained by Sun *et al.* 8 for Laplacian distribution in MPEG-4 FGS coding. Finally, with the same heuristic method in developing the GGD R-Q model, (20) is modified into:

$$D(\Delta) = \beta^2\left[1 + \frac{e^{-z(\sqrt{2}\Delta/\beta)^\alpha}}{2(1 - e^{-(\sqrt{2}\Delta/\beta)^\alpha})}\left[(\sqrt{2}\Delta/\beta)^{2\alpha}(1 - 2z) - 2(\sqrt{2}\Delta/\beta)^\alpha\right]\right] . \qquad (22)$$

Equation (22) is the proposed D-Q model for GGD under the general form of DZ+UTSQ/NURQ scheme, providing different GGD parameters and dead-zone ratios to adapt to various source signals and quantizer implementations.

5   Simulation Experiments

To rigorously confirm the effectiveness of our findings, the theoretical GGD rate and distortion are used as the benchmark to evaluate different R-Q and D-Q models under DZ+UTSQ/NURQ scheme. The theoretical GGD rate is simulated using (4) and (5), and the theoretical GGD distortion is simulated using (6) and (7), where GGD shape α is chosen from 0.5 to 1.0 and β is chosen from 2 to 10 to conform to the statistics of the actual DCT coefficients [8]. The DZ+UTSQ/NURQ is set by adjusting z within the range $1/2 < z \leq 1$ [7], which represents the efficient DZ+UTSQ/NURQ design criteria used in practice. Therefore, the theoretical GGD rate and distortion can well reflect the actual coding properties of the practical video coding systems, which guarantee the validity of the simulation experiments.

For comparison, the Cauchy model [2] and the Quadratic model [3] are selected to represent the analytical and heuristic modeling methods. For each model, the nonlinear data fitting method is employed and the quantization step ranges from 0.625 to 208, corresponding to the quantization parameter (QP) of H.264/AVC from 0 to 50. For $\alpha = 0.5$ and $\alpha = 1.0$, related results of R-Q models comparing with the theoretical GGD rate are illustrated in Fig. 3 and Fig. 4. It should be noted that the entropy rate is measured in bits/sample, which means that even slight difference can result in significant prediction error in the bit rate of the whole system. Therefore, it is clearly exhibited from global perspective that the proposed R-Q model matches the theoretical rate of the practical video coding systems perfectly from low to high bit rate, while other two models are both of some deviations. Similarly, related results of D-Q models comparing with the theoretical GGD distortion are illustrated in Fig. 5 and Fig. 6, from which it is well demonstrated that the proposed D-Q model matches the theoretical distortion of the practical systems accurately from low to high bit rate while other two models are both of some deviations. In Fig. 5, it is observed that the quadratic D-Q model is also a good approximation of the GGD distortion when $\alpha = 0.5$. However, for GGD shape 1.0, obvious prediction error of the quadratic model is shown in Fig. 6, which indicates the effectiveness of this model is unable to be guaranteed for various source signals. It should also be noted that the proposed model is derived under URQ which is a sub-optimal case of NURQ for GGD, thus in high bit rate, slightly greater distortion than the theoretical data is obtained when $p \neq 0$.

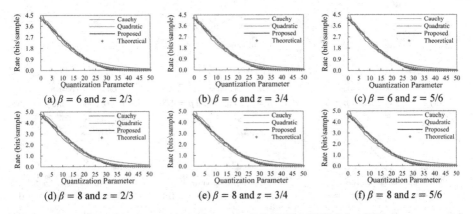

Fig. 3. Comparison of the three R-Q models under DZ+UTSQ/NURQ for GGD source $\alpha = 0.5$

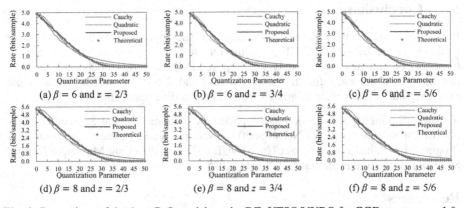

Fig. 4. Comparison of the three R-Q models under DZ+UTSQ/NURQ for GGD source $\alpha = 1.0$

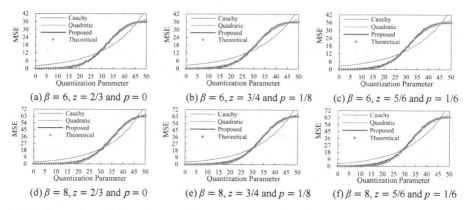

(a) $\beta = 6$, $z = 2/3$ and $p = 0$ (b) $\beta = 6$, $z = 3/4$ and $p = 1/8$ (c) $\beta = 6$, $z = 5/6$ and $p = 1/6$

(d) $\beta = 8$, $z = 2/3$ and $p = 0$ (e) $\beta = 8$, $z = 3/4$ and $p = 1/8$ (f) $\beta = 8$, $z = 5/6$ and $p = 1/6$

Fig. 5. Comparison of the three D-Q models under DZ+UTSQ/NURQ for GGD source $\alpha = 0.5$

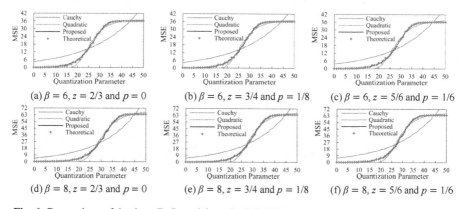

(a) $\beta = 6$, $z = 2/3$ and $p = 0$ (b) $\beta = 6$, $z = 3/4$ and $p = 1/8$ (c) $\beta = 6$, $z = 5/6$ and $p = 1/6$

(d) $\beta = 8$, $z = 2/3$ and $p = 0$ (e) $\beta = 8$, $z = 3/4$ and $p = 1/8$ (f) $\beta = 8$, $z = 5/6$ and $p = 1/6$

Fig. 6. Comparison of the three D-Q models under DZ+UTSQ/NURQ for GGD source $\alpha = 1.0$

6 Conclusion

In this paper, we combine the analytical and heuristic approaches to perform the R-D modeling of the DZ+UTSQ/NURQ scheme for GGD. Compared with the representative heuristic and analytical models, the proposed R-Q and D-Q models achieve superior estimation accuracy for the simulated rate and distortion, which promises the efficiency and effectiveness to guide various video coding applications.

Acknowledgement. This work was supported by National Natural Science Foundation of China under contract No. 60902004, National Basic Research Program (973 Program) of China under contract No.2009CB320907 and Beijing Municipal Natural Science Foundation under contract No. 4102025.

References

1. Hang, H.-M., Chen, J.-J.: Source model for transform video coder and its application. I. Fundamental theory. IEEE Trans. Circuits Syst. Video Technol. 7(2), 287–298 (1997)

2. Kamaci, N., Altunbasak, Y., Mersereau, R.: Frame bit allocation for the H.264/AVC video coder via Cauchy-density-based rate and distortion models. IEEE Trans. Circuits Syst. Video Technol. 15(8), 994–1006 (2005)
3. Chiang, T., Zhang, Y.-Q.: A new rate control scheme using quadratic rate distortion model. IEEE Trans. Circuits Syst. Video Technol. 7(1), 246–250 (1997)
4. Ribas-Corbera, J., Lei, S.: Rate control in DCT video coding for low-delay communications. IEEE Trans. Circuits Syst. Video Technol. 9(1), 172–185 (1999)
5. Sun, J., Gao, W., Zhao, D., Li, W.: On rate-distortion modeling and extraction of H.264/SVC fine-granular scalable video. IEEE Trans. Circuits Syst. Video Technol. 19(3), 323–336 (2009)
6. Sullivan, G.J., Sun, S.: On dead-zone plus uniform threshold scalar quantization. In: Proc. SPIE Visual Commun. Image Process (VCIP), Beijing, China (July 2005)
7. Duan, Y., Sun, J., Liu, J., Guo, Z.: Efficient dead-zone plus uniform threshold scalar quantization of generalized Gaussian random variables. In: Proc. IEEE Visual Commun. Image Process (VCIP), Tainan, Taiwan (November 2011)
8. Sun, J., Gao, W., Zhao, D., Huang, Q.: Statistical model, analysis and approximation of rate-distortion function in MPEG-4 FGS videos. In: Proc. SPIE Visual Commun. Image Process (VCIP), Beijing, China (July 2005)
9. Sullivan, G.J.: Efficient scalar quantization of Exponential and Laplacian random variables. IEEE Trans. Inform. Theory 42(5), 1365–1374 (1996)

A   Appendix

In the appendix, we provide the proof of:

$$\sum_{k=1}^{+\infty} k\mu\Delta e^{-k\mu\Delta} = \frac{\mu\Delta e^{-\mu\Delta}}{(1-e^{-\mu\Delta})^2} \ . \tag{A1}$$

Let $u_k(x) = e^{-k\mu x}$, here we have $u_k(\Delta) = e^{-k\mu\Delta}$. We use $(0, \Delta_s)$ to specify the range of Δ with arbitrary $\Delta_s > 0$. It is easy to know that $\sum_{k=1}^{+\infty} u_k(\Delta)$ is convergent in $(0, \Delta_s)$. Let $u'_k(\Delta)$ be the derivative function of $u_k(\Delta)$ on Δ, and $\sum_{k=1}^{+\infty} u'_k(\Delta)$ makes the new series. It is also easy to prove the continuity of $u'_k(\Delta)$ as well as the uniform convergence for $\sum_{k=1}^{+\infty} u'_k(\Delta)$ in the range of $(0, \Delta_s)$. We specify the sum function $S(\Delta)$ for $u_k(\Delta)$ with $S(\Delta) = \sum_{k=1}^{+\infty} u_k(\Delta) = e^{-\mu\Delta} / (1-e^{-\mu\Delta})$, using the theory of derivation one by one in the series of function, in $(0, \Delta_s)$ we have:

$$S'(\Delta) = \sum_{k=1}^{+\infty} u_k'(\Delta) \ . \tag{A2}$$

Finally with (11) and (A2), we come into the conclusion that

$$\sum_{k=1}^{+\infty} k\mu\Delta e^{-k\mu\Delta} = \sum_{k=1}^{+\infty} -(e^{-k\mu\Delta})' = -\sum_{k=1}^{+\infty} u_k'(\Delta) = -S(\Delta)'$$

$$= -(\frac{e^{-\mu\Delta}}{1-e^{-\mu\Delta}})' = \frac{\mu\Delta e^{-\mu\Delta}}{(1-e^{-\mu\Delta})^2}. \tag{A3}$$

Flexible Presentation of Videos
Based on Affective Content Analysis

Sicheng Zhao, Hongxun Yao, Xiaoshuai Sun, Xiaolei Jiang, and Pengfei Xu

School of Computer Science and Technology, Harbin Institute of Technology,
No.92, West Dazhi Street, Harbin, P.R. China, 150001
{zsc,h.yao,xiaoshuaisun,xljiang,pfxu}@hit.edu.cn

Abstract. The explosion of multimedia contents has resulted in a great demand of video presentation. While most previous works focused on presenting certain type of videos or summarizing videos by event detection, we propose a novel method to present general videos of different genres based on affective content analysis. We first extract rich audio-visual affective features and select discriminative ones. Then we map effective features into corresponding affective states in an improved categorical emotion space using hidden conditional random fields (HCRFs). Finally we draw affective curves which tell the types and intensities of emotions. With the curves and related affective visualization techniques, we select the most affective shots and concatenate them to construct affective video presentation with a flexible and changeable type and length. Experiments on representative video database from the web demonstrate the effectiveness of the proposed method.

Keywords: Video presentation, affective analysis, emotion space, HCRFs.

1   Introduction

The explosion of multimedia contents has resulted in a great demand of video presentation. On one hand, viewers need to get a gist of video content, watch video highlights due to time limit and then make the decision to view the entire video (e.g. a movie) or not. On the other hand, video broadcast platforms, especially television stations, have to check substantial videos and select legal and valuable ones to play, which is a time-consuming and tedious task. Thus, effective video presentation techniques can make video reviewers' work more convenient and efficient.

Most previous works on content-based video presentation focused on certain type of videos, such as sports videos, home videos, or summarizing videos by event detection [1-5]. Liu *et al.* [1] proposed a novel flexible racquet sports video content summarization framework, by combining the structure event detection method with the highlight ranking algorithm. Zhao *et al.* [2] proposed a novel system of highlight summarization in sports videos based on replay detection. Based on videos' three properties: emotional tone, local main character and global main character, Xiang and Kankanhalli [3] employed affective analysis to automatically create adaptive presentations from home videos for three types of social groups: family, acquaintance

S. Li et al. (Eds.): MMM 2013, Part I, LNCS 7732, pp. 368–379, 2013.

and outsider. Wang *et al.* presented an approach for event driven web video summarization by tag localization and key-shot mining in [4], and turned a movie clip to comics automatically in [5].

Meanwhile, affective image and video content analysis has been paid much attention recently. Generally, all these works are based on two kinds of emotion models: categorical (discrete) emotion states, or dimensional (continuous) emotion space. In categorical (discrete) models [3, 6-8], emotions are usually considered to be one of a few basic categories, such as *fear, anger, happy, sad, etc.* Machajdik and Hanbury [6] exploited theoretical and empirical concepts from psychology and art theory to extract image features that are specific to the domain of artworks with emotional expression, and classified an image into eight affective categories: *amusement, awe, contentment, excitement, anger, disgust, fear* and *sad.* Kang [7] trained two Hidden Markov Models (HMMs) to detect affective states in movies. Xu *et al.* [8] utilized fuzzy clustering and HMMs to classify video shots in movies into five emotion types, including *fear, anger, sad, happy,* and *neutral.* Dimensional or continuous models mostly employ the 3-D Valence-Arousal-Control (VAC) emotion space [9] or 2-D Valence-Arousal (VA) emotion space for affective representation and modeling [10-13]. Hanjalic and Xu [10] modeled arousal and valence separately using linear feature combinations, and drew arousal curve and valence curve or combined affect curve. A novel computation method for affect-based video segmentation, which is designed based on the Pleasure-Arousal-Dominance (P-A-D) emotion model, was introduced in [11]. Zhang *et al.* [12] proposed affective information based movie browsing using affective visualization techniques. Nicolaou *et al.* [13] proposed a multi-layer hybrid framework for emotion classification that is able to model inter-dimensional correlations. Both emotion models have their advantages. The former one is easier for users to understand and label, while the latter one is more flexible and richer in descriptive power.

Some researchers have also considered viewers' affective reactions to video contents for video analysis [14-18]. Ma *et al.* [14] presented a generic framework of video summarization based on the modeling of viewer's attention. Joho *et al.* [15, 16] proposed a new approach for personal highlights detection and video summarization based on the analysis of facial activities of viewers, using the model of pronounced level and expression's change rate. Zhao *et al.* [17,18] presented a novel method to classify, index and recommend videos based on affective analysis, mainly on facial expression recognition of viewers.

However, few works have considered the affective analysis of video content for general video presentation of different genres, while affective information in videos is closely related with users' experiences and preferences [12]. In this paper, we propose a novel method to present general videos of different genres based on affective content analysis. The contributions lie in three aspects: 1. We propose an improved categorical emotion space, by combining the advantages of categorical emotion states and dimensional emotion space; 2. We propose a novel video presentation method with a flexible and changeable type and length; 3. Two HCRFs are trained to compute the labels of types and intensities of emotions, respectively.

The rest of this paper is organized as follows. Section 2 describes the proposed method in detail. We conduct relative experiments and analyze the results in Section 3. Finally, conclusion and future work are discussed in Section 4.

2   The Proposed Method

The framework of the proposed method is shown in Figure 1. First, we segment each video into video shots. Second, affective audio-visual features are extracted and selected. Then affective models are trained to map selected features into affective states in an improved categorical emotion space using HCRFs. Finally, we draw affective curves, visualize affective states, select the most affective shots, and cluster these shots to construct video presentation according to time order. Further, we can change the type and length of video presentation based on users' feedback.

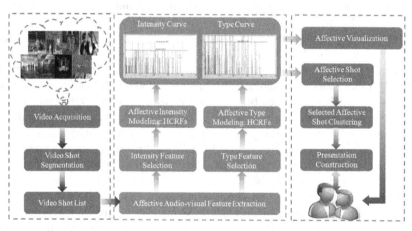

Fig. 1. The framework of the proposed method

2.1   An Improved Categorical Emotion Space and Related Affective Curves

An emotion can be evoked in a user when watching affective videos. Different scenes of different videos result in different emotion types and intensities of viewers. On one hand, categorical emotion states are easier for users to label and understand, while dimensional emotion space can express the intensity of emotions. On the other hand, the former one is computed in a classification framework, while the latter one is done regressively, which makes the computation more complex. By combining their advantages, we propose an improved categorical emotion space named IT emotion space, in which "I" represents the intensity or degree of emotion and "T" stands for the type of emotion. That is, we add discrete intensity dimension to the original categorical emotion states. Assume that the number of T is m, the number of I is n, and the total number of affective states is $N = m \times n$. If T contains the emotion type of *'neutral'*, *N* turns to $m \times n - m + 1$, as *'neutral'* does not have different intensities. For instance, in this paper, we assume that

TE{*'neutral'*, *'sadness'*, *'anger'* , *'disgust'*, *'fear'*, *'surprise'*, *'happiness'*} and IE{*'very low'*, *'low'*, *'mid'*, *'strong'*, *'very strong'*}.

Totally we can express 31 types of different affective states, as shown in Figure 2(a). For simplicity, we use TE{1, 2, 3, 4, 5, 6, 7} and IE{1, 2, 3, 4, 5} to represent different types and intensities of emotions respectively.

In this case, type and intensity of emotion are the functions of different audio and visual features. We model type and intensity time curve as

$$T(k) = F(f^k_{T\_audio}, f^k_{T\_visual}), I(k) = G(f^k_{I\_audio}, f^k_{I\_visual}) \qquad (1)$$

where k, $f^k_{T\_audio}$, $f^k_{T\_visual}$, $f^k_{I\_audio}$, $f^k_{I\_visual}$ stand for the kth short, audio and visual features for type and intensity respectively.

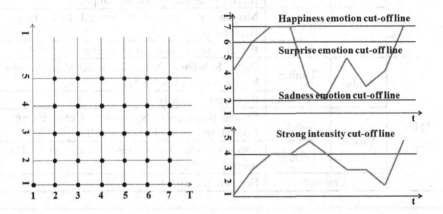

Fig. 2. (a) Illustration of the improved (b)Related affective time curves with cut-off categorical emotion space   lines

2.2   Affective Feature Extraction

Based on related works of affective feature extraction [3, 6, 8, 10, 11], we extract 11visual features and 13 audio features, as shown in Table 1.

The *Audio Intensity Features* typically describe the intensity of the audio in video clips. They are closely related with the intensity of emotion and have been frequently used in audio affective analysis. *Mel Frequency Ceptral Coefficients (MFCC)* work well for excited and non-excited intensity detection. The *Audio Timbre Features* are frequently used to distinguish different types of sound production. *Rhythm* is an important tool used by artists to express their emotions, based on Onset detection and Drum detection. The *visual clues* in videos are also important in describing the affective states, and are commonly utilized to render the emotions and evoke certain moods in audience. We extract color and texture visual features for emotion type. After the feature extraction, we employ Gaussian Normalization to normalize these features into [0, 1]. Besides, we also recognize facial expressions and estimate the expression intensities [17-19], as facial expression and its intensity tell the emotion type and intensity of actors to a great extent.

Table 1. Extracted affective features for emotions' type and intensity estimation

	Category		Name
Intensity	Audio	Intensity	Zero Crossing Rate (ZCR), Short Time Energy (STE) , MFCC
		Timbre	Sub-band Peak, Sub-band Valley, Sub-band Contrast
		Rhythm	Tempo, Rhythm Strength, Rhythm Contrast, Rhythm Regularity, Drum Amplitude
	Visual		Motion Intensity, Short Switch Rate, Frame Brightness
	Content		Facial expression intensity estimation
Type	Audio	Timbre	Pitch, Sub-band Peak, Sub-band Valley, Sub-band Contrast, Pitch STD
		Rhythm	Rhythm Regularity
	Visual	Color	Frame Brightness, Saturation, Color Energy, Color Emotion, Colorfulness, Hue
		Texture	Tamura, Wavelet textures, Gray-level Co-occurrence Matrix
	Content		Facial expression recognition

2.3   Affective Labeling

We utilize HCRFs to compute the type and intensity of emotion, based on the extracted audio and visual features. In nature, the main idea behind HCRFs is to enrich CRFs by adding hidden states to capture complex dependencies or implicit structures in the training samples [20, 21].

Suppose $F = \{f(y, h, x; \theta)\}$ be a set of feature functions, and $H = \{h_1, h_2, \cdots, h_m\}$ be a set of hidden variables. Equation (2) and (3) list the probability and potential function.

$$P(y \mid x, \theta) = \frac{\sum_h \exp(\varphi(y, h, x; \theta))}{\sum_{y'} \sum_h \exp(\varphi(y', h, x; \theta))} \qquad (2)$$

$$\varphi(y, h, x; \theta) = \sum_j f(x, j) \cdot \theta(h_j) + \sum_j \theta(y, h_j) + \sum_{(j,k) \in E} \theta(y, h_j, h_k), f(x, j) \qquad (3)$$

$$= [f_{j-1}^T, f_j^T - f_{j-1}^T]^T$$

Based on previous works on CRFs and HCRFs [20, 21], we use the following objective function for training the parameters:

$$\sum_i \log P(y_i \mid x_i, \theta) - \frac{1}{2\sigma^2} \parallel \theta \parallel^2 \tag{4}$$

Given a test shot with observations $x' = \{x_1, x_2, \cdots, x_n\}$ and trained parameters θ^* of the HCRFs, the label y' of this shot is recognized as follows:

$$y' = \arg \max_{y \in Y} P(y \mid x', \theta^*) \tag{5}$$

2.4 Affective Presentation Construction

Based on the computed affective type and intensity time curves, we are able to present videos flexibly. As affective presentation should contain those shots with high intensity levels, we use a cut-off line to segment the highlight shots. In this paper, we utilize 5 different intensity types, so in the intensity time curve, we draw a constant function $i=level$, $level \in \{1, 2, 3, 4, 5\}$. And the shots between the intersection points (the constant function and the intensity time curve, that is, $i(t)=level$) and the local maximums of the intensity time curve ($i(t)=5$) are the highlights. Further, we can change the value of $level$ or cluster the selected shots using k-mean clustering to fit the demanded length of the video presentation according to users' time arrangement.

On the other hand, different people at different time have different moods, and want to see video presentation of different types. In this case, we provide a user feedback interface considering users' interests and moods. Users can choose the emotion types in which they are interested, and then we draw a constant function $v=value$, $value \in \{1, 2, 3, 4, 5, 6, 7\}$. And the shots at the intersection points of the constant function and the type time curve ($v(t)=value$) are the selected ones. Further, we can change $level$ and $value$ simultaneously to specify emotion type and intensity. One example of cut-off line in emotion transitions is given in Figure 2(b).

3 Experiments

In this section, we introduce the creation and labeling of the used dataset, conduct relative experiments, analyze our results and compare them with latest research.

3.1 Dataset and Labels

We select 20 representative videos (*Titanic, Avatar, Kung Fu Panda 2, The Ring, Got the Money Anyway* and so on) to create our database. They are of different video types, including *comedy, tragedy, horror, sport videos, etc.* The total length is about 40 hours and the average length is about 2 hours. As far as we know, our database is currently one of the most representative video databases [3, 7, 9-11].

We invite 20 volunteers (12 male, 8 female; 10 postgraduates, 10 undergraduates; 15 computer related major, 5 art related major) to label the affective ground truths. They are required to choose emotion type and intensity values (intensity in {1, 2, 3, 4, 5}, type in {1, 2, 3, 4, 5, 6, 7}) to describe each shot's affective states when they are watching the videos. As emotions evoked by videos are strongly influenced by different educations, cultures, backgrounds, *etc*, different participants may have different emotion types and intensities. The final evaluation value of emotion type and intensity for a shot is the latest integer of the average score of all viewers. In order to reduce the interference and confusion in subjective labeling, we provided a standard explanation of each emotion type and intensity. After watching one video, viewers can have one break to lower the impact of fatigue. Totally about 12000 shots' affective ground truths are labeled.

3.2   Results of Affective Labeling

We select six representative video genres to test our emotion type and intensity labeling accuracy: *comedy*, *tragedy*, *action*, *horror*, *exciting*, and *sports*. We separate the data into a training and testing set using K-fold Cross Validation (K=5). As different videos genres have different affective shot types, we test them respectively. In experiment, the number of hidden states of HCRF is selected 3. The experimental results are presented in Table 2, in which "—" represents that there is no such emotion type or intensity manual labels in related video genres. As T=1 stands for the emotion type of *neutral* (no emotion), we don't test its accuracy for simplicity.

From Table 2, it is clear to see that *horror* videos get the highest emotion type accuracy, *comedy* and *tragedy* lowest, and that the emotion intensity accuracy of *comedy*, *exciting*, and *sports* is higher. This is reasonable because the emotion types in *comedy* and *tragedy* are more likely expressed by conversations between characters, while the emotion types in *horror* videos are mainly expressed by our selected audio-visual features and the emotion intensities in *comedy*, *exciting*, and *sports* videos are commonly produced by motion and sound. The comparison of our method and [12] is shown in Figure 3, from which we can conclude our performance is better than [12], because we select more affective features, especially visual color and texture features, and facial expression and its intensity, and use HCRFs to capture the latent information of affective features and emotional labels.

Table 2. Emotion type and intensity estimation accuracy on the testing set (%)

	type of emotion						intensity of emotion				
	2	3	4	5	6	7	1	2	3	4	5
comedy	—	—	70.0	—	—	76.5	—	66.6	78.6	78.0	60.0
tragedy	75.0	60.0	60.0	50.0	65.0	73.3	60.0	60.0	77.5	73.3	60.0
action	—	75.0	75.0	70.0	77.5	75.0	70.0	75.0	82.5	80.0	70.0
horror	85.7	—	—	86.2	81.2	—	—	70.0	72.0	70.0	73.3
exciting	—	63.6	61.5	77.7	76.2	72.7	60.0	63.6	71.9	76.2	70.0
sports	—	71.4	66.6	—	81.6	70.6	—	72.7	72.0	82.2	75.0

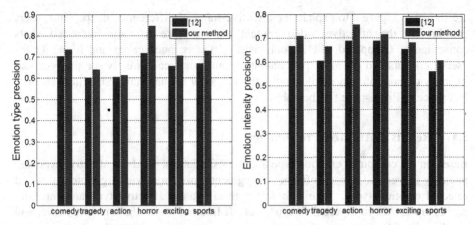

Fig. 3. Comparison of the proposed method and [12] in emotion type and intensity precision

3.3 Affective Visualization

Similar to [12], we use an affective visualization technique to make users understand videos' affective states better. We employ commonly used color to represent affective states on time axis. As we use the proposed IT emotion space, containing 7 emotion types and 5 levels of emotion intensity, we simply use 7 and 5 different colors to represent relevant type and intensity, respectively. A progress bar and two color bars

(a)Different colors for different emotion types and intensities

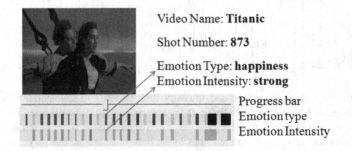

Video Name: **Titanic**

Shot Number: **873**

Emotion Type: **happiness**
Emotion Intensity: **strong**

Progress bar
Emotion type
Emotion Intensity

(b) An instance of affective visualization on time axis

Fig. 4. Illustration of easy-understanding affective visualization results

are shown in Figure 4 to explain the affective states of different contents. Thus, users can browse the affective video content and navigate to their interested shots conveniently. Compared to [12], our visualization is much easier and friendlier for users to understand the affective content, because they can directly see the type and intensity of emotion they are dealing with, instead of the hard-understanding values of VA space or PAD space for the public.

3.4   Results of Affective Presentation

Based on the computed emotion type and intensity labels, we draw affective type and intensity curves and use cut-off lines to segment those shots with high intensities and selected emotion types. Take *Titanic* for instance, the affective type and intensity curves are shown in Figure 5. We use cut off-line $i=4$ to cut the intensity curve, and get the related key frames and corresponding emotion types, partly shown in Figure 6. Similar result of comedy *Got the Money Anyway* is shown in Figure 7. We can also use emotion type cut off-line to get the shots with the interested moods, such as $t=6$ in Figure 5.

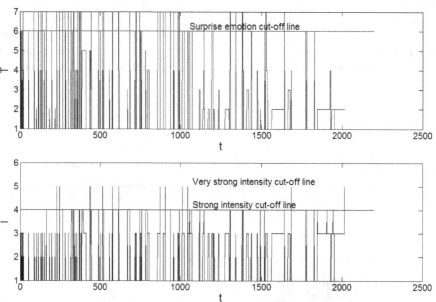

Fig. 5. Illustration of emotion type and intensity time curves and segmentation with cut-off lines. The magenta line in the above figure is the surprise emotion cut-off line. The interaction points of the magenta line and the blue type curve correspond to the shots which evoke surprise emotion in viewers. The yellow and red lines in the bottom figure are the very strong intensity cut-off line and strong intensity cut-off line. The interaction points of the yellow line and the black intensity curve correspond to the shots which evoke very strong emotion intensities in viewers. The interaction points of the red line and the black intensity curve correspond to the shots which evoke strong emotion intensities in viewers.

Fig. 6. Examples of some affective key frames of tragedy <Titanic>, and related emotion types

Fig. 7. Examples of some affective key frames of comedy <Got the Money Anyway>, and related emotion types

In order to evaluate the effectiveness of the proposed method, we use the subject agreement (SAR). We asked the 20 volunteers and another 50 users to watch the constructed video presentations and the provided affective video visualization, and score them with a ten-level Likert scale (1-10, 1: not satisfied at all; 10: very satisfied). Then we compute the average score of the two groups respectively. The satisfaction score are 8.86 and 8.25. Through the results of user study, we can see that the proposed presentation method is able to satisfy users' demand.

4   Conclusion and Future Work

In this paper, we propose a novel general video presentation method based on affective content analysis. According to computed affective labels of emotion type and intensity in the proposed easy-understanding emotion space, we draw affective curves which tell the types and intensities of emotion changes and transitions. With the curves and related affective visualization techniques, we select the most affective shots and concatenate them to construct a flexible affective video presentation according to time order. The presentation results are satisfying.

As viewers of different ages and different races differ in affective semantics understanding, how to present videos affectively according to viewers' experiences is worth studying. How to tackle the case that different viewers may have different emotion types and intensities is a challenging topic. Maybe assigning one shot with different emotion types and intensities in an emotion vector is an appropriate answer. Combining video's affective contents and viewers' affective reactions to them effectively can present better. Further, how to map low-level affective audio-visual features into affective states more precisely (such as affective audio-visual words and latent topic driving model [22]) is also our future research task.

Acknowledgments. The work was supported by the National Natural Science Foundation of China (No. 61071180) and Key Program (No.61133003).

References

1. Liu, C., Huang, Q., Jiang, S., et al.: A framework for flexible summarization of racquet sports video using multiple modalities. Computer Vision and Image Understanding 113(3), 415–424 (2009)
2. Zhao, Z., Jiang, S., Huang, Q., Zhu, G.: Highlight summarization in sports video based on replay detection. In: IEEE International Conference on Multimedia & Expo, pp. 1613–1616 (2006)
3. Xiang, X., Kankanhalli, M.: Affect-based adaptive presentation of home videos. In: ACM Multimedia, pp. 553–562 (2011)
4. Wang, M., Hong, R., Li, G., Zha, Z., Yan, S., Chua, T.: Event Driven Web Video Summarization by Tag Localization and Key-Shot Identification. IEEE Transactions on Multimedia 14(4), 975–985 (2012)
5. Wang, M., Hong, R., Yuan, X., Yan, S., Chua, T.: Movie2Comics: Towards a Lively Video Content Presentation. IEEE Transactions on Multimedia 14(3), 858–870 (2012)
6. Machajdik, J., Hanbury, A.: Affective image classification using features inspired by psychology and art theory. In: ACM Multimedia, pp. 83–92 (2010)
7. Kang, H.: Affective Content Detection Using HMMs. In: ACM Multimedia, pp. 259–262 (2003)
8. Xu, M., Jin, J., Luo, S.: Hierarchical Movie Affective Content Analysis Based on Arousal and Valence Features. In: ACM Multimedia, pp. 677–680 (2008)
9. Schlosberg, H.: Three Dimensions of Emotion. Psychological Review 61(2), 81–88 (1954)
10. Hanjalic, A., Xu, L.: Affective Video Content Representation and Modeling. IEEE Transactions on Multimedia 7(1), 143–154 (2005)
11. Arifin, S., Cheung, P.Y.K.: A Computation Method for Video Segmentation Utilizing the Pleasure-Arousal-Dominance Emotional Information. In: ACM Multimedia, pp. 68–77 (2007)
12. Zhang, S., Tian, Q., Huang, Q., Gao, W., Li, S.: Utilizing affective analysis for efficient movie browsing. In: IEEE International Conference on Image Processing, pp. 1853–1856 (2009)

13. Nicolaou, M.A., Gunes, H., Pantic, M.: A Multi-layer Hybrid Framework for Dimensional Emotion Classification. In: ACM Multimedia, pp. 933–936 (2011)
14. Ma, Y.-F., Lu, L., Zhang, H.-J., Li, M.: A User Attention Model for Video Summarization. In: ACM Multimedia (2002)
15. Joho, H., Staiano, J., Sebe, N., Jose, J.M.: Looking at the viewer: analysing facial activity to detect personal highlights of multimedia contents. Multimedia Tools Application 51(2), 505–523 (2011)
16. Joho, H., Jose, J.M., Valenti, R., Sebe, N.: Exploiting facial expressions for affective video summarization. In: ACM International Conference on Image and Video Retrieval (2009)
17. Zhao, S., Yao, H., Sun, X., Xu, P., Liu, X., Ji, R.: Video Indexing and Recommendation Based on Affective Analysis of Viewers. In: ACM Multimedia, pp. 1473–1476 (2011)
18. Zhao, S., Yao, H., Sun, X.: Video Classification and Recommendation Based on Affective Analysis of Viewers. Neurocomputing (to appear, 2012)
19. Yang, P., Liu, Q., Metaxas, D.N.: RankBoost with l_1 regularization for Facial Expression Recognition and Intensity Estimation. In: IEEE International Conference on Computer Vision, pp. 1018–1025 (2009)
20. Lafferty, J., McCallum, A., Pereira, F.: Conditional random fields: Probabilistic models for segmenting and labeling sequence data. In: IEEE International Conference on Machine Learning (2001)
21. Quattoni, A., Collins, M., Darrell, T.: Conditional random fields for object recognition. In: Neural Information Processing Systems (2004)
22. Irie, G., Satou, T., Kojima, A., Yamasaki, T., Aizawa, K.: Affective Audio-Visual Words and Latent Topic Driving Model for Realizing Movie Affective Scene Classification. IEEE Transactions on Multimedia 16(2), 523–535 (2010)

Dynamic Multi-video Summarization of Sensor-Rich Videos in Geo-Space

Ying Zhang, He Ma, and Roger Zimmermann

School of Computing, National University of Singapore, Singapore 117417
{yingz118,mahe,rogerz}@comp.nus.edu.sg

Abstract. User generated videos are much easier to be produced today due to the progress in camera technology on mobile devices. The ubiquitous built-in sensors in digital devices greatly enrich these videos with sensor descriptions, especially geo-spatial properties. A repository of such sensor-rich videos can be a great source of information for prospective tourists when they plan to visit a city and would like to get a preview of its main areas.

In this study we propose an interactive geo-video search system. When a user specifies a start point and a destination (*e.g.*, on a map), the system dynamically retrieves a video summarization along the path between the two points. Moreover, the query can be interactively updated during the video playback, by changing either the tour path or the target destination. The main features of our technique are, first, that it is fully automatic and leverages sensor meta-data information which is acquired in conjunction with videos. Second, the system dynamically adapts to query updates in real-time, and no prior knowledge is required by users. Third, a concise but comprehensive summarization from multiple user generated videos is proposed for any queried route. Finally, the system incrementally adapts to the latest contributions to the video repository.

1 Introduction

Nowadays people are very familiar with trip planning via such sites as Google Maps where they designate a start location (*e.g.*, address) and an end location and the map system creates a short and effective route. Additionally, users can use Street View to get a visual impression of a location. Inspired by this concept, our system implements a similar idea, but with major difference of leveraging crowd-sourced user generated videos (UGV) to provide the "lifeliness" of a place, rather than static contents such as Google's panoramic imagery. Additionally, UGVs are potentially much more up-to-date to reflect the latest view of an area, especially for the popular places.

This paper describes the detailed techniques of our previous demonstration system [17] with the overall architecture in Figure 1. To satisfy the real-time property of an interactive query, the system is divided into an **online** query response component and an **offline** summarization preparation. In the online part, given the user's designated start point and destination, a *Query Builder*

S. Li et al. (Eds.): MMM 2013, Part I, LNCS 7732, pp. 380–390, 2013.
© Springer-Verlag Berlin Heidelberg 2013

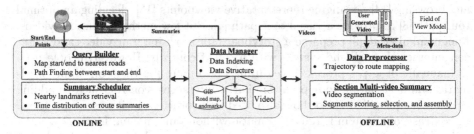

Fig. 1. Proposed system architecture

decides a tour path which may be long and composed of multiple *sections*, the atomic logical units for which we pre-process small summarizations. Secondly, for each section, a *Summary Scheduler* contacts the data management repository and retrieves a profile about its surrounding summaries. The *Data Manager* maintains an index structure, containing the profiles for each section in geo-space. The final video is an ordered combination of all its section summarizations. Offline summarization selects informative sub-parts from a big set of UGVs, collected through specially designed recording applications for smartphones with a stream of sensor values from built-in GPS and compass devices automatically. The *Data Preprocessor* maps videos trajectories to roads and only the pieces containing sufficient information are retained. Then, for each route section, we retrieve all its companion video segments, and extract useful pieces to produce a summary. A profile containing the summary meta-data is retained and indexed by the data manager. The highlights of our work are summarized as follows.

- All processes are automatic with light-weighted computation utilizing sensor meta-data only.
- A summary quality is formulated, including a set of metrics, to produce a single summary from *multiple* videos.
- An efficient index is built to retain the relationship between the sections and their corresponding summaries. The average query process time is around 0.043 seconds per section which provides a good foundation for a real-time query system.
- The system can be incrementally updated so that when new videos are inserted, the most update to date results will be produced.

In the rest of this paper, Section 2 describes the related work and background information. Sections 3 and 4 detail the video segmentation and summarization strategies, separately. Section 5 introduces the online query design. An experimental evaluation is presented in Section 6 and conclusions in Section 7.

2 Related Work and Preliminaries

2.1 Related Work

Rich web-sources facilities are used by a lot of visitors as travel assistant services. Most of existing researches focus on photos such as through photo organization and browsing [14], arranging representative viewpoints [13], deciding an optimal tour path [8] or provide a visual tour path [4]. But few methods consider videos. Pongnumkul *et al.* [9] prepared a storyboard for browsing tour videos and a user can interact through the system with a set of controls. However, people need to manually drag pre-processed key-frames to a map, pin them on proper landmarks, and designate a tour path. Similarly, interactive controls such as adding or removing provided video shots also need to be processed with user knowledge. In contrast, our system is fully automatic without any manual feedback.

For the existing video summarization techniques, only a few studies target multiple video summarization. Li *et al.* [6,7] presents a series of work to globally compare aural, visual and hybrid features of frames from all videos and select the most distinguishing ones as key frames in the final summary. Similarly, Shao *et al.* [10] proposed a cluster-based keyframe selection using visual and textual features and Wang *et al.* [15] assign each frame a score proportional to its globally covered information using visual features. Aside from the computational complexity, all of these works intend to choose key frames in a set of videos. In our work, we choose a set of informative video segments fully leveraging the sensor meta-data attached to the videos to achieve a light-weight computation.

2.2 Terminology and Symbolic Notations

For a video $v \in V$, its frames are denoted by f_i at time t_i. $|v|$ counts the number of frames in a video and $|V|$ is the size of the video database. At each time index t_i, meta-data from camera-attached sensors is recorded, including GPS location $\mathcal{L}$, and view direction θ. The scene for each frame is modeled as a spatial object,

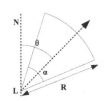

Fig. 2. Illustration of 2D FOV

which can be represented by a 2D *field-of-view* (*FOV*) with four parameters: $FOV \equiv \langle \mathcal{L}, \theta, \alpha, R \rangle$. (1) The camera position $\mathcal{L}$ consists of the latitude and longitude coordinates read from GPS. (2) Camera direction θ is obtained from a digital compass. (3) R is the maximum visible distance at which an object within the camera's *FOV* can be recognized. (4) The viewable angle α is obtained based on the camera and lens properties for the current zoom level [3]. In 2D space, the *FOV* of the camera at time t_i forms a pie-slice-shaped area as in Fig. 2. We use *FOV* to represent each frame's coverage scene and 𝔽𝕆𝕍 for a segment that sums up all its frames' viewable scenes.

3 Video Segmentation

The process of finding and extracting salience from original video contents is termed as video segmentation. For geo-tagged videos, salience refers to objects that attract users to capture them and we term them *landmarks* (denoted by $H = \{h_i\}$, $i = 1$ to $|H|$). Landmark position is pre-known from external GIS database. In our system, we would like to include the video segments that cover as many popular landmarks in a queried region as possible so only the video pieces containing rich landmark information are retained as candidates in the final summary. The informativeness of a video clip depends on both the popularity of its covered landmarks and how integrate the landmarks are described in videos.

3.1 Landmark Popularity

A set of UGVs are recorded by many individual users and usually people would like to capture what mostly attract them within the geographically surrounded area. If a landmark h located at $\mathcal{L}_h$ is captured in more videos, a higher popularity is assigned: $P(h) = \frac{|\{v|v \in V, \mathcal{L}_h \in \text{FOV}_v,\}|}{|V|}$. Secondly, the longer a landmark is shown in a video, the higher popularity the landmark is: $Q(h) = \frac{|\{f|f \in v, \mathcal{L}_h \in \text{FOV}_f\}|}{\sum_{\{v|\mathcal{L}_h \in \text{FOV}_v\}} |v|}$. So the popularity score u_h is proportional to both factors $u_h \propto P(h) \times Q(h)$.

3.2 Landmark Completeness

If a landmark is described more integrally in a video clip, this clip is more informative. Usually, (1) the central an object is captured in a screen or (2) the nearer the object is captured from the camera location, it's of larger possibility for the object to be presented completely. Denote o_f as the direction from camera location $\mathcal{L}_f$ to landmark position $\mathcal{L}_h$ and a visible distance d_h within which the landmark can be observed. For the first factor, the less camera direction θ derives from o_f, the higher score this frame gains. $A(f) = \sum_{\{h|\mathcal{L}_h \in \text{FOV}_f\}} w_h(\frac{\|\theta_f, o_f\|}{\alpha/2})^{-1}$, w_h is a weighted factor. Similarly for the second factor, $D(f) = \sum_{\{h|\mathcal{L}_h \in \text{FOV}_f\}} w_h(\frac{\|\mathcal{L}_f, \mathcal{L}_h\|}{d_h})^{-1}$.

Frame informativeness is $\mathscr{S}(f) = A(f) \cdot D(f)$, and take $w_h = u_h$. To select a set of video segments for summarization, we adopt a sliding window based solution by averaging scores from all containing frames, and a boundary is decided only when the score falls below a threshold. Such solution reduces the influence from occasional sensor data noises. To further improve video quality with less discontinuities, we remove the segments with a duration falling below an acceptable observable duration and recover a combined segment from multiple pieces with trivial gaps.

4 Summarization

To prepare video answers for different user queries when multiple route sections are involved, the sections with more popular surrounding landmarks will have a larger possibility to be chosen for a detailed version of summarization. This detail-level is measured by a length threshold. In this section, we design a score mechanism to retrieve summaries of various durations.

4.1 Scoring Metrics

COVERAGE: An object can be observed through multiple viewpoints, from either distance or angular perspective. Each of them presents the object with certain semantics. From a distance aspect, we divide the distance between the camera position and the landmark position within its visible distance into multiple ranges $\mu(h)$, so the distance coverage of a video v for a single landmark is formulated $\mathscr{D}(h) = |\bigcup\{\mu_f(h)|f \in v, \mathcal{L}_h \in FOV_f\}|$. Similarly, each frame f captures the covered landmark h with a specific heading direction and a well presented summary should cover many view angles for all visible landmarks. The angular coverage for a segment is the sum from all its frames $\mathscr{A}(h) = |\bigcup\{o_f(h)|f \in s, \|\mathcal{L}_f, \mathcal{L}_h\| \le d_h\}|$. The final coverage score is a combination of these two factors over all their surrounding landmarks H, $\mathbf{C} = \sum_{h \in H} |\mathscr{D}(h))| + |(\mathscr{A}(h)|$.

VISIBILITY QUALITY: This metric describes how well the surrounding landmarks are recorded in a video. Objects in the central view [12,5], or with nearer distance [11] are usually better than on the boundary, or from far away. For a video v, its visibility quality is the average informativeness score of all its frames as described in Section 3: $\mathbf{Q} = \frac{\sum_{\{f|f \in v\}} \mathscr{S}(f)}{|\{f|f \in v\}|}$.

REDUNDANCY: A well-prepared summary should avoid too much shared video contents from different segments. We formulate redundancy R as the total similarities between any pair of summary segments. Given two segments s_i and s_j, similarity is measured as the total number of their shared distance ranges and angular perspectives over all surrounding landmarks $M_{\mathscr{X}}(s_i, s_j) = \sum_{h \in H} |\mathscr{X}_i(h) \cap \mathscr{X}_j(h)|$, where $\mathscr{X} \in \{\mathscr{A}, \mathscr{D}\}$. Final redundancy score is formulated as: $\mathbf{R} = \sum_{s_i \in v} \sum_{s_j \in v, i \ne j} \mathbf{W}[M_{\mathscr{A}}(s_i, s_j), M_{\mathscr{D}}(s_i, s_j)]^T$ where W is a weighted factor.

Finally, the overall scoring function of a summary S' is a combination of these three factors $score(S') = \mathbf{C}(S') + \mathbf{Q}(S') + \mathbf{R}(S')$ and a higher score indicates better quality.

4.2 Summarization Selection

Given a set of video segments $S = \{s_i\}, i = 1, \ldots, n$, we need to determine a minimal subset, $S' = \arg\max_{S' \subseteq S} score(S')$, where $\sum_{s \in S'} |s| \le Th_{time}$. The global optimization problem is solved by a greedy algorithm:

1. The summary S is initialized with an empty set.
2. Select the segment $s = \arg\max\{score(s \cup S) || s \cup S| <= Th_{time}\}$.
3. $S = S \cup s$. Go to step 2.

Cornuejols et al. [1] proved that such a greedy algorithm has quality at least $\frac{e-1}{e}$ times the optimal solution, where e is the base of the natural logarithm.

5 Online Query Process

After given a start and end point, the system computes a path and returns a video summarization along it. The initial path can be realized using any existing algorithms (*e.g.*, A*). Our focus is on how a system can fast and interactively decide which summarization should be scheduled to capture the most interesting part along the path. With no prior knowledge of a user's preference, the initial result may not well satisfy their queries. So during the video playback, the system can further allow interaction with the user by providing the functionality to alter the desired path, either through changes of the path directly via mouse manipulations or by choosing another destination. A quick adaption to such a dynamic update should be taken into consideration for the system design. Usually, a travel path R may be long, composed of multiple lower level units, named *sections* r_i, $R = \{r_i\}, i = 1, \ldots, n$. From maps such as TIGER/Line or OpenStreetMap, each road consists of piece-wise linear sections and it is unlikely for videos along two sections to share similar contents due to reasons such as blocked views by buildings. Hence, we define the summarization on one section (*section summarization*) as atomic, primitive unit S_i' and a path summarization is an ordered set of independent sub-summarizations for each section, $S' = \langle S_1', S_2', \ldots, S_n' \rangle$.

5.1 Data Structure Design

In a large-scale application with a big video repository and its associated sensor meta-data, to retrieve the needed information quickly and shorten users wait time is a critical issue. An elegant data structure design, capturing the essential information, usually assists with fast video summarization preparations.

Two categories of information are retained during the off-line section summarization. The first is an overview of its surrounding landmarks which are captured by at least one video. The other includes the summarization information of multiple video segments. With different time allocations, multiple versions of summaries can quickly be scheduled. When a new query arrives and the path has been computed, the system needs to determine which landmarks should be included and how many segments should be scheduled in real time. A sample profile to record the above information is shown in Table 2a.

Table 1.

Name	Description	Name	Description
id	Section ID	id	Segment ID
position	Section start/end point's position	vid	Video ID
landmarks	IDs of covered Landmark	ts/te	Start/end timestamp
videos	Descriptions for each segment	previous	ID of previous segment
	in summarization	landmarks	IDs of covered Landmarks

(a) Sample Summarization Profile.   (b) Segment Description.

5.2 Data Indexing

To manage the profiles of sections, we require an index to store these geo-characteristics. There exist data-driving structures (*e.g.*,R-tree, R^+tree, R^*) and space-driven (*e.g.*, Grid Files, Voronoi Diagrams) structures for spatial data.

However, the second category takes spatial object as points which does not fit our *FOV* model. So we adopt an R-tree index and insert the Minimal Boundary Rectangle (MBR) for each section r into it. Because each section is represented as a straight line section in our database, the MBR is completely determined by the start and end points.

The index I is constructed as follows:

1. For a new arrival video segment, map it to a section map R.
2. For each section $r \in R$, apply a range query over its MBR to get all its indexed profiles.
3. If no record is found, it means no section summarization was conducted before, insert section MBR to I. Otherwise go to step 4.
4. Recompute the summary, update the profile and the index I.

Note that in step 4, if two segments have the same contribution and a similarity score of one, then the one with the latest time-stamp will be selected. This ensures that our system adapts well to newly inserted videos and the most recent clips are used for presentation.

If an incremental update to the road map occurs we need to modify the overall index, but such changes happen with very low frequency. By setting the load factor in the R-tree index suitably, the tree structure remains the same when new sections are inserted. For improved scalability with video databases covering large geo-spaces, multiple indices for sub-spaces can be built and processed in parallel.

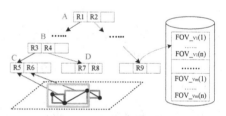

Fig. 3. Index structure overview: the profile of each primitive unit (section) is indexed by its MBR in an R-tree

5.3 Online Query Processing

The index is maintained by a data manager and a series of services will be triggered when new query arrives:

1. Once start and destination points are determined, a path $R = \{r_i\}, i = 1, \ldots, n$ is calculated.
2. For each section r_i, conduct a range query on the index and retrieve corresponding profile in order $P = p_i, i = 1$ to n.
3. By analyzing the covered landmarks from all profiles, the sections surrounded by more popular landmarks will be allocated with more detailed summarizations.
4. Schedule the video segments accordingly for the user to watch.

We designed a new backtracking range query mechanism to speed up the node retrieval during query processing. A single path query consists of multiple sections, each of which corresponds to a range query on the R-tree index. These sections have tight continuity due to their geo-proximity which indicates that several consecutive range queries have a large probability to be contained in the same MBR of a non-leaf node. A traditional R-tree range query travels from the root top-to-down until an overlapping MBR is found or no records remain in the index. Hence we cache a trace file in main memory of the traversal path for the last query. When a new query arrives, the check will start from the leaf node in the trace file and step back to the parent level by level if no match is found, until a parent node's MBR has overlap with the current query. Then a traditional range query is conducted from this node top-to-down until another leaf node is reached. For example, in Fig. 3, $A \rightarrow, \ldots, B \rightarrow C$ is the trace of the last query, so if the next query targets leaf D, the process only needs to step back to B, instead of starting a new search from A.

6   Experimental Evaluation

6.1   Offline Video Summarization

Data Set. To evaluate the performance of our proposed framework for multi-video summarization for a desired section or a path composed of multi-sections, we conducted experiments on a real data set of 72 videos. These videos were recorded from December 19, 2010 to March 28, 2011 in the downtown area of Singapore and mainly four local attractions (the Merlion statue, the Marina Bay Sands (MBS), the Esplanade, and the Singapore Flyer) are captured. These videos were acquired with Samsung Galaxy S and Motorola Droid smartphones running Android 2.2 with resolutions post-processed to be 720×480 pixels, and their frame rate is 24 or 30 frames per second. The minimum sampling rate for the location and orientation information is around 5 samples per second.

Offline Summarization. OpenStreetMap is taken as an external GIS database which includes all roads' information. We divided all paths that have videos associated into totally 96 sections with average path length around 100 meters per section. The summary is conducted for each of them. To evaluate our strategy, we compare our results (DVS) with both the popular K-Means clustering and the Video-MMR solutions [6] for four randomly chosen sections. As both these two solutions work for only key-frame selection, we uniformly select a set of frame from segments candidate and take them as our algorithm input. We only use the video track for analysis as audio in these UGVs usually come from environmental noises. User-made summaries are collected from 12 people, each of which selects 5 to 10 keyframes from each frame set. User-selected frames $B = b_i, i = 1$ to $|B|$ are taken and a baseline and summarization quality is formulated to be the visual similarities with it $Q = \frac{1}{|B|} \Sigma_{i=1}^{|B|} \max_{f \in S, b \in B} sim(f, b_i)$. This method extracts SIFT features and their descriptors from each video frame, converts them to word of bag using K-Means clustering and the final similarity is the cosine distance between their visual word histograms.

Figure. 4 illustrates similarity quality scores with different numbers of selected frames for a randomly selected frame set. The higher the score is, the better the results are. As more frames are included in the summary, our methods presents smaller difference with the manually selected results.

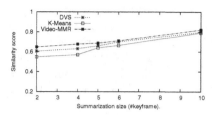

Fig. 4. Summarization comparison

Table 3a presents the scores with same number of selected frames for four randomly chosen sets. We find that our strategy produces a reasonably comparable result, 3.7% difference with best of the three solutions on average. However, it greatly reduces the computation workload as shown in Table 3b. The processing time of our strategy between two frames is 67 times faster than Video-MMR and computation increases intensively when more frames are included.

Table 2. Comparison of (a) summarization quality scores and (b) total processing time (seconds) among three different summarization strategies

	Set 1	Set 2	Set 3	Set 4
DVS	0.58	0.65	0.64	0.76
KMeans	0.50	0.59	0.74	0.72
V-MMR	0.66	0.65	0.68	0.75

(a)

#frames	2	4	8	16
DVS	0.13	0.52	1.37	3.88
KMeans	3.77	18.39	64.01	470.44
V-MMR	9.43	37.88	201.4	1458

(b)

6.2 Online Query

Simulation and Data Setup. To test the performance of our query system, we prepare a simulation to obtain the average query time and the results when query updates. We constructed a local MySQL database and stored all the FOV meta-data and the index tables. All the experiments were conducted on a PC with 2 Quad Intel® Core™ Q9550 2.83 GHz CPUs and 3 GB of RAM running Mircosoft Windows XP Professional. A geo-region around 20.42 km × 21.89 km of the downtown area of Singapore is chosen, which contains 3,705 route sections with total length 199.57 km. The path finding implementation is from Marcus Wolschon [16]. For each route section, we randomly generated at most 50 different video trajectories with average FOV sampling rate around 5/meter (only the meta-data of longitude and latitude). In total, 94,350 videos with 12.97 million FOVs were produced. There exist totally 30 landmarks in this geo-region and we assigned a random popularity score for each and allocated a random number of landmarks to each section. The R-tree index was constructed with the implementation of Greg Douglas [2]. For the R-tree, each parent had at most eight children with a loading factor of 50%. Both the page size and the buffer size were set to 4,096. The total construction time for a four-level R-tree based index was around 106 milliseconds.

Performance Results and Discussion. For each query, the processing time T includes three parts. (1) The query build time (T_1) includes finding the nearest query points on the road network and path finding between the start and the destination. (2) The summary decision time (T_2) denotes the profile retrieval time for all routes on a path and analysis of the time distribution. (3) The

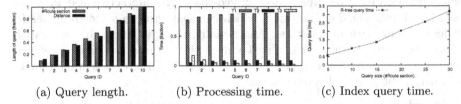

| (a) Query length. | (b) Processing time. | (c) Index query time. |

Fig. 5. Online processing time analysis

summarization schedule time (T_3) represents the time to compute the allocated video segments for each route section.

We randomly generated a pair of start and end points as an initial query and took its path as baseline (98 route sections, 4.7 km). Then we gradually increased the size of the query region by decreasing/increasing the min/max point of the base path with step 0.0001 degrees (110 m). Ten queries are selected to illustrate the experiments and the comparison of their path length is presented in Fig. 5(a), both in the number of sections and the distance in meters. Each bar-cluster corresponds to one query, the x-axis is the query ID and the values of the y-axis are fractional values between the length of a query and the largest one, *e.g.*, 0.8 means the path length is 0.8 times the 10th query. In the simulation, we randomly allocated at most five landmarks for each section as we observed that a single section is relevant to few local attractions.

Table 3. Simulation time

Name	Time ($ms/section$)	Percentage (%)
T1	26.7	63.2
T2	0.57	1.3
T3	15	35.5
total	42.27	100

The three components in the whole processing time are illustrated in Fig. 5(b), with the y-axis representing the fractional value between each time component and the total service time, $\frac{T_i}{\sum_{i=1}^{3} T_i}, i = 1, \ldots, 3$. As seen from the figure, the major service time for each query is spent on path finding. T_3 is generally steady due to the fact that the summary scheduler selects the most essential sections, so the overall IO time to read the profiles will not be significant. Fig 5(c) shows the increasing trend of T_2 for six queries, with total section number from 5 to 30. The time is linearly increasing as more sections are included. However, usually a meaningful tour path is within a limited geo-region, so it is less possible for hundreds of sections to be taken in, leading the overall query time to remain at a level of second. This time can be reduced by storing multiple R-tree nodes in a single access page in our future work. Processing time for each component are listed in Table 3. The average processing time is a fraction of seconds which is a good foundation for real-time and interactive queries.

7   Conclusions

We propose novel techniques that fully leverages sensor-equipped smartphones and geo-information database to automatically generate user specified summarization from multiple videos in geo-space. An interactive system is prepared with efficient data indexing, which can fast retrieve a video answer for a desired tour path in real-time. Additionally, the system can quickly satisfy query update to produce personalized and satisfied results, as well as an elegant adaption when new videos are included in database. All processes are automatic and with light-weighted computation.

Acknowledgements. This research has been supported by the Singapore National Research Foundation under its International Research Centre @ Singapore Funding Initiative and administered by the IDM Programme Office.

References

1. Cornuejols, G., Fisher, M.L., Nemhauser, G.L.: Location of bank accounts to optimize float: an analytic study of exact and approximate algorithms. Management Science (1977)
2. Douglas, G.: R-Tree, Templated C++ Implementation (2010),
   `http://superliminal.com/sources/RTreeTemplate.zip`
3. Graham, C.H., Bartlett, N.R., Brown, J.L., Hsia, Y., Mueller, C.C., Riggs, L.A.: Vision and Visual Perception. John Wiley & Sons, Inc. (1965)
4. Jing, F., Zhang, L., Ma, W.Y.: Virtualtour: an online travel assistant based on high quality images. In: 14th ACM MM (2006)
5. Judd, T., Ehinger, K., Durand, F., Torralba, A.: Learning to predictwhere humans look. In: 11th IEEE ICCV (2009)
6. Li, Y., Merialdo, B.: Multi-video summarization based on Video-MMR. In: WIAMIS (2010)
7. Li, Y., Merialdo, B.: Video Summarization Based on Balanced AV-MMR. In: Schoeffmann, K., Merialdo, B., Hauptmann, A.G., Ngo, C.-W., Andreopoulos, Y., Breiteneder, C. (eds.) MMM 2012. LNCS, vol. 7131, pp. 370–382. Springer, Heidelberg (2012)
8. Lu, X., Wang, C., Yang, J.M., Pang, Y., Zhang, L.: Photo2trip: generating travel routes from geo-tagged photos for trip planning. In: 18th ACM MM (2010)
9. Pongnumkul, S., Wang, J., Cohen, M.: Creating map-based storyboards for browsing tour videos. In: 21st ACM UIST (2008)
10. Shao, J., Jiang, D., Wang, M., Chen, H., Yao, L.: Multi-video summarization using complex graph clustering and mining. In: ComSIS (2010)
11. Shen, Z., Arslan Ay, S., Kim, S.H., Zimmermann, R.: Automatic tag generation and ranking for sensor-rich outdoor videos. In: ACM Multimedia (2011)
12. Snavely, N., Garg, R., Seitz, S.M., Szeliski, R.: Finding paths through the world's photos. In: ACM TOG (2008)
13. Snavely, N., Seitz, S.M., Szeliski, R.: Photo tourism: exploring photo collections in 3D. In: ACM TOG (2006)
14. Toyama, K., Logan, R., Roseway, A.: Geographic location tags on digital images. In: 11th ACM MM (2003)
15. Wang, F., Merialdo, B.: Multi-document video summarization. In: IEEE ICME (2009)
16. Wolschon, M.: Traveling salesman API,
    `http://wiki.openstreetmap.org/wiki/Traveling_salesman`
17. Zhang, Y., Zimmermann, R.: DVS: A dynamic multi-video summarization system of sensor-rich videos in geo-space. In: 20th ACM MM (2012)

Towards Automatic Music Performance Comparison with the Multiple Sequence Alignment Technique

Chih-Chin Liu

Department of Bioinformatics, Chung Hua University, Hsin-Chu City, Taiwan
ccliu@chu.edu.tw

Abstract. In this paper, we propose an approach towards automatic music performance comparison based on the multiple sequence alignment technique. In this approach, the onset detection technique is first applied to the multi-version recordings of the same music work. The signal between two adjacent onsets is represented with its corresponding chroma feature vector and symbolized as a chroma symbol. Thus a piece of music signal can be transformed into its associated chroma string. The progressive multiple sequence alignment technique is applied to these chroma strings to find a global alignment for multiple performances. After these chroma strings are aligned, dynamics and tempo comparisons among the multi-version performances can be carried out in various scale such as a note, a phrase, or the whole song. Nine versions of CD recordings on Sonatas and Partitas for Violin Solo, composed by Johann Sebastian Bach, are selected as the data set for the experiments. A phynogenetic tree for the nine performances can be automatically generated based on the distance matrix of their aligned chroma strings.

Keywords: Progressive multiple sequence alignment, chroma strings, music performance comparison, music interpretation, content-based music analysis.

1 Introduction

The analysis of music expression is one of the most important topics in traditional music research. In recent years, with the advancement in content-based analysis of multimedia data, it has become a new research issue to apply information technology to analysis of music expression.

Traditionally, the music signals are aligned parewisely, which limits the scale of performace comparisons. While the techniques to align a set of DNA or protein sequences have been highly developed in the field of molecular biology, these techniques can be employed to align multiple performances of the same music works. Also, the notion of the phynogenetic tree is a valuable tool to provide a global landscape for the music performances to be compared. The techniques presented in this paper can be used in the following music applications.

- Comparative analysis of musical works: Through comparative and computational analysis of the performances of the same musical works, we can discover the

S. Li et al. (Eds.): MMM 2013, Part I, LNCS 7732, pp. 391–402, 2013.

variations in timing, tempo, dynamics, harmony, articulation, and timbre, among these performances.

- Understanding music interpretation: To illustrate the characteristics of a specific performance, a global alignment can provide many quantitative descriptors about the performer in dealing with each piece of music, such as tempo, dynamics, harmony, and musical interpretation of the tension and emotion.
- Music education: In addition to the basic playing skills a music student can learn, by quantitatively analyzing the differences between the expressive factors performed by teachers and students, students can understand how to practically improve their performances to satisfy the teacher's requirements on musical interpretation.

The rest of this paper is organized as follows. Related works on sequence alignments are reviewed in Section 2. The overall architecture of the proposed music performance analyzing system is explained in Section 3. Section 4 discusses the techniques to segment a piece of music signal. The semantics and the computation of the chroma features, and the chroma string representation for music performances are explained in Section 5. The progressive multiple sequence alignment algorithm with necessary modifications for the chroma strings is explained in detail in Section 6. In Section 7, experiments are performed to show the effectiveness of the techniques proposed. Finally, Section 8 concludes this paper and presents our future research directions.

2   Related Works

The related works on sequence alignments can be largely divided into two categories: music pairwise alignment (in the music analysis field) and multiple sequence alignment (in the bioinformatics field), as follows.

Music alignment is the basic technique required in applications such as music information retrieval [12] and query-by-humming systems. Orio and Schwarz [14] defined the term music alignment as *"the association of events in a musical score (in our case, notes) with points in the time axis of an audio signal."* Basically, there are two approaches to align music signals with their scores, i.e., the *hidden Markov model (HMM)* [15] and the *dynamic time warping (DTW)* [11]. Early studies on score following try to align a music signal with its corresponding score. Later research showed that score following can be efficiently done by both the HMM [9][16] and the DTW [14] techniques. Raphael proposed an HMM-based model for segmenting a piece of music signal into a sequence of regions corresponding to the notes and rests in the score [16]. Both the melodies in a music database and the hummed queries are treated as time series. Therefore, the query-by-humming problem can be transformed into the similarity matching problem in time series databases. The index scheme proposed for time series database can thus be employed to solve the query-by-humming problem [9].

Early studies on the *multiple sequence alignment (MSA)* problem were largely done in the field of molecular biology. Since the MSA of a set of DNA or protein sequences may reveal the evolutionary history of these sequences, many research works were conducted on developing computational methods for MSA. In [13], based on the concept of *dynamic programming*, an approach known as *Needleman-Wunsch algorithm* was proposed to find a global alignment for the animo acid sequences of two proteins. Feng and Doolittle proposed the *progressive multiple sequence alignment algorithm* [8] by iteratively applying the Needleman-Wunsch algorithm for building pairwise alignments for every pair of multiple protein sequences. The distance matrix for the protein sequences can be found and a phylogenetic tree can be constructed to show the evolutionary relationships among these protein sequences.

To improve the sensibility of the progressive MSA algorithm, Thomson *et al.* proposed an enhanced dynamic programming alignment algorithm for finding the optimal alignment scores [20]. Based on this improvement and the idea of residue-specific gap penalties, a widely used open system called *Clustal W* was developed [21]. A heuristic was proposed by Carrillo and Lipman [4], which can find a close to the optimal solution of MSA.

3 System Overview

Figure 1 shows the overall architecture of the proposed music performance analyzing system. For a set of multi-versions of music performances with or without the score information, the onset of every pitch event is detected first. Every interval between two adjacent onsets, known as *IOI (inter-onset interval)*, is derived and acts as the basic unit for music performance comparison. The music signals of IOIs are symbolized into theirs corresponding chroma strings. The chroma strings of various versions of music performances are aligned with multiple sequence alignment technique. Based on the aligned results, a phylogenetic tree which shows the similarity among these versions of music performances can be derived.

For example, the right-top of Figure 1 shows the first measure of the Andante movement from Bach's Sonata No.2 for solo violin. There are eight pitch events in this measure. Assume the eight onsets for these pitch events are detected. Then, the chroma feature vectors of the music signals in this measure are computed and every IOI is transformed into its corresponding chroma symbol. In this example, the chroma string with respected to the eight IOIs is "E-E-C-E-F-E-D-C." Assume there are five versions of music performances for this piece of Bach's work, which are played by Enescu, Heifetz, Milstein, Mintz, and Szeryng. The corresponding chroma strings are automatically computed as "EECEFEDC", "CECCFCCC", "EECEFECC", "ECCCFEDC", and "CECCFCDC." The five chroma strings are aligned with the progressing sequence alignment technique, which will be explained in section 6. Based on the alignment result, a distance matrix which represents the similarities of all pairs of music performances is derived. Finally, a phylogenetic tree which visually shows the overall performance similarities is plotted. In this example, Milstein's version is more similar to that of Enescu, and Heiftz's version is more similar to that of Szeryng. The global evolutionary relationships among these five versions of music performances can be inferred and illustrated.

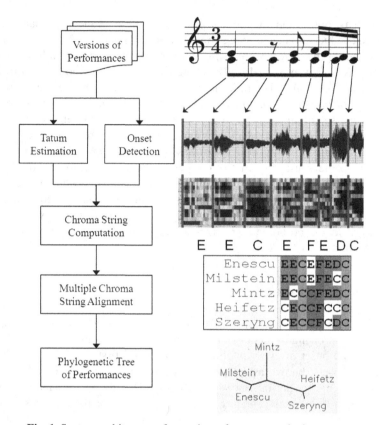

Fig. 1. System architecture of a music performance analyzing system

4   Segmentation of Music Signals

The music signal should be segmented into a sequence of pitch events before the alignment procedure can be carried on. In this section, two methods for segmenting music signals are provided.

The performances we studied are extracted from commercially available CD recordings. The frame size is set to 50 ms and the stepping time is set to 10 msec. Hamming window is used. The features we used for onset detection are F_1 = spectral flux, F_2 = phase deviation, and F_3 = weighted sum of spectral powers. These onset features are added with weights for continuous k frames.

$$F_{i,k}(t) = \sum_{j=1}^{k} \alpha_j \left(F_i(t+j) - F_i(t+j-1) \right) \tag{1}$$

Then peaks $F_{i,k}(t)$ which are a local maximum within $\pm m$ frames (m is set to 5) and exceed the threshold P_i are picked as the candidate onsets.

$$Peaks\left(F_{i,k}(t),m,P_i\right)=\begin{cases}1,F_{i,k}(t)>P_i\\0\end{cases} \qquad (2)$$

The global peaks are picked by the weighted sum of these local peaks.

$$Onset(t)=\sum_{i=1}^{3}\beta_i Peaks\left(F_{i,k}(t),m,P_i\right) \qquad (3)$$

The performance of the onset detection technique is affected by the instruments (pitched/non-pitched, percussion/non-percussion) and the music genres. Typically, the actual onsets detected are more than we expected, which decreases the quality of alignment. Therefore, another segmentation strategy is proposed based on the notion of metrical structure. The metrical structure of music consists of a hierarchy of points in time. The basic units that make up a metrical hierarchy are *beats* (or *tactus*), which are often referred to as foot tapping upon hearing a song. The grouping of strong and weak beats forms a higher level of unit called *meter*. For lower level of metrical unit, Bilmes defined the notion of *tatum* as: "When we listen to or perform music, we often perceive a high frequency pulse, frequently a binary, trinary, or quaternary subdivision of the musical tactus (the stroke of the hand or baton in conducting)." This pulse is termed tatum [3].

Although theoretically the value of tatum in the whole song should remain stable, humans are not machines. Even if the tempo markings such as Accelerando and Ritardando are absent from a score, typically the tatum values in a real performance are normally distributed. Similarly, the time intervals of all beats in a real performance are normally distributed. Therefore, the distribution of all IOIs I of a given piece of music can be fitted with a one-dimensional *Gaussian mixture model (GMM)* of k-components.

$$hist(I)\approx GMM(t)=\sum_{i=1}^{k}w_i N(t\mid\mu_i,\sigma_i) \qquad (4)$$

where w_i and $N(x\mid\mu_i,\sigma_i)$ are the weight and the i-th component with mean μ_i and standard deviation σ_i, respectively. The number of components and the parameters of these components can be estimated by using the *Expectation Maximization (EM) algorithm*. The *Akaike Information Criterion (AIC)* [1] is used in the estimating procedure. Since the tatum is defined as the smallest metrical unit of music, the time interval of a beat should be an integer multiple of that of a tatum. We can utilize this characteristic in discovering the tatum value and the beat value of a given music performance. Assume there are N components in an estimated GMM for fitting the distribution of a set of IOIs in the performance. The value of the tatum μ_{tatum} can be estimated by

$$\underset{\mu_{tatum}}{\arg\min}\left[T_{error}=\sum_{i=1}^{N}\left|\underset{k\in\{1,2,3,4,6,8\}}{\min}(\mu_i-k\mu_{tatum})\right|\right] \qquad (5)$$

where μ_i is the mean of the i-th component; $k \in \{1, 2, 3, 4, 6, 8\}$ is the common integer ratio of μ_{beat} to μ_{tatum}. After the tatum value of a piece of music is estimated, it can be segmented into a sequence of fixed time intervals $\Delta T = k\mu_{tatum}$, where $k = 1, 2, 1/2,$ or $1/4$.

5   Chroma String Representation

Early studies in music psychology suggested that the perception of pitch should be better represented by a two-dimensional model [6][17][19], rather than by a unidimensional model. The first dimension of the two-dimensional model is called the *tone height*, and the other dimension is called the *chroma*. As suggested by [6][17], the two-dimensional pitch model could be visualized as a spiral line, called the *pitch spiral*. The vertical and angular dimensions of the pitch spiral represent the tone height and the chroma, respectively. For example, since *C1*, *C2*, ..., and *C8* have the same chroma, all *Cs* lie at $0°$ azimuth. However, the tone heights of all *Cs* are different.

Bartsch and Wakefield [2] formulated the pitch spiral as

$$p = 2^{c+h} \tag{6}$$

where $c \in [0,1]$ and $h \in Z$ denote the chroma and the tone height of the pitch p (in Hz) of a given tone, respectively. The chroma c of the given tone can thus be calculated using

$$c = \log_2 f - \lfloor \log_2 f \rfloor \tag{7}$$

The domain of the chroma is typically divided into 12 equal parts which can be associated with the 12 *pitch classes {C, C#, D, D#, E, F, F#, G, G#, A, A#, B}* in the equal-tempered chromatic scale. In many music applications, the music signal can be divided into 88 frequency bands with semi-tone bandwidth, which correspond to the standard 88 keys (A0~C8, 27.5~4186 Hz) on a piano. For a given short-time music signal, the spectral energy for each of the 88 frequency bands is computed. The spectral energy of the subbands corresponding to the same pitch class, e.g., C1, C2, ..., and C8, is added up to form a 12-dimensional vector $\vec{v} = (v_1, v_2, \cdots, v_{12})$, called the *chroma feature vector*

$$v_c = \sum_i S(2^{\frac{c-1}{12}+i} * 16.35), \quad c = 1, 2, \cdots, 12 \tag{8}$$

where $S(f)$ is the spectral energy of the frequency band corresponding to pitch f with the i-th octave and the pitch class associated with c.

The chroma feature vectors were proved to perform very well in many music analyzing tasks, such as audio thumbnailing [2], music retrieval [9], and audio synchronization [12]. Since the chroma feature is a compact representation for both melodic and harmonic information, it can also capture the variances among multi-versions of music performances. We apply the chroma feature to automatic music performance comparison. Before the performance comparison can be processed, the chroma feature vectors corresponding to the different versions of music performances must be aligned first. The alignments of chroma feature vectors can be done using signal processing techniques or symbolic sequence matching algorithms. In this paper, the multiple sequence alignment method is applied. Therefore, the chroma feature vectors associated with a piece of music signal should be symbolized. We defined a *chroma string* S of length n to be a symbolic representation for a given piece of music signal with n pitch events performed by a certain musician.

$$S = s_1 s_2 \cdots s_n, \ s_i = \alpha(\vec{v}_i), \ s_i \in \Sigma \tag{9}$$

where $\vec{v}_i$ is the chroma feature vector of the i-th pitch event; s_i is the *chroma symbol* corresponding to $\vec{v}_i$; Σ is the *alphabet* for all chroma symbols; and α is a *symbolizing function* which maps a chroma feature vector into its corresponding chroma symbol. In this paper, the *alphabet of the prominent pitch strings*, Σ_{pitch}, is defined to be the twelve pitch classes.

$$\Sigma_{pitch} = \{C, \ C\#, \ D, \ D\#, \ E, \ F, \ F\#, \ G, \ G\#, \ A, \ A\#, \ B\} \tag{10}$$

For a given chroma feature vector $\vec{v} = (v_1, v_2, \cdots, v_{12})$, the *symbolizing function for a prominent pitch strings* $\alpha_{pitch} : R^{12} \rightarrow \Sigma_{pitch}$ is defined to be

$$\alpha_{pitch}(\vec{v}) = s_i \in \Sigma_{pitch}, \ i = \underset{i=1,2,\cdots,12}{\arg \max}(v_i) \tag{11}$$

6   Multiple Sequence Alignment

After the signals of the multi-versions of music performances are symbolized into their corresponding chroma strings, multiple sequence matching techniques can be applied to align these strings for drawing a phylogenetic tree to show the *kinships* (or *evolutionary relationships*) among these performances and for further comparison. In this paper, the progressive multiple sequence alignment algorithm [8] proposed by Feng and Doolittle in the field of bioinformatics is employed. The procedure of the algorithm with necessary modifications for aligning chroma strings is explained as follows.

Step 1. Pairwise alignment of the Chroma Strings: Given a set of chroma strings $S=\{s_1, s_2, ..., s_n\}$, the *distance matrix* D for all pair of chroma strings is calculated using *Needleman-Wunsch algorithm* [13]. For two chroma string $s_A = \alpha_1 \alpha_2 \cdots \alpha_n$ and $s_B = \beta_1 \beta_2 \cdots \beta_m$, the score $C_{i,j}$ when the alignment is partially done at position i in string s_A and position j in string s_B can be computed recursively

$$C_{i,j} = \max \begin{pmatrix} C_{i-1,j-1} + M(\alpha_i\beta_j), \\ \max_{x\geq 1}(C_{i-x,j} - \varepsilon_x), \\ \max_{y\geq 1}(C_{i,j-y} - \varepsilon_y) \end{pmatrix} \tag{12}$$

where $M(\alpha_i\beta_j)$ is the corresponding score in the *similarity matrix* (also called the *substitution matrix* or the *score matrix*) for matching the two chroma symbols α_i and β_j; ε_x is the insertion cost; ε_y is the deletion cost. And the distance between s_A and s_B will be $d_{AB} = C_{n,m}$. The similarity function can be a simple editing distance function or more complicated tonal distance functions.

Step 2. Generating the Guide Tree: According to the distance matrix D for all pair of chroma strings, we can build an unrooted, bifurcating tree, called the phylogenetic tree, with branch lengths proportional to performance divergence along each branch. Based on the goal to minimize the total branch lengths in the phylogenetic tree, the *neighbor-join method* [18] can be applied. Among all pairs of chroma strings $S=\{s_1, s_2, \ldots, s_n\}$, the one that gives the minimum sum of branch lengths C_{ij} in the phylogenetic tree is chosen and joined. The two chroma strings "*joined*" is then regarded as one chroma string to proceed. The neighbor-join procedure is done repeatedly until all chroma strings are joined. After the unrooted tree is generated, the corresponding *guide tree* can be derived by place the root at the "mid-point" position of the unrooted tree which the branch lengths are equal in either branches of the root [20].

Step 3. Progressive Alignment: According to the branching information specified in the guide tree, a series of pairwise chroma string alignments are performed in a bottom-up manner, *i.e.*, from the most closely related groups to most distantly related groups. To align a chroma string with a group of aligned chroma string, each gap (insertion error or deletion error) in the aligned chroma string is replaced with a neutral symbol X with the highest similarity score. The score for matching two aligned symbols is the average of all pairwise scores.

Step 4. Drawing the phylogenetic tree: After the progressive alignment is done, a new distance matrix for all pair of chroma strings can be derived. The distances in the guide tree are revised to reflect the global alignment result. The final version of the phylogenetic tree can be drawn.

7   Experiments

Following the data sets used in [5], the Andante (third) movement from Bach's Sonata No. 2 in A minor BWV 1003 for solo violin is selected due to its regular pulse. We have collected nine performances for this classical masterpiece from commercially available CD recordings. The nine versions were interpreted by Enescu, Grumiaux, Heifetz, Menuhin, Milstein (1956, 1975), Mintz, Szeryng, and Szigeti.

The score and phrasing information for this Sonata are available in [5]. A reference version is generated from the MIDI sequence downloaded from the Mutopia Project (http://www.mutopiaproject.org). The source code of Clustal-W [21] is modified by disabling all gap penalty settings and replacing the alphabet and the similarity matrix for the chroma strings.

The first experiment illustrates the effectiveness of the fixed interval approach for music segmentation described in section 4. According to the experiment results, all Gaussian distributions of tatum values are correctly estimated by equation (5) for the 10 versions of Bach's Sonata. However, since the signals of the ten performances are segmented into a sequence of fixed time intervals $\Delta T = k\mu_{tatum}$, where $k = 1, 2, 1/2$, or 1/4, there may exist an *onset error* Δt_{onset} between every manually labeled onset point and the boundary of the nearest fixed time interval to the onset. We measure the onset error Δt_{onset} for every manually labeled onset point in the 10 versions with various ΔT settings. The average values of the onset errors are shown in Table 1. The average onset error decreases if smaller fixed time interval ΔT is set. If the error tolerance of a onset point is set to 50 msec, which is suggested in many onset detecting studies, $\Delta T = \dfrac{1}{4}\mu_{tatum}$ will be a feasible setting.

Table 1. The average onset error Δt_{onset} with various ΔT settings

Performer	Δt_{onset}, when $\Delta T =$			
	$2\mu_{tatum}$	μ_{tatum}	$\dfrac{1}{2}\mu_{tatum}$	$\dfrac{1}{4}\mu_{tatum}$
Enescu	0.399	0.192	0.107	0.052
Gramiaux	0.353	0.141	0.072	0.042
Heifetz	0.416	0.197	0.098	0.050
Menuhin	0.386	0.205	0.104	0.048
Milstein 56	0.457	0.244	0.107	0.050
Milstein 75	0.288	0.141	0.079	0.041
Mintz	0.429	0.210	0.104	0.053
Szeryng	0.418	0.217	0.103	0.049
Szigeti	0.416	0.198	0.105	0.055
Average	0.396	0.194	0.098	0.049

The second experiment compares the alignment accuracy of the two segmentation approaches for generating the chroma strings. We measure the error between every manually labeled onset point to the start point of the aligned time interval that contains this onset point pluses the onset error Δt_{onset}, the sum is the *onset alignment error* after the sequence contains this onset is aligned, which is denoted as $\Delta t'_{onset}$. By adding all $\Delta t'_{onset}$ of every onset point in a music phrase, we get the *phrase alignment error* Δt_{phrase} for this phrase. The average phrase alignment errors for the 10 performances are shown in Table 2. In general, the tatum method performs better than the onset method in music segmentation for alignment.

The third experiment illustrates the overall results of the multiple chroma string alignment for the 10 performances. Based on the average distance matrix of the 10 performances, a phylogenetic tree can be plotted to illustrate the overall similarity and the inferred evolutionary relationships among these performances. The PHYLIP package [7] is used to draw the phylogenetic tree corresponding to the average phrase distance matrix of the 10 performances. As shown in Figure 2, similar pairs of performances are (Milstein 55, Milstein 75), (Gramiaux, Mintz), and (Szigeti, Heifetz).

Table 2. The average phrase alignment errors for the 10 performances

Performer	Average Phrase Alignment Error (sec)	
	Onset Method	Tatum Method
Score	1.233	0.854
Enescu	0.613	1.398
Gramiaux	1.101	0.868
Heifetz	2.566	0.875
Menuhin	1.432	0.753
Milstein 56	3.512	1.491
Milstein 75	2.509	0.720
Mintz	0.503	0.961
Szeryng	1.639	1.050
Szigeti	1.233	0.854
Average	1.679	0.997

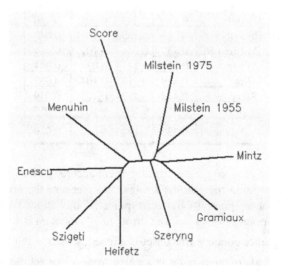

Fig. 2. The phylogenetic tree for the 10 versions of Bach's Sonata for Violin No. 2

8   Conclusion

The chroma strings are used to represent the tonal information in music performances. The progressive multiple sequence alignment algorithm is applied on these strings for performance comparison. In the future, three research issues will be addressed. First, other possible symbolic representations such as dynamics strings and tempo strings for performance comparison will be studied. Second, the index schemes proposed in time series databases can be applied to improve the performance of alignment. Finally, issues about the music performance similarity matrix will be investigated.

Acknowledgments. This work was partially supported by the National Science Council under Contract No. NSC 101-2221-E-216-032.

References

1. Akaike, H.: A New Look at the Statistical Model Identification. IEEE Trans. on Automatic Control 19(6), 716–723 (1974)
2. Bartsch, M.A., Wakefield, G.H.: Audio Thumbnailing of Popular Music Using Chroma-Based Representations. IEEE Trans. on Multimedia 7(1), 96–104 (2005)
3. Bilmes, J.A.: Techniques to Foster Drum Machine Expressivity. In: Proc. Int. Comp. Music Conf., pp. 276–283 (1993)
4. Carrillo, H., Lipman, D.: The Multiple Sequence Alignment Problem in Biology. SIAM Journal on Applied Mathematics 48(5), 1073–1082 (1988)
5. Cheng, E., Chew, E.: Quantitative Analysis of Phrasing Strategies in Expressive Performance: Computational Methods and Analysis of Performances of Unaccompanied Bach for Solo Violin. Journal of New Music Research 37(4), 325–338 (2008)
6. Deutsch, D.: Music Recognition. Psychological Review 76(3), 300–307 (1969)
7. Felsenstein, J.: PHYLIP - Phylogeny Inference Package. Version 3.2, Cladistics 5(2), 163–166 (1989)
8. Feng, D.-F., Doolittle, R.F.: Progressive Sequence Alignment as a Prerequisiteto Correct Phylogenetic Trees. Journal of Molecular Evolution 25(4), 351–360 (1987)
9. Hu, N., Dannenberg, R.B., Tzanetakis, G.: Polyphonic Audio Matching and Alignment for Music Retrieval. In: Proc. IEEE Workshop on Applications of Signal Processing to Audio and Acoustics, pp. 185–188 (2003)
10. Lipman, D.J., Altschul, S.F., Kececioglu, J.D.: A Tool for Multiple Sequence Alignment. Proc. Nail. Acad. Sci. 86, 4412–4415 (1989)
11. Myers, C., Rabiner, L., Rosenberg, A.: Performance Tradeoffs in Dynamic Time Warping Algorithms for Isolated Word Recognition. IEEE Trans. on Acoustics, Speech, and Signal Processing 28(6), 623–635 (1980)
12. Müller, M.: Information Retrieval for Music and Motion, 2nd edn. Springer Publisher (2010)
13. Needleman, S.B., Wunsch, C.D.: A General Method Applicable to the Search for Similarities in the Amino Acid Sequence of Two Proteins. Journal of Molecular Biology 48(3), 443–453 (1970)
14. Orio, N., Schwarz, D.: Alignment of Monophonic and Polyphonic Music to a Score. In: Proc. of the Intl. Computer Music Conf., pp. 129–132 (2001)

15. Rabiner, L.: A Tutorial on Hidden Markov Models and Selected Applications in Speech Recognition. Proc. IEEE 77(2), 257–286 (1989)
16. Raphael, C.: Automatic Segmentation of Acoustic Musical Signals Using Hidden Markov Models. IEEE Trans. on Pattern Analysis and Machine Intelligence 21(4), 360–370 (1999)
17. Revesz, G.: An Introduction to the Psychology of Music. University of Oklahoma Press, Norman (1954)
18. Saitou, N., Nei, M.: The Neighbor-Joining Method: A New Method for Reconstructing Phylogenetic Trees. Mol. Biol. Evol. 4(4), 406–425 (1987)
19. Shepard, R.N.: Circularity in Judgments of Relative Pitch. J. Acoust. Soc. Amer. 36, 2346–2353 (1964)
20. Thompson, J.D.: Introducing Variable Gap Penalties to Sequence Alignment in Linear Space. CABIOS 11(2), 181–186 (1995)
21. Thompson, J.D., Higgins, D.G., Gibson, T.J.: CLUSTAL W: Improving the Sensitivity of Progressive Multiple Sequence Alignment through Sequence Weighting, Position-Specific Gap Penalties and Weight Matrix Choice. Nucleic Acids Research 22, 4673–4680 (1994)

Multi-frame Super Resolution Using Refined Exploration of Extensive Self-examples*

Wei Bai, Jiaying Liu, Mading Li, and Zongming Guo**

Institute of Computer Science and Technology, Peking University,
Beijing, P.R. China 100871

Abstract. The multi-frame super resolution (SR) problem is to generate high resolution (HR) images by referring to a sequence of low resolution (LR) images. However, traditional multi-frame SR methods fail to take full advantage of the redundancy in LR images. In this paper, we present a novel algorithm using a refined example-based SR framework to cope with this problem. The refined framework includes two innovative points. First, based upon a thorough study of multi-frame and single frame statistics, we extend the single frame example-based scheme to multi-frame. Instead of training an external dictionary, we search for examples in the image pyramids of the LR inputs, *i.e.*, a set of multi-resolution images derived from the input LRs. Second, we propose a new metric to find similar image patches, which not only considers the intensity and structure features of a patch but also adaptively balances between these two parts. With the refined framework, we are able to make the utmost of the redundancy in LR images to facilitate the SR process. As can be seen from the experiments, it is efficient in preserving structural features. Experimental results also show that our algorithm outperforms state-of-the-art methods on test sequences, achieving the average PSNR gain by up to 1.2dB.

1 Introduction

The super resolution process of a LR image is inverse to the LR imaging process, during which partial high-frequency information is lost. And for the SR problem, such loss leads to non-unique solutions, making the problem ill-posed. Multi-frame SR methods make use of their redundancy in LR information to constrain the solution space. The problem is to reconstruct a HR image from a sequence of low-resolution images. In the literature, these LR images can either be images acquired from the same scene but slightly differ in viewing angles, or a sequence of consecutive video frames. A key point in the former situation is that the LR images should be sub-pixel aligned. Integer pixel alignment is meaningless because that supplies the same amount of information as the single image. In consecutive video frames, information not included in one frame may

---

* This work was supported by National Natural Science Foundation of China under contract No.61071082, National Basic Research Program (973 Program) of China under contract No.2009CB320907 and Doctoral Fund of Ministry of Education of China under contract No.20110001120117.
** Corresponding author.

S. Li et al. (Eds.): MMM 2013, Part I, LNCS 7732, pp. 403–413, 2013.

appear in adjacent frames. In a word, multi-frame SR approaches exploit redundancy in input LRs for extra information and exchange temporal resolution for spatial resolution.

As current SR methods become more and more sophisticated, people demand more details restored from the LR input. Details are high frequency information. To achieve this, contemporary SR methods either apply pre-defined priori knowledge or refer to a learned dictionary for external information. For example, Tsai and Huang [1] proposed the frequency domain method by transforming the LR image data into the discrete Fourier transform (DFT) domain. Here the relationship between DFT coefficients of LR and HR images can be considered as priori knowledge. Farsiu *et al.* [2] also proposed an l_1-norm regularization method based on a bilateral total variation. However, the performance of these methods degrades rapidly when applied with a large magnification factor, which constrains their application. In recent years, example-based SR methods draw tremendous attention around the world because of their simplicity and potential to break the upper bound of magnification factor compared with the aforementioned algorithms. They usually depend on extra information provided by an image database. Freeman *et al.* [3] estimated high frequency details from a large training set of HR images that encode the relationship between HR and LR images. Glasner *et al.* verify the internal similarity in natural images statistically in [4], meaning that it is feasible to perform the example-based SR from a single image. However, these methods fail to take into consideration the structural features while exploiting examples for extra information.

The foundation of example-based SR is the recurrence of small image patches in different resolutions. These HR/LR pairs indicate how the input LR patches be super resolved. A typical example-based SR algorithm [4] is presented in the following steps:

– Let L be the LR image while H is the HR image. B is the blur kernel. Thus the imaging model is formulated as:

$$L_j(p) = H * B_j = \sum_{q_i \in S_{B_j}} H(q_i) \cdot B_j(q_i - q),\tag{1}$$

where p is a pixel in the LR image, corresponding to q in the HR image. And the HR pixels q_i belong to the support of kernel B, which centers at q.
– Down-sample the input image L to a cascade of scales, comprising an image pyramid with multiple resolutions.
– For each patch in L, search for similar patches in the above image pyramid. If the similar patches are in lower scales than the input scale, their parent patches are of higher resolution, thus providing examples for the upsampling of the LR patch. Similar patches of the same scale as the input scale also count. By fitting these parent patches or similar patches in the imaging model, the solution space of (1) is constrained.

In this scheme, apparently, it is very important to find plausible examples with enough information, especially when we are handling the multi-frame SR problem. Similar patches recur in multiple frames, more frequent than in a single frame according to the statistics, making example-based method a proper choice. Grounding on this, we propose a new metric to search for reliable examples whereas lower the search cost at the same time. Our method, REESE (refined exploration of extensive self-examples), takes into account of the structure feature and intensity feature of a local image patch and search its nearest neighbors in a multi-frame and multi-resolution image set, which is derived totally from the input images. With all the similar patches constraining the final HR result, we utilize numerical method to solve the least square problem. Experiments show the superiority of our method, which not only magnifies images efficiently but also preserves the structure details better.

The rest of this paper is organized as follows: Section 2 describes each part of the proposed algorithm in detail. Experimental results are shown in Section 3. Finally, concluding remarks are given in Section 4.

2 Proposed Multi-frame SR Algorithm

2.1 Motivation and Algorithm Framework Overview

According to the prior work that patches recur in a image pyramid, we can easily deduce that multiple frames can provide more similar patches. Fig.1 demonstrates this assumption by comparing the possibilities to find a certain number of similar patches in a single frame and multi-frame at different scales. The horizontal axis shows the number of similar patches found in a frame or frames while the vertical axis indicates the corresponding percentage of input patches. Thus we can come to the aforementioned conclusion that it's reasonable to introduce the self-similarity property to solve the multi-frame SR problem.

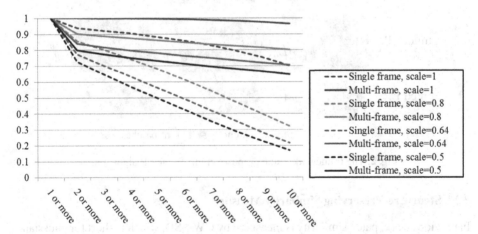

Fig. 1. Comparison of the amount of similar patches found in single frame and multi-frame, searched at different scales

Motivated by the existence of extensive examples in multiple frames, we propose an algorithm which follows the process illustrated in Fig.2. For the convenience of expression, the input LR images are represented as $\{..., L_0^{n-1}, L_0^n, L_0^{n+1}, ...\}$, and L_0^n is the to-be-magnified image. The down-sampled images are denoted as $\{..., L_{-k}^{n-1}, L_{-k}^{n}, L_{-k}^{n+1}, ...\}$, $k = 1, 2, 3, ...$, where the down-sampling scale is $\alpha^{-k}, \alpha > 1$. In order to get the corresponding HR image, L_k^n, we need to solve a least square problem. Generally, it can be presented as:

$$min(B * L_k^n \downarrow - L_0^n)^2, \qquad (2)$$

where $\downarrow$ denotes the down-sampling process.

At essence, the fundamental idea is for the patch at every pixel of L_0^n, we search for its similar patches in both $\{..., L_0^{n-1}, L_0^n, L_0^{n+1}, ...\}$ and $\{..., L_{-k}^{n-1}, L_{-k}^{n}, L_{-k}^{n+1}, ...\}$. Theoretically, a global search for similar patches needs to be done at each pixel in the to-be-super-resolved image, L_0^n. However, it is a huge amount of calculation to do a global search, which is very time-consuming. So we can take a pre-processing step before the search part. This method is elaborately described in [5], using a motion estimation process to relocate the search window, thus reducing search cost. Then the similar patches in $\{..., L_0^{n-1}, L_0^n, L_0^{n+1}, ...\}$ and the parent patches of those in $\{..., L_{-k}^{n-1}, L_{-k}^{n}, L_{-k}^{n+1}, ...\}$, as the blue line show in the diagram, will render constraints on the final HR image in (2), resulting in an optimal solution.

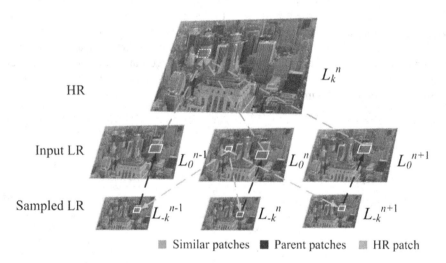

Fig. 2. The framework of the proposed multi-frame SR algorithm

2.2 Structure-Preserving Similarity Measure

In previous work, patch similarity is measured by GW-SSD, which is short for gaussian-weighted SSD (sum of squared difference). It assigns greater weight to pixels that are closer to the center of the patch. However, this is not enough to find the most structurally similar patches. GW-SSD just discriminates centering pixels from peripheral pixels, ignoring different structural features of each patch. Thus, we introduce a new similarity measure to define the distance between two patches. It takes both intensity similarity and structure similarity into consideration, corresponding to the two terms in the following formulation, respectively.

$$dist(P_{x_1,y_1}, P_{x_2,y_2}) = \lambda \cdot d_i(P_{x_1,y_1}, P_{x_2,y_2}) + (1 - \lambda) \cdot d_s(P_{x_1,y_1}, P_{x_2,y_2}), \quad (3)$$

where P_{x_1,y_1} and P_{x_2,y_2} are two patches centered at the pixel (x_1, y_1) and (x_2, y_2). $dist(P_{x_1,y_1}, P_{x_2,y_2})$ represents the distance between the two patches. d_i is the intensity term, while d_s is the structure term. λ is a parameter to weight the two terms which depends on the complexity of the image to be super resolved. The conventional GW-SSD is utilized to calculate the intensity distance:

$$d_i(P_{x_1,y_1}, P_{x_2,y_2}) = \sum_{x,y} G_\sigma \cdot (P_{x_1,y_1}(x,y) - P_{x_2,y_2}(x,y))^2, \quad (4)$$

where G_σ stands for the gaussian kernel, which is the same size as the patches.

As to the structure similarity measure, we utilize the covariance matrix to extract the local structure feature of the patches. A covariance matrix can reflect to what extent neighboring pixels rely on each other. In other words, it implicitly indicates the structure of a local area. The method of calculating the covariance matrix is stated below.

1. Let $P_{x,y}$ denote a patch centered at the point (x, y) in L_0^n. For each pixel in $P_{x,y}$, we build a sample vector V_i. Take point (x, y) (the 3-tagged pixel) for example, as shown in Fig.3, the pixels tagged from number 1 to 5 compose the elements of the sample vector. Note that, V_i is a 1×5 row vector.

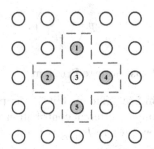

Fig. 3. Formation of sample vectors

2. Assume the patch size to be $N * N$, i.e. N^2 pixels. Their sample vectors are denoted as $V_i, i = 1, 2, 3, ..., N^2$, so the patch vector $V_{x,y}$ can be written as:

$$V_{x,y} = \begin{pmatrix} V_1 \\ V_2 \\ ... \\ V_{N^2} \end{pmatrix} \quad (5)$$

3. Then we calculate the self-covariance matrix of $P_{x,y}$'s sample vector $V_{x,y}$, represented as $C_{x,y}$. Finally the structure similarity measure is formulated as:

$$d_s(P_{x_1,y_1}, P_{x_2,y_2}) = \sum_{x,y} G_\sigma \cdot (C_{x_1,y_1}(x,y) - C_{x_2,y_2}(x,y))^2. \qquad (6)$$

After we get the intensity and the structure term, the weight parameter λ is applied as mentioned before. In practice, the distance changes as the value of λ varies, making a fixed λ inappropriate. Thus, considering the impact of parameter selection, we adaptively choose λ based upon the smoothness of an image patch. If the patch is not so smooth, *i.e.*, there are plenty of textures, the weight of the structure term should be magnified. In [6], the gradient strength in two perpendicular directions is used to measure the smoothness of local image patches.

$$S_i = \frac{1}{n_i} \sum_{(x,y) \in b_i} S(x,y) = \frac{1}{n_i} \sum_{(x,y) \in b_i} \nabla L(x,y) \cdot \nabla L(x,y)^T, \qquad (7)$$

In the above equation, n_i stands for the number of pixels in block b_i. Let $\lambda_1^{(i)}$ and $\lambda_2^{(i)}$ denote the eigenvalues of matrix S_i, which represent the gradient strength then the smoothness s_i of a block b_i is defined as:

$$s_i = |\lambda_1^{(i)}| + |\lambda_2^{(i)}|, \qquad (8)$$

Thus, the weight parameter λ can be formulated as follows with μ a constant.

$$\lambda = exp\left(-\frac{s_i}{2\mu^2}\right), \qquad (9)$$

2.3   Solving the Weighted Least Square Problem

We use the metric described in Sec.2.2 to exploit similar patches. These patches differ slightly in content, thus resulting in different contribution to the eventual HR patch. So we give them weights as follows, where P_{x_1,y_1} is the reference patch in L_0^n and σ is a parameter.

$$weight_{P_{x,y}} = G_\sigma * exp\left(-\frac{dist(P_{x_1,y_1}, P_{x,y})}{2\sigma^2}\right), \qquad (10)$$

Considering the difference between similar patches, the aforementioned least square problem is further extended to a weighted least square problem, shown as below:

$$min\frac{1}{2}(B * L_k^n \downarrow - L_0^n)^T W (B * L_k^n \downarrow - L_0^n), \qquad (11)$$

where W is a diagonal matrix, composed of each patch's weight calculated by eq.(10). Thus, we see that we have weighted constraints at the unknown pixels of the images in the higher resolution image L_k^n and we can use gradient descent or other numerical methods to solve the weighted least square problem.

3   Experimental Results

The experiments are performed using MATLAB R2010a on Intel Core CPU 2.4GHz Microsoft Windows platform. All the test image frames are blurred and decimated by a factor of 1:2 (in each axis), and then contaminated by an additive noise with standard deviation 2.

To demonstrate the validity of our proposed method, we first test on single images to perform single image SR. The size of the patch is set as 5*5, and the cascade of low-resolution images is simplified to a 1/2 size reduced one of the input image. In Table 1, we compare our result (S-REESE, single image REESE) with those obtained by Bicubic, [4], and [7], which are relatively state-of-the-art methods that can be found in the literature.

Specifically, since we utilize a similar self-similarity framework with the one in [4], we compare our results with Glasner's to verify that the proposed new metric for similarity does work. We zoom to see the details of HR images super resolved by Glasner's method and the proposed method, as Fig.5 shows. We see that by our algorithm the jaggy effect is reduced and the edges are clearer.

Fig.4 presents another group of results obtained by KR, Jurio's [8] and the proposed algorithm S-REESE. When we zoom in to see the details (best viewed on screen), for example, as presented in *Lena*, we can see that KR over-smoothes the image in contrary to Jurio's, which over-sharps the image even to render jaggies. Our method seeks to balance between smoothness and sharpness.

Then we implement the complete version of the proposed algorithm, *i.e.*, for the input video sequence, we enhance each frame's resolution with reference to its adjacent frames. The efficiency of the proposed multi-frame super resolution algorithm is evaluated both objectively and subjectively. In the objective part, we compare the PSNRs (average of 5 frames in the experiment) of different image sequences obtained by

Table 1. PSNR (dB) Comparison of Different Methods on Single Image Implementation

Images	Bicubic	KR [7]	Glasner's [4]	S-REESE (vs. Glasner's)
House	31.29	32.00	31.82	**32.85(+1.03)**
Cameraman	25.17	25.46	26.18	**26.36(+0.18)**
Elaine	32.20	31.89	31.92	**32.51(+0.59)**
Boat	28.80	29.16	29.16	**30.05(+0.90)**
Bridge	25.78	25.59	26.03	**26.69(+0.66)**
Car	29.58	30.01	29.99	**30.91(+0.92)**
Clock	28.81	29.36	29.78	**30.37(+0.59)**
Peppers	29.44	30.14	30.10	**31.17(+1.07)**
Ship	29.42	30.23	30.29	**31.04(+0.75)**
Window	21.74	21.69	22.62	**22.84(+0.22)**
Average	28.22	28.55	28.79	**29.48(+0.69)**

Table 2. Average PSNR (dB) Comparison of Test Sequences

Sequences	Bicubic	3DKR	S-REESE	REESE
Ice	30.58	31.12	33.24	**33.43**
Soccer	27.85	28.25	29.17	**29.26**
Harbour	23.35	23.86	24.63	**24.70**
City	26.82	27.43	27.87	**27.90**
Foreman	31.36	32.68	33.46	**33.62**
Crew	30.16	30.80	32.40	**32.45**
Average	28.35	29.02	30.13	**30.23**

different methods. To add, 3DKR [7] is the mutli-frame version of KR and the single frame REESE is also listed to demonstrate the superiority of multiple frames, as indicated below in Table 2.

Fig.6 are interpolated results on the video by different methods. For the *Foreman* sequence, our algorithm obtains better result visually with a PSNR of 34.18 dB, achieving a 1.2 dB gain compared with the KR algorithm while for *City*, our algorithm also obtains better result visually with a PSNR of 27.94 dB, achieving a 1.07 dB gain.

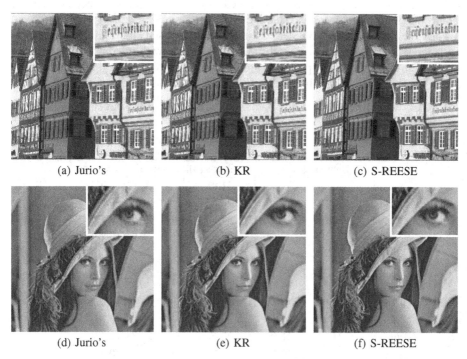

(a) Jurio's   (b) KR   (c) S-REESE

(d) Jurio's   (e) KR   (f) S-REESE

Fig. 4. Results of our algorithm compared with other methods

(a) Bicubic   (b) Glasner's   (c) S-REESE

(d) Bicubic   (e) Glasner's   (f) S-REESE

(g) Bicubic   (h) Glasner's   (i) S-REESE

(j) Bicubic   (k) Glasner's   (l) S-REESE

Fig. 5. Comparison of results obtained by Glasner's and S-REESE on test images

(a) Original

(b) Bicubic, PSNR=31.68dB

(c) 3DKR, PSNR=32.98dB

(d) REESE, PSNR=34.18dB

(e) Original

(f) Bicubic, PSNR=26.87dB

(g) 3DKR, PSNR=27.51dB

(h) REESE, PSNR=27.94dB

Fig. 6. Comparison of different SR methods on sequence *Foreman*

4   Conclusions

In this work, based on example-based SR framework, we focus on how to find more reliable similar patches. Considering the inadequacy of contemporary example-based methods, we propose a novel method for similar patch exploration. Another contribution of this work is we incorporate multiple frames to enrich the details of the HR images while keeping computing cost as low as possible. Experimental results are free of jaggies and of high quality.

References

[1] Tsai, R., Huang, T.: Multiple frame image restoration and registration. In: Advances in Computer Vision and Image Processing, vol. 1, pp. 317–339 (1984)

[2] Farsiu, S., Robinson, D., Elad, M., Milanfar, P.: Fast and robust multi-frame super-resolution. IEEE Transactions on Image Processing 13, 1327–1344 (2003)

[3] Freeman, W., Jones, T., Pasztor, E.: Example-based super-resolution. IEEE Computer Graphics and Applications 22(2), 56–65 (2002)

[4] Glasner, D., Bagon, S., Irani, M.: Super-resolution from a single image. In: IEEE International Conference on Computer Vision, pp. 349–356 (2009)

[5] Zhuo, Y., Liu, J., Ren, J., Guo, Z.: Nonlocal based super resolution with rotation invariance and search window relocation. In: IEEE International Conference on Acoustics, Speech, and Signal Processing, pp. 853–856 (2012)

[6] Su, H., Tang, L., Wu, Y., Tretter, D., Zhou, J.: Spatially adaptive block-based super-resolution. IEEE Trans. on Image Processing 21(3), 1031–1045 (2012)

[7] Takeda, H., Milanfar, P., Protter, M., Elad, M.: Super-resolution without explicit subpixel motion estimation. IEEE Transactions on Image Processing 18(9) (2009)

[8] Jurio, A., Pagola, M., Mesiar, R., Beliakov, G., Bustince, H.: Image magnification using interval information. IEEE Trans. on Image Processing 20(11), 3112–3123 (2011)

Iterative Super-Resolution for Facial Image
by Local and Global Regression

Fei Zhou[1,2], Biao Wang[1,2], Wenming Yang[1], and Qingmin Liao[1,2]

[1] Visual Information Processing Lab., Graduate School at Shenzhen,
Tsinghua University, Shenzhen, China
[2] Department of Electronic Engineering, Tsinghua University, Beijing, China
{flying.zhou,yangelwm}@163.com, wangbiao08@mails.thu.edu.cn,
liaoqm@sz.tsinghua.edu.cn

Abstract. In this paper, we propose an iterative framework to super-resolve the facial image from a single low-resolution (LR) input. To retrieve local and global information, we first model two linear regressions for the local patch and global face, respectively. In both regression models, we restrict the responses of the regressors under the considerations of facial property and discriminability. Since the responses estimated from the LR training samples can be directly applied to the (high-resolution) HR training ones, the restricted linear regressions essentially describe the desired output. More specifically, the local regression reveals the facial details, and the global regression characterizes the features of overall face. The final results are obtained by alternately using two regressions. Experimental results show the superiority of the proposed method over some state-of-the-art methods.

Keywords: Super-resolution (SR), face hallucination, linear regression.

1   Introduction

In many application based on videos and images, the details of some objects in images are usually of significance, such as the faces in surveillance video and visual telephone. Nevertheless, the image details are usually unsatisfying due to the limited resolution of the interested objects. Image super-resolution (SR) technique is expected to solve this problem by generate a high resolution (HR) image or object from the low resolution (LR) input. Generally speaking, existing SR methods retrieve high-frequency information from complementary multi-images [1] or learning samples [4], [5], [6]. In multi-images SR, the performances highly depend on the accuracy of subpixel registration. However, it is generally difficult to achieve a robust registration for local objects with arbitrary transformations. Although a robust registration method is proposed in [2] to align license plates in sub-pixel accuracy, the accurate and robust registration of local objects that are non-grid and non-planar, e.g. faces, is still a challenge. Moreover, systematic errors are inevitable in the real-physical imaging if the registration is performed on LR images [3]. Recently, learning-based SR has become popular [4], [5], [6]. In [4], SR is performed in a patch-based fashion. The

S. Li et al. (Eds.): MMM 2013, Part I, LNCS 7732, pp. 414–424, 2013.

image is regarded as a Markov network with the patches as its nodes. HR images are inferred through Bayesian belief propagation. In [5], the principle of locally linear embedding (LLE) is employed: The linear relation is reserved among neighbors under decimation. Thus, the HR patch is constructed by a linear combination of several neighbors. In [6], sparse representation is successfully applied to patch-based SR, where an over-complete dictionary is learned from pairs of HR patches and their LR counterparts. However, most SR methods are designed for the SR problem of general images.

To process the specific objects in images, we can take advantage of the knowledge about the specific domain. Hence, better performances are expected to be obtained. The SR methods developed for a specific domain are known as domain specific SR [7]. In this paper, we focus on the SR problem of facial images, which is also known as face hallucination [8]. A classic work for face SR is based on eigen transformation (ET) [9], where principal component analysis (PCA) is used to fit the input LR image as a linear combination of the LR images in the training set. Accordingly, the output HR image is obtained by the same combination of the corresponding HR images in the training set. This work provides a theoretical basis for the example-based reconstruction method of global faces. However, this method does not pay attention to the local details. Consequently, some methods, e.g., [10] and [11], integrate global reconstruction and residue compensation in a two-step framework. In [10], the global reconstruction is performed using the Gaussian model of PCA coefficients while residue compensation is achieved by a nonparametric Markov network. The method in [11], which is known as locality preserving hallucination (LPH) method, follows the similar framework as [10], but prefer to project the conventional face subspace to another subspace. Specifically, locality preserving projections is adopted in [11] to remain the linear relations between the neighbors information in a more robust manner. Nevertheless, the subspace transformation might reduce the discrimination of hallucinated faces. Furthermore, the local residue compensations in [10] and [11] do not taken the facial property into account. More recently, the work in [12] demonstrates that patch-based method could achieve plausible results without the residue compensation. In fact, the step of residue compensation is only indispensable for the methods with dimension reduction. In [12], HR patches are inferred using the training patches of the same position and the final result is obtained by stitching these over-lapped patches together. Unlike the SR in pixel domain, the method in [13] builds patch-based SR in (discrete cosine transform) DCT domain. The direct current (DC) coefficients are estimated by an interpolation while the alternating current (AC) coefficients are inferred through a simplified inference model. However, the natures of patches are not taken into full consideration in [12] and [13], e.g., the paradigm of the fixed positions cannot model the expressional variations. Moreover, since there is no guarantee for the retrieve of global characteristics, the discriminability of hallucinated faces would be impaired.

In this paper, we do not ignore either local or global information. The former is modeled by local regression while the latter is included in global regression. For the local regression, the natures of patches in context of facial images are used to restrict the responses. For the global regression, in order to preserve facial discriminability, the responses are also constrained without any dimension reduction method. The HR output is reconstructed by alternately adopting two regressions. Experiments demonstrate that our method outperforms the compared methods in the aspects of visual quality, signal fidelity, and recognition rate.

The remainder of the paper is organized as follows: In Section 2, the regression models for both the local patches and global faces are presented. Section 3 provides the implementation of the proposed face SR method. Experimental results are given in Section 4, and Section 5 concludes the paper.

2   Regression Model

Regression is a statistical analysis for estimating the relationships among variables. Although types of regression are available, linear regression is used as the prototype in our work. The reasons are twofold: First, it is one of the most simple and widely-used regression. Secondly, the linearity between either the local patches [5], [12] or the global facial images [7], [9] has been proven. The linear regression model takes the form of

$$\mathbf{a} = w_1 \cdot \mathbf{b}_1 + \cdots + w_i \cdot \mathbf{b}_i + \cdots + w_N \cdot \mathbf{b}_N + \varepsilon \ , \tag{1}$$

where $\mathbf{a}$ is known as regressand, $\mathbf{b}_i$ ($1 \leq i \leq N$) is called regressor, w_i is the responses of $\mathbf{b}_i$, N is the number of regressors, and ε is the intercept. In the context of SR, the regressand is the input or output, the regressors are the training samples, and w_i can be regarded as the weights of the training samples. In our processing, all the data are set to zero-mean by subtracting the mean at beginning and adding it at the end. Thus, the intercept ε equals zero. In the rest of this paper, the intercept is ignored. To finalize the regression, we need to estimate the weights or responses w_i.

2.1   Local Regression

For the local regression, the regressands and regressors are the LR patches cropped from the input and LR training images. The traditional method for estimating the responses is merely based on the minimization of reconstruction errors. However, this method is relative naive for face domain. Therefore, we should constrain the value of the responses. Essentially, the responses represent the extent to which the associated regressors contribute to the regressand.

Our insight is that the LR training patches with similar natures to the input patch should make more contributions. In other words, the regressand should have larger responses on the similar regressors. Although the dissimilar ones may help to get lower square errors, the dissimilar natures of them and their HR counterpart would produce artifacts in the SR results. To measure the similarity between the regressand and regressor, we make use of both the pixel intensities and spatial positions. The pixel intensities can express the texture features, which is one of the most important natures of patches. In additional, the human faces are highly structured. Moreover, the structures of various faces are the same substantially. Hence, the local characteristics on faces are dependent on the spatial positions. In this work, the similarity s of patch $\mathbf{x}$ and $\mathbf{y}^1$ is defined as

$$s(\mathbf{x}, \mathbf{y}) = \exp\left(-\frac{1}{M \cdot \sigma_1^2}\|\mathbf{x} - \mathbf{y}\|_2^2\right) \cdot \exp\left(-\frac{1}{\sigma_2^2}\|p(\mathbf{x}) - p(\mathbf{y})\|_2^2\right) , \tag{2}$$

where M is the number of pixels in an LR patch, $p(\bullet)$ is the operation of obtaining the position in the HR grid, $\|\bullet\|_2$ represents the L2-norm, and σ_1 and σ_2 are the photometric spread and geometric spread, respectively. The photometric spread σ_1 can be set according to the tolerable noises. Larger σ_1 means that the training patches with larger disparity of intensities are incorporated. Likewise, the geometric spread σ_2 can be set based on the probable variations of expression. Larger σ_2 mean that the training patches from more distant positions are included. The principle of (2) is similar to that of bilateral filtering [14], which is performed on pixels to smooth images with edge preserving. It is worthwhile to notice that the patch-based similarity defined in (2) is only suitable for the highly structured images, e.g., facial images.

As mentioned above, the similarity defined in (2) provides a constraint for the responses of the local regression: larger similarity should result in larger responses. Suppose that similarity is proportional to the response, we can further obtain the ratio ϕ by the minimization of square errors:

$$\hat{\phi}(\mathbf{x}_L) = \arg \min_{\phi(\mathbf{x})} \left\| \mathbf{x}_L - \phi(\mathbf{x}_L) \sum_i s(\mathbf{x}_L, \mathbf{y}_{Li}) \cdot \mathbf{y}_{Li} \right\|_2^2 , \tag{3}$$

where $\mathbf{x}_L$ is the regressand (LR input patch), $\mathbf{y}_{Li}$ is the i-th regressor (LR training patch). The solution of (3) can be readily derived as

$$\phi(\mathbf{x}_L) = \frac{\mathbf{x}_L^T \cdot \left(\sum_i s(\mathbf{x}_L, \mathbf{y}_{Li}) \cdot \mathbf{y}_{Li}\right)}{\left(\sum_i s(\mathbf{x}_L, \mathbf{y}_{Li}) \cdot \mathbf{y}_{Li}\right)^T \cdot \left(\sum_i s(\mathbf{x}_L, \mathbf{y}_{Li}) \cdot \mathbf{y}_{Li}\right)} . \tag{4}$$

From (3) and (4), we can conclude two points: First, the ratio ϕ is associated with the regressand and independent with the regressors. Secondly, the data consistency

---

[1] In this paper, all the patches and images are denoted in lexicographic order.

constraint is also satisfied. After getting φ, we can calculate the responses in the local regression as follows:

$$w_i\left(\mathbf{x}_L, \mathbf{y}_{Li}\right) = \phi\left(\mathbf{x}_L\right) \cdot s\left(\mathbf{x}_L, \mathbf{y}_{Li}\right) . \tag{5}$$

To simplify the local regression, the regressors with very small similarities are excluded by setting the associated responses to zero. Since the responses in the local regression are restricted by (2), we can regard it as a restricted linear regression.

2.2   Global Regression

As [7] and [9] indicate, any face can be approximated by a linear combination of the prototype faces in the training data. That is, for the global regression, the regressand and regressors are the LR input and the LR training images, respectively. However, the regression merely based on the minimization of reconstruction errors could reduce the facial discriminability. For example, the resulting eyes of the hallucinated face may appear similarly for different person. More faces contributing to the final SR result imply lower discriminability of hallucinated face. Similar to the idea in previous sub-section, we only consider similar faces in the training data as effective regressors.

A straightforward way is to select the similar faces as the k nearest neighbors (KNN) faces. However, KNN selects the samples in an independent manner. Therefore, a selected training face may be highly correlated with the other selected ones. Consequently, the information provided by KNN may be redundant in some parts as well as insufficient in the other parts. From the analysis above, we can find the expected properties of selected regressors are as follows: First, the number effective regressors, i.e., the regressors with non-zero responses, should be limited. Secondly, the regressors should be incoherent with each other. Thirdly, the regressand should be well approximated by the regressors. Inspired by these properties, we employ L1-norm to restrict the global regression:

$$\left\{r_j \left| 1 \le j \le N\right.\right\} = \underset{\left\{r_j \left|1 \le j \le K\right.\right\}}{\arg\min} \left\| \mathbf{I}_L - \sum_{j=1}^{N} r_j \cdot \mathbf{J}_{Lj} \right\|_2 \tag{6}$$
$$\text{subject to } \sum_{j=1}^{N} \left\| r_j \right\|_1 \le n$$

where $\mathbf{I}_L$ is the LR input image, $\mathbf{J}_{Lj}$ is the j-th LR training image, r_j is the response of $\mathbf{J}_{Lj}$, $\|\bullet\|_1$ represents the L1-norm, and $n \ge 0$ is a tuning parameter to control the number of non-zero responses. The responses in (6) are sparse, i.e., the number of non-zero responses is small. Hence, (6) is also known as sparse representation or compressed sensing. Since the ability of sparse representation to discriminate faces has been demonstrated in [15], the global regression can preserve facial discriminability in SR process. To solve (6), we adopt the feature-sign search algorithm [16]. Some examples of the global regression are illustrated in Fig. 1. It is worth to note that the constraint defined in the local regression can also guarantee the responses are sparse. Therefore, both the regressions can be regarded as sparse linear regressions.

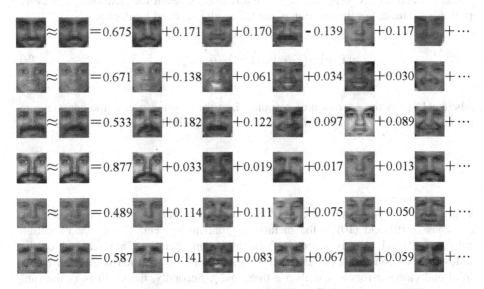

Fig. 1. Illustrations of the global regression

3   SR Implementation

Since the responses, i.e., the weights, remain the same under decimation [6], [12], the desired local HR detail $\mathbf{x}_H$ for the input patch $\mathbf{x}_L$ is given by

$$\mathbf{x}_H = \sum_i w_i (\mathbf{x}_L, \mathbf{y}_{Li}) \cdot \mathbf{y}_{Hi} \,, \tag{7}$$

where $\mathbf{y}_{Hi}$ is the HR counterpart of $\mathbf{y}_{Li}$, and w_i is defined in (5). The position of $\mathbf{x}_H$ in the HR grid can be readily calculated using the position of $\mathbf{x}_L$ in the LR grid and the amplification factor. We denote the desired HR detail at the position $\mathbf{p}$ as $\mathbf{x}_{H,\mathbf{p}}$. Likewise, the desired global HR image $\mathbf{I}_H$ for the input image $\mathbf{I}_L$ can be calculated as

$$\mathbf{I}_H = \sum_j r_j \mathbf{J}_{Hj} \,, \tag{8}$$

where $\mathbf{J}_{Hi}$ is the HR counterpart of $\mathbf{J}_{Li}$, and r_j is the solution of (6). In fact, (7) indicates the desired local information while (8) represents the desired global information. However, it is impossible that the result from (8) is exactly in accord with that from (7). To make a tradeoff between them, we make use of an iteratively strategy to alternate revise the current SR result.

Let the current SR result be $\mathbf{I}_r$. We first revise the result for local details at every probable position. For the position $\mathbf{p}$, the SR result is revised locally by:

$$c(\mathbf{I}_r,\mathbf{p}) = (1-\alpha)\cdot c(\mathbf{I}_r,\mathbf{p}) + \alpha\cdot\mathbf{x}_{H,\mathbf{p}} \ , \tag{9}$$

where $c(\mathbf{I}_r, \mathbf{p})$ is the operation of cropping $\mathbf{I}_r$ at the position $\mathbf{p}$, α is the parameter to control the step of local revision. Then, the result is revised globally by:

$$\mathbf{I}_r = (1-\beta)\cdot\mathbf{I}_r + \beta\cdot\mathbf{I}_H \ , \tag{10}$$

where β is the parameter to control the step of global revision. Essentially, the revisions in (9) and (10) is the method of gradient descent. The cyclic execution between (9) and (10) is stopped if the maximum iterations are reached or the following two conditions are satisfied: First, the total error which is the sum of the local and global errors is less than a threshold t. Secondly, the result is in the sub-space of HR facial images. To judge whether a hallucinated result is in the face sub-space, we resort to PCA. PCA is performed on the HR images in the training data to calculate the eigenvectors and associated eigenvalues. The eigenvectors represent the orientations of principal components while eigenvalues describe the reasonable variation ranges on the associated eigenvectors. Therefore, in this work, a hallucinated face is believed to be in the face sub-space, if its projections on each eigenvector is no more than the twice of the associated eigenvalue. Otherwise, we consider it as an image out of the face sub-space. The phenomenon that a hallucinated result is not included in the face sub-space mainly appears when blocking artifacts are produced by patch-based operations. Hence, once the first condition is satisfied but the second one is still not satisfied, we halve the value of α in (9).

4   Experiments

In many practical applications, e.g., surveillance video [17], faces only contain a small part of pixels in the whole image. Although face detection is an important issue therein, it is beyond the scope of this paper. In this work, we assume that faces have been well detected and aligned.

4.1 Experimental Data and Settings

Our experiments are conducted on publicly available face databases, known as FERET [18], which involves a gallery set as well as a test set. We randomly select 300 facial images from the test set and consider them as the ground truth in our experiments. The decimated versions of them serve as the input LR images. The training samples include the associated images from the gallery set and extra 200 facial images. Thus, the number of HR training images is 300 + 200 = 500. The LR training images are the decimated counterpart of HR ones. All the images are manually aligned and cropped to 126×126 pixels. The amplification/decimation factor in our experiments is 7×7, thus the LR input images are of size 18×18 pixels. The LR patches are of size 3×3 pixels with one pixel overlapped with adjacent patches. Consequently, the size of HR patches is 21×21 pixels. The photometric spread σ_1 and the geometric spread σ_2 in (2) are set to 50 and 7 for all the testing inputs, respectively. The step α and β in (9) and (10) are initialized as 0.3. The initial $\mathbf{I}_r$ is obtained through bicubic interpolation (BI).

To verify the performances, our method is compared with BI and some state-of-the-art methods, including the Eigen Transformation (ET) method [9], the LPH method [11], and the DCT domain method [13]. These compared SR methods are based on the global image, two-step, and image patches, respectively. In order to provide fair comparisons, all the data and settings remain the same for all these methods.

4.2 Intuitive Comparisons

In this sub-section, we provide the intuitive comparisons. In Fig. 2, some examples of hallucinated faces are exhibited. It can be observed from Fig. 2 that the proposed method results in the sharpest hallucinated faces with the least artifacts among all the SR methods. More specifically, the results from BI and [13] suffer from the problem of blurring to varying degrees. Some details that are sensitive to visual perception and discriminability, e.g., the region of mouths and eyes, are not well generated in the results of [9] and [11].

Table 1. Quantitative comparisons

Method	MSE	SSIM	Recognition Rate (LBP) %	Recognition Rate (UP) %
BI	7.791	0.7129	42.33	69.33
ET [9]	6.104	0.7201	72.67	75.67
LPH [11]	5.980	0.8075	73.00	60.67
DCT domain [13]	5.810	0.8431	66.00	79.67
Our method	**5.567**	**0.8634**	**82.33**	**85.33**
Original HR	0	1	96.00	93.22

4.3   Quantitative Comparisons

In comparison with intuitive comparisons, quantitative comparisons are more objective and convincing. To measure the SR results, we first employ a traditional criterion, known as mean square error (MSE). However, MSE has been reported to be poor in visual tasks in literatures. Therefore, we further utilize the structural similarity (SSIM) index [19], which is believed to be consistent with human visual perception, to measure the quality of SR results. Lower MSE and higher SSIM imply better SR results. Besides MSE and SSIM, we also use face recognition rate to evaluate the performances. Better SR results should have higher recognition rate.

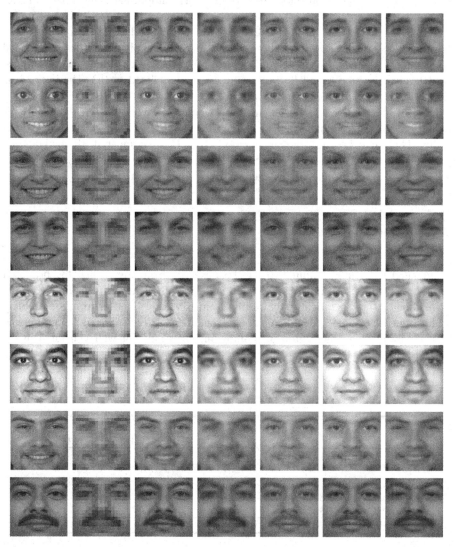

Fig. 2. Visual comparisons. From left to right are the HR original images (ground truth), LR inputs, the results of our method, BI, ET [9], LPH [11], and DCT domain [13], respectively.

The quantitative comparisons are shown in Table 1, where the quantitative results of original HR images are also provided as the benchmarks. The values of MSE and SSIM in Table 1 are the average values over all the 300 testing images. From MSE and SSIM, we can conclude that the proposed method obtains the most similar image quality to the original HR images. In Table 1, two face recognition technologies are used to demonstrate the superiority of our method. One is known as the method of local binary pattern (LBP) [20], the other one is the recently proposed uniform pursuit (UP) approach [21]. The former is a local-based face recognition method while the latter prefers to measure the similarity of the global faces. Both the methods of the local and global recognition indicate that the proposed method achieves the closest recognition rate to that of original HR images. It means that our method can well retrieve local details as well as holistic facial characteristics.

5   Conclusion

In this paper, an SR method is proposed to infer facial images from a single LR input by using the local and global regression. Both regressions are designed by restricting the linear regression in the context of facial characteristics and discriminability. For learning-based SR problem, the LR input and training samples are regarded as the regressand and regressors, respectively. Our aim is to regularize the responses. In local regression, the intensities and positions of patches contribute to the final responses of the regressors. In the global regression, the responses are restricted in L1-norm so that the facial discriminability can be preserved. An iterative procedure is also presented to obtain the final results with the help of both the regression. The experimental results demonstrate the superiority of the proposed method over some state-of-the-art methods via various criteria.

Acknowledgments. The research leading to this work was supported by the National Natural Science Foundation of China under Grant 61271393.

References

1. Zhou, F., Yang, W., Liao, Q.: Interpolation-Based Image Super-Resolution Using Multisurface Fitting. IEEE Transactions on Image Processing 21(7), 3312–3318 (2012)
2. Zhou, F., Yang, W., Liao, Q.: A Coarse-to-Fine Subpixel Registration Method to Recover Local Perspective Deformation in the Application of Image Super-Resolution. IEEE Transactions on Image Processing 21(1), 53–66 (2012)
3. Wang, Z., Qi, F.: On Ambiguities in Super-Resolution Modeling. IEEE Signal Processing Letter 11(8), 678–681 (2004)
4. Freeman, W.T., Jones, T.R., Pasztor, E.Z.: Example-Based Super-Resolution. IEEE Computer Graphics and Applications 22(2), 56–65 (2002)
5. Chang, H., Yeung, D.-Y., Xiong, Y.: Super-Resolution through Neighbor Embedding. In: IEEE Conference on Computer Vision and Pattern Recognition, pp. 275–282 (2004)
6. Yang, J., Wright, J., Ma, Y., Huang, T.S.: Image Super-Resolution via Sparse Representation. IEEE Transactions on Image Processing 19(11), 2861–2873 (2010)

7. Park, J.-S., Lee, S.-W.: An Example-Based Face Hallucination Method for Single-Frame, Low-Resolution Facial Images. IEEE Transactions on Image Processing 17(10), 1838–1857 (2008)

8. Baker, S., Kanade, T.: Hallucinating Faces. In: IEEE International Conference on Automatic Face and Gesture Recognition, pp. 83–88 (2000)

9. Wang, X., Tang, X.: Hallucinating Faces by Eigentransform. IEEE Transactions on Systems, Man, and Cybernetics - Part C: Applications and Reviews 35(3), 425–434 (2005)

10. Liu, C., Shum, H.Y., Freeman, W.T.: Hallucinating Faces: Theory and Practice. International Journal of Computer Vision 75(1), 115–134 (2007)

11. Zhuang, Y., Zhang, J., Wu, F.: Hallucinating Faces: LPH Super-Resolution and Neighbor Reconstruction for Residue Compensation. Pattern Recognition 40(11), 3179–3194 (2007)

12. Ma, X., Zhang, J., Qi, C.: Hallucinating Face by Position-Patch. Pattern Recognition 43(6), 2224–2236 (2010)

13. Zhang, W., Cham, W.-K.: Hallucinating Face in the DCT Domain. IEEE Transactions on Image Processing 20(10), 2769–2779 (2011)

14. Tomasi, C., Manduchi, R.: Bilateral Filtering for Gray and Color Images. In: IEEE International Conference on Computer Vision, pp. 839–846 (1998)

15. Wright, J., Yang, A.Y., Ganesh, A., Sastry, S.S., Ma, Y.: Robust Face Recognition via Sparse Representation. IEEE Transactions on Pattern Analysis and Machine Intelligence 31(2), 210–227 (2009)

16. Lee, H., Battle, A., Raina, R., Ng, A.Y.: Efficient Sparse Coding Algorithms. In: Advances in Neural Information Processing Systems, pp. 801–808 (2007)

17. Yang, W., Zhou, F., Liao, Q.: Object Tracking and Local Appearance Capturing in a Remote Scene Video Surveillance System with Two Cameras. In: Boll, S., Tian, Q., Zhang, L., Zhang, Z., Chen, Y.-P.P. (eds.) MMM 2010. LNCS, vol. 5916, pp. 489–499. Springer, Heidelberg (2010)

18. Phillips, P.J., Moon, H., Rizvi, S.A., Rauss, P.J.: The FERET Evaluation Methodology for Face Recognition Algorithms. IEEE Transactions on Pattern Analysis and Machine Intelligence 22(6), 1090–1104 (2000)

19. Wang, Z., Bovik, A.C., Sheikh, H.R., Simoncelli, E.P.: Image Quality Assessment: From Error Visibility to Structural Similarity. IEEE Transactions on Image Processing 13(4), 600–612 (2004)

20. Ahonen, T., Hadid, A., Pietikäinen, M.: Face Description with Local Binary Patterns: Application to Face Recognition. IEEE Transactions on Pattern Analysis and Machine Intelligence 28(12), 2037–2041 (2006)

21. Deng, W., Hu, J., Guo, J., Cai, W., Feng, D.: Robust, Accurate and Efficient Face Recognition from a Single Training Image: A Uniform Pursuit Approach. Pattern Recognition 43(5), 1748–1762 (2010)

Stripe Model: An Efficient Method to Detect Multi-form Stripe Structures

Yi Liu[1,2], Dongming Zhang[1], Junbo Guo[1], and Shouxun Lin[1]

[1] Advanced Computing Research Laboratory, Beijing Key Laboratory of Mobile Computing and Pervasive Device, Institute of Computing Technology,
Chinese Academy of Sciences, Beijing 100190, China
[2] Graduate University of Chinese Academy of Sciences, Beijing 100049, China
{liuyi,dmzhang,guojunbo,sxlin}@ict.ac.cn

Abstract. We present a general mathematical model for multiple forms of stripes. Based on the model, we propose a method to detect stripes built on scale-space. This method generates difference of Gaussian (DoG) maps by subtracting neighbor Gaussian layers, and reserves extremal responses in each DoG map by comparing to its neighbors. Candidate stripe regions are then formed from connected extremal responses. After that, approximate centerlines of stripes are extracted from candidate stripe regions using non-maximum suppression, which eliminates undesired edge responses simultaneously. And stripe masks could be restored from those centerlines with the estimated stripe width. Owing to the ability of extracting candidate regions, our method avoids traversing to do costly directional calculation on all pixels, so it is very efficient. Experiments show the robustness and efficiency of the proposed method, and demonstrate its ability to be applied to different kinds of applications in the image processing stage.

Keywords: stripe model, scale-space, difference of Gaussian, non-maximum suppression.

1 Introduction

The stripe, curvilinear structure of a certain width, is a fundamental component of many objects, such as character strokes, balustrades, roads, pipelines etc. So stripe detection plays an important part in the image processing stage of computer vision applications, e.g., it could be used to extract strokes in text detection as preprocessing, find roads in aerial images, or detect tubular objects in medical and industrial fields.

Early methods detect stripes by considering the gray values of the image only and using purely local criteria [1, 2]. They aim at stripes of a very small width, i.e. lines, and their performance is deeply constrained by stripe direction [1]. In order to detect wide curving stripes, they have to zoom out the image and combine results in all possible directions, which leads to high computational cost and low robustness.

Moreover, the stripe body could even consist of dense subcomponents as shown in Fig. 1, and they could be blurred due to the compression in real images and videos.

S. Li et al. (Eds.): MMM 2013, Part I, LNCS 7732, pp. 425–435, 2013.
© Springer-Verlag Berlin Heidelberg 2013

Under such a condition, a group of generalized methods based on derivatives of Gaussian kernels [3, 4] could be applied.

Fig. 1. Different stripe forms from left to right: narrow, different brightness, wide, blur, and dense subcomponent

Koller et al. proposed a multi-scale filter [3] that detects both the left and right edge responses of stripes, combines them nonlinearly and iterates in scale-space to detect stripes of arbitrary widths. The advantage of this approach is that, since the particular nonlinear scheme is used, it can subtly resolve the side effects of edges, as occurs for every linear filter based on derivatives of Gaussian kernels. However, due to its inability of locating candidate stripe regions, it has to apply the whole procedure on every pixel while extending 1D theory to 2D situation, which is redundant and takes high computational cost.

Steger gave a comprehensive elucidation on behavior of stripes under derivatives of Gaussian kernels and introduced an unbiased detector [4] building upon previous works on multi-scale ridge detection [3, 5, 6, 7]. The detector locates stripes according to the first directional derivative and the Laplacian of Gaussian (LoG) response, and it generates meaningful results in consideration of asymmetry of edges. But it is also computationally expensive to traverse all pixels the same as [3].

Most recently, Epshtein et al. put forward the stroke width transform (SWT) to detect character strokes [8], which uses Canny filter to find edges and retrieves the stripe width information by concatenating edge points along the gradient direction. Though making a great progress on the text detection task in nature scenes, it is unfit for noisy or blurred condition as reported, because SWT heavily depends on the edge detection.

We give a summary in Table 1 to compare the ability of three popular methods in different condition.

Table 1. Ability comparison among different methods

Method	Width selection	Blur condition	Subcom-ponents	Less sensitive to noise	Efficiency
Koller [3]	√	√	√	√	×
Steger [4]	×	√	√	√	×
Epshtein [8]	√	×	×	×	√

Consequently, there lacks of a method that is efficient and can fulfill multi-form stripes. In this paper, by taking advantage of most recent achievements, we define the stripe structure and build a mathematical model to illustrate its characteristics. Then, we propose an efficient and robust scheme to detect stripes of diverse forms.

Comparing to the analogue and most popular ones [3, 4, 8], our method does the directional calculation only on candidate stripe regions which indeed speeds up the processing, and the scale-space based scheme makes it flexible to screen or select stripes of certain widths.

2   Stripe Detection

2.1   Stripe Model

Our stripe model is motivated by the characteristics of the difference of Gaussian (DoG) response of an edge, which we will illustrate with 1D signal, and then extend to 2D. The DoG function provides a close approximation to the scale normalized LoG, $\sigma^2\nabla^2G$ [9], as studied by Lindeberg, and Lowe use it to build the image pyramid [10], which is so efficient and will be utilized in our scheme later.

Here, we define the stripe structure as the curvilinear one of a certain width, and the ideal 1D profile is given by

$$f(x) = au(x+\frac{w}{2})+bu(-x+\frac{w}{2})+c \tag{1}$$

where $u(x)$ is the unit step signal, w is the stripe width, $a,b,c \in R$ and $ab > 0$.

For $u(x)$ could be seen as the profile of an ideal edge in 1D, to exemplify the features of the edge response under DoG filter, we construct $g(x) = au(x)$ with amplitude a and convolve it with DoG function

$$h_{step}(x;\sigma) = DoG(x;\sigma) * g(x) = a\int_{-\infty}^{x} DoG(t;\sigma)dt \tag{2}$$

where

$$DoG(x;\sigma) = G(x;k\sigma)-G(x;\sigma) \tag{3}$$

$$G(x;\sigma) = \frac{1}{\sqrt{2\pi}\sigma}e^{-x^2/2\sigma^2} \tag{4}$$

$k > 1$ is a constant multiplicative factor of two nearby scales.

Take the derivative of $h_{step}(x;\sigma)$ with respect to x and set it to zero, giving

$$x_1 = k\sigma\sqrt{\frac{2\ln k}{k^2-1}}, \quad x_2 = -k\sigma\sqrt{\frac{2\ln k}{k^2-1}} \tag{5}$$

In consideration of the second derivative of $h_{step}(x;\sigma)$, we can conclude the offsets of the extremal DoG responses of a step signal are x_1 and x_2, which only depend on scale σ but are irrelevant to amplitude a, shown in Fig. 2.

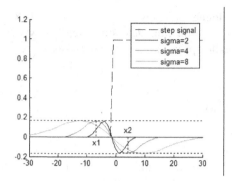

Fig. 2. DoG responses of a step signal in multi-scale with $a = 1$

Proposition 1. The offsets of the extremal DoG responses of a step signal only depend on the scale of DoG filter but are irrelevant to the signal amplitude.

Bring x_1 back to Eq. 2, we get the extremal response

$$\hat{h}_{step}(x_1;\sigma) = \frac{a}{2}[\phi(\sqrt{\frac{\ln k}{k^2-1}}) - \phi(k\sqrt{\frac{\ln k}{k^2-1}})] \tag{6}$$

where

$$\phi(x) = \frac{2}{\sqrt{\pi}}\int_0^x e^{-t^2}dt \tag{7}$$

So, we find $\hat{h}_{step}$ only depends on amplitude a and is irrelevant of scale σ.

Proposition 2. The values of the extremal DoG responses of a step signal only depend on the signal amplitude but are irrelevant to the scale.

Now, given a square signal, the 1D profile of an ideal stripe structure with a width w, it could be seen as a linear combination of two shifted step signals, as Eq. 1. For DoG filter is a linear operator, the final response of the square signal is also a linear combination of that two step signal responses

$$h_{square}(x,\sigma) = DoG(x;\sigma) * f(x)$$
$$= aDoG(x;\sigma) * u(x+\frac{w}{2}) + bDoG(x;\sigma) * u(-x+\frac{w}{2}) \tag{8}$$

Take the result into account along with proposition 1 and 2, we infer that there must be a scale σ, at which the DoG response in the center of the square signal reaches the extremum as shown in Fig. 3. And bring $x_1 = w/2$, the extremum offset, back to (5), we get the proper scale

$$\hat{\sigma} = \frac{w}{2k} \sqrt{\frac{k^2 - 1}{2 \ln k}} \qquad (9)$$

At the same time, the values in the center neighborhood also reach their extrema due to the continuity property of the response. So we can summarize the feature of the stripe model in 1D as:

Proposition 3. Given a square signal of the width w, DoG responses in the center and its neighborhood reach extrema at scale $\sigma = \hat{\sigma}$.

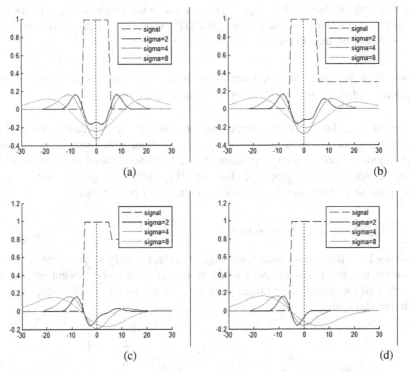

Fig. 3. DoG responses of a square signal in multi-scale with different a and b configurations. (a) shows that symmetrical edges generate corresponding responses. When $a \ne b$, though do not affect the extrema, they turns biased (b, c). (d) gives a limit case that $b = 0$ which degrades to the step signal situation.

Now, we will discuss the blurred condition. As mentioned before, this model fulfills blurred stripes which could be seen as a Gaussian smoothing ones

$$f_s(x; \sigma_s) = G(x; \sigma_s) * f(x) \qquad (10)$$

According to the semi-group property of Gaussian kernel [11], the convolution of two Gaussian kernels is the same as that of one kernel with different scale

$$G(\cdot\,;\sigma)*G(\cdot\,;\sigma_s)=G(\cdot\,;\sigma+\sigma_s) \qquad (11)$$

So the final response turns to be $h_{square}(x;\sigma+\sigma_s)$ in accordance with Eq. 11, and all properties described above still hold with the corresponding scale rising to $\sigma+\sigma_s$ concomitantly.

Since the previous analysis is based on 1D, we need extend to 2D to apply to images. Using the separability property of 2D Gaussian kernels, convolving the stripe with a 2D Gaussian kernel is the same as convolving it with a 1D one in the stripe direction first and then in the perpendicular direction. For the convolution in the stripe direction seldom changes the stripe itself, and that in the perpendicular direction could be seen as convolving a square signal, the 1D profile of an ideal stripe structure, we can easily extend the 1D theory to a 2D version:

Proposition 4. Given a stripe structure of a width w, DoG responses in the centerline and its neighborhood reach extrema at scale $\sigma=\hat{\sigma}$.

The analysis so far has been carried out for a method to generate maps containing extremal responses without any directional calculation, which supplies the candidate stripe regions to reduce the computation cost and get accelerated as mentioned before. However, not only stripe responses but edge (Fig. 3d) and edge-side ones (Fig. 3a-c) exist in the same map, so we need to screen out the stripe responses.

2.2   Edge Responses Elimination

Applying DoG filter to an image, at a certain scale, not only stripes generate extremal responses, but also edges and edge sides. Moreover, edge responses appear as extrema even though the whole scale range, so it is necessary to separate the stripe responses from the edge ones.

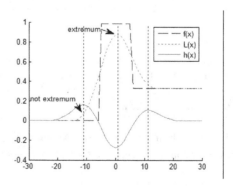

Fig. 4. The response profiles of the edge and stripe in 1D

We draw $f(x)$, $L(x;\sigma)$ and $h_{square}(x;\sigma)$ in Fig. 4 to illustrate their mutual relationships where

$$L(x;\sigma) = G(x;\sigma) * f(x) \tag{12}$$

In Fig. 4, when $h_{square}(x;\sigma)$ reaches local extremum on both sides, $L(x;\sigma)$ is still far from its peak, but when $h_{square}(x;\sigma)$ reaches extremum inside, $L(x;\sigma)$ reaches its peak nearby. This discovery constitutes the foundation of distinguishing the extremal responses of edges and stripes.

Now we will examine the peak location of $L(x;\sigma)$. There is a bias problem while stripe edges are asymmetry [4] that the peak of $L(x;\sigma)$ will shift from the extremum of $h_{square}(x;\sigma)$ when $a \neq b$, as also shown in Fig. 4. And the offset, i.e. the root of the first derivative of $L(x;\sigma)$ is

$$l = \frac{\sigma^2}{w}\ln(\frac{a}{b}) \tag{13}$$

If l belongs to the $h_{square}(x;\sigma)$ extremal interval, we could easily estimate the center by seeking the extremum of $L(x;\sigma)$ in that interval, corresponding to the candidate region in 2D. We name them salient stripes that satisfy this condition and have extremal responses beyond a certain threshold.

However, it is hard to determine the extremal interval analytically, so we compute a/b and its generated offset ratio $2l/w$ numerically to give some perceptual knowledge.

Table 2. a/b and its generated offset ratio when $k = 1.8$

a/b	40	6	3	2	1.5	1
$2l/w$	1.00	0.48	0.30	0.19	0.11	0

Table 2 shows that when $a/b \leq 3$, the extremum of $L(x;\sigma)$ is very close to that of $h_{square}(x;\sigma)$, i.e. it generates salient stripes. And the condition $a/b \leq 3$ is so loose that it could be satisfied in most common situations.

By shifting along the direction perpendicular to the candidate stripe regions and do non-maximum suppression, we can get the approximate centerlines of stripes.

An equivalent way to carry out the non-maximum suppression is to check zero-crossing. Assuming $l(x)$ indicates the value of $L(\mathbf{x},\sigma)$ that x pixels away from $\mathbf{x}$, shifted along the unit direction $\mathbf{e}$, and $\mathbf{x}_0$ is the sample point, we get

$$l(x) = L(\mathbf{x}_0 + x\mathbf{e})$$

$$\approx L(\mathbf{x}_0) + x\frac{\partial L(\mathbf{x}_0)}{\partial \mathbf{x}}^T \mathbf{e} + \frac{x^2}{2}\mathbf{e}^T\frac{\partial^2 L(\mathbf{x}_0)}{\partial \mathbf{x}^2}\mathbf{e} \tag{14}$$

The location of the extremum, is estimated by taking the derivative of this function with respect to x and setting it to zero, giving

$$\hat{x} = (\mathbf{e}^T \frac{\partial^2 L(\mathbf{x}_0)}{\partial \mathbf{x}^2} \mathbf{e})^{-1} \frac{\partial L(\mathbf{x}_0)}{\partial \mathbf{x}}^T \mathbf{e} \tag{15}$$

Consequently, we can discriminate between stripe responses and edge ones by simply checking whether the estimated $\hat{x} \in [-\frac{1}{2}, \frac{1}{2}]$ meets, and the approximate centerline map is obtained.

2.3   Stripe Detection Scheme

Based on the stripe model and false response elimination algorithm, we begin to set up the whole stripe detection scheme.

First, we choose a target stripe width w and build the scale-space around the scale σ according to Eq. 9. The parameter k should be tuned according to different requirements, because it affects the precision of stripe width selection and determines whether the centerlines stay inside the candidate stripe regions along with a/b. A bigger k results in less selection precision and more stay-inside possibility. In general, a recommended value $k = 1.8$ could fulfill most common needs and we use it as default in Sec. 3.

Second, we build the scale-space and generate the DoG maps by subtracting neighbor Gaussian layers. The extremum maps that contain candidate stripe regions are generated by comparing pixel responses to its neighbor DoG maps and keeping the extrema above a certain screen threshold. We could set the screen threshold to select stripes with prominent edges in consideration of the edge contrast as used in [4]. E.g., if stripes with edge contrasts of $a, b \geq 70$ are to be selected, they will have extremal responses about -22.5 from Eq. 6. Therefore, the upper threshold for the absolute value of the response is set to 20. If intersection operation is needed, the lower threshold is set to 3.6 in order to collect adequate responses.

Finally, edge responses are eliminated by the method described in Sec 2.2, so that only approximate one-pixel-width centerlines remain. And we can also estimate stripe masks from centerlines with dilate operation according to the target stripe width.

The flowchart of the whole stripe detection scheme is shown in Fig. 5.

Fig. 5. The flowchart of the proposed scheme

In addition, if stripe is too wide that will add a heavy burden to Gaussian convolution due to the filtering window size, we can use Lowe's work [10] to generate image pyramid and calculate on a smaller image to save computing cost.

3   Experiments and Results

In this section, we compare the proposed method to Koller [3], Steger [4], and Epshtein [8], regarding to the quality, width selection and efficiency.

First, we compare the detection results in an aerial image as Fig. 6a. We find that Koller, Steger and the proposed methods generate similar results. However, the effect of Steger is a little worse due to its inability of width selection, so roads and other components with narrower width all emerge. In order to reduce these noises to the same level of Koller and the proposed for comparison, we have to tune screen threshold and some fractions of roads disappear. Epshtein is deeply trapped by those noises, and it is hard to distinguish roads from them without other cues.

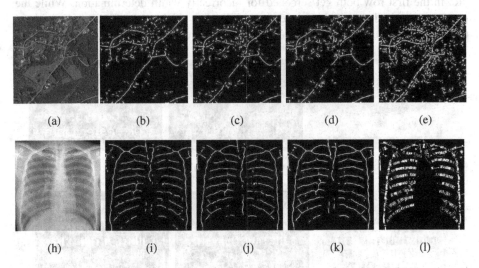

Fig. 6. Results from an aerial image (a) and a medical one (h) by the proposed (b, i), Koller (c, j), Steger (d, k) and Epshtein (e, l)

Table 3. The time cost (s) comparison on Fig. 6a, h

Method	Koller	Steger	Epshtein	Proposed
Fig. 6a	47.2	17.4	2.0	1.63
Fig. 6h	10.8	4.0	0.7	0.8

Fig. 6h gives another example on medical image. Due to the smoothness property of Gaussian kernel, Koller, Steger and the proposed method are applicable in blurred or low edge contrast condition. While Epshtein, depending on edge detection badly, misses some stripes.

Table 3 shows time cost of each method on original image of Fig. 6a with size 803×596 and Fig. 6h with size 365×300. All methods are carried out in matlab on a PC with 2.4GHz CPU and 3GB memory. The proposed method runs fastest on Fig. 6a and the second fastest on Fig. 6h. The main reason lies in fewer pixels to calculate, for the proposed method only does calculation on pixels within candidate stripe regions comparing to the traversal scheme used by Koller and Steger. Epshtein is also fast in both examples, and its efficiency is directly related to the number of edge pixels extracted by Canny filter.

In Fig. 7, another common task, scene text detection [12], is examined. As Koller and Steger generate similar results as the proposed, we only compare to Epshtein.

Epshtein aims at scene text detection by stroke transformation and rule based assembly. However, in complex background like trees and overlapped objects, the transformation catches a lot of noisy responses (Fig. 7c) or split the stroke into pieces (Fig. 7f) which changes the actual stroke width. Furthermore, it fails to accurately determine the width of crossing strokes when strokes are thick, for it is determined by the median width of a scanning ray. E.g., in Fig. 7f, left strokes of the letter "D" and "R" in the first row both get screened for incorrectly width determination. While the proposed method uses width selection ability to shield the effects of trees and overlapped objects and gains a better result (Fig. 7b, e).

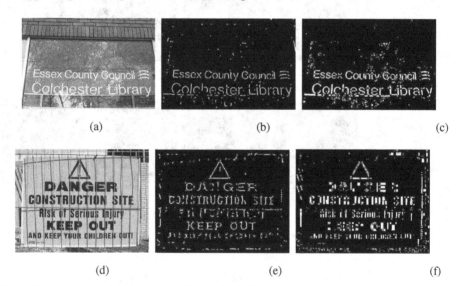

(a)   (b)   (c)

(d)   (e)   (f)

Fig. 7. Comparing results of scene text image (a, d) between the proposed (b, e) and Epshtein (c, f). For we aim at distinguishing thick stripes from the thin balustrade in (d), those thin stroke characters also get screened in (e, f).

4   Conclusion

In this paper, we studied current methods which detect multiple forms of stripes, and introduce a mathematical model to define the common stripe structures by exploiting the characteristics of the edge DoG response. Then, based on the model, we present a scale-space based method to estimate the stripe centerlines and masks efficiently. Due to the peculiar property of the scale-space and the scale specification of extremal maps, our method extracts candidate stripe regions to get accelerated and can detect stripes of diverse forms. Furthermore, we show the method works well on practical images comparing to other popular methods in experiments.

Our method only gives approximate results of centerlines and stripe masks concentrating on detection efficiency. If more precise stripe boundary and bias removal are needed, detailed analysis of asymmetrical stripe profiles [4] may be helpful, and could be integrated into our scheme.

In the future, we will integrate GPU based parallel computing scheme [13, 14] to make the method processing much faster.

Acknowledgements. This work is supported by National Nature Science Foundation of China (61273247, 61271428), National Key Technology Research and Development Program of China (2012BAH39B02), and Co-building Program of Beijing Municipal Education.

References

1. Gonzalez, R.C., Woods, R.E.: Digital image processing, 3rd edn., pp. 719–722. Prentice-Hall Publisher (2008)
2. Hough, P.V.C.: Method and means for recognizing complex patterns. U.S. Patent 3,069,654 (December 18, 1962)
3. Koller, T.M., Gerig, G., Szekely, G., Dettwiler, D.: Multiscale detection of curvilinear structures in 2-D and 3-D image data. In: Proc. ICCV, pp. 864–869 (1995)
4. Steger, C.: An unbiased detector of curvilinear structures. IEEE Trans. PAMI 20(2), 113–125 (1998)
5. Eberly, D., Gardner, R., Morse, B., Pizer, S., Scharlach, C.: Ridges for image analysis. J. Math. Imaging Vis. 4(4), 353–373 (1994)
6. Koenderink, J.J., van Doorn, A.J.: Two-plus-one-dimensional differential geometry. Pattern Recognition Letters 15(5), 439–444 (1995)
7. Lindeberg, T.: Edge detection and ridge detection with automatic scale selection. In: Proc. CVPR, pp. 465–470 (1996)
8. Epshtein, B., Ofek, E., Wexler, Y.: Detecting text in natural scenes with stroke width transform. In: Proc. CVPR, pp. 2963–2970 (2010)
9. Lindeberg, T.: Scale-space theory: a basic tool for analyzing structures at different scales. J. Appl. Stat. 21(2), 224–270 (1994)
10. Lowe, D.G.: Distinctive image features from scale-invariant keypoints. Int. J. Comput. Vis. 60(2), 91–110 (2004)
11. Lindeberg, T.: Generalized Gaussian scale-space axiomatics comprising linear scale-space, affine scale-space and spatio-temporal scale-space. J. Math. Imaging Vis. 40(1), 36–81 (2011)
12. Shahab, A., Shafait, F., Dengel, A.: ICDAR 2011 robust reading competition challenge 2: reading text in scene images. In: Proc. ICDAR, pp. 1491–1496 (2011)
13. Wang, W.Y., Zhang, D.M., Zhang, Y.D., Li, J.T., Gu, X.G.: Robust spatial matching for object retrieval and its parallel implementation on GPU. IEEE Trans. on Multimedia 13(6), 1308–1318 (2011)
14. Xie, H.T., Gao, K., Zhang, Y.D., Tang, S., Li, J.T., Liu, Y.Z.: Efficient feature detection and effective post-verification for large scale near-duplicate image search. IEEE Trans. on Multimedia 13(6), 1319–1332 (2011)

Saliency-Based Content-Aware Image Mosaics

Dongyan Guo, Jinhui Tang, Jundi Ding, and Chunxia Zhao

School of Computer Science and Engineering,
Nanjing University of Science and Technology
dongyan.guo@gmail.com

Abstract. In this paper, we propose a novel content-aware image mosaic approach based on image saliency. The image saliency is used in the whole process of creating an image mosaic with variable size tiles, while a novel energy map is proposed by combining Neighborhood Inhomogeneity Factor and Graph-Based Visual Saliency. The target image is divided into small tiles with variable sizes based on the energy map. Image retrieval is introduced to choose the tile images from a certain database. Considering the specialization of tile image retrieval, we propose a new feature representation called brightness distribution vector, which indicates the image global brightness distribution. Extensive experiments are conducted to show that the proposed approach creates better mosaics in visual aspect than the conventional methods.

1 Introduction

The traditional sense of the mosaic is a design in which a large image is formed by small pieces of colored glass, stone or ceramic tiles. The image mosaic is the digital refinement of the traditional mosaic, in which the tiles are small images

| (a)Target image | (b)Conventional method | (c)Proposed method |

Fig. 1. Image mosaics of "Panda": in (b) and (c) each rectangle is a tile and filled with a tile image. Clearly, the proposed approach gives a batter mosaic result in visual aspect.

S. Li et al. (Eds.): MMM 2013, Part I, LNCS 7732, pp. 436–444, 2013.
© Springer-Verlag Berlin Heidelberg 2013

called tile images [1]. An example of an image mosaic is shown in Fig. 1. In an image mosaic the tile images are arranged in such a way: when they are seen separately, each of them is a unique image; when they are seen together from a distance, they suggest a whole large image. The technology of combining groups of images in this way has application in the artistic field. There are also some other potential applications, such as encoding extra information in an image for transmission or security, new forms of half-toning screens for printing, and association of images for advertising. For example, people can synthesize their portraits by a collection of images which record their personal growth history.

The tiles of conventional image mosaics tend to have the same size and shape [1,2,3]. They are usually obtained by applying a uniform grid to the target image. How to choose the appropriate size for image tiles is a problem. If the tiles are big, the large image is blocky, while if the tiles are small, the image tiles themselves are hardly discernible. To solve the problem, Achanta et al. [5] introduced the image saliency to image mosaics and assigned variable sized image tiles based on an energy map decomposition technique. They recursively subdivided the target image into four equal quadrants, subject to the energy contained within each quadrant exceeding a predefined threshold.

The approach in [5] gives us a lot of inspiration. However, it leads to another question that worth considering. What kind of energy map can give better mosaic results? For an image mosaic with variable-size tiles, the important areas should be divided into smaller tiles to prevent the details of the target image, while the unimportant areas should be divided into larger tiles to better display appearances of tile images themselves. To meet this goal, we apply the Neighborhood Inhomogeneity Factor (NIF) [4] to image saliency to get a novel energy map for variable-size tile decomposition. This content-aware decomposition technique better preserves the details of important areas than the uniform grid decomposition method [1,2,3].

After decomposition, the obtained tiles are filled with corresponding tile images. These tile images are chosen from a certain image database through image retrieval [10,11]. Considering the specialization of tile image retrieval that the searched image and the tile should be visually consistent, we propose a new feature representation called brightness distribution vector which denotes the image global brightness distribution. Another two features are also introduced into the image retrieval step: the saliency weighted color histogram and the saliency weighted texture histogram [9]. The tile image retrieval method ensures the good performance of image mosaics. Fig. 2 illustrates the process of our mosaic approach.

The rest of this paper is organized as follows. Section 2 presents the proposed approach for content aware image decomposition. Section 3 details our method for saliency based tile image retrieval. The mosaic results of the proposed approach are given in Section 4, followed by the conclusions in Section 5.

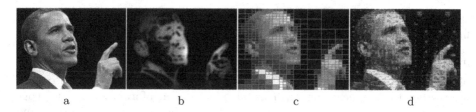

$$\text{a} \qquad\qquad \text{b} \qquad\qquad \text{c} \qquad\qquad \text{d}$$

Fig. 2. Process of the proposed image mosaic: (a) Target image; (b)Energy map computation; (c) Variable-Sizes tile decomposition; (d) Saliency-weighted image retrieval for mosaics

2   Content-Aware Image Decomposition

2.1   The Novel Energy Map

For an image mosaic with variable-size tiles, the important areas should be divided into smaller tiles to prevent the details of the target image, while the unimportant areas should be divided into larger tiles to show the tile images their own appearances. Therefore, we need an energy map to indicate the importance of different areas, while the energy map for decomposition should meet the following two requirements: first, the map should highlight the significant areas, and second, the map should keep details of the significant areas. Here we combine two measures to compute the energy map: Neighborhood Inhomogeneity Factor (NIF) [4] and Graph-Based Visual Saliency (GBVS) [7].

The NIF denotes the inhomogeneity of an image. It is computed through the square neighborhood of an arbitrary pixel p in the image. The NIF value of each pixel is up to the texture variation of its surrounding neighborhood. Therefore the NIF map not only highlights the significant areas but also part of the background areas which are inhomogeneous. The details of computing NIF map can be found in [4].

The GBVS is a bottom-up visual saliency model which shows a remarkable consistency with the attention deployment of human subjects. The method uses a novel application of ideas from graph theory to concentrate mass on activation maps, and to form activation maps from raw features. It first forms activation maps on certain feature channels, and then normalizes them in a way which highlights conspicuity and admits combination with other maps. The GBVS map robustly highlights salient regions even far away from object borders. In other words, the GBVS can highlight the whole object region, including both borders and internal areas. That is the reason why we choose GBVS map as the saliency map here. The method of computing GBVS map can be referred to [7].

By combining the two measures, the details of the significant regions are well preserved and the background noises are removed. Let I denote the input image, we define the energy map as

$$E(I) = E_{nif} \times E_{gbvs}. \tag{1}$$

where E_{nif} is the NIF map and E_{gbvs} is the GBVS map.

An example is shown in Fig. 3. The NIF map highlights both the significant object and the background structure in the image. By multiplying with the GBVS map the highlighted part of the background is filtered out.

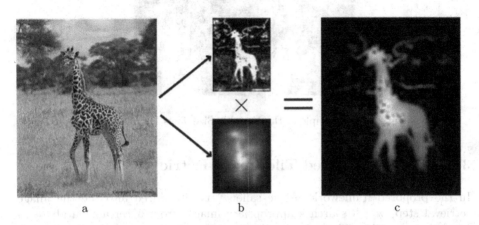

a   b   c

Fig. 3. The proposed energy map is the product of the NIF measure and the GBVS measure. (a) Original image; (b) top: NIF map; bottom: GBVS map; (c) Energy map.

2.2 Variable-Size Tiles Decomposition

For an image I, when obtaining its energy map $E(I)$, we can divide I into small tiles with variable sizes by using the energy map decomposition technique [5]. The input image is recursively subdivided into four equal quadrants, subject to the cumulative energy M for each quadrant exceeding a threshold T_e. The cumulative energy of quadrant Q_j is computed by:

$$M_j = \sum_{p \in Q_j} E(I(p)). \tag{2}$$

Suppose the initial quadrant is Q, the recursive function can be described as follows:

Use Eq.(2) to compute the cumulative energy M_Q of Q. If $M_Q \leq T_e$, terminate the recursion;

Otherwise, perform the following recursive body:

1. Divide Q into four equal quadrants, Q_1, Q_2, Q_3 and Q_4;
2. Call the recursive function to subdivide Q_1;
3. Call the recursive function to subdivide Q_2;
4. Call the recursive function to subdivide Q_3;
5. Call the recursive function to subdivide Q_4.

A decomposition example is shown in Fig. 4.

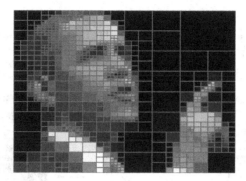

Fig. 4. An Example of the Variable-Size Tiles Decomposition

3 Saliency-Weighted Tile Image Retrieval

In the proposed framework, image saliency is introduced into the tile image retrieval step, which searches appropriate images from a certain database to fit the mosaic tiles. The color similarity between the image and the tile has a strong influence on the mosaic result. Therefore, we use the frequency tuned (FT) saliency detection method [8] to compute the saliency map. The FT method directly defines pixel saliency on pixel color difference. It calculates distances of pixels in a Gaussian blurred version of the input image to the average color of the same image in the *Lab* color space to obtain the saliency map.

Taking into account the particularity of image retrieval for mosaics, three image features are defined for tile fitting: two saliency weighted features based on the saliency map and one proposed brightness distribution vector based on the grayscale image. After extracted features, the Euclidean distance between texture vectors are used to measure the difference between the tile and a certain image from the database.

3.1 Image Feature Extraction

Three image features are denoted for image fitting: the saliency weighted color histogram, the saliency weighted texture histogram and the brightness distribution vector. Let I denotes the input image, S denotes the FT saliency map of I, $sum(S)$ denotes the sum of all elements in S, the three features can be described as follows:

Saliency Weighted Color Histogram. Global color histogram is a kind of classic color feature. However, simple color histograms are poor abstractions of images as two visually dissimilar images can have the same histogram. The saliency weighted color histogram alleviates this problem. We compute the histogram in HSV color space in which the colors are quantized to 256 level colors in proportion to the criterion $H : S : V = 16 : 4 : 4$. The saliency weighted color histogram is then defined as:

$$H_{SWC}(k) = \frac{sum(S(color(k)))}{sum(S)}. \tag{3}$$

where $color(k)$ is the k-th color, $sum(S(color(k)))$ is the sum of the elements in S whose corresponding pixels in the HSV space have the k-th color value.

Saliency Weighted Texture Histogram. Similarly to the saliency weighted color histogram definition, the saliency weighted texture histogram is defined based on a texture map. We use Local binary pattern (LBP) [12] to extract the image texture feature and get a LBP map. The values in LBP map are quantized to 64 bins. The saliency weighted texture histogram is then defined as:

$$H_{SWT}(k) = \frac{sum(S(LBP(k)))}{sum(S)}. \tag{4}$$

where $LBP(k)$ is the k-th value, $sum(S(LBP(k)))$ is the sum of the elements in S whose corresponding pixels in the LBP map have the k-th LBP value.

Brightness Distribution Vector. Each tile of the mosaic has its own brightness distribution, for example, the left part of the tile may be darker and the right part may be lighter. During the image matching, it is better to find the image with the same brightness distribution. Therefore we propose a brightness distribution vector to describe the image brightness distribution. Firstly the original image is averagely divided into N blocks. Then the N dimensional brightness distribution vector can be obtained by computing the average brightness of each block. For simplicity, the image is resized to 256×256 and divided into 64 blocks each with 32×32 size. The vector is simplify defined as:

$$V(j) = \frac{sum(B(j))}{32 \times 32}. \tag{5}$$

where $B(j)$ is the j-th block and $sum(B(j))$ is the sum of all elements in $B(j)$.

Image Feature Fitting. The Euclidean distance between feature vectors are introduced to measure the difference between the tile and a certain image from the database. Let F^t denotes the tile feature and F^q denotes the image feature, the difference between the two is defined as:

$$d(F^t, F^q) = \alpha D(H_{SWC}^t, H_{SWC}^q) + \beta D(H_{SWT}^t, H_{SWT}^q) + \gamma D(V^t, V^q). \tag{6}$$

where $\alpha \in [0,1]$, $\beta \in [0,1]$, $\gamma \in [0,1]$ and $\alpha + \beta + \gamma = 1$. The function $D(T,Q)$ is the Euclidean distance:

$$D(T,Q) = \| T \cdot Q \|. \tag{7}$$

In addition, if the mosaic tile is small, the image with similar brightness distribution is apt to be found; if the mosaic tile is large, the image with similar texture is apt to be found. Therefore during the realistic computation, the values of α, β, γ are dynamically changed corresponding to each tile size.

4   Experimental Results

We choose 5000 images from the Corel database to form a representative database. The mosaics shown in this paper all consist of images retrieved from this database.

The process of the proposed approach is presented in Fig. 2. Fig. 2(a) shows the target image. Fig. 2(b) shows the novel energy map obtained by combining NIF map and GBVS map. Fig. 2(c) shows the divided tiles with variable sizes, which are obtained through the energy map based decomposition technique. Fig. 2(d) shows the final mosaic result by filling each tile with retrieved images.

A comparison between the conventional image mosaic and the proposed image mosaic is shown in Fig. 1. The target panda image is shown in Fig. 1(a). Fig. 1(b) shows the conventional mosaic result with the same size tiles, while Fig. 1(c) shows the proposed mosaic result with variable-size tiles. The panda's eyes and nose are distorted in Fig. 1(b), but they are well presented in Fig. 1(c).

Face Segmentation Constraint. For human faces, the facial skin texture is relatively homogeneous and the neck skin is similar to the facial skin. Therefore, doing mosaic to the human portrait may lead to distortion in the face and neck, such as the mosaic shown in Fig. 5(a). To handle this problem, we apply a face segmentation constraint to the proposed mosaic approach. The segmented face region is used to weight the energy map. The weighted energy map highlights the human face so that it can be divided into more detailed tiles. Fig. 5(b) shows the result with face segmentation constraint. In Fig. 5(b) the face region is represented in detail and separated from the neck.

Color Correction. The mosaic composed of chosen tile images may sometimes not match the target image in color. In this situation, we need to alter the two to be more uniform in color. The objective is to better match the color of the tile

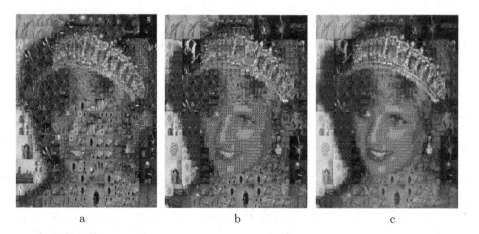

a   b   c

Fig. 5. Different image mosaics. (a) Mosaic without face segmentation constraint; (b) Mosaic with face segmentation constraint; (c) Mosaic with Color Correction.

images to the color of the region in the target image that is covered by the tile. A simple method is using the weighted average method. The chosen tile images and the target image are weighted averaged to get the final mosaic result. Fig. 5(c) shows a more living result obtained by applying color correction to the mosaic in Fig. 5(b).

5   Conclusions

In this paper we propose a content aware image mosaic approach based on saliency. Image saliency is used both in the target image decomposition step and the tile image retrieval step. Different saliency maps are chosen for different purposes. The graph-based visual saliency map is combined with the NIF map to generate a novel energy map for variable-size tiles decomposition. The FT saliency map is used in image retrieval for tile image choosing. The proposed approach creates better mosaics in visual aspect than that of the conventional methods. In current mosaic results, the tile images are simply arranged together, each tile is a rectangle. In the next work, we can create a collage from the tile images, which can give better visual aspect in mosaics. Until now there is no clear definition or measure to evaluate the performance of image mosaic. In future we may try to make a criterion.

Acknowledgments. This work is supported in part by 973 Program under Grant 2012CB316304, in part by the NSFC under Grant 61103059 and 61173104, in part by the NSF of Jiangsu Province under Grant BK2011700 and Open Project Program of the National Laboratory of Pattern Recognition (NLPR).

References

1. Finkelstein, A., Range, M.: Image Mosaics. In: Hersch, R.D., André, J., Brown, H. (eds.) RIDT 1998 and EPub 1998. LNCS, vol. 1375, pp. 11–22. Springer, Heidelberg (1998)
2. Silvers, R., Hawley, M.: Photomosaics. Henry Holt, New York (1997)
3. Tran, N.: Generating photomosaics: an empirical study. In: Proceedings of the Symposium on Applied Computing (1999)
4. Ding, J., Shen, J., Pang, H., Chen, S., Yang, J.: Exploiting Intensity Inhomogeneity to Extract Textured Objects from Natural Scenes. In: Zha, H., Taniguchi, R.-i., Maybank, S. (eds.) ACCV 2009, Part III. LNCS, vol. 5996, pp. 1–10. Springer, Heidelberg (2010)
5. Achanta, R., Shaji, A., Fua, P., Susstrunk, S.: Image Summaries using Database Saliency. In: ACM SIGGRAPH ASIA (2009)
6. Achanta, R., Susstrunk, S.: Saliency Detection for Content-aware Image Resizing. In: IEEE Conf. on Image Processing (2009)
7. Harel, J., Koch, C., Perona, P.: Graph-based visual saliency. In: NIPS (2006)
8. Achanta, R., Hemami, S., Estrada, F., Susstrunk, S.: Frequency-tuned salient region detection. In: CVPR (2009)

9. Zou, G., Ma, R., Ding, J., Zhong, B.: Image retrieval based on saliency weighted color and texture. Journal of Shandong University (Natural Science) (2010) (in Chinese)
10. Wang, M., Hua, X., Hong, R., Tang, J., Qi, G., Song, Y.: Unified Video Annotation Via Multi-Graph Learning. IEEE Transactions on Circuits and Systems for Video Technology, 733–746 (2009)
11. Wang, M., Hong, R., Li, G., Zha, Z., Yan, S., Chua, T.: Event Driven Web Video Summarization by Tag Localization and Key-Shot Identification. IEEE Transactions on Multimedia, 975–985 (2012)
12. Ojala, T., Pietikainen, M., Harwood, D.: A Comparative Study of Texture Measures with Classification Based on Feature Distributions. Pattern Recognition 29, 51–59 (1996)

Combining Visual and Textual Systems within the Context of User Feedback

Leszek Kaliciak[1], Dawei Song[2], Nirmalie Wiratunga[1], and Jeff Pan[3]

[1] The Robert Gordon University, Aberdeen, UK
[2] The Open University, Milton Keynes, UK
[3] Aberdeen University, Aberdeen, UK
{l.kaliciak,n.wiratunga}@rgu.ac.uk, Dawei.Song@open.ac.uk,
jeff.z.pan@abdn.ac.uk

Abstract. It has been proven experimentally, that a combination of textual and visual representations can improve the retrieval performance ([20], [23]). It is due to the fact, that the textual and visual feature spaces often represent complementary yet correlated aspects of the same image, thus forming a composite system.

In this paper, we present a model for the combination of visual and textual sub-systems within the user feedback context. The model was inspired by the measurement utilized in quantum mechanics (QM) and the tensor product of co-occurrence (density) matrices, which represents a density matrix of the composite system in QM. It provides a sound and natural framework to seamlessly integrate multiple feature spaces by considering them as a composite system, as well as a new way of measuring the relevance of an image with respect to a context. The proposed approach takes into account both intra (via co-occurrence matrices) and inter (via tensor operator) relationships between features' dimensions. It is also computationally cheap and scalable to large data collections. We test our approach on ImageCLEF2007photo data collection and present interesting findings.

Keywords: Visual and Textual Systems' Combination, Visual Features, Textual Features, User Feedback, Tensor Product, Density Matrix, Expectation Value.

1 Introduction

It has been proven experimentally (i.e. the annual imageCLEF competition results) that a combination of textual and visual representations can improve the retrieval performance ([20], [23]). It is due to the fact, that the textual and visual feature spaces often represent complementary yet correlated aspects of the same image, thus forming a composite system. This, in turn, presents an opportunity to utilize this complemetarity by combining the systems in order to improve their performance.

Visual and textual systems can be combined within the context of image retrieval or automatic image annotation. The latter exploits the relationships between the features' dimensions to automatically annotate images that do not have textual descriptions. However, even after auto-annotating the images, the retrieval system often (apart from some projection based methods, i.e. LSI) needs to combine the features in a meaningful way in order to utilize the complementarity of the aforementioned feature spaces

S. Li et al. (Eds.): MMM 2013, Part I, LNCS 7732, pp. 445–455, 2013.

to improve the retrieval. Some of these combination methods can be modified to incorporate the user feedback.

This paper focuses on the combination of the systems within the context of image retrieval, and to be more precise - the context of a user feedback. The data collection that we conduct our experiments on, ImageCLEF2007photo, is a fully annotated one (albeit the description field which was present in the ImageCLEF2006 collection is now unavailable).

Thus, most approaches that combine visual and textual features in content based image retrieval systems are fusion methods that would:

1. pre-filter the data collection by visual content and then re-rank the top images by text ([4]);
2. pre-filter the data collection by text and then re-rank the top images by visual content ([5]);
3. pre-filter the data collection by visual (textual) content and then aggregate the scores of the textual (visual) representations of the top retrieved images (transmedia pseudo-relevance mechanism [6]);
4. fuse the representations (early fusion [7]);
5. fuse the scores or ranks (late fusion [8]).

This paper is organized as follows: Section 2 presents work related to the combination of visual and textual features in general. Section 3 describes the theoretical model for combination of visual and textual systems in the context of user feedback. The experimental setup and results with their discussion forms the next section, Section 4. Finally, Sections 5 and 6 are devoted to conclusions and future work, respectively.

2 Related Work

In this work, we modify the existing models (that combine visual and textual features) in order to incorporate user feedback. Thus modified approaches will serve as our comparison baselines.

Pre-filtering by text and re-ranking by visual content is usually a well performing method. However, the main drawback of this approach is that the images without the textual description will never be returned by the system (although one could try to auto-annotate the collection beforehand). Moreover, this type of pre-filtering relies heavily on the textual features and the assumption that the images are correctly annotated.

The most common early fusion technique is concatenation of visual and textual representations. Some recently proposed models incorporate the tensor product to combine the systems [9]. The aforementioned tensor product presents a sound fusion technique as it takes into account all of the combinations of different features' dimensions. The main drawback of the early fusion approach, however, is the well known curse of dimensionality. Later in the paper we show, that the curse of dimensionality can often be avoided as the similarity between the fused representations may be characterized as the combinations of similarities computed on individual feature spaces.

In case of the late fusion, the most widely used method is the arithmetic mean of the scores, their sum (referred to as CombSUM), or their weighted linear combination. One of the best performing systems on the ImageCLEF2007 data collection, XRCE [10], utilizes both (for comparison purposes) early (concatenation of features) and late (an average of scores) fusion approaches. Another common combination method, referred to as CombPROD in the literature, is the square of the geometric mean of the scores - their product. It has been argued, that the major drawback of the late fusion approaches is their inability to capture the correlation between different modalities [11]. However, later in the paper we show, that in some cases the late fusion can be represented as early fusion.

Other features' combination methods involve a combination of late fusion and image re-ranking [12] (because the first step is the pre-filtering of the collection by text, the model is called semantic combination). Some researchers [9] experimented with ten-soring of the representations and modeling the inherent dependencies between features' dimensions (although the incorporation of dependencies did not improve the retrieval effectiveness and the model was not scalable to large image collections due to its high computational cost).

The fusion approach that can be easily modified to incorporate the user feedback is based on the transmedia pseudo-relevance mechanism. This so-called inter-media feedback query expansion is based on textual query expansion in most of the papers ([13],[14]). Typically, textual annotations from the top visually-ranked images (or from a mixed run) are used to expand a textual query.

There is a proliferation of other models that utilize user feedback (mono-modal) in order to improve the retrieval. In this paper, however, we focus on the issue of com-bining the visual and textual features in the context of user feedback, therefore we are interested in hybrid approaches that combine the visual and textual features, and also hybrid approaches that combine them within the context of user feedback.

Our main contribution is the proposed model for combining visual and textual sys-tems within the context of user feedback. The model was inspired by the expectation value of the measurement utilized in quantum mechanics and the tensor product of the density matrices of the systems (that results in a density matrix of the composite system). It was designed to capture both intra-relationships between features' dimen-sions (visual and textual correlation matrices) and inter-relationships between visual and textual representations (tensor product). The model provides a sound and natural framework to seamlessly integrate multiple feature spaces by considering them as a composite system, as well as a new way of measuring the relevance of an image with respect to a context by applying quantum-like measurement. It opens a door for a series of theoretically well-founded further exploration routes, e.g. by considering the inter-ference among different features.

3   Combining Visual and Textual Features within the Context of User Feedback

Modern retrieval systems allow the users to interact with the system in order to narrow down the search. This interaction takes the form of implicit or explicit feedback. The representations of the images in the feedback set are often aggregated or concatenated (or co-occurrence matrices may be aggregated to represent i.e. probability distribution matrix). The information extracted from the feedback set is utilized to expand the query or re-rank the top images returned in the first round of the retrieval.

Here, we are going to introduce our model for visual and textual systems' combination within the context of a user feedback. The proposed model was inspired by the measurement used in quantum mechanics, which is based on an expectation value, predicted mean value of the measurement

$$\langle A \rangle = tr\left(\rho A\right) \tag{1}$$

where tr denotes the trace operator, ρ represents a density matrix of the system and A is an observable. We can also represent an observable A as a density matrix (corresponding to the query or an image in the collection). For more information on the analogies between quantum mechanics and information retrieval the curious reader is referred to [17].

We are going to use the tensor operator $\otimes$ to combine the density matrices corresponding to visual and textual feature spaces. In quantum mechanics, the tensor product of density matrices of different systems represents a density matrix of the combined system (see [15]).

Thus, the proposed measurement is represented by

$$tr\left((M_1 \otimes M_2) \cdot \left((a^T \cdot a) \otimes (b^T \cdot b)\right)\right) \tag{2}$$

where M_1, M_2 represent density matrices (co-occurrence matrices) of the query and images in the feedback set corresponding to visual and textual spaces respectively, a and b denote vectors representing visual and textual information for an image from the data collection, and T is a transpose operation. We would perform this measurement on all the images in the collection, thus re-scoring the dataset based on the user feedback.

Assuming that the systems were prepared independently (otherwise we would have to try to model a concept analogous to entanglement [18]), we get

$$tr\left((M_1 \otimes M_2) \cdot \left((a^T \cdot a) \otimes (b^T \cdot b)\right)\right) =$$
$$tr\left((M_1 \cdot (a^T \cdot a)) \otimes (M_2 \cdot (b^T \cdot b))\right) =$$
$$tr\left(M_1 \cdot (a^T \cdot a)\right) \cdot tr\left(M_2 \cdot (b^T \cdot b)\right) =$$
$$\langle M_1 | a^T \cdot a \rangle \cdot \langle M_2 | b^T \cdot b \rangle \tag{3}$$

where $\langle \cdot | \cdot \rangle$ denotes an inner product operating on a vector space.

Let q_v, q_t denote the visual and textual representations of the query, c^i, d^i denote visual and textual representations of the images in the feedback set, r_1, r_2 denote the weighting factors (constant, importance of query and feedback density matrices respectively), and n denote the number of images in the feedback set. Then, we define M_1 and M_2 as weighted combinations of co-occurrence matrices (a subspace generated by the query vector and vectors from the feedback set)

$$M_1 = r_1 \cdot D_q^v + \frac{r_2}{n} \cdot D_f^v =$$
$$r_1 \cdot q_v^T \cdot q_v + \sum_i \left(\frac{r_2}{n} \cdot (c^i)^T \cdot c^i\right) \tag{4}$$

and

$$M_2 = r_1 \cdot D_q^t + \frac{r_2}{n} \cdot D_f^t =$$

$$r_1 \cdot q_t^T \cdot q_t + \sum_i \left(\frac{r_2}{n} \cdot \left(d^i \right)^T \cdot d^i \right) \tag{5}$$

Co-occurrence matrices are quite often utilized in the Information Retrieval (IR) field. Because they are Hermitian and positive-definite, they can be thought of as density matrices (probability distribution). The common way of co-occurence matrix generation is to multiply the term-document matrix by its transpose (rows of the matrix represent the documents $d_1, \ldots d_m$), that is $D = M^T \cdot M$. Notice, that this is equivalent to $D = \sum_{i=1}^n d_i^T \cdot d_i$.

This observation, due to the properties of the inner product, will allow us to further simplify our model

$$\langle M_1 \otimes M_2 | \left(a^T \cdot a \right) \otimes \left(b^T \cdot b \right) \rangle = \langle M_1 | a^T \cdot a \rangle \cdot \langle M_2 | b^T \cdot b \rangle =$$

$$\left\langle r_1 \cdot q_v^T \cdot q_v + \sum_i \left(\frac{r_2}{n} \cdot \left(c^i \right)^T \cdot c^i \right) | a^T \cdot a \right\rangle \cdot$$

$$\left\langle r_1 \cdot q_t^T \cdot q_t + \sum_i \left(\frac{r_2}{n} \cdot \left(d^i \right)^T \cdot d^i \right) | b^T \cdot b \right\rangle =$$

$$\left(\langle r_1 \cdot q_v^T \cdot q_v | a^T \cdot a \rangle + \sum_i \frac{r_2}{n} \left\langle \left(c^i \right)^T \cdot c^i | a^T \cdot a \right\rangle \right) \cdot$$

$$\left(\langle r_1 \cdot q_t^T \cdot q_t | b^T \cdot b \rangle + \sum_i \frac{r_2}{n} \left\langle \left(d^i \right)^T \cdot d^i | b^T \cdot b \right\rangle \right) =$$

$$\left(r_1 \cdot \langle q_v | a \rangle^2 + \frac{r_2}{n} \cdot \sum_i \langle c^i | a \rangle^2 \right) \cdot \left(r_1 \cdot \langle q_t | b \rangle^2 + \frac{r_2}{n} \cdot \sum_i \langle d^i | b \rangle^2 \right) \tag{6}$$

Notice, that the model breaks down into the weighted combinations of individual measurements. The squares of the inner products come from the correlation matrices and can play an important role in the measurement. Later in the paper, we are going to justify this claim.

We can consider a variation of the aforementioned model, where just like in the original one $M_1 = r_1 \cdot D_q^v + \frac{r_2}{n} \cdot D_f^v$ and $M_2 = r_1 \cdot D_q^t + \frac{r_2}{n} \cdot D_f^t$. We can decompose (eigenvalue decomposition) the density matrices M_1, M_2 to estimate the bases[1] (p_i^v, p_j^t) of the subspaces generated by the query and the images in the feedback set. Now, let us consider the measurement

$$\langle P_1 \otimes P_2 | \left(a^T a \right) \otimes \left(b^T b \right) \rangle \tag{7}$$

---

[1] It has been highlighted [19] that the orthogonal decomposition may not be the best option for visual spaces because the receptive fields that result from this process are not localized, and the vast majority do not at all resemble any known cortical receptive fields. Thus, in the case of visual spaces, we may want to utilize decomposition methods that produce non-orthogonal basis vectors.

where P_1, P_2 are the projectors onto visual and textual subspaces generated by query and the images in the feedback set $(\sum_i (p_i^v)^T p_i^v, \sum_j (p_j^t)^T p_j^t)$, and a, b are the visual and textual representations of an image from the data set. Because the tensor product of the projectors corresponding to visual and textual Hilbert spaces (H_1, H_2) is a projector onto the tensored Hilbert space $(H_1 \otimes H_2)$, the measurement (7) can be interpreted as probability of relevance context, the probability that vector $a \otimes b$ was generated within the subspace (representing the relevance context) generated by $M_1 \otimes M_2$. Hence

$$\langle P_1 \otimes P_2 | \left(a^T a\right) \otimes \left(b^T b\right) \rangle =$$
$$\langle P_1 | a^T a \rangle \cdot \langle P_2 | b^T b \rangle =$$
$$\left\langle \sum_i (p_i^v)^T p_i^v | a^T a \right\rangle \cdot \left\langle \sum_j (p_j^t)^T p_j^t | b^T b \right\rangle =$$
$$\sum_i \langle p_i^v | a \rangle^2 \cdot \sum_j \langle p_j^t | b \rangle^2 =$$
$$\sum_i Pr_i^v \cdot \sum_j Pr_j^t =$$
$$\left\| \left(\langle p_1^v | a \rangle, \ldots, \langle p_n^v | a \rangle\right) \otimes \left(\langle p_1^t | b \rangle, \ldots, \langle p_n^t | b \rangle\right) \right\|^2 \tag{8}$$

where Pr denotes the projection probability and $\|\cdot\|$ represents vector norm.

We can see, that this measurement is equivalent to the weighted combinations of all the probabilities of projections for all the images involved. In quantum mechanics, the square of the absolute value of the inner product between the initial state and the eigenstate is the probability of the system collapsing to this eigenstate. In our case, the square of the absolute value of the inner product can be interpreted as a particular contextual factor influencing the measurement.

In this paper, we are going to experimentally test the model based on the expectation value of the measurement and the tensor product of density matrices. The proposed model can incorporate both implicit (i.e. query history) and explicit (i.e. relevance data) forms of user feedback.

4   Experiments and Discussion

We evaluate the proposed model on ImageCLEFphoto 2007 data collection [20]. ImageCLEFphoto2007 consists of 20000 everyday real-world photographs. It is a standard collection used by Information Retrieval (IR) community for evaluation purposes. This allows comparison with published results. There are 60 query topics that do not belong to the collection.

Because of the abstract semantic content of many of the queries, ImageCLEFphoto 2007 data collection is considered to be very difficult for retrieval systems. For example, the topic "straight road in the USA" could be very difficult for visual features whereas "church with more than two towers" could render the textual features helpless. That is why the hybrid models should play an important role in modern retrieval systems.

4.1 Experimental Setup

We test our model (expectation value with a tensor product of density matrices) within a simulated user feedback framework. First, we perform the first round retrieval for a topic from the query set based on the visual features only (we retrieve 1000 images). We use the visual features only because in the real life scenario many images would not have textual descriptions. We also do not combine the features in the first round retrieval as this would represent a different task. In this work we want to focus on testing the features' combination models within the user feedback framework.

Next, we identify 1, 2 and 3 relevant images respectively from the highest ranked images based on the ground truth data (starting from the most similar). Thus obtained images simulate the user feedback and are utilized in the proposed model to re-score the data collection. For each query topic (60 in total) we calculate mean average precision (MAP) for the top 20 retrieved images, as most users would only look at this number of documents. We set the weights r_1, r_2 to 1 and 0.8 respectively (standard weights' values for query and its context as in the classic Rocchio algorithm, for example).

The visual features used in the experiment are based on the Bag of Visual Words framework (see [21] for a detailed description). They are regarded as a mid-level representation.

The textual features were obtained by applying the standard Bag of Words technique, with Porter stemming, stop words removal, and term frequency - inverse document frequency weighting scheme.

4.2 Experimental Results and Discussion

As aforementioned, we modify existing models in order to incorporate the user feedback. We use several baselines for comparison purposes.

Thus, early fusion is represented by a modified Rocchio algorithm (*earlyFusion*). The only difference between this variation and the classic model is that we apply it to concatenated visual and textual vectors, as opposed to visual or textual representations only. Let $\oplus$ denote the concatenation operation (other notation as in the previous sections). Then, this model modify the query in a following way

$$newQuery = q_v \oplus q_t + \frac{0.8}{n} \sum_i (c_i \oplus d_i) \tag{9}$$

After the query modification the scores are recomputed.

Another baseline, which we will refer to as *lateFusion* will be represented as a combination of all the scores

$$sim(q_v, a) + \frac{0.8}{n} \sum_i sim(c_i, a) + sim(q_t, b) + \frac{0.8}{n} \sum_i sim(d_i, b) \tag{10}$$

where sim denotes the similarity between given vectors. In this work sim is an inner product between two vectors.

Our third baseline *rerankText* denotes the re-ranking of the results obtained from the first round retrieval based on the aggregated textual representations of the feedback images. Similarly, $rerankVis$ represents re-ranking of the top retrieved images based on the aggregated visual representations of the images from the feedback set.

Next model $trMedia$ represents, as the label suggests, inter-media feedback query modification. Here, textual annotations from the feedback images (identified by visual features) are used to expand a textual query.

The system performance without simulated feedback will be denoted as $noFeedback$ and the proposed model for combination of visual and textual features within the context of simulated relevance feedback will be denoted as $prMeanMeasure$.

Table 1 presents the obtained results.

Table 1. Simulated Relevance Feedback, ImageCLEF2007photo results (MAP)

	1 Feedback Image	2 Feedback Images	3 Feedback Images
$noFeedback$	0.013	0.013	0.013
prMeanMeasure	**0.079**	**0.094**	**0.11**
$earlyFusion$	0.066	0.082	0.085
$lateFusion$	0.066	0.082	0.085
$rerankText$	0.055	0.069	0.075
$rerankVis$	0.034	0.036	0.031
$trMedia$	0.061	0.078	0.081

From the experimental results we can see, that the best performing model is based on the proposed predicted mean value of the measurement ($prMeanMeasure$) with the density matrix of the composite system (tensor product of the subspaces). The difference (in terms of means) between $prMeanMeasure$ and the rest of the baselines is statistically significant (paired t-test, $p < 0.05$). The inter-media feedback query expansion ($trMedia$) also performed well, albeit worse than early and late fusion ($earlyFusion$, $lateFusion$). In general, all the models' performance suggests that they are quite effective in utilizing users' feedback.

An interesting observation is that both early ($earlyFusion$, modified Rocchio) and late fusion strategies ($lateFusion$, combination of scores) show exactly the same performance. It is because

$$newQuery = q_v \oplus q_t + \frac{0.8}{n} \sum_i (c_i \oplus d_i)$$

$$imagesInDataset = a \oplus b \quad forAll \ a, b \in Dataset \tag{11}$$

$$\langle newQuery | imagesInDataset \rangle =$$

$$\left\langle q_v \oplus q_t + \frac{0.8}{n} \sum_i (c_i \oplus d_i) \, | a \oplus b \right\rangle =$$

$$\langle q_v \oplus q_t | a \oplus b \rangle + \frac{0.8}{n} \sum_i \langle c_i \oplus d_i | a \oplus b \rangle =$$

$$\langle q_v | a \rangle + \frac{0.8}{n} \sum_i \langle c_i | a \rangle + \langle q_t | b \rangle + \frac{0.8}{n} \sum_i \langle d_i | b \rangle \tag{12}$$

Thus, in our case the early and late fusion strategies (modified Rocchio algorithm operating on concatenated representations and weighted linear combination of scores) are interchangeable. We are going to address this interesting discovery in our future work.

We observe, that even one feedback image can help to narrow down the search, thus increasing the match between user's preferences (in this case, a human expert who assesed the relevance of images in ground truth data). Let us assume, that the visual query pictures a person wearing sunglasses. In the first round retrieval, the system may recognize (return more images of) a concept representing sunglasses without a person present on the picture. However, the human assesor might have deemed an image relevant only if both concepts were present in the image. A user feedback can then reinforce the subjective (perceived) relevance of the query to the retrieved images. In case of using the visual representations only in the user feedback ($rerankVis$), more images in the feedback set can sometimes confuse the visual features (especially if they significantly differ in terms of colour, texture, viewpoint or illumination). Thus, approaches like $rerankVis$ may strongly depend on the type of visual features used (while visual features A may be suitable for the particular feedback set C, visual features B may not work so well on C and vice versa).

In this work, the MAP is calculated for 20 top images only as this is a more realistic scenario (especially for user simulation/user feedback context). However, for 1000 top and 3 feedback images, the system performace is approximately $MAP \approx 0.206$. If we consider the ImageCLEF2007photo results of other systems (the best models utilize both visual and textual information) which can be found on the ImageCLEF website [23], the proposed model places itself among the best performing approaches. However, it must be noted that our model combines visual and textual features within the context of user feedback framework (different task).

We also need to take into consideration the disadvantages of automatic evaluation methods. The ultimate test for every retrieval system (especially for user simulation/user feedback context) should be the real user evaluation (although it is a time consuming task). The relevance of an image is a highly subjective concept and the automatic evaluation seems to fail to address this problem. Moreover, there is a glitch in the trec-eval evaluation software, that can bring the reported results into question. To be more specific, if some images obtain the same similarity score, they will be re-ordered by the software. The result is that two identical submissions may get different performance scores.

5   Conclusions

In this paper, we have presented the model for visual and textual features' combination within the context of user feedback. The approach is based on mathematical tools also used in quantum mechanics - the predicted mean value of the measurement and the tensor product of the density matrices, which represents a density matrix of the combined systems. It was designed to capture both intra-relationships between features' dimensions (visual and textual correlation matrices) and inter-relationships between visual and textual representations (tensor product). The model provides a sound and natural framework to seamlessly integrate multiple feature spaces by considering them as a composite system, as well as a new way of measuring the relevance of an image with respect to a context by applying quantum-like measurement. It opens a door for a series of theoretically well-founded further exploration routes, e.g. by considering the interference among different features. It is easily scalable to large data collections as it is general and computationally cheap. The results of the experiment conducted on ImageCLEF data collection show the significant improvement over other baselines.

6 Future Work

The future work will involve testing different notions of correlation within the proposed framework (we can construct correlation matrices in such a way that they can be regarded as density matrices). In this paper, we incorporate document/image level correlations only. However, in case of textual representations, we can also experiment with Hyperspace Analogue to Language (HAL). In the aforementioned approach, the context is represented by a sliding window of a fixed size (while in document level correlation the context is represented by the whole document). We can also consider a visual counterpart to HAL, where a window of a fixed size (e.g. square, circular) is shifted from one instance of a visual word to another. Then, the number of instances of visual words that appear in the proximity of the visual word on which the window is centered can be calculated. In case of a dense sampling, the window would be shifted analogously to HAL in text IR. If the sparse sampling was utilized, however, the window would shift from one instance of a visual word to another.

References

1. Zhao, R., Grosky, W.I.: Narrowing the semantic gap-improved text-based web document retrieval using visual features. IEEE Transactions on Multimedia 4, 189–200 (2002)
2. Ferecatu, M., Sahbi, H.: TELECOM ParisTech at Image Clef photo 2008: Bi-modal text and image retrieval with diversity enhancement. In: Working Notes of CLEF (2008)
3. Martinez-Fernandes, J.L., Serrano, A.G., Villena-Roman, J., Saenz, V.D.M., Tortosa, S.G., Castagnone, M., Alonso, J.: MIRACLE at ImageCLEF 2004. In: Working Notes of CLEF (2004)
4. Yanai, K.: Generic image classification using visual knowledge on the web. In: Proceedings of the 11th ACM International Conference on Multimedia, pp. 167–176 (2003)
5. Tjondronegoro, D., Zhang, J., Gu, J., Nguyen, A., Geva, S.: Integrating Text Retrieval and Image Retrieval in XML Document Searching. In: Fuhr, N., Lalmas, M., Malik, S., Kazai, G. (eds.) INEX 2005. LNCS, vol. 3977, pp. 511–524. Springer, Heidelberg (2006)
6. Maillot, N., Chevallet, J.P., Valea, V., Lim, J.H.: IPAL Inter-media pseudo-relevance feedback approach to ImageCLEF 2006 photo retrieval. In: CLEF Working Notes (2006)
7. Rahman, M.M., Bhattacharya, P., Desai, B.C.: A unified image retrieval framework on local visual and semantic concept-based feature spaces. J. Visual Communication and Image Representation 20(7), 450–462 (2009)
8. Simpson, M., Rahaman, M.M.: Text and content-based approaches to image retrieval for the ImageCLEF 2009 medical retrieval track. In: Working Notes for the CLEF 2009 Workshop (2009)

9. Wang, J., Song, D., Kaliciak, L.: Tensor product of correlated text and visual features: a quantum theory inspired image retrieval framework. In: AAAI-Fall 2010 Symposium on Quantum Information for Cognitive, Social, and Semantic Processes, pp. 109–116 (2010)

10. Mensink, T., Csurka, G., Perronnin, F.: LEAR and XRCE's participation to visual concept detection task - ImageCLEF 2010. In: Proceedings of the 14th Annual ACM International Conference on Multimedia, pp. 77–80 (2006)

11. Mensink, T., Verbeek, J., Csurkay, G.: Weighted transmedia relevance feedback for image retrieval and auto-annotation. Technical Report Number 0415 (2011)

12. Clinchant, S., Ah-Pine, J., Csurka, G.: Semantic combination of textual and visual information in multimedia retrieval. In: ACM International Conference on Multimedia Retrieval, ICMR (2011)

13. Depeursinge, A., Muller, H.: Fusion techniques for combining textual and visual information retrieval. In: ImageCLEF. The Springer International Series on Information Retrieval, vol. 32, pp. 95–114 (2010)

14. Chang, Y.-C., Chen, H.-H.: Increasing Precision and Diversity in Photo Retrieval by Result Fusion. In: Peters, C., Deselaers, T., Ferro, N., Gonzalo, J., Jones, G.J.F., Kurimo, M., Mandl, T., Peñas, A., Petras, V. (eds.) CLEF 2008. LNCS, vol. 5706, pp. 612–619. Springer, Heidelberg (2009)

15. Combining systems: the tensor product and partial trace, http://www.quantum.umb.edu/Jacobs/QMT/QMT-AppendixA.pdf

16. Li, Y., Cunningham, H.: Geometric and quantum methods for information retrieval. SIGIR Forum 42(2), 22–32 (2008)

17. van Rijsbergen, C.J.: The geometry of information retrieval. Cambridge University Press (2004)

18. Bruza, P.D., Kitto, K., Nelson, D., McEvoy, C.L.: Entangling words and meaning. In: Proceedings of the 2nd Quantum Interaction Symposium, pp. 118–124 (2008)

19. Olshausen, B.A., Field, D.J.: Emergence of simple-cell receptive field properties by learning a sparse code for natural images. Nature 381, 607–609 (1996)

20. Grubinger, M., Clough, P., Hanbury, A., Muller, H.: Overview of the ImageCLEF 2007 photographic retrieval task. In: Working Notes of the 2007 CLEF Workshop (2007)

21. Kaliciak, L., Song, D., Wiratunga, N., Pan, J.: Novel local features with hybrid sampling technique for image retrieval. In: Proceedings of Conference on Information and Knowledge Management (CIKM), pp. 1557–1560 (2010)

22. Nowak, E., Jurie, F., Triggs, B.: Sampling Strategies for Bag-of-Features Image Classification. In: Leonardis, A., Bischof, H., Pinz, A. (eds.) ECCV 2006. LNCS, vol. 3954, pp. 490–503. Springer, Heidelberg (2006)

23. ImageCLEF website, http://www.imageclef.org

24. Grubinger, M., Clough, P., Hanbury, A., Müller, H.: Overview of the ImageCLEFphoto 2007 Photographic Retrieval Task. In: Peters, C., Jijkoun, V., Mandl, T., Müller, H., Oard, D.W., Peñas, A., Petras, V., Santos, D. (eds.) CLEF 2007. LNCS, vol. 5152, pp. 433–444. Springer, Heidelberg (2008)

25. Chen, Z., Liu, W., Zhang, F., Li, M.J., Zhang, H.J.: Web mining for web image retrieval. Journal of the American Society for Information Science and Technology 52(10), 831–839 (2001)

A Psychophysiological Approach to the Usability Evaluation of a Multi-view Video Browsing Tool

Carmen Martinez-Peñaranda[1], Werner Bailer[2], Miguel Barreda-Ángeles[1], Wolfgang Weiss[2], and Alexandre Pereda-Baños[1]

[1] Barcelona Media, Barcelona, Spain
maikaranda@gmail.com,
{miguel.barreda,alexandre.pereda}@barcelonamedia.org
[2] JOANNEUM RESEARCH – DIGITAL, Graz, Austria
{werner.bailer,wolfgang.weiss}@joanneum.at

Abstract. The aim of this study is to investigate the usability of a video browsing tool. The tool aims at facilitating content navigation and selection in media post-production, supporting also multi-view content. Psychophysiological measures such as skin conductance level are used to measure cognitive effort. Objective measures based on content retrieval tasks as well as self-report measures of usability are also reported. Results indicate the differential effect of introducing specific support for multi-view content in the browsing tool, and encourage further research on the use of psychophysiological techniques in usability evaluations.

Keywords: psychophysiology, video browsing, usability, interactive search.

1   Introduction

With the increasing amount of multimedia data to be handled in production and post-production, there is growing demand for more efficient ways of supporting exploration and navigation of multimedia data. In post-production environments, users typically deal with large amounts of audiovisual material, such as newly shot scenes, archive material and computer generated sequences. A large portion of the material is unedited and often very redundant, e.g. containing several takes of the same scene shot by a number of different cameras. Typically only few metadata annotations are available (e.g. which production, which camera, which date). The goal is to support the user in navigating and organizing these audiovisual collections, so that unusable material can be discarded, yielding a reduced set of material from one scene or location available for selection in the post-production steps. Recently, boosted by 3D cinema and 3DTV, content is increasingly shot with stereo cameras or even more views, recording multiple views of one scene. While this increases the freedom in post-production (e.g., for inserting or moving objects, or adjusting the depth level), multi-view content increases the problem of content management in post-production, as even more (and more redundant) items need to be handled.

S. Li et al. (Eds.): MMM 2013, Part I, LNCS 7732, pp. 456–466, 2013.

Multimedia content abstraction methods such as browsing tools are complementary to search and retrieval approaches, as they allow for exploration of an unknown content set, without the requirement to specify a query in advance. This is relevant in cases where only few metadata are available for the content set, and where the user does not know what to expect in the content set, so that she is not able to formulate a query. In order to enable the user to deal with large sets of content, it has to be presented in a form which facilitates its comprehension and allows judging the relevance of segments of the content set. In order to support users in post-production, media content abstraction methods shall (i) support the user in quickly gaining an overview of a known or unknown content set, (ii) organize content by similarity in terms of any feature or group of features, and (iii) select representative content for subsets of the content set that can be used for visualization. The focus of this work is on assessing the added value of support for multi-view content in order to support users in media post-production.

In this paper we present an approach for evaluating a video browsing tool for multi-view content collections using psychophysiological methods. In Section 2, we give an overview of related work on evaluation approaches. Section 3 discusses the video browsing tool to be evaluated and the task design. We present the results of the experiments in Section 4 and draw conclusions in Section 5.

2   Related Work

2.1   Evaluation of Video Browsing and Interactive Video Search Tools

Most of the literature on evaluation of exploratory search deals with text documents. In the multimedia domain evaluation approaches for summarization and skimming systems often deal only with single multimedia documents, rather than with collections. The following classes of evaluation approaches have been proposed (of course combinations of the methods from different classes are sometimes used).

Survey (Self Report). The users are asked about their experience with the tool, their satisfaction with the results and the relevance of certain features of the tool (e.g., [10]). This type of evaluation does not require any ground truth or specific preparation of a data set.

Analysis of System Logs. This approach uses either server-side logs [2] or specific client applications that log user actions [3]. The main advantage is that evaluation does not interfere with the user's work with the system and that the approach can be used for long-term studies. However, comparison across different types of tasks and systems might be difficult.

Question Answering. Users are asked fact finding questions about the content in order to evaluate whether they have found the correct segment of content or were able to extract information from the collection of multimedia documents (e.g., [4]). The questions can be open or in the form of a multiple choice test (quiz). The correctness of answers to open questions needs to be checked by a human, while multiple choice tests can be very efficiently evaluated once the ground truth for a specific data set has been created.

Indirect Evaluation. The user performs a task using the tool or system. Based on the success of this task the effectiveness of the tool can be measured. The task can for example be a content retrieval task [1] or gathering information from a meeting archive browser [7]. Once ground truth for these tasks has been created the answers can be checked automatically.

In this paper, indirect evaluation using content search tasks is performed, in combination with psychophysiological methods and a set of open questions.

2.2   Subjective and Objective Evaluation Methods

As the terminology in the area is often used confusingly, let's clarify that by subjective methods we imply here self-reported qualitative measures, as those obtained by employing interviews, questionnaires or whatever method in which the user is directly asked about some aspect of his experience with a certain content and/or technology. On the other hand, objective evaluation is meant here to refer to the collection of indirect measures of users' experience, that often reveal reactions that are not consciously available to the user (such as variations in performance, motor behavior, emotional reaction and so on) and that therefore cannot be captured by self-report measures. A common occurrence is to refer to the former as observational evaluations and to the latter as experimental evaluations, though in principle, nothing prevents both of these methodologies to work with either objective or subjective data [9].

In self report, the user informs directly about his conscious perceptions, and therefore is "cognitively mediated" [12], and if users cannot report what they cannot perceive consciously, self-report measures are inherently flawed [6]. What is needed is to adopt a dimensional approach to the concept of user experience measurement, where the different aspects that contribute to an engaging experience are analyzed in consonance with their own characteristics; such aspects are as varied as perceptual quality, comfort, emotional reactions, attentional load, etc. The concept of quality of experience is a complex one, and any evaluation activity will need to take into account which of these aspects is a key in the experience delivered to the user. As regards the work discussed here, we consider that the key aspect to measure here is cognitive effort, we combine qualitative methods with indirect psychological measures, namely, performance measures and psychophysiological measures.

2.3   Psychophysiological Evaluation Techniques

Psychophysiological techniques are a perfectly suited methodology for testing cognitive effort in reaction to audiovisual media tasks in an indirect fashion, as this is the branch of psychology that deals with the physiological basis and indicators of psychological processes. Historically, the term has been reserved for the study of the responses of the autonomic nervous system. However, techniques such as EEG (electroencephalography, the recording of electrical cerebral activity) or fMRI (functional magnetic resonance imaging, the measure of changes in cerebral blood flow) also allow observing the activity of the central nervous system. Furthermore, other techniques that cannot be considered as psychophysiological measures (but also let to obtain information from users in a indirect way) such as eye-tracking, have been used on usability research [15]. In any case, for the study of cognitive effort, and certain attentional reactions such us orientation responses, this is still the method of choice for many researchers in the area. Traditionally, registering of autonomic activity is performed over three main measures:

- *Electrodermal activity (EDA)*: reflects changes in the electrical conductance of the skin due to the activity of the sweating glands induced by the sympathetic system, whose activity is related to the degree of emotional activation and it can also be an indicator of cognitive effort.
- *Electromyographyc activity (EMG)*: It measures muscular activity and is often employed to measure facial muscle activity. Depending on the registered muscle and stimuli conditions it can indicate the emotional valence (positive/negative valuation), discomfort and attentional capture.
- *Cardiac and respiratory rhythms*: indicators of phasic attentional responses as well as general stress levels.

These variables are registered by means of the polygraph, an instrument allowing capturing, modulating, amplifying and graphically registering these physiological systems. Regarding cognitive activation measurement, and our research objectives, EDA measures, are often used as cognitive activation and attentional indicators of the level of cognitive effort induced by audiovisual material [11]. Variations in the EDA levels reflect variations in arousal [8], which can signal increases in emotional activation or cognitive effort, depending on the stimulation conditions. In our research we focus on exploring the capacity of these technique for measuring cognitive effort.

3   Video Browsing Tool Evaluation

3.1   Tool Description

The developed video browsing tool performs automatic content analysis of newly ingested data. Currently, camera motion estimation, visual activity estimation, extraction of global color features and estimation of object trajectories are performed. In order to select content, the users follow an iterative selection process, consisting of alternating steps of clustering and selecting subsets of the current data set. Importantly, a new feature was introduced in the latest version enabling to handle multi-view content. This allows users, especially in 3D productions, to reduce the material they have to deal with, and therefore facilitating the search process by reducing the attentional load of the task. The tool does not support text-based queries. The central component of the video browsing tool's user interface is a light table, which shows the current content set and cluster structure using a number of representative key frames for each of the clusters, visualized by colored image borders. The following clustering features are implemented in the tool.

- *Motion activity:* clusters by the global movement in a given clip
- *Color layout:* clusters global color distribution of key frame
- *Media item:* shows all the different video files
- *Multi-view camera:* shows all videos shot with each of the possible cameras
- *Multi-view media Item:* shows all videos, but each view (camera) separately

The browsing workflow starts with selecting a dataset. Then, by selecting one of the available features the content is clustered according to this feature. Depending on the current size of the content set and the available space in the browser window, a fraction of the segments (mostly a few percent or even less) is selected to represent a cluster. The user can then decide to select a subset of clusters that seems to be relevant and discard the others, or repeat clustering on the current content set using another feature. In the first case, the reduced content set is the input to the clustering step in the next iteration. If no key frame of the selected view is available then an alternative view is selected where the user gets informed by displaying the name of the alternative view. On the left side of the application window the history, the similarity search and the result list are arranged. The history feature automatically records all clustering and selection actions done by the user. By clicking on one of the entries in the history, the user can jump back to the selected point. The browsing path can be branched at this point and explored using alternative cluster features without losing previous iterations. To execute the similarity search, the user drags the desired key frame into the marked area for the similarity search. Then, the application displays all available search options to allow the user to further proceed with the search task. The result list is used to memorize video segments and to extract segments of videos for further video editing, for example, as edit decision list (EDL). The user drags relevant key frames into the result list at any time, thus adding the corresponding segment of the content to it. Specific clustering features and further options are offered to the user via a context menu of the key frames where the users can select and discard a whole cluster or select and discard a single video. Furthermore, the metadata of the selected key frame can be displayed.

3.2   Evaluation Measures

The design of the series is based on item retrieval tasks, where the participant has to locate a given item (known or unknown) in a dataset, and performance of the participants is observed under conditions in which the new multi-view feature is used or not. The measures obtained in this tests are the performance measures employed earlier in [1], namely precision, recall, and F1 (the harmonic mean of precision and the recall) of the retrieval task; a psychophysiological measurement of mental activity related responses (EDA, as described earlier). A questionnaire was also filled by every participant in order to gather qualitative reports of their experience with the tool.

3.3   Task Design

For this test we devised four tasks which, we predicted, would vary in difficulty due to the number and the length of the available results. Participants were instructed to use the browsing tool to locate all segments that matched the given textual description, and to note that for two of the four tasks they were not allowed to use the "multiview" clustering feature. For every task, they had a time limit of ten minutes. The two a priori easy tasks where task A (look for segments with various people walking around) and task C (look for segments showing empty tables with no people present on the scene). The first one had eleven possible solutions with large average segment durations, whereas the second had five solutions but still long segment durations. The tasks where selected also to be more amenable to clustering (by motion activity in the case of task A and by color layout in the case of task C). The two a priori difficult tasks where task B (look for segments showing a man flashing a light in his hand), and task D (look for segments showing two people standing on a platform that falls through an abyss), these had respectively six and one matching segment and the event searched was always of a very short duration. The clustering features were described to the participants and they were introduced to the workflow with the browsing tool. All the participants performed the four tasks, and order of presentation of the task and the presence of multiview condition were counterbalanced between participants. The main condition in the test was the use or not of multi-view clustering, though other variables of interest collected in the questionnaire are the familiarity with the dataset, with web/video browsing general, and with this browsing tool in particular.

4   Results

4.1   Retrieval Performance

We report here the results from the fourteen participants (28 years old average, standard deviation (SD) = 2.7) in the test. Regarding precision and recall measures, overall mean precision was .81 (SD = .33), whereas overall mean recall was .53 (SD = .33). The precision and recall results by task are depicted in Fig. 1. Mean F1 was .61 (SD = .32).

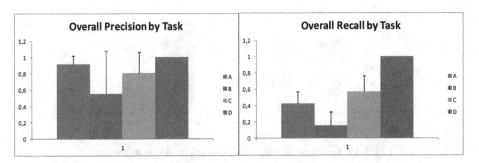

Fig. 1. Precision and recall for the four tasks. Error bars represent standard error

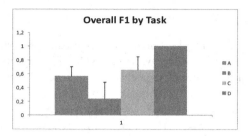

Fig. 2. F1 scores for the four tasks. Error bars represent standard error

From here on we report only the results of the F1 measure. Fig. 2, depicts the F1 score for each task. As can be seen, the only task that did not match our predictions regarding difficulty was task D, due to the fact that, despite having a single solution, the key frame containing the solution appeared repeatedly represented in various clusters and was easily localizable. Therefore, given that there is no variability associated to that task, we restrict the following analyses to tasks A, B and C. As regards these three tasks, the reasons used to predict their difficulty a priori are supported by the performance data. Of the four grouping variables collected in the questionnaire, F1 scores showed a slight trend to be higher only with high levels of general familiarity with video browsing (see Fig. 3). Regarding the main condition on the test, the use of multi-view clustering (see Fig. 4), there was a clear trend towards better F1 scores when multi-view was allowed for the two harder tasks, but the opposite was observed for the easier task. It could be interpreted that for the easier task, the introduction of a further clustering feature hampers rather than improves performance, though this interpretation must be taken with care, given that the difference in difficulty (as reflected by the F1 score) is not too large between tasks A and C.

Regarding task order, a learning effect was not observed on F1 scores, although for the harder task B, there was a clear difference, given that no participant was able to provide a single correct answer when this task was presented first. Evidently, as participants work with the tool, they grow increasingly familiarized with the materials increasing the chances of encountering the relevant segments.

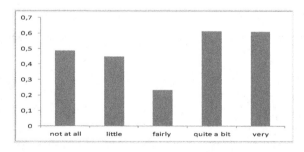

Fig. 3. F1 Scores by familiarity with video browsing in general

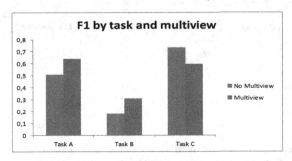

Fig. 4. F1 by task and multi-view

4.2 Electrodermal Activity Data

EDA measures were also registered in order to obtain a further indicator of the amount of mental effort associated with the browsing activity. As described in above, the increase in arousal signaled by variations in EDA levels has been used as an indicator of mental effort when the emotional significance of the experimental material is controlled as proposed initially by [5]. Thus, we calculated the mean EDA activity for the different tasks and both levels of the multi-view condition. Z-scores (standardized scores) of EDA values for each participant in each task were calculated from the raw data by subtracting the mean EDA value of all tasks to the value of each task, and dividing the results by the standard deviation of all task of this participant. Regarding the different tasks there were no significant differences between the levels of EDA for the different tasks ($F<1$, see Fig. 5a), nor for the order of the tasks, that is, during the approximate one hour session, the levels did not vary with time ($F<1$, see Fig. 5b).

Regarding the multi-view condition though, a clear trend was observed for a higher activity in conditions where multi-view was allowed ($F = 2.64$, $p < .078$, see Fig. 6a), which was mostly due to differences between tasks B and C (respectively $F = 2.8$, $p < .20$, and $F = 2.72$, $p < .118$, see Fig. 6b). The fact that the comparisons fell short of significance is clearly due to the few participants tested and a few participants more will surely provide significant results.

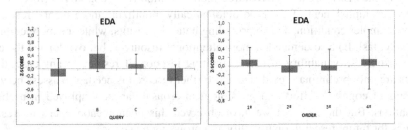

Fig. 5. (a) EDA values by task (query), (b) EDA values by order of presentation. Error bars represent standard error.

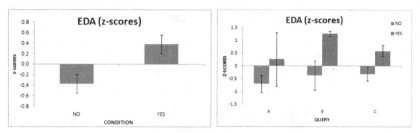

Fig. 6. (a) EDA average values per multi-view condition, (b) EDA average values per multi-view condition and task Error bars represent standard error

It is evident that the EDA data further corroborates the results from the performance test, and this with just an average analysis of tonic activity. These results are a further proof of the impact that indirect measures of user's reaction might have for usability tests. The fact that the EDA activity is higher for the condition where multi-view was allowed might seem counterintuitive at first, if we think that a better performance should be related to less mental effort. But the concept of mental effort is not necessarily negative when performing a task. More complex tasks require more mental resources (i.e. attention, or cognitive effort) to be processed [5], and it can be reflected on an increased arousal. It does not mean that the mental processing of the more complex task has to be hindered respect to a simpler one. For example, Lang et al. [13] found that increasing the structural complexity of television messages (by increasing the number of edits on them) resulted on increased arousal on viewers, more attention paid (measured by a decrease on heart rate) and even a better memory for the content of the messages. These authors explained that more structurally complex stimuli elicit automatic orientation of attention [14], which, in turn, increases the automatic allocation of mental resources to the task processing, and thus improving it. However, if the complexity of the stimulus is excessive, the mental resources required to its processing exceed the mental capacity of the viewer, resulting in a poor processing and a bad recall of the content. Some similar could be happened at our experiment: The inclusion of the multiview condition (i.e. the increase on the complexity of the task) might have required more attention (i.e. "cognitive effort") to the correct accomplishment of the task, automatically mobilizing more mental resources than the simpler condition, and so providing better F1 results, while the more complex tasks (e.g. task B) also demanded more attentional resources, but overloaded the cognitive resources availability of users, resulting in poor F1 results. In any case, these results are only preliminary evidence and further research is needed to assess how the concepts of cognitive effort and physiological arousal can be employed in usability evaluations. But the fact that we have observed this phenomenon opens interesting avenues for future research on usability and browsing behavior.

4.3   Verbal Reports

Participants suggested a series of features for improving the usability of the tool, for example, customizable window size, showing frames on the timeline of the video player or the possibility to select various videos at once. Participants did not like that in occasions the clusters are automatically modified when changing frame size, some did not like the clustering features available, or the overlapping between the preview and video player windows. As it is well known from evaluations of previous versions of the tool, the fact that all participants find unanimously distracting is that the key frames for a given clip are not constant, but depends on the feature used for clustering. We found that we had to emphasize the use of discard and select clusters function in the context menu. A participant suggested making these actions more evident by putting them on the above menu bar. Turning finally to the positive feedback, those who were not familiar with this kind of search liked the idea of searching videos by visual features and, as one participant put it; its potential utility in editing works where the content of the clips is not directly relevant and search by primitive visual features suffices. Also, they all unanimously agreed that they were more comfortable with using the similarity search than clustering.

5   Conclusion

The first thing to note is that participants tended not to produce false alarms, as reflected by the asymmetry between high precision and low recall. The explanation probably lies in the moderate size of the dataset. The F1 score behaved as expected in terms of the predicted difficulty of the queries, and the main result to emerge from this study is of course the differential effect of introducing the multi-view condition, which was further confirmed by the EDA data, although more detailed analyses could shed more light on the exact causes of these effects. For example, looking at phasic EDA responses, which would need an event-related approach to the design of the task, was not possible with the present material. It is interesting to note how the EDA measure was sensible to the manipulation of the search processed introduced by allowing the use of multi-view, but that this was not the case for the manipulations of task difficulty in terms of number and size of the relevant segment. This could be interpreted as a further confirmation that the variations in EDA level observed are due to variations in the attentional load of the task, rather than an increased emotional activation due to the difficulty in finding relevant results, though again, an empirical confirmation of the latter would require a more controlled task design. Finally, regarding the experience of the participants with this kind of software, only the previous experience with video browsing in general seemed to have an effect on performance, though a bigger pool of participants would be needed to extract firm statistical conclusions in this regard. It is also very interesting to note that the verbal feedback provided by the participants after performing the task made few direct references to the effect of having or not the possibility of using multi-view clustering, despite the fact that EDA and precision/recall measures indicate that this is a key variable affecting their performance. This is an encouraging result in terms of bringing this kind of measurement into evaluating tools professionals' use in their daily work.

Acknowledgements. The research leading to these results has received funding from the European Union's Seventh Framework Programme (FP7/2007-2013) under grant agreement no. 215475, "2020 3D Media – Spatial Sound and Vision".

References

1. Bailer, W., Rehatschek, H.: Comparing fact finding tasks and user survey for evaluating a video browsing tool. In: Proc. of ACM Multimedia, Beijing, CN (October 2009)
2. Fissaha Adafre, S., de Rijke, M.: Exploratory search in Wikipedia. In: SIGIR Workshop on Evaluating Exploratory Search Systems (2006)
3. Jansen, B.J., Ramadoss, R., Zhang, M., Zang, N.: Wrapper: An application for evaluating exploratory searching outside of the lab. In: SIGIR EESS Workshop (2006)
4. Jijkoun, V.B., de Rijke, M.: A pilot for evaluating exploratory question answering. In: SIGIR Workshop on Evaluating Exploratory Search Systems (2006)
5. Kahneman, D.: Attention and effort. Prentice-Hall, Englewood Cliffs (1973)
6. Knoche, H., De Meer, H.G., Kirsh, D.: Utility curves: Mean Opinion Scores considered biased. In: Proc. of 7th Intl. Workshop on Quality of Service, London (June 1999)
7. Kraaij, W., Post, W.: Task based evaluation of exploratory search systems. In: SIGIR Workshop on Evaluating Exploratory Search Systems (2006)
8. Ravaja, N.: Contributions of psychophysiology to media research: Review and recommendations. Media Psychology 6, 193–235 (2004)
9. Jumisko-Pyykkö, S., Strohmeier, D.: Report on research methodologies for the experiments. Technical report of the MOBILE3DTV Project (2008)
10. Qu, Y., Furnas, G.W.: Model-driven formative evaluation of exploratory search: A study under a sensemaking framework. Inf. Process. Manage. 44(2), 534–555 (2008)
11. Sánchez-Vives, M.V., Slater, M.: From presence to consciousness through virtual reality. Nature Reviews Neuroscience 6, 333–339 (2005)
12. Wilson, G.M., Sasse, M.A.: Listen to your heart rate: counting the cost of media quality. Affective Interactions Towards a New Generation of Computer Interfaces (2000)
13. Lang, A., Zhou, S., Schwartz, N., Bolls, P.D., Potter, R.F.: The effects of edits on arousal, attention, and memory for television messages: When an edit is an edit can an edit be too much? Journal of Broadcasting & Electronic Media 44, 94–109 (2000)
14. Turpin, G.: Effects of stimulus intensity on automatic responding: The problem of differentiating orienting and defense reflexes. Psychophysiology 23, 1–14 (1986)
15. Cooke, L.: Eye tracking: How it works and how it relates to usability. Technical Communication 52, 456–463 (2005)

Film Comic Generation with Eye Tracking

Tomoya Sawada, Masahiro Toyoura, and Xiaoyang Mao

University of Yamanashi

Abstract. Automatic generation of film comic requires solving several challenging problems such as selecting important frames well conveying the whole story, trimming the frames to fit the shape of panels without corrupting the composition of original image and arranging visually pleasing speech balloons without hiding important objects in the panel. We propose a novel approach to the automatic generation of film comic. The key idea is to aggregate eye-tracking data and image features into a computational map, called iMap, for quantitatively measuring the importance of frames in terms of story content and user attention. The transition of iMap in time sequences provides the solution to frame selection. Word balloon arrangement and image trimming are realized as the results of optimizing the energy functions derived from the iMap.

Keywords: Visual attention, eye-tracking, film comic, video processing, frame selection, image trimming, balloon arrangement.

1 Introduction

Film comic, also called cine-manga, is a kind of art medium created by editing the frames of a movie into a book in comic style. It is loved by a wide range of readers from little babies to comic manias around the world. Traditionally, film comics are manually created by professional editors who are familiar with comic styles. The frames of movies are manually selected, trimmed and arranged into the comic-like layouts. Verbal information of the movie, such as dialogues and narrations, are inserted into the correspondent panels as word balloons. Since a movie usually consists of a huge number of frames, those editing tasks can be very tedious and time consuming.

Recently, several research works have been conducted trying to provide computational support to the generation of film comic [1-4]. The work by R. Hong *et al.* [3] first succeeded in automating the whole procedure of film comic by employing modern computer vision technologies for extracting key frames and for allocating the word balloons. However, their method replies on existing human face and speaking lip detecting technologies for identifying speakers, and are not applicable to cartoon animations, which is the main target of film comics, because the appearance of characters in cartoon animation is usually very different from that of human face. Actually, many issues of automatic film comic generation cannot be solved with naive image/video processing or even by combining text analysis. For example, a subtle

S. Li et al. (Eds.): MMM 2013, Part I, LNCS 7732, pp. 467–478, 2013.

change of character's posture which causes no remarkable change to image features may be very critical in conveying the story. When placing a word balloon, only avoiding occluding the speaker is usually not enough since an object about which the speaker is talking may also be very important. All these problems are story or even user dependent.

In this paper, we present a new approach of using eye-tracking data to automate the whole process of film comic creation. It is known that eye movement and gaze information provides good cues to the interpretation of scenes by viewers [5]. Existing studies [6-8] showed the effectiveness of applying visual attention model for measuring the significance of images in creating video summarization. Those techniques use the computation model of visual attention which is mainly based on the low level features of images and motions. The real attention of viewers, however, is largely dependent on the content of a story as well as the personal background of the viewers, such as their age, sex and cultural background. By using the eye-tracking data, we can provide a better prediction to the attention of viewers. Furthermore, by using subjects from the assumed population of readers, e.g. using the eye-tracking data of children for creating a film comic mainly targeting at children, we can even create film comics adapted to a particular population.

The major contribution of this paper can be summarized as follows:

1. Proposal of a new framework for automatically converting a movie into a comic based on eye-tracking data.
2. Proposal of iMap, a new computational map which combines eye-tracking data and image feature for quantitatively measuring the local informativeness in a movie frame.
3. Proposal of a new algorithm for detecting critical frames based on iMap .
4. Proposal of the optimization scheme for achieving desirable image trimming and balloon arrangement.

The remainder of the paper is organized as follows: Section 2 reviews the related work. Section 3 gives the overview of the proposed method and Section 4 introduces iMap together with its construction algorithm. The detailed algorithms of frame selection, trimming, balloon allocation are described in Section 5. Section 6 is about implementation and experiment and Section 7 concludes paper.

2   Related Works

A pilot work on film comic generation was done by W.I. Hwang et al. [1]. Assuming the correspondence between a balloon and a speaker is known, their method can automatically allocate balloons following the rules of comic, but the major tasks of film comic generation such as the selection of frame images were done manually. Preuß et al. [2] proposed an automatic system for converting movies into comics, but their method assumes that the screenplay of the movie is available and hence limits its application to a very special case. More recently, R. Hong et al. [3] presented a full automatic system which employs face detector, lip motion analysis, and motion analysis to realize the automatic script-face mapping and key-scene extraction. But as mentioned in the previous section, their technique cannot be used for cartoon animations. M. Toyoura *et al.*[4] proposed a technique to automatically detect typical camera works, such as zooming, panning, fade-in/fade-out, and represent them in a particular comic style. The same function is supported in our system.

Video summarization [9] is the field most closely related to film comic generation. Both video summarization and film comic generation share the two major technique issues: how to select appropriate images from a video/movie and how to arrange the images to effectively depicting the contents of the video/movie. Video summarization has been well studied in the past decade and researchers have challenged the problem through various approaches ranging from computer vision [9-10], text mining [11], to fMRI [12]. Several researchers proposed to use human visual attention model for video segmentation and skimming [6-7]. Those methods were further enhanced through combining multiple visual cues [8] as well as high level semantic and cognitive information [13]. W. Peng et al. proposed an approach using both eye-tracking and facial expression detection for selecting important frames in home video summarization [14]. Comic style was also explored in the field of video summarization for achieving an effective representation of video summaries [15]. While a video summary is mainly for providing the overview of a video or serving as the index enabling users to quickly browse the required information in the original video, film comic is an alternative of movie for storytelling and hence should enable readers to easily understand the whole story without refereeing to the original movie. Therefore, many technologies of video summarization cannot be applied to film comic generation directly. For example, W. Peng et al. selected the frame sequence with long gaze fixation as the important frames to be included in the summary [14]. But from the viewpoint of story development, a frame causing the change of attention should be more important. We succeeded in selecting such frames by detecting the significant transition of iMap.

3   Framework

Figure 1 depicts the framework of the proposed technique. Given a movie including subtitle information, our technique converts it into a comic in following 10 steps. First, sample viewers are invited to watch the movie and their eye positions are recorded with an eye-tracker. At the same time SURF image feature detection is performed for each frame of the movie (Step 2). Then an iMap is generated from the eye-tracking data and SURF feature (Step 3). Based on iMap, the frame sequence of the movie is segmented into groups of frames called *shots* (Step 4). Those shorts are subjected to camera-works detection (Step 5). Then for each shot, the first frame, the frames with speech line and the frames for representing camera works are selected (Step 6). Based on the number of selected frames and the information on camera works, panel layout of the comic is generated (Step 7). The information required for trimming the frames and for placing the balloons are computed by referring to iMap (Step 8&9). Finally a comic book is created by arranging the trimmed frames into the comic layout and balloons with speech lines are inserted to the frames (Step 10). We use existing technologies for Step 5 and 7, and the other steps are realized with newly developed techniques, whose details are described in the next section.

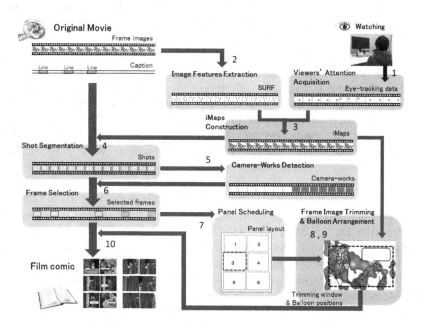

Fig. 1. The framework of film comic generation with proposed technique

4   Film Comic Generation with iMap

4.1   Construction of iMap

A good film comic should contain the frames that are important for understanding the story and each of the frames should be represented in a way well preserving the original information while being visually pleasing as a comic. Those requirements give rise to the issue of measuring how informative a frame or a region in a frame is in conveying the story. We address the issue by combing eye-tracking data and image features into iMap, a probability distribution map providing a quantitative measure to the local informativeness in each frame. It is known that eye movement plays an important role in event perception [5]. Our eyes are likely to be drawn to a region which is most relevant to the story development and a big eye movement usually provides a cue to the change of events. However, eye-tracking data usually consists of noise. It is also difficult to distinguish an unintentional or meaningless fixation from a deliberate or purposeful one. To solve the problem, we first convert the raw eye-tracking data of multiple viewers into an *attention map,* a probability distribution of attention estimated with Gaussian kernel. At the same time we generate a *feature map* which is a probability distribution estimated from SURF (Speeded Up Robust Feature) [16]. We adopt SURF for taking its advantage in speed and robustness in tracking features across frames. Then we construct iMap as the aggregation of the attention map and the feature map. The significance of combing eye-tracking data and image feature lies in the fact that a fixation at a position with little image feature such as edges and corners is likely to be a meaningless one and the resulting iMap gives a more reliable prediction to how informative the position is.

A feature map M^t_f representing the probability distribution of image features in t^{th} frame is calculated as follows:

$$M^t_f = f^t_{SURF}(x,y) \otimes G(0,\sigma^2_f) \qquad f^t_{SURF}(x,y) = \begin{cases} 1 & if\ (x,y)\ is\ SURF \\ 0 & if\ (x,y)\ is\ not\ SURF \end{cases} \tag{1}$$

where $G(0,\sigma^2_f)$ is a 2D Gaussian kernel with average 0 and variance σ^2_f. $\otimes$ is convolution. If pixel (x,y) is SURF, its contribution to the probability distribution of image feature is estimated with the Gaussian kernel.

Similarly, an attention map M^t representing the distribution of n sample viewer's attention is calculated as follows:

$$M^t = \sum_{i=1}^{n} f^t_{i\,EYE}(x,ty) \otimes G(0,\sigma^2_a) \qquad f^t_{i\,EYE}(x,y) = \begin{cases} 1 & if\ (x,y)\ is\ an\ eye\ position \\ 0 & if\ (x,y)\ is\ not\ an\ eye\ position \end{cases} \tag{2}$$

iMap M^t of tth frame is constructed from maps M^t_f and M^t_a as follows:

$$M^t = M^t_f * M^t_a. \tag{3}$$

Since M^t is the product of M^t_f and M^t_a, a high value of M^t indicates an informative region in terms of both image features and viewers' attention. When there is a significant change of M^t in time sequences, there should be a transition of either/both of image features or/and viewers' attention. By detecting such transition, the movie can be segmented into the shots consisting of frames of the same contents.

Although M^t gives a good solution to the detection of content transition among frames, it does not contain enough information for performing frame image trimming and balloon arrangement, which requires measuring the local informativeness in the context given by neighboring frames also. We further extend iMap to encounter the probability distribution in time dimension. An iMap representing the spatial-temporal informativeness is calculated from the neighboring frames of tth frame as follows:

$$M^{Nt} = \sum_{s \in N(t)} G(t,\sigma^2_t)(s) * M^s, \tag{4}$$

where $N(t)$ is the set of neighboring frames of tth frame and the contribution of each neighboring frames is given by Gaussian function $G(t,\sigma^2_t)$ (s) with highest value at t^{th} frame.

4.2   Frame Selection

To select frames which are important for conveying the story of the movie, we first segment the movie into shots by detecting significant change of iMaps across adjacent frames. The set of transition frames $\Omega(t)$ are detected as

$$\Omega(t) = \left\{ t \mid \| M^{Nt+1} - M^{Nt} \| > T \right\} \qquad T = \iint_M \| G(x,y) - G(x - x_d, y - y_d) \| \qquad (5)$$

Where $G(x,y)$ is the Gaussian filter used for computing iMap, with $\|*\|$ denotes L_2 norm, $d = \sqrt{x_d + y_d}$ is the possible distance between the eye positions at adjacent frames when a saccade occurs. A saccade is a fast movement of an eye between fixations. The central part of our retina, known as the fovea, plays a critical role in resolving objects. We move our eyes quickly toward a target so as to sense it with greater resolution. Therefore, if a saccade is detected, we can say that the region of interest has changed, which indicates a possible transition of story contents.

After transition frames are detected, the movie is segmented into the group of frames, say shots, at the transition frames. Then the first frame of each shot, which is the transition frame, is selected for being included in the film comic. All the frames with speech lines are also selected. If a shot contains special camera-works, multiple frames are selected depending on the type of camera-works.

4.3   Image Trimming

If the shape and size of a selected frame image are different from those of the corresponding panel, the frame image is trimmed and resized to fit into the panel. As our current implementation uses rectangular panels only, trimming is necessary only when the frame image and its corresponding panel have different aspect ratios. A good trimming should meet two requirements: First, the area being cropped should be less informative. Second, the composition of original frame should be preserved as possible. We achieve the best trimming by minimizing the following cost computed from iMap M^{Nt}:

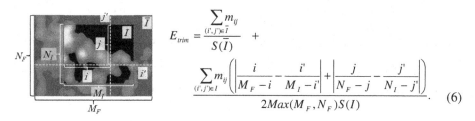

$$E_{trim} = \frac{\sum_{(i',j') \in \bar{I}} m_{ij}}{S(\bar{I})} + \frac{\sum_{(i',j') \in I} m_{ij} \left(\left| \dfrac{i}{M_F - i} - \dfrac{i'}{M_I - i'} \right| + \left| \dfrac{j}{N_F - j} - \dfrac{j'}{N_I - j'} \right| \right)}{2Max(M_F, N_F)S(I)}. \qquad (6)$$

Fig. 2. Image trimming by referring to iMap M^{Nt}

As shown in Figure 2, F denotes the original frame image, I the trimmed one and $\bar{I}$ the set of the pixels being cropped away with the trimming. (i',j') is the position in original frame image F for the pixel in I or $\bar{I}$. $S(I)$ and $S(\bar{I})$ are the total number of pixels (i,j) in I and $\bar{I}$, M_F, M_I, N_F, N_I are the width and height of the frame image F and trimmed image I, respectively. The first term in Eq (6) measures how less informative the region being cropped is and the second term evaluates whether the relative location of a pixel in original frame is preserved in the trimmed image. By multiplying the cost with the value in M^{Nt}, we give those informative pixels a higher priority for preserving their relative positions.

4.4   Arranging Balloons

A balloon is placed into the panel of a selected frame with corresponding speech lines. Balloons should be placed in a way to meet 3 requirements: First, it should not occlude an important region. Second, it should be in a visually pleasing shape. Third, it should be spatially close to the speaker. For the first requirement, iMap provides the information about the importance of regions. In our current implementation, we use oval shaped balloons, which is one of the most popular shapes found in comics. To avoid generating long narrow shaped balloons, we use the ratio of the two axes of the ellipse as the measure to the aesthetic quality of the balloon. To measure the distance from the balloon to the speaker, the position of the speaker should be given. We have tried out existing face detector and speaking lip detector, but found all of them failed to produce good results. The main reason is that the appearance of characters in cartoon animation are usually very unique, which makes it difficult to apply the existing face detector. Another reason is that a cartoon animation usually contains exaggerated motions throughout the movie and this makes it difficult to detect small motions such as a speaking lip. Put all together, a desirable balloon arrangement can be obtained by minimizing the following energy function:

$$E_{ballon} = a\frac{\sum\limits_{(i,j)\in I_{balloon}} m_{ij}}{S(I_{balloon})} + b\frac{\left\|P_{balloon} - P_{I_{speaker}}\right\|}{M^2 + N^2} + c\frac{L_{short}}{L_{long}}, \tag{7}$$

where $I_{balloon}$ is the region occluded by a balloon, $P_{balloon}$, $P_{I_{speaker}}$ are the gravity center of balloon and speaker region, L_{short}, L_{long} are the short and long axes of balloon, respectively. a, b, c are user given parameters for controlling the weight of each term. The values of L_{short}, L_{long} are dependent to the length of speech lines. L_{short} and L_{long} take discrete values to avoid breaking a line at the middle of a word. Given a speech line, we first build a look up table containing all the possible pair of L_{short}, L_{long} and loop through all the entries of the table during optimization.

After the shape and location of a balloon is decided, a tail is attached to the balloon and headed toward the speaker.

5   Experiments

We invited sample viewers (3 male university students in their 20s) to watch the beginning part of "Pinocchio" and "Snow White and the Seven Dwarfs" by Disney Inc. and a part of "Roman Holiday". The movies were displayed on a 1920x1200 monitor and EMR-AT VOXER of Nac Image Technology was used for tracking subject's eye positions at 60Hz. Frame rate of the movies was 29.97fps. Figure 8 are some example pages of the resulting film comics and Table 1 shows some statistics of the results for the two cartoon movies.

Figure 3 and Figure 4 compare the results of frame selection by the existing color histogram based video segmentation [10] and by our method. Figure 3 is the scene the cricket sneaks into the house in the freezing night trying to warm his hip up with the burnt stone from the fireplace. Since the color histogram does not change much through the scene, the result by existing work as shown in Figure 3(a) failed to detect enough frames required for representing such mise-en-scene. Figure 4 is a part of "Roman Holiday" where Mr. Joe and Princess Ann try to insert their hands into the mouth of truth. Princess Ann gets frightened and moves her hand toward the mouth timidly. Such process could not be visualized with the result of existing technique in Figure 4(a) because of the little change of color histogram. In Figure 4(b), our proposed method succeeded in catching such subtle but important transition of contents.

Figure 5(a) shows an example of image trimming. In the scene, the Prince exchanges bows with Snow White. Both of the Prince and Snow White are considered to be informative and should be included in the result of trimmed image. Figure 5(b) is the iMap of the frame, and the lines indicated the trimmed area by energy optimization. The yellow regions indicate high values in the iMap. We can recognize that they are the regions corresponding to the Prince and Snow White. Moreover, the highest position is Prince's face. It demonstrates that our proposed method could estimate the important positions with the use of multiple viewers' eye movement data and SURF feature tracking.

Table 1. Experimental Results

	Length	Panels	Ballons	Position error	Tail error
Pinocchio	20m 00s	352	254	10 (3.9%)	32 (12.5%)
Snow White	11m 10s	200	94	6 (6.3%)	12 (12.7%)
Roman Holiday	1m 24s	38	19	3 (15.7%)	1 (5.2%)

(a) By existing color histogram based technique

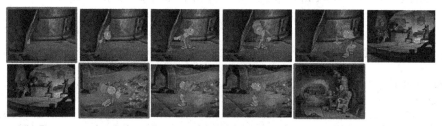

(b) By proposed technique. Images with red frames are corresponding to the ones of (a)

Fig. 3. Comparison of frame selection results for "Pinocchio"

(a) By existing color histogram based technique

(b) By proposed technique. Images with red frames are corresponding to the ones of (a)

Fig. 4. Comparison of frame selection results for "Roman Holiday"

Figure 6(a) shows a result of balloon arrangement. The blue box in Figure 6(b) is the resulting balloon position given by energy minimization. By placing the balloon at this position and directing the tail to the highest area of the iMap, the important area in the frame remains not occluded and the tail is directed to the speaker.

The number of position error in Table 1 is the number of balloons that partially occlude a character or an object which is related to the story development. The number of tail error is the number of tails not heading to the speaker. Figure 7(a) and 7(b) are the examples of position error and tail error, respectively. In Figure 7(a), the balloon has been placed on the body of Wicked Queen since the surrounding areas have more SURF features than the body area. We expect to solve this kind of problem by using other higher level image features such as those for measuring objectiveness. In Figure 7(b), the tail of the balloon has a wrong direction since the speaker is not included in the scene. We expect to reduce such errors by combining eye-tracking data with some advanced motion detection techniques.

   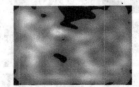

(a) Original frame image   (b) iMap corresponding to (a)

Fig. 5. An example of image trimming by referring to iMap

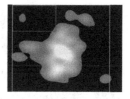

(a) Composited balloon

(b) Optimizing the position of a balloon on iMap

Fig. 6. Arranging a balloon by referring to iMap

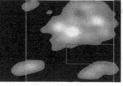

(a) Balloon arrangement failure   (b) Tail attaching failure

Fig. 7. Example of balloon arrangement and tail attaching failure

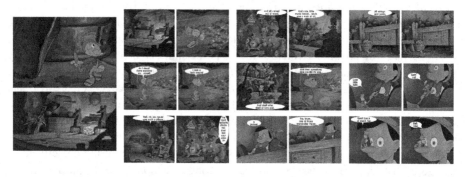

(a)   Pinocchio

(b)   Show White and the Seven Dwarfs

Fig. 8. Comic pages generated with Proposed technique

(c)   Roman Holidays

Fig. 8. (*Continued*)

6   Conclusions and Feature Works

This paper presented a novel technique for generating a film comic from a movie automatically based on the eye-tracking data of sample viewers. We believe that the concept of iMap, together with the optimization scheme for obtaining desirable image trimming and balloon arrangement, has a very high potential for being used in other image/video processing applications. Also we believe that with the advancement of eye-tracking technology as well as the cost down of eye tracking devices in the future, the proposed technique will enable casual users to easily enjoy movie in the style of comic.

Acknowledgement. This work was supported by KAKENHI 21300033 Grant-in-Aid for Scientific Research (B), Japan Society for the Promotion of Science (JSPS).

References

1. Hwang, W.I., Lee, P.J., Chun, B.K., Ryu, D.S., Cho, H.G.: Cinema comics: Cartoon generation from video stream. In: International Conference on Computer Graphics Theory and Applications, pp. 299–304 (2006)
2. Preuß, J., Loviscach, J.: From movie to comics, informed by the screenplay. In: ACM SIGGRAPH, Poster (2007)
3. Hong, R., Yuan, X.T., Xu, M., Wang, M., Yan, S., Chua, T.S.: Movie2comics: a feast of multimedia artwork. In: Proceedings of the International Conference on Multimedia, pp. 611–614 (2010)
4. Toyoura, M., Kunihiro, M., Mao, X.: Film Comic Reflecting Camera-Works. In: Schoeffmann, K., Merialdo, B., Hauptmann, A.G., Ngo, C.-W., Andreopoulos, Y., Breiteneder, C. (eds.) MMM 2012. LNCS, vol. 7131, pp. 406–417. Springer, Heidelberg (2012)
5. Smith, T.J., Whitwell, M., Lee, J.: Eye movements and pupil dilation during event perception. In: Proceedings of the Eye Tracking Research and Applications Conference (2006)

6. Ma, Y., Hua, X., Lu, L., Zhang, H.: A generic framework of user attention model and its application in video summarization. IEEE Trans. on Multimedia 7(5), 907–919 (2005)

7. Li, K., Guo, L., Faraco, C., Zhu, D., Deng, F., Zhang, T., Jiang, X., Zhang, D., Chen, H., Hu, X., Miller, S., Liuothers, T.: Human-centered attention models for video summarization. In: International Conference on Multimodal Interfaces and the Workshop on Machine Learning for Multimodal Interaction, vol. 27. ACM (2010)

8. You, J., Liu, G., Sun, L., Li, H.: A multiple visual models based perceptive analysis framework for multilevel video summarization. IEEE Transactions on Circuits and Systems for Video Technology 17(3), 335–342 (2007)

9. Money, A.G., Agius, H.: Video summarisation: A conceptual framework and survey of the state of the art. Journal of Visual Communication and Image Representation 19(2), 121–143 (2008)

10. Porter, S.V.: Video Segmentation and Indexing using Motion Estimation. PhD thesis, University of Bristol (2004)

11. Chen, B., Wang, J., Wang, J.: A novel video summarization based on mining the story structure and semantic relations among concept entities. IEEE Transactions on Multimedia 11(2), 295–312 (2009)

12. Hu, X., Deng, F., Li, K., Zhang, T., Chen, H., Jiang, X., Lv, J., Zhu, D., Faraco, C., Zhang, D., et al.: Bridging low-level features and high-level semantics via fmri brain imaging for video classification. In: Proceedings of the International Conference on Multimedia, pp. 451–460. ACM (2010)

13. Liu, A., Yang, Z.: Watching, thinking, reacting: A human-centered framework for movie content analysis. International Journal of Digital Content Technology and its Applications 4(5), 23–37 (2010)

14. Peng, W.-T., Huang, W.-J., Chu, W.-T., Chou, C.-N., Chang, W.-Y., Chang, C.-H., Hung, Y.-P.: A User Experience Model for Home Video Summarization. In: Huet, B., Smeaton, A., Mayer-Patel, K., Avrithis, Y. (eds.) MMM 2009. LNCS, vol. 5371, pp. 484–495. Springer, Heidelberg (2009)

15. Boreczky, J.S., Girgensohn, A., Golovchinsky, G., Uchihashi, S.: An interactive comic book presentation for exploring video. In: ACM Conference on Computer-Human Interaction (CHI), pp. 185–192 (2000)

16. Bay, H., Ess, A., Tuytelaars, T., Gool, L.V.: Speeded-up robust features (surf). Computer Vision and Image Understanding 110(3) (2008)

Quality Assessment of User-Generated Video Using Camera Motion

Jinlin Guo[1], Cathal Gurrin[1], Frank Hopfgartner[1],
Zhenxing Zhang[1], and Songyang Lao[2]

[1] CLARITY and School of Computing, Dublin City University
Glasnevin, Dublin 9, Dublin, Ireland
{jinlin.guo,cgurrin,frank.hopfgartner,zzhang}@computing.dcu.ie
[2] School of Information System and Management
National University of Defence Technology, Changsha, Hunan, China
laosongyang@vip.sina.com

Abstract. With user-generated video (UGV) becoming so popular on the Web, the availability of a reliable quality assessment (QA) measure of UGV is necessary for improving the users' quality of experience in video-based application. In this paper, we explore QA of UGV based on how much irregular camera motion it contains with low-cost manner. A block-match based optical flow approach has been employed to extract camera motion features in UGV, based on which, irregular camera motion is calculated and automatic QA scores are given. Using a set of UGV clips from benchmarking datasets as a showcase, we observe that QA scores from the proposed automatic method and subjective method fit well. Further, the automatic method reports much better performance than the random run. These confirm the satisfaction of the automatic QA scores indicating the quality of the UGV when only considering visual camera motion. Furthermore, it also shows that the UGV quality can be assessed automatically for improving the end users quality of experience in video-based applications.

Keywords: Quality Assessment, User-generated Video, Irregular Camera Motion, Optical Flow.

1 Introduction

As the proliferation of Web 2.0 applications, user-generated video (UGV) [1] is poised to inundate the Internet. Recent statistics show that, on the primary video sharing website, YouTube[1], 48 hours of video are uploaded every minute by users, resulting in nearly 8 years of content uploaded every day. Furthermore, over 800 million unique users visit YouTube each month, and more than 3 billion hours of video are watched on YouTube[2]. This has increased the requirement on video websites to match the video quality expectation of the end users and viewers,

---

[1] http://www.youtube.com
[2] According to http://www.youtube.com/t/press_statistics

S. Li et al. (Eds.): MMM 2013, Part I, LNCS 7732, pp. 479–489, 2013.

such as users always prefer to viewing the best-quality of video among many clips captured in a same concert using personal capturing devices. Therefore, reliable quality assessment (QA) of UGV plays an important role in providing good quality of service (QoS), in improving the end users' quality of experience (QoE) and in managing such a large amount of video data.

Reliable QA of video has attracted a lot of research interest, and numerous of video QA methods and measurements have been proposed over the past years with varying focuses on objective or subjective QA of video. Previous approaches to video QA have the following characteristics.

(1) Many aspects may affect the quality of video including, but not limited to, acquisition, process, compression, transmission, display and reproduction systems. Most of existing approaches to QA of video focus on the distortion caused by compression [2,3] and transmission [4]. The quality of video itself is not assessed when it's captured. However, capturing conditions such as irregular camera motion (IRRCM) can degrade the perceived video quality.

(2) The commonly-used video QA measurements such as signal-to-noise ratio (SNR), peak-signal-to-noise ratio (PSNR) and mean squared error (MSE) [5], are computationally simply, however, they disregard the characteristics of human visual perception.

(3) A lot of research interest has been focused on objective video QA [2,6], however, methods to assess the visual quality of digital video as perceived by human observer are becoming increasingly important, due to the large number of applications that target humans as the end users of video. The only reliable method to assess the video quality is to ask human subjects for their own opinions, which is termed subjective video QA. The subjective methods are based on groups of trained or untrained users viewing the video content, and rating for quality [5,7]. It is impractical for most applications, and also time consuming, laborious and expensive, due to the human involvement in the process. However, subjective QA studies provide the means to evaluate the performance of objective or automatic technologies of QA. Combination of objective and subjective QA, which means objective QA methods should produce video QA scores that highly correlate with the subjective assessments provided by human evaluators, will likely be a trend of future research.

Moreover, the traditional QA methods are usually performed on broadcast video which has been professionally preproduced [8]. Compared to broadcast video such as news and sports video, UGV is usually of lower quality, due to the uncontrolled capturing conditions and various types of capture devices. For example IRRCM and fuzzy backgrounds are very common in UGV [9].

It should be noted that in [10], Wu et al. analyzed the IRRCM in home videos, and proposed a segmentation algorithm for home videos based on the categorization of camera motion. By support vector machines (SVMs), the effects caused by the camera motion were classified into four types: Blurred, Shaky, Inconsistent and Stable according to the changes of camera motion in speed, direction and acceleration. Finally, video sequence were segmented, and each segment was labeled as as one of the four camera motion effects. However, in this paper, we employ IRRCM to assess the visual quality of UGV. The rationale is that IRRCM or camera shaking is so commonplace in UGV and it causes the degradation of perceived quality of UGV. Our contributions in this work are that 1)firstly, we propose a automatic approach to perform QA of UGV using IRRCM, and 2) subjective QA of UGV is conducted and compared with the proposed automatic QA method.

The remaining parts of this paper are organized as follows. In Section 2, we analyze the IRRCM feature in UGV and describe the approach to UGV IRRCM extraction and scoring proposed in this work. In Section 3, experimental results based on a set of UGV clips from benchmarking datasets are represented, and also compared with the results from the subjective assessment and random run. Finally, we give our conclusions and outline future work in Section 4.

2 IRRCM Extracting and Scoring

It is hard to define the relationship between the video visual quality and camera motion. However, if one video clip contains more IRRCM, the visual quality is generally perceived as of being lower in subjective assessment.

2.1 Camera Motion Analysis in UGV

Camera motion is an important factor affecting visual quality of video. The visual quality of UGV is highly relevant to three properties of camera motion [10], that is *speed*, *direction* (*orientation*) and *acceleration*. These three properties affect UGV quality in different ways. As shown in Fig. 1, the classification of effects caused by the change of three properties of camera motion can be represented as a decision tree [10].

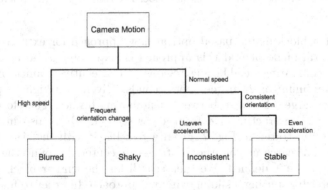

Fig. 1. Classification of camera motion effects in UGV

If the speed of camera motion is high, the captured frames will be blurred. When the speed is normal, but the orientation of camera motion changes frequently, namely, the camera moves back and forth repeatedly, the captured videos are regarded as shaky. When speed is normal and orientation is consistent, but the accelerations of camera motion in consecutive-extracted frames are uneven, that is, the variance of acceleration is large, the captured videos are inconsistent. The normal camera motion with rare orientation changes and even accelerations lead to stable motion.

In this paper, we name IRRCM to be the effects caused by the changes of acceleration and orientation. In the quality assessment, we jointly weight the acceleration and orientation changes. The acceleration change is measured by the magnitude change of two consecutive-extracted motion vectors, whereas the orientation change is calculated by angle between these two vectors. In the following parts of this section, we will describe our QA of UGV method in detail.

2.2   Extraction of Background Camera Motion

We use a two-parameter motion model to deal with camera motion (X and Y). X depicts horizontal movement to the left and right, commonly referred to as X transition and pan. Y depicts vertical movements (up and down), referred to as Y transition and tilt. This provides relative computational efficiency.

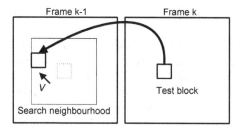

Fig. 2. Neighborhood search for similar blocks (Origin is at bottom left corner)

We adopt a block-match based optical flow approach for extracting camera motion. The rationale behind this approach is that camera movement can be detected by comparing neighboring regions of consecutive frames. Each video clip has a minimum of 24 frames per seconds. Given this high frequency, we do not expect large differences between neighboring frames. Therefore, we limit our approach by extracting five frames per second, which also improves the efficiency. As illustrated in Fig. 2, given a test block with the size of S in the current frame, a search neighborhood of $3S$-size, centered around the test block in preceding frame is defined. We then search for the similar block within the search neighborhood using a sliding window approach. In order to find the most similar block in the search neighborhood, the *Maximum Matching Pixel Count* (MMPC) is determined as follows:

$$D\left(x_t, y_t; x_p, y_p\right) = \begin{cases} 1 & if \sum_{c\in\{R,G,B\}} |P_c(x+i, y+j)- \\ & Q_c(x+d_x+i, y+d_y+j)| \leq T \\ 0 & else \end{cases} \tag{1}$$

$$(x'_p, y'_p) = \underset{(x_p, y_p)}{argmax} \sum_{i=1}^{S} \sum_{j=1}^{S} D(x_t, y_t; x_p, y_p) \qquad (2)$$

Where (x_t, y_t) is the center of a test block. (x_p, y_p) defines the center of searched block located in the search neighborhood. P_c, is the color value of the pixel in the current frame, and Q_c is the color value of the pixel in the previous frame, (x, y) is the coordinate of the bottom left corner of the test block. The displacement between the two centers is defined as $d_x = x_p - x_t$ and $d_y = y_p - y_t$. T is the threshold. S is the size of the test block. Therefore, the displacement vector $\mathbf{v}$ (optical flow) is given by $v_x = x'_p - x_t$, $v_y = y'_p - y_t$. v_x and v_y are the X and Y motion component, respectively.

However, there are many uniform-texture areas in the video frame images that make the detected motion vector $\mathbf{v}$ unreliable. We checked the number of similar blocks within the neighborhood, if there are more than N blocks, this indicates a uniform-texture area, and the motion vector $\mathbf{v}$ is unreliable. In this case, we set $\mathbf{v} = 0.$.

This process is repeated over the whole image to obtain a optical flow for each block. Large homogeneous regions, typically half of the image, are considered to be the background. The camera motion between frame $k - 1$ and frame k is determined by computing the average motion vector $\overline{\mathbf{v}}_{k-1,k}$ of the blocks within this background region. Specifically, all the detected motion vectors are grouped into a histogram with eight bins by *orientation assignment*, each of which represents one orientation as shown in Fig. 3.

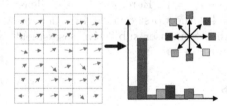

Fig. 3. Orientation assignment of optical flows for finding the camera motion

2.3 Scoring Irregular Camera Motion

After extracting all the camera motions between each consecutive pairwise-frame, we can get the changes of a camera motion by considering both acceleration and orientation change. For two consecutive camera motion vectors (corresponding to three consecutive-extracted frames $k - 1$, k, $k + 1$), the acceleration a, and the orientation change $\theta_{k-1,k,k+1}$ both describe the IRRCM. They are computed as follows:

$$a = \left\| \overline{\mathbf{v}}_{k-1,k} - \overline{\mathbf{v}}_{k,k+1} \right\| / \Delta t = m_{k-1,k,k+1} / \Delta t \doteq m_{k-1,k,k+1} \qquad (3)$$

$$\theta_{k-1,k,k+1} = arccos \left(\frac{\overline{\mathbf{V}}_{k-1,k} \cdot \overline{\mathbf{V}}_{k,k+1}}{\|\overline{\mathbf{V}}_{k-1,k}\| \, \|\overline{\mathbf{V}}_{k,k+1}\|} \right) \quad (4)$$

where Δt is time interval between two consecutive-extracted frames. Since we sample the frames uniformly (five frames per second), that is, Δt is a constant, the acceleration measurement is equal to the magnitude of the difference of two consecutive motion vectors, $m_{k-1,k,k+1}$. For a whole video clip, the final IRRCM is represented by the average acceleration (AA) $\overline{m}$ and average orientation change (AOC) $\overline{\theta}$, which are calculated by the average of all magnitude changes and orientation changes.

In order to give a QA score to a video clip based on how much IRRCM it contains, we firstly consider to assess the AA $\overline{m}$ and AOC $\overline{\theta}$, respectively. A quality grade system is firstly built on a training set by quantifying all the AAs into five levels. Given a test video clip and its $\overline{m}$, its AA rank r_m can be obtained by:

$$r_m = \left\lfloor 5 * \frac{\overline{m}}{m_{max}} \right\rfloor \quad (5)$$

where m_{max} is the maximum AA extracted in the training set (it keeps $\frac{\overline{m}}{m_{max}} < 1$). $\lfloor \cdot \rfloor$ is the *Floor Function*. The AOC rank r_θ can be obtained by the same way as the AA rank. The final rank r indicating the IRRCM in this clip can be described as:

$$r = \lfloor \omega_m * r_m + \omega_\theta * r_\theta \rfloor \quad (6)$$

where ω_m and ω_θ are the weights for the AA rank r_m and AOC rank r_θ respectively, which show the importance attached to the AA and AOC respectively by the observer, and $\omega_m + \omega_\theta = 1$. Here, $r = 0$ means least IRRCM, that is best quality, whereas, $r = 4$ is the worst quality. In order to compare the QA score obtained here with that from user subjective assessment, we set the final QA score of the video clip to $5 - r$.

3 Experiments

3.1 Experimental Setup

We conduct our experiments using the Internet video collection of the NIST TRECVid [11] 2011 Multimedia Event Detection (MED) task[3]. This dataset consists of publicly available UGV posted to various Internet video hosting sites. For this evaluation, we randomly select a subset of 1000 video clips (88 hours playing time) from the dataset and split it into a training (700 video clips) and a smaller-size testing set (300 videos clips) since subjective assessment is very time and labor consuming. The training set is used for training the related thresholds and aforementioned parameters. Based on the preliminary analysis of a preceding experiment, we chose the following settings: test block size $S = 10$, similarity threshold $T = 10$, threshold for number of similar blocks $N = 4$, and $\omega_m = \omega_\theta = 0.5$

---

[3] http://www.nist.gov/itl/iad/mig/med11.cfm

Fig. 4. User assessment interface. A video is displayed on the left, Users were asked to score the video quality by choosing 1-5 stars shown on the right.

We conducted user subjective QA for evaluating the performance of the proposed method in this paper. In total, ten human evaluators, all postgraduate students from our research center, were asked to assess the quality of all video clips in the testing set. Fig. 4 displays the user-assessment interface. This interface allows the users to play the video and to assess its quality on a Five Point Likert scale, ranging from very serious IRRCM (1) to few IRRCM (5). Before the evaluation, the users were given example video clips for each category that allowed them to familiarize themselves with this task and the expected quality of the video clips. For each video, we receive ten subjective scores from ten independent assessors, which we average to receive the final user assessment score s_u:

$$s_u = \begin{cases} \lfloor \bar{s} \rfloor & if\ \bar{s} - \lfloor \bar{s} \rfloor < 0.5 \\ \lfloor \bar{s} \rfloor + 1 & else \end{cases} \tag{7}$$

where $\bar{s}$ is the average of all the independent scores from the subjects. Furthermore, results from a random run are also reported.

3.2   Results and Analysis

We firstly analyze the factors that affect the camera motion detection. There are commonly three different identifiable categories of motion in video sequences, namely, background or camera motion, the foreground object motion and shot or scene change. The shot or scene change is of different origin, namely external manual editing influences. The motion caused by the shot or scene change is easily removed since there are less matched pairwise-blocks in two consecutive frames that separately belong to two shots or scenes. Video clips showing foreground object motion under static camera are generally assessed as the best quality, with score 5. In this case, the detection method is especially effective if the object (or person) occupies a small part of the background. However, if an object is very close to a camera and moving irregularly, it displays the same visual effect as that caused by an IRRCM. In the case that both the object (person) and camera move, which is more complicated hinder the effectiveness of the detection method, we choose to process more frame images to overcome it compromisingly.

Now, we summarize the QA results from three methods. Table 1 compares the subjective scores, the determined scores using our proposed method and the random scores, respectively. As shown by the distribution of the scores, IRRCM is common in UGV.

Table 1. Number of UGV clips for each score grade from three methods

Method	Score				
	1	2	3	4	5
User #	10	31	62	79	118
Proposed #	16	49	59	55	121
Random #	53	64	59	68	56

A common challenge in user-based evaluation is the subjective nature of the assessment. Given a video clip, different subjects may give varying scores, which may be attributed to the subjective reasons, such as underestimating or overestimating. Fig. 5 depicts the Top 10 video clips with the highest standard deviation of user assessment scores. Moreover, the figure depicts the final subjective score, our automatic score and random score (shown in brackets above the box plot for each video clip). As can be seen, the subjective scores cover a wide range (from score 1 to 5) for these video clips, which most likely is due to subjective reasons or misuse. Nevertheless, the triplewise-score listed in the brackets show that our automatic method reports nearly the same scores as the subjective scores, and outperforms the random run.

In order to compare the differences between the proposed method or random run and the subjective assessment in detail, here, we take the scores from subjective assessment as the ground truth of QA results. We define three match levels between the automatic or random score and the subjective score for each video clip:

$$
\begin{aligned}
&Exact\ Match,\ if\ |s_{pg} - s_u| == 0\ \ or\\
&Close\ Match,\ if\ |s_{pg} - s_u| == 1\ \ or\\
&Dismatch,\ \ \ if\ |s_{pg} - s_u| >= 2
\end{aligned}
\tag{8}
$$

where, s_{pg} is the score from our proposed method or random run.

The pie graph in Fig. 6a shows the proportions of the three match levels. Overall, the results are very encouraging, more than 3/5 of the video clips with their two scores matched exactly, and about 32% of them achieved a close match. However, match results shown in Fig. 6b indicate that random run reports much worse results. Table 2 shows the differences between the automatic and subjective score pairs and the random-subjective score pairs in more detail. The results suggest that the proposed method performs well on the video clips without or

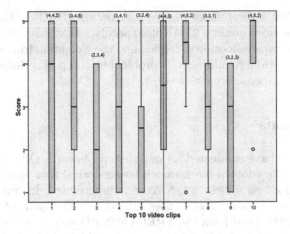

Fig. 5. Top 10 video clips with maximum standard deviation of user scores

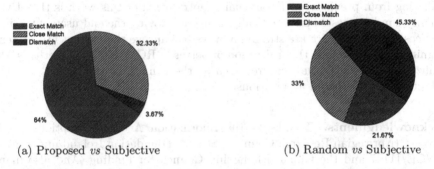

(a) Proposed *vs* Subjective   (b) Random *vs* Subjective

Fig. 6. Proportions of three match levels for proposed method *vs* subjective assessment, and random run *vs* subjective assessment

with very few camera motion since both methods scored nearly the same numbers of video clips with score 5 (118 *vs* 121). In total, 192 video clips receive the same scores from the proposed and subjective methods, 73 clips got lower scores from our method, whilst 35 clips were overestimated by our method. In contrast, random method only achieves 21.76% exact match and reports larger deviation to the subjective scores. This may be explained by the following facts: 1) Given

Table 2. Numbers of video clips for each score difference

	Score Difference ($s_{pg} - s_u$)							
	-4	-3	-2	-1	0	1	2	3
Proposed #	–	1	9	63	192	34	1	–
random #	23	44	38	62	65	37	22	9

488 J. Guo et al.

a video clip, only several short parts of it contain IRRCM. The users easily overestimate the video quality; 2) We also speculate that this may be introduced by the imperfect camera motion detection and the weighting framework in Eq. 6. As future work, we aim to address this problem by comparing different weighting frameworks to compensate for this effect.

4 Conclusions

In this paper, we have conducted an initial study towards QA of UGV by analyzing IRRCM. We adopt a block-match based optical flow approach to detect the IRRCM, based on which, a QA score is determined. In order to evaluate this quality score, we conducted a user subjective assessment of UGV quality. Using a set of UGV clips from the TRECVid MED task as a showcase, our results suggest that QA scores from proposed and subjective methods fit well. And the proposed method reports much better performance than the random run. Differing from previous work, our main contribution of this work is that UGV quality can be assessed automatically, hence improving the end users' QoE in a low-cost manner. There are still many ways to improve the current QA system. Significant ones include the utilization of better IRRCM detection methods, the utilization of other visual features such as the camera motion speed, and the adoption of UGV-based applications.

Acknowledgments. Thanks to the Information Access Disruptions (iAD) Project (Norwegian Research Council), Science Foundation Ireland under grant 07/CE/I1147 and the China Scholarship Council for funding. And also many thanks to the HMA group[4] for their help in the assessment effort.

References

1. Guo, J., Gurrin, C.: Short user-generated videos classification using accompanied audio categories. In: The First ACM International Workshop on Audio and Multimedia Methods for Large-Scale Video Analysis (AMVA), Nara, Japan (2012)
2. Olsson, S., Stroppiana, M., Baina, J.: Objective methods for assessment of video quality: state of the art. IEEE Transactions on Broadcasting 43(4), 487 (1997)
3. Farias, Q., Carli, M., Neri, A., Mitra, S.K., Barbara, C.S., Barbara, S., Tre, R., Navale, V.: Video quality assessment based on data hiding driven by optical flow information. In: Proceedings of SPIE, vol. 5294, pp. 190–200 (2004)
4. Van der Auwera, G., Reisslein, M.: Implications of smoothing on statistical multiplexing of h.264/avc and svc video streams. IEEE Transactions on Broadcasting (2009)
5. Seshadrinathan, K., Soundararajan, R., Bovik, A., Cormack, L.: Study of subjective and objective quality assessment of video. IEEE Transactions on Image Processing (2010)

---

[4] http://hma.dcu.ie/HMA/Home.html

6. Chikkerur, S., Sundaram, V., Reisslein, M., Karam, L.: Objective video quality assessment methods: A classification, review, and performance comparison. IEEE Transactions on Broadcasting (2011)
7. Internatinal Telecom Union: Rec. ITU-R BT.500-11. Technical report
8. Staelens, N., Moens, S., Van den Broeck, W., Mariën, I., Vermeulen, B., Lambert, P., Van de Walle, R., Demeester, P.: Assessing quality of experience of iptv and video on demand services in real-life environments. IEEE Transactions on Broadcasting (2010)
9. Guo, J., Scott, D., Hopfgartner, F., Gurrin, C.: Detecting complex events in user-generated video using concept classifiers. In: The 10th Workshop on Content-Based Multimedia Indexing (CBMI), pp. 1–6 (2012)
10. Wu, S., Ma, Y.-F., Zhang, H.-J.: Video quality classification based home video segmentation. In: ICME, pp. 217–220 (2005)
11. Smeaton, A.F., Over, P., Kraaij, W.: Evaluation campaigns and trecvid. In: MIR 2006: 8th ACM Int. Workshop on Multimedia Information Retrieval, pp. 321–330. ACM Press, New York (2006)

Multiscaled Cross-Correlation Dynamics on SenseCam Lifelogged Images

N. Li[1,2], M. Crane[1], H.J. Ruskin[1], and Cathal Gurrin[2]

[1] Centre for Scientific Computing & Complex Systems Modelling
[2] CLARITY: Centre for Sensor Web Technologies
School of Computing, Dublin City University, Ireland
na.li@dcu.ie, {mcrane,hruskin,cgurrin}@computing.dcu.ie

Abstract. In this paper, we introduce and evaluate a novel approach, namely the use of the cross correlation matrix and Maximum Overlap Discrete Wavelet Transform (MODWT) to analyse SenseCam lifelog data streams. SenseCam is a device that can automatically record images and other data from the wearer's whole day. It is a significant challenge to deconstruct a sizeable collection of images into meaningful events for users. The cross-correlation matrix was used, to characterise dynamical changes in non-stationary multivariate SenseCam images. MODWT was then applied to equal-time Correlation Matrices over different time scales and used to explore the granularity of the largest Eigenvalue and changes, in the ratio of the sub-dominant Eigenvalue spectrum dynamics, over sliding time windows. By examination of the eigenspectrum, we show that these approaches can identify "Distinct Significant Events" for the wearers. The dynamics of the Eigenvalue spectrum across multiple scales provide useful insight on details of major events in SenseCam logged images.

Keywords: Lifelogging, SenseCam, Images, equal-time Correlation Matrices, Maximum Overlap Discrete Wavelet Transform.

1   Introduction

In *Lifelogging*, the subject typically wears a device to record episodes of their daily lives. This concept has been pioneered to the extent of including a wearable computer, camera and viewfinder with wireless internet connection in order to capture personal activities through the medium of images or video. Developed by Microsoft Research in Cambridge, UK, SenseCam [1] is such a camera worn around the neck to capture images and other sensor readings automatically, in order to record the wearer's every moment. Such images and other data can be periodically reviewed to refresh and strengthen the wearer's memory. The device takes pictures at VGA resolution, (480x640 pixels), and stores these as compressed JPEG files on internal flash memory. SenseCam can collect a large amount of data, even over a short period of time, with a picture typically taken every 30 seconds. Hence there are about 4,000 images captured in any one day, or of the order of 1 million images captured per year.

Although research shows that the SenseCam can be an effective memory-aid device [2,3], as it helps to improve retention of an experience, wearers seldom wish to review

S. Li et al. (Eds.): MMM 2013, Part I, LNCS 7732, pp. 490–501, 2013.

life events by browsing large collections of images manually [4,5,6,7]. The challenge then is to manage, organise and analyse these large image collections in order to automatically highlight *key episodes* and, ideally, classify these in order of importance to the wearer. Previously, the lifelog of SenseCam images has been segmented into approximately 20 distinct events in a wearer's day, or about 7,000 events per year [8], but this large collection of personal information still contains a significant percentage of routine events. The challenge is to determine which events are the most important or unusual to the wearers.

In recent years, the behaviour of the largest Eigenvalue of a cross-correlation matrix over small windows of time, has been studied extensively, e.g. for financial series [9,10,11,12,13,14,15,16,17,18,19,20], electroencephalographic (EEG) recordings [21,22], magnetoencephalographic (MEG) recordings [23] and a variety of other multivariate data. In this paper, we investigate the same approach to analysis SenseCam lifelog data streams. We aim to apply the multiscaled cross-correlation matrix technique to study the dynamics of the SenseCam images, where this time series should exhibit *atypical* or *non-stationary*, characteristics, which highlight "Distinct Significant Events" in the data. We also evaluate our approach by identifying the boundaries between different daily events, which might include working at the office, walking outside, shopping etc. We found that different distinct events or activities can be detected at different scales.

This paper is organized as follow: in Section 2, we review the methods, in Section 3 we describe the data used, while Section 4 details with results obtained. Conclusions are given in Section 5.

2 Methods

Our previous research [24] has shown that SenseCam image time series reflect strong *long-range correlation*, indicating that the time series is not a random walk, but is cyclical, with continuous low levels of background information picked up constantly by the device. In this section, we first use equal-time cross-Correlation Matrices to characterise dynamical changes in non-stationary multivariate SenseCam time-series. The Maximum Overlap Discrete Wavelet Transform (MODWT) is then used to calculate equal-time Correlation Matrices over different time scales. This enables exploration of details of the Eigenvalue spectrum and in particular, examination of whether specific events show evidence of distinct *signatures* at different time scales.

2.1 Correlation Dynamics

The equal-time cross-correlation matrix can used to characterise dynamical changes in non-stationary multivariate time series. Before examining the image time series in detail, it is important to introduce the *gray scale pixel values* concept. In a gray scale image, a pixel with a value of 0 is completely black and a pixel with a value of 255 is completely white. While images captured from SenseCam are coloured, these are converted to gray-scale images in order to simplify the calculation. To reduce the size of the calculation further and the amount of memory used, we first adopt an averaging method to decrease image size from 480x640 pixels to 6x8 pixels. Hence the correlation matrix is made up of 48 time series for over 10,260 images. Given pixels $G_i(t)$, of a collection of images, we normalize G_i within each window in order to standardize the different pixels for the images as follows:

$$g_i(t) = \frac{G_i(t) - \overline{G_i(t)}}{\sigma_{(i)}} \tag{1}$$

where $\sigma_{(i)}$ is the standard deviation of G_i for image numbers $i=1,...,N$, and $\overline{G_i}$ is the time average of G_i over a time window of size T. Then the equal-time cross-correlation matrix may be expressed in terms of $g_i(t)$

$$C_{ij} \equiv \langle g_i(t)g_j(t) \rangle \tag{2}$$

The Eigenvalues λ_i and eigenvectors $\overline{v}_i$ of the correlation matrix C are found from the Eigenvalue equation $C\overline{v}_i = \lambda_i \overline{v}_i$.

2.2   Wavelet Multiscale Analysis

The Maximum Overlap Discrete Wavelet Transform (MODWT) [25,26,27,28], is a linear filter that transforms a series into coefficients related to variations over a set of scales. It produces a set of time-dependent wavelet and scaling coefficients with basis vectors associated with a location t and a unitless scale $\tau_j=2^{j-1}$ for each decomposition level $j=1,...,J_0$. Unlike the DWT, the MODWT, has a high level of redundancy. However, it is *non-orthogonal* and can handle any sample size N, whereas the DWT restricts the sample size to a multiple of 2^j. MODWT retains downsampled [1] values at each level of the decomposition that would be discarded by the DWT. This reduces the tendency for larger errors at lower frequencies when calculating frequency dependent variance and correlations, as more data are available. For MODWT the jth level equivalent filter coefficients have a width $L_j=(2^j-1)(L-1)+1$, where L is the width of the $j=1$ base filter.

Decomposing a signal using the MODWT to J levels theoretically involves the application of J pairs of filters. The filtering operation at the jth level consists of applying a rescaled father wavelet to yield a set of detail coefficients

$$\tilde{D}_{j,t} = \sum_{l=0}^{L_j-1} \tilde{\varphi}_{j,l} f_{t-l} \tag{3}$$

and a rescaled mother wavelet to yield a set of scaling coefficients

$$\tilde{S}_{j,t} = \sum_{l=0}^{L_j-1} \tilde{\phi}_{j,l} f_{t-l} \tag{4}$$

---

[1] Downsampling or decimation of the wavelet coefficients retains half of the number of coefficients that were retained at the previous scale. Downsampling is applied in the Discrete Wavelet Transform.

for all times $t = ..., -1, 0, 1, ...$, where f is the function to be decomposed [29]. The rescaled mother, $\tilde{\varphi}_{j,t} = \frac{\varphi_{j,t}}{2^j}$, and father, $\tilde{\phi}_{j,t} = \frac{\varphi_{j,t}}{2^j}$, wavelets for the jth level are a set of scale-dependent localized differencing and averaging operators and can be regarded as rescaled versions of the originals. The jth level equivalent filter coefficients have a width $L_j = (2^j - 1)(L - 1) + 1$, where L is the width of the $j = 1$ base filter. In practice the filters for $j > 1$ are not explicitly constructed because the detail and scaling coefficients can be calculated, using an algorithm that involves the $j = 1$ filters operating recurrently on the jth level scaling coefficients, to generate the $j + 1$ level scaling and detail coefficients [29].

The wavelet variance $\nu_f^2(\tau_j)$ is defined as the expected value of $\tilde{D}_{j,t}^2$ if we consider only the non-boundary coefficients. An unbiased estimator of the wavelet variance is formed by removing all coefficients that are affected by boundary conditions and is given by

$$\nu_f^2(\tau_j) = \frac{1}{M_j} \sum_{t=L_j-1}^{N-1} \tilde{D}_{j,l}^2 \qquad (5)$$

where $\tilde{D}_{j,l}$ is a rescaled father wavelet, which yields a set of scaling coefficients, $M_j = N - L_j + 1$ is the number of non-boundary coefficients at the jth level.

The wavelet covariance between functions $f(t)$ and $g(t)$ is similarly defined to be the covariance of the wavelet coefficients at a given scale. The unbiased estimator of the wavelet covariance at the jth scale is given by

$$\nu_{fg}(\tau_j) = \frac{1}{M_j} \sum_{t=L_j-1}^{N-1} \tilde{D}_{j,l}^{f(t)} \tilde{D}_{j,l}^{g(t)} \qquad (6)$$

The MODWT estimate of the wavelet cross-correlation between functions $f(t)$ and $g(t)$ may be calculated using the wavelet covariance and the square root of the wavelet variance of the functions at each scale j. The MODWT estimator, of the wavelet correlation is thus given by

$$\rho_{fg}(\tau_j) = \frac{\nu_{fg}(\tau_j)}{\nu_f(\tau_j)\nu_g(\tau_j)} \qquad (7)$$

where $\nu_{fg}(\tau_j)$ is the covariance between $f(t)$ and $g(t)$ at scale j, $\nu_f(\tau_j)$ is the variance of $f(t)$ at scale j and $\nu_g(\tau_j)$ is the variance of $g(t)$ at scale j.

The multiscaled cross-correlation matrix technique is adopted in order to help highlight non-stationary events (in SenseCam lifelog data streams), which could be of importance.

3   Data

In this study, the data were generated from one person wearing the SenseCam over a six day period, from a Saturday to a Thursday. These particular days were chosen in order to include a weekend, where normal home activity varies in comparison to events on weekdays or a working week. Forming a total lifelog of 10,260 images, with average wearing time varying from about 11 hours on Saturday to about 6 hours on Tuesday.

Saturday involved the subject walking to the nearest bus stop from home, a bus journey to the city centre, walking through local streets as well as a visit to a shopping centre. This day also involved dinner with a friend and a bus journey back to the original bus stop. Over the next five days, these images described a typical day for the subject: sitting in the office, talking with a colleague and sharing lunch in the cafeteria, the journey from the office to home, and the next morning from home to their office and so on. Figure 1 shows some examples of SenseCam images. Data statistics are reported in Table 1. To create a ground truth, the user reviewed her collection and manually marked the boundary image between all events.

Fig. 1. Example of SenseCam Images

Table 1. Data Statistics

User	Events Catalogue	Groundtruthed Events	Images
1	Working	15	6146
1	Walking Outside	32	1494
1	Shopping	12	826
1	Eating	3	658
1	Taking Bus	2	297
1	Others	5	839
		Total: 69	Total: 10,260

4   Results

4.1   Dynamics of the Largest Eigenvalue for Different Window Sizes

In financial data, it has been known for some time that the largest Eigenvalue (λ_1) contains information on risk associated with the particular assets of which the covariance matrix is comprised, (i.e. the 'market' factor) [30]. Similarly we would expect the largest Eigenvalue to present information from the image that reflects the largest change in the SenseCam recording.

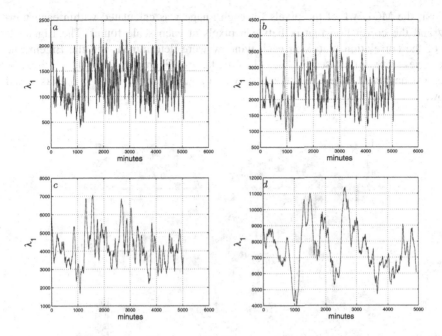

Fig. 2. Total 10260 Images (size: 6x8 pixels) largest Eigenvalue Distribution using a sliding window of 50 Images(*a*), 100 Images(*b*), 200 Images(*c*) and 400 Images(*d*)

Typically SenseCam captures two images every minute, so we can measure wavelet Eigenvalue dynamics in time (minutes). Figure 2 shows the time series of the largest Eigenvalue for different window sizes. From these, we note the following features:

- With increased window size comes increased smoothing – as expected. This removes some of the high frequency small-scale changes, typically associated with noise.
- As the window size is increased, the peaks in the series become more pronounced. These peaks reflect large changes in greyscale of the images.

4.2 Wavelet Analysis

For the present study, we selected the least asymmetric (LA) wavelet, (known as the Symmlet, [31]), which exhibits near symmetry about the filter midpoint. LA filters are defined in even widths and the optimal filter width is dependent on the characteristics of the signal and the length of the data series. The filter width chosen for this study was the LA8, (where 8 refers to the width of the scaling function), since this enables accurate calculation of wavelet correlations to the 10th scale, which is appropriate given the length of data series available. Although the MODWT can accommodate any level, J_0, the largest level, is chosen in practice, so as to prevent decomposition at scales longer than the total length of the data series, (hence the choice of the 10th), while still containing enough detail to capture subtle changes in the signal, [29].

First, the MODWT of the pixels for each image was calculated within each window and the correlation matrix between pixels at each scale found. The Eigenvalues of the correlation matrix in each window were determined, and the Eigenvalue time series were normalised in time. Then the largest Eigenvalue for different window sizes was analysisd. These results are shown in a heat map in Figure 3 and discussed below.

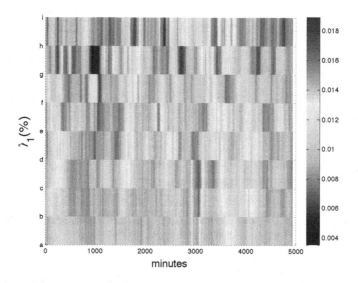

Fig. 3. Heatmap diagram showing the dynamics of the largest Eigenvalue λ_1 across 9 wavelet scales. Scales 1 *(a)* to 9 *(i)* correspond to a 1-2 minute period, a 2-4 minute period, a 4-8 minute period, a 8-16 minute period, a 16-32 minute period, a 32-64 minute period, a 64-128 minute period, a 128-256 minute period, a 256-512 minute period.

Dynamics of the Largest Eigenvalue at Various Wavelet Scales. Figure 3 shows the time series of the largest Eigenvalue dynamics across different wavelet scales. Some peaks are consistently captured by the SenseCam at certain scales, such as a peak around 3000 minutes, (captured by wavelet scales 1, 2, 3 and corresponding to a 1-2 minute period, a 2-4 minute period and a 4-8 minute period). These peaks should help us to identify major events or activities in the data. The different features, found at various scales, suggest that the correlation matrix captured different major events with different time horizons. This will be examined in more detail in the next subsection.

The Largest Eigenvalue λ_1 Compared with the Ratio of λ_1/λ_2 Dynamics. We also wish to ascertain whether the sub-dominant Eigenvalues λ_2 hold further information on the key sources or major events and what information these contribute additionally to the images. The dynamics of the largest Eigenvalue and changes in the ratio of the largest Eigenvalues were examined from a MODWT analysis. Here, we detail several scenarios for the peaks in the largest Eigenvalue and the ratio Eigenvalues for a window size of 400 images. We have tried to identify the position and nature of peak sources or major events from the real images generated from SenseCam collections.

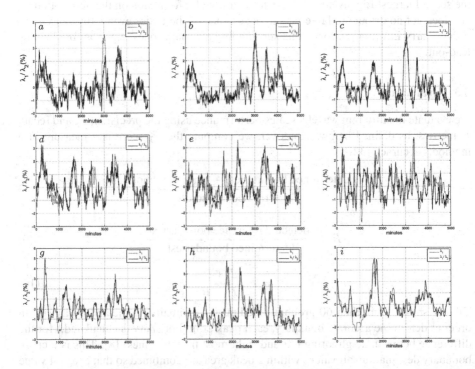

Fig. 4. The largest Eigenvalue λ_1 *(red)* and the ratio of λ_1/λ_2 *(black)* dynamics across 9 wavelet scales. The Top 3 figures *(a-c)* are for scales 1 to 3 corresponding to a 1-2 minute period, a 2-4 minute period, a 4-8 minute period, Middle 3 figures *(d-f)* are for scales 4 to 6 corresponding to a 8-16 minute period, a 16-32 minute period, a 32-64 minute period and Bottom 3 figures *(g-i)* are for scales 7 to 9 corresponding to a 64-128 minute period, a 128-256 minute period, a 256-512 minute period.

In Figure 4, the dynamics of the series for the largest Eigenvalue and changes of the Eigenvalue ratio λ_1/λ_2 were examined for the MODWT analysis. The technique gives a clear picture of the movements in the image time series by reconstructing them using each wavelet component. MODWT captured the particular marked features at specified scales. A number of features from the image are reproduced and can be examined by studying these Eigenvalue series.

We have detailed the scenario for each peak in Figure 4. The subject was sitting in front of her laptop, with laptop, lights and seating position unchanged over on extended period, contributing high pixel values in a sequence of images. This typical case was always marked by a peak in the SenseCam signal. The signal fluctuation is caused by changes, such as the subject moving from indoors to outdoors, light level alternatives, the subject changing position from sitting to moving, movement increase and more people joining in the scene. Note that movement increase or multiple person interactions can be captured by specific scales, using the MODWT method. The ratio analysis strongly reinforces observations on the largest Eigenvalue over time. The ratio of λ_1/λ_2 has smaller variation compared to that for the largest Eigenvalue λ_1. This implies that the second largest Eigenvalue (λ_2) carries additional information on the description of, but does not contribute in large part to occurrence of the major events for SenseCam. It does carry information for events surrounding the major ones, e.g. possible lead-in, lead-out.

4.3 Evaluation

We evaluate the different wavelet scales performance using the precision, P, and recall, R metrics, as defined below. Moreover, we compute the F_1 score is a measure of a method's accuracy [32].

$$P = \frac{|\text{determined boundaries}| - |\text{wrong boundaries}|}{|\text{determined boundaries}|} \tag{8}$$

$$R = \frac{|\text{detected reference boundaries}|}{|\text{determined boundaries}|} \tag{9}$$

$$F_1 = 2 * \frac{P * R}{P + R} \tag{10}$$

Table 1 shows more than 60 groundtruthed events manually segmented by a user. In order to determine accurate boundary, each peak point boundary is calculated, (for the difference between neighboring left and right most trough values) [8]. This is a crude boundary designation, all values within a peak area are combined so that a signal value is less informative. Significant peaks are determined (distinct events or activities) by λ_1/λ_2 percentage pixel values are larger than zero. Table 2 shows the precision, recall and F_1 measure for λ_1/λ_2 at different wavelet scales. As we can see, all scales appear with high precision but very low recall. The main weakness as well as strength

Table 2. Precision, Recall and F_1 measures for MODWT method

Wavelet Scales	λ_1/λ_2		
	P	R	F_1
Scale1 (1-2 minute period)	0.3929	0.4058	0.3992
Scale2 (2-4 minute period)	0.7857	0.2029	0.3225
Scale3 (4-8 minute period)	0.5000	0.3188	0.3894
Scale4 (8-16 minute period)	0.4783	0.3333	0.3929
Scale5 (16-32 minute period)	0.5238	0.3043	0.3850
Scale6 (32-64 minute period)	0.5789	0.2754	0.3732
Scale7 (64-128 minute period)	0.7333	0.2174	0.3354
Scale8 (128-256 minute period)	0.9167	0.1739	0.2924
Scale9 (256-512 minute period)	1	0.1594	0.2750

for wavelet scales is that different scales highlight different distinct events dependent on the time horizons. Some events at certain scales will be missed, so that the overall recall values are low for this approach. In addition, some activities such as working in front of the laptop last for several hours. In manually segmenting 69 events of 10,260 images only, the detection probability for a given event is quite low. In consequence this approach is quite crude and suggest that further modification needed, such as incorporating other than peak distance an weighting scale combinations.

5   Conclusions

The Maximum Overlap Discrete Wavelet Transform (MODWT) method, calculating equal-time Correlation Matrices over different time scales, was used to investigate the largest Eigenvalue and the changes in the sub-dominant Eigenvalue ratio spectrums. As shown in Figure 3, the different features, found at various scales, suggest that the correlation matrix captured different major events with different time horizons. We note that these "jitters" may contain additional information surrounding the major events. This suggests that the correlation matrix for different information captured from SenseCam can be filtered by different time horizons. Those consistently occurring peaks should help us to identify major events captured by the SenseCam. By examining the behaviour of the largest Eigenvalue and the change in Eigenvalue ratios over time, the Eigenvalue ratio analysis confirmed that the largest Eigenvalue carries most of the major event information, whereas subsequent Eigenvalues carry information on supporting or lead in/ lead out events. On analyzing major events, (with a sliding window set to 400 images), we identified the light level as a major event delineator during static periods of image sequence. While the methods set out in this paper appear to perform quite well for precision value, they do quite poorly for the recall value at different time scales. The value in the method lies in the fact that prior information about events can be used as an additional filter. An example of this is that sitting in front of a laptop is likely to show up in the higher scaled wavelet scales. Overall the MODWT method provides a powerful tool for examination of the nature of the captured SenseCam data for certain categories of users.

Future work includes checking larger datasets and multiple users to help to confirm initial findings which are clearly capable of refinement, as mentioned Section 4. Studying the multi-scaled correlation dynamics over multiple sensors may be useful to detect distinct events or activities and could also be worthwhile for classification of event type in SenseCam data.

Acknowledgments. NL would like to acknowledge generous support from the Sci-Sym Centre Small Scale Research Fund, as well as additional support from the School of Computing and CLARITY Centre.

References

1. Hodges, S., Williams, L., Berry, E., Izadi, S., Srinivasan, J., Butler, A., Smyth, G., Kapur, N., Wood, K.: SenseCam: A Retrospective Memory Aid. In: Dourish, P., Friday, A. (eds.) UbiComp 2006. LNCS, vol. 4206, pp. 177–193. Springer, Heidelberg (2006)

2. Harper, R., Randall, D., Smyth, N., Evans, C., Heledd, L., Moore, R.: Thanks for the Memory. In: Proceedings of the 21st BCS HCI Group Conference, HCI 2007, Lancaster, U.K. (2007)

3. Harper, R., Randall, D., Smyth, N., Evans, C., Heledd, L., Moore, R.: The Past is a Different Place: They Do Things Differently There. In: Designing Interactive Systems, Cape Town, South Africa, pp. 271–280 (2008)

4. Ashbrook, D., Lyons, K., Clawson, J.: Capturing Experiences Anytime, Anywhere. IEEE Pervasive Computing (2006)

5. Bell, G., Gemmell, J.: A Digital Life. Scientific American (2007)

6. Lee, M.L., Dey, A.K.: Providing good memory cues for people with episodic memory impairment. In: Assets 2007: Proceedings of the 9th International ACM SIGACCESS Conference on Computers and Accessibility, Tempe, Arizona, USA, pp. 131–138 (2007)

7. Lin, W.H., Hauptmann, A.: Structuring Continuous Video Recordings of Everyday Life Using Time-Constrained Clustering. In: Multimedia Content Analysis, Management and Retieval SPIE-IST Electronic Imaging, vol. 6073, San Jose, California, USA, pp. 111–119 (2006)

8. Doherty, A.R., Smeaton, A.F.: Automatically Segmenting Lifelog Data into Events. In: Proc. WIAMIS 2008, pp. 20–23 (2008)

9. Laloux, L., Cizeau, P., Bouchaud, J.-P., Potters, M.: Noise dressing of financial correlation matrices. Phys. Rev. Lett. 83(7), 1467–1470 (1999)

10. Plerou, V., Gopikrishnan, P., Rosenow, B., Nunes Amaral, L.A., Stanley, H.E.: Universal and non-uiversal properties of cross-correlations in financial time series. Phys. Rev. Lett. 83(7), 1471–1474 (1999)

11. Laloux, L., Cizeau, P., Bouchaud, J.-P., Potters, M.: Random matrix theory and financial correlations. Int. J. Theoret. Appl. Finance 3(3), 391–397 (2000)

12. Plerou, V., Gopikrishnan, P., Rosenow, B., Nunes Amaral, L.A., Guhr, T., Stanley, H.E.: A random matrix approach to cross-correlations in financial data. Phys. Rev. E 65, 066126/1–066126/18 (2000)

13. Gopikrishnan, P., Rosenow, B., Plerou, V., Eugene Stanley, H.: Indentifying business sectors from stock price fluctuations. Phys. Rev. E 64, 035106R/1–035106R/4 (2001)

14. Utsugi, A., Ino, K., Oshikawa, M.: Random matrix theory analysis of cross-correlations in financial markets. Phys. Rev. E 70, 026110/1–026110/17 (2004)

15. Bouchaud, J.P., Oshikawa, M.: Theory of Financial Risk and Derivative Pricing. Cambridge University Press (2003)

16. Wilcox, D., Gebbie, T.: On the analysis of cross-correlations in South African market data. Physica A 344(1-2), 294–298 (2004)

17. Sharifi, S., Crane, M., Shamie, A., Ruskin, H.J.: Random matrix theory for portfolio optimization: A stability approach. Physica A 335(3-4), 629–643 (2004)

18. Conlon, T., Ruskin, H.J., Crane, M.: Random matrix theory and fund of funds portfolio optimisation. Physica A 382(2), 565–576 (2007)

19. Conlon, T., Ruskin, H.J., Crane, M.: Wavelet multiscale analysis for Hedge Funds: Scaling and Strategies. Physica A, 5197–5204 (2008)

20. Podobnik, B., Stanley, H.E.: Detrended cross-correlation analysis: A new method for analysing two non-stationary time series. Phys. Rev. Lett. 100(8), 084102/1–084102/11 (2008)

21. Schindler, K., Leung, H., Elger, C.E., Lehnertz, K.: Assessing seizure dynamics by analysing the correlation structure of multichannel intracranial EEG. Brain 130, 65–77 (2007)

22. Schindler, K., Elger, C.E., Lehnertz, K.: Increasing synchronization promote seizure termination: Evidence from status epilepticus. Clin. Neurophysiol. 118(9), 1955–1968 (2007)

23. Kwapien, J., Drozda, S., Ionannides, A.A.: Temporal correlations versus noise in the correlation matrix formalism: An example of the brain auditory response. Phys. Rev. E 62, 5557–5564 (2000)
24. Li, N., Crane, M., Ruskin, H.J.: Automatically detecting significant events on sensecam. ERCIM News (87) (2011)
25. Peng, C.-K., Buldyrev, S.V., Havlin, S., Simons, M., Stanley, H.E., Golderberger, A.L.: On the mosaic organization of DNA sequences. Phys. Rev. E 49, 1685–1689 (1994)
26. Matos, J.A.O., Gama, S.M.A., Ruskin, H.J., Sharkasi, A.A., Crane, M.: An econophysics approach to the Portuguese Stock Index-PSI-20. Physica A 342, 665–676 (2004)
27. Buldyrev, S.V., Goldberger, A.L., Havlin, S., Mantegna, R.N., Matsa, M.E., Peng, C.-K., Simons, M., Stanley, H.E.: Long-range correlation properties of coding and noncoding DNA sequences: GenBank analysis. Phys. Rev. E 51, 5084–5091 (1995)
28. Heneghan, C., McDarby, G.: Establishing the relation between detrended fluctuation analysis and power spectral density analysis for stochastic processes. Phys. Rev. E 62, 6103–6110 (2000)
29. Percival, D.B., Walden, A.T.: Wavelet methods for time series analysis. Cambridge University Press (2000)
30. Sharkasi, A., Crane, M., Ruskin, H.J., Matos, J.A.: The reaction of stock markets to crashes and events: A comparison study between emerging and mature markets using wavelet transforms. Physica A 368, 511–521 (2006)
31. Burrus, C.S., Gopinath, R.A., Gao, H.: Introduction to wavelets and wavelets transforms. Prentice Hall (1997)
32. Misra, H., Hopfgartner, F., Goyal, A., Punitha, P., Jose, J.M.: TV News Story Segmentation Based on Semantic Coherence and Content Similarity. In: Boll, S., Tian, Q., Zhang, L., Zhang, Z., Chen, Y.-P.P. (eds.) MMM 2010. LNCS, vol. 5916, pp. 347–357. Springer, Heidelberg (2010)

Choreographing Amateur Performers
Based on Motion Transfer between Videos

Kenta Mizui[1], Makoto Okabe[2,3], and Rikio Onai[2]

[1] Department of Informatics, The University of Electro-Communications, 2-11, Fujimicho, Chofu, Tokyo, 182-0033, Japan
[2] Department of Computer Science, The University of Electro-Communications, 2-11, Fujimicho, Chofu, Tokyo, 182-0033, Japan
[3] JST PRESTO
mizui@onailab.com, m.o@acm.org, onai@cs.uec.ac.jp

Abstract. We propose a technique for quickly and easily choreographing a video of an amateur performer by comparing it with a video of a corresponding professional performance. Our method allows the user to interactively edit the amateur performance in order to synchronize it with the professional performance in terms of timings and poses. In our system, the user first extracts the amateur and professional poses from every frame via semi-automatic video tracking. The system synchronizes the timings by computing dynamic time warping (DTW) between the two sets of video-tracking data, and then synchronizes the poses by applying image deformation to every frame. To eliminate unnatural vibrations, which often result from inaccurate video tracking, we apply an automatic motion-smoothing algorithm to the synthesized animation. We demonstrate that our method allows the user to successfully edit an amateur's performance into a more polished one, utilizing the Japanese sumo wrestling squat, the karate kick, and the moonwalk as examples.

Keywords: shape deformation, video editing.

1   Introduction

We may imagine ourselves as professional dancers or action stars, but since our physical skills are usually inadequate, the resulting performances are poor. This situation is commonplace in the movie industry. When an actor's physical abilities are limited, a professional (or stunt) performer is often substituted, but the face must be carefully concealed. To solve this problem, we are interested in developing a technique for editing a performer's motions to make him or her appear more dexterous (Fig. 1).

A number of existing approaches for editing a video of a human performance are based on three-dimensional (3D) reconstruction of the performer. For example, Jain *et al.* proposed a method for reconstructing the shape of a performer's body by fitting it to a morphable body model, which can be edited interactively (e.g., to be thinner or fatter). However, our goal is to edit the speed and pose of

S. Li et al. (Eds.): MMM 2013, Part I, LNCS 7732, pp. 502–512, 2013.

(a) An amateur performance   (b) A professional performance   (c) The choreographed performance

Fig. 1. Our method uses a pair of videos of a physical activity, such as a karate kick, as input; one is the target video of an amateur performance (a), and the other is a video of a corresponding professional performance (b). We extract timing and pose information from the professional performance and use it to choreograph the amateur performance (c).

a performance more dynamically. The video-based character method is closely related to our technique, but synthesizes a video sequence of a new performance of a pre-recorded character [2]. Although this allows the user to freely edit a 3D human performance and move the camera at will, expensive equipment, including a special studio and multiple cameras, is required.

Since performance editing in 3D space is powerful but expensive, two-dimensional (2D) video editing is a reasonable and popular alternative. Video texture approaches allow an input video or its internal objects to be edited to play in a seamless infinite loop, or with a user-specified temporal behavior [3, 4]. However, since these techniques rearrange only the temporal order of the video frames, it is impossible to change the speed or pose of an object. The method of Scholz *et al.* allows a video object to be moved, rotated, or bent, but only low-dynamic edits have been demonstrated [5]. The method of Weng *et al.* is closely related to our work [6], but the motion of a video object is edited only in user-specified key frames: i.e., the deformations are applied in a sparse set of frames, and the system then propagates the deformations to the rest of frames. Their method requires contour tracking of a video object, a time-consuming task for the user, which our technique avoids. Also, our method is example-based, allowing the user to quickly specify a desired motion.

To make the awkward motions of an amateur performer appear more polished, we propose a method for quickly and easily choreographing a video of the amateur performer by utilizing a video of a corresponding professional performance. Our method allows an amateur performance to be edited by transferring the timings and poses extracted from the professional performance. Our technique is purely 2D, and enables the user to edit human performances at low cost, without the large datasets of 3D models, a special studio, or multiple cameras. We successfully applied our method to amateur performances involving the Japanese sumo wrestling squat, the karate kick, and the moonwalk, making them all look more polished.

2   System Overview

Figure 2 shows an overview of our system. The system utilizes videos of an amateur performance (Fig. 2a) and a corresponding professional performance (Fig. 2b) as inputs. The karate kick is used in this example. In our experiments, we assume that each video is captured with a single, fixed camera, and the background image is known. Our method consists of the following four stages. First, the user extracts the trajectories from the two input videos (Fig. 2c). Our system tracks a motion and associates its trajectory with a user-specified body part. Since the automatic tracking is incomplete, and often generates incorrect trajectories, our system supports a user interface to interactively adjust them. Second, the system applies dynamic time warping (DTW) to synchronize the amateur's motions with the professional's (Fig. 2d). DTW finds temporal correspondences between the amateur and professional trajectories, and temporally aligns video frames. Third, the system deforms the amateur's pose in every frame, so that the trajectories of both performers are spatially matched (Fig. 2e). Finally, when significant unnatural vibrations appear in the choreographed animation, the user can optionally request the system to apply automatic smoothing. Our optimization algorithm removes the differences between the motions of choreographed and professional animations (Fig. 2f). The tracking and DTW stages are semi-automatic, while the deformation and smoothing stages are fully automatic.

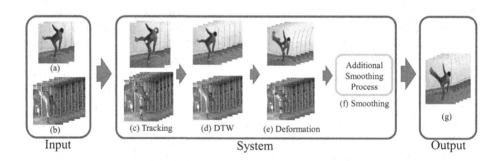

Fig. 2. System overview

3   Semi-automatic Video Tracking

We first analyze the motions of both the amateur and professional, and extract the trajectories of the corresponding body parts (Fig. 3). In this example, we are tracking seven body parts: the head, back, hip, left and right knees, and left and right feet. We rely on semi-automatic video tracking because an automatic algorithm will not always provide correct results. Even the latest techniques, such as the particle video method [7], compute inaccurate, noisy trajectories for fast motions. On the other hand, as we shall demonstrate, only a small amount of user interaction is necessary to efficiently modify such trajectories. Since we

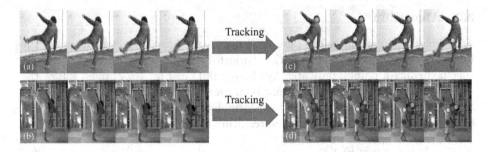

Fig. 3. (a) Amateur frames. (b) Professional frames. (c and d) The colored dots represent the tracking of each body part.

track the center of each body part instead of its contour [8], the user's burden is smaller than in methods based on rotoscoping.

3.1 User Interface

Figure 4 shows the flow of our semi-automatic video tracking. The user begins to track the left foot by clicking on it (Fig. 4a). The system automatically computes the trajectory, indicated by the blue dots in the top part of Fig. 4, based on an optical flow method [9]. However, the blue dots gradually slide outside the foot area. To correct this, the user drags and modifies the last tracker position, indicated by the red dot in Fig. 4b. The system then updates the trajectory, indicated by the orange dots in the bottom part of Fig. 4, using our optimization algorithm. In this way, the user can efficiently and accurately extract a trajectory with minimal burden.

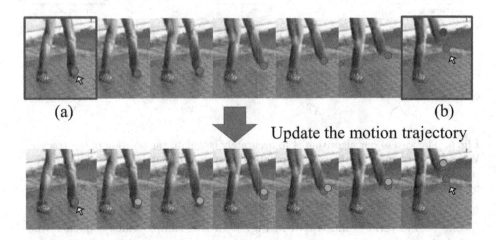

Fig. 4. User interface for the semi-automatic video tracking

3.2   Algorithm

To develop the user interface described above, we based our algorithm on the following three policies. The resulting trajectory 1) must pass through the user-specified control points, indicated by the red dots in Fig. 4b, 2) must follow the sampled optical flow under the trackers in every frame as much as possible, and 3) must be as smooth as possible. We formulate this optimization problem in a least-squares sense: we calculate the resulting trajectory R from

$$R = \arg \min_{X} E(X), \tag{1}$$

$$E(X) = \lambda \sum_{i \in I^c} (x_i - x_i^c)^2 + \sum_{j \in I^c} \sum_i^n w_i^j (x_i - t_i^j)^2 + \sum_i^{n-1} w_i^a (v_i - v_{i-1})^2, \tag{2}$$

$$v_i = x_{i+1} - x_i, \tag{3}$$

where $X = \{x_1, \ldots, x_n\}$ is the set of tracker positions, I^c is the set of indices of the frames with user-specified control points, $X^c = \{i \in I^c | x_i^c\}$ is the set of user-specified control point positions, and $T^j = \{t_1^j, \ldots, t_n^j\}(j \in I^c)$ is the set of tracker positions, calculated at each control point. The first, second, and third terms correspond to the three policies. The first term relates to the user-specified control points, the second term compels the trackers to try to follow the underlying optical flow, and the third term is the smoothness term. λ, w_i^c, and w_i^a are the respective weights for the terms. λ is set equal to 10,000 in our experiment.

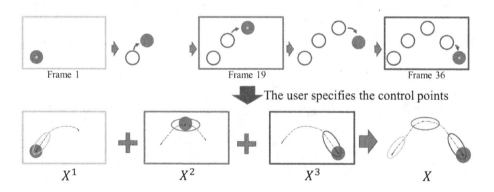

Fig. 5. The idea of our algorithm is to accurately calculate the position of the tracker. When the user specifies the control points, our system calculates the position of the tracker from each control point (bottom). We obtain the precise position of the tracker by mixing these trajectories.

The second term is illustrated in Fig. 5. In this sample video sequence, the red circle moves along an arc, as shown in the top part of Fig. 5. The user specifies three control points, indicated by the orange dots in Frame 1, Frame

19, and Frame 36. Our system then calculates the trajectories T^1, T^{19}, and T^{36}, starting from each control point and sampling the underlying optical flow. The calculated positions of the trajectories are accurate and reliable near the control points around the ellipse in each figure, but inaccurate and unreliable far from the control points. Therefore, we want to merge the reliable parts of those trajectories and obtain accurate trajectories for the trackers. To achieve this, we set w_i^j to be inversely proportional to the distance from each control point.

Without the third term, the first and second terms introduce unnatural vibrations into the trajectory (Fig. 6a), resulting in awkward motions in the final synthesized animation. Thus, we introduce the third term to smooth the trajectory. We want to keep the accelerations as small as possible, while preserving the significantly large accelerations (Fig. 6b). w_i^a is designed according to Eq. 4.

$$w_i^a = e^{-|(x_{i+1}-x_i)-(x_i-x_{i-1})|^5/10000} \tag{4}$$

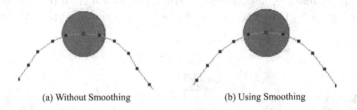

(a) Without Smoothing   (b) Using Smoothing

Fig. 6. Tracking results w/o smoothness term

4   Motion Transfer and Rendering

Synchronization of Motion Speed via DTW. After extracting the trajectories, we applied DTW between the amateur and professional trajectories and synchronized the amateur's motions with those of the professional [10]. For example, Fig. 7 shows amateur and professional videos of the Japanese sumo wrestling squat, with durations of 84 frames and 96 frames, respectively. Using seven trackers, we obtained seven motion trajectories. Since each tracker represents a 2D position, we obtained (2×7)-dimensional signals for the videos, with durations of 96 and 84 frames, respectively. After applying DTW between these signals, we obtained the lengthened (i.e., slowed down) synchronized amateur video shown in Fig. 7. DTW usually succeeded, but often included poor synchronization in the results. In such a case, we used the "time remapping" user interface of Adobe After Effects CS5 to modify it.

Deformation. After synchronizing the motions, we deformed the amateur's pose frame-by-frame to match the professional's pose. We applied feature-based image warping [11] to accomplish this. As Fig. 8 shows, we created a skeleton for each frame by connecting the neighboring trackers, and then applied the algorithm between the amateur and professional skeletons. In the resulting image, the amateur's leg was as high as that of the professional.

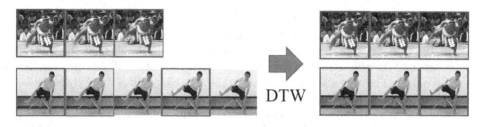

Fig. 7. Synchronization of the amateur's motions with those of the professional

We also tried as-rigid-as-possible (ARAP) shape manipulation [12]. ARAP is effective for preserving the original shape and area of the input image. However, we often had to deform the image drastically, and in such a case, ARAP sometimes yielded awkward results because of the rigidity constraints (Fig. 8d). On the other hand, feature-based image warping employs several parameters, which enabled us to control the rigidity of the deformation and obtain better results (Fig. 8c).

Fig. 8. (a) Amateur frame. (b) Professional frame. (c) Deformation results of feature-based image warping. (d) Deformation results of ARAP.

Smoothing Unnatural Vibration. Inaccurate motion trajectories cause visible artifacts, which take the form of unnatural vibrations in the synthesized animation. This effect was especially conspicuous in our moonwalk results. However, it is almost impossible for the user to create perfect trajectories in the semi-automatic video tracking process. Therefore, we introduced an automatic optimization technique to remove such unnatural vibrations and refine the resulting motions. We based our algorithm on the assumption that both the choreographed amateur and professional should have the same motions. Hence, if the choreographed amateur exhibits unnatural vibrations, these can be obtained as motion differences between the choreographed amateur and professional. We therefore calculated these differences, and then deformed each frame of the choreographed amateur video until the difference disappeared. Figure 9 shows the details of our smoothing process. The choreographed amateur's ear moved downward, but the professional's ear moved to the right (Figs. 9a, 9b, and 9c). The computed optical flow is shown in Fig. 9d. Our algorithm computed the difference between the optical flow vectors (the red arrow in Fig. 9e), and deformed the choreographed amateur's frame in accordance with it (Fig. 9f). We repeated this procedure for every frame, using all the trackers defined in the video tracking process. In this procedure, we used image deformation based on moving least squares (MLS) [13], which is computationally efficient and effective for preserving the original shape.

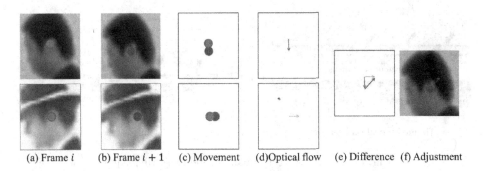

(a) Frame i (b) Frame $i + 1$ (c) Movement   (d)Optical flow   (e) Difference   (f) Adjustment

Fig. 9. Smoothing unnatural vibration

Rendering. The deformation algorithm not only deforms the performer but also distorts the background (Fig. 10a). To recover the correct background, we applied the alpha-blending method to create the final composite. Since the background image was given in advance (Fig. 10b), we first performed the background subtraction and obtained a mask for the performer. We then applied the same deformation to the mask (Fig. 10c). Finally, we computed a trimap from the mask and applied the Poisson matting method [14] to create the composite (Fig. 10d).

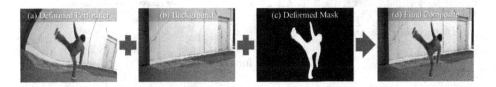

(a) Deformed Performer   (b) Background   (c) Deformed Mask   (d) Final Composite

Fig. 10. Rendering the final result with background subtraction and alpha blending

5   Result

We applied our technique to 1) the Japanese sumo wrestling squat, 2) the karate kick, and 3) the moonwalk. For each of these, we prepared a pair of videos, one amateur and one professional. We captured the former ourselves, and obtained the latter from YouTube (http://www.youtube.com/). The results are shown in our website (https://sites.google.com/a/onailab.com/mizui/english).

The Japanese Sumo Wrestling Squat. The amateur performer could not raise his leg high enough because his hip joint was not as flexible as that of the professional (Fig. 11). Our method successfully deformed the amateur's pose and raised his leg as high as that of the professional. However, we also acquired a visible artifact: the deformed leg grew a bit thicker. This is because our image-warping technique is not effective for preserving area.

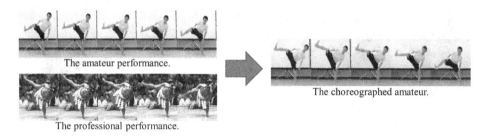

The amateur performance.

The choreographed amateur.

The professional performance.

Fig. 11. Results for the Japanese sumo wrestling squat

The Karate Kick. Our method successfully edited the low, slow karate kick of the amateur (Fig. 12) to create a high, quick karate kick video. The side-by-side comparison in our supplementary video confirms its similarity to the professional performance.

The amateur performance.

The choreographed amateur.

The professional performance.

Fig. 12. Results for the karate kick

The Moonwalk. Our method was also used to modify an amateur's moonwalk performance, which initially looked as if he was just walking slowly backwards (Fig. 13). After motion synchronization and pose deformation, the resulting animation looked awkward, and visible artifacts were apparent: the choreographed performance had unnatural vibrations throughout the video sequence. Our smoothing algorithm successfully removed these vibrations and synthesized a smooth animation. The comparison in our website video confirms this.

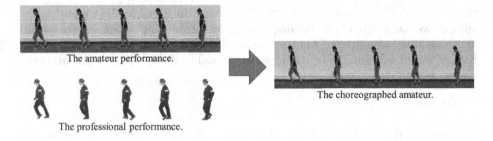

The amateur performance.

The professional performance.

The choreographed amateur.

Fig. 13. Results for the moonwalk

5.1   User Study

We conducted a subjective evaluation to investigate whether or not subjects could find visible artifacts created by our system. We asked 15 subjects to evaluate the system, all of them university students who were not familiar with our technique. We showed them the original videos and choreographed amateur performances, and each subject rated the quantity of visible artifacts via a seven-point Likert scale, where 1 meant "no artifact" and 7 meant "a lot of artifacts". In the case of the karate kick, the original video was rated at 2.20 $\pm$ 1.83, and the choreographed video at 5.27 $\pm$ 1.44, where $\pm$ denotes the standard deviation. This shows that our finished video still had visual artifacts, but 2 out of 15 subjects rated the choreographed video as having no visible artifacts. 14 out of 15 subjects noticed that the choreographed performer kicked higher and more quickly than in the original video. Also, some subjects pointed out that the leg had grown thicker and the foot was slipping.

5.2   Limitation and Future Work

The quality of the video tracking is important for the success of our method. Several sophisticated 3D video tracking techniques have been proposed recently [15, 16]. Since these might reduce the user burden and also improve the quality of the resulting animation, we intend to try them in the future.

Since our video editing procedure is strictly 2D, we cannot handle occlusions in the video. Figure 14 shows a sample situation. The two legs are overlapped and we must edit each leg independently. However, our method does not support

Fig. 14. We cannot separate the two overlapped legs

this, resulting in the awkward deformation shown in Fig. 14c. To support such editing, we must extend our technique, e.g., by introducing image segmentation and hole-filling algorithms. Another possibility is to use a depth sensor such as Kinect, which will provide depth information and facilitate image segmentation.

References

[1] Jain, A., Thormählen, T., Seidel, H.-P., Theobalt, C.: Moviereshape: Tracking and reshaping of humans in videos. ACM Trans. Graph. 29(5) (2010)

[2] Xu, F., Liu, Y., Stoll, C., Tompkin, J., Bharaj, G., Dai, Q., Seidel, H.-P., Kautz, J., Theobalt, C.: Video-based characters: creating new human performances from a multi-view video database. In: ACM SIGGRAPH 2011 Papers, pp. 32:1–32:10 (2011)

[3] Schödl, A., Szeliski, R., Salesin, D., Essa, I.: Video textures. In: Proceedings of ACM SIGGRAPH 2000. Annual Conference Series, pp. 489–498. ACM SIG-GRAPH (2000)

[4] Schödl, A., Essa, I.: Controlled animation of video sprites. In: Proc. of the 2002 ACM SIGGRAPH/Eurographics Symposium on Computer Animation, pp. 121–127 (2002)

[5] Scholz, V., El-Abed, S., Seidel, H.-P., Magnor, M.A.: Editing object behavior in video sequences. CGF 28(6), 1632–1643 (2009)

[6] Weng, Y., Xu, W., Hu, S., Zhang, J., Guo, B.: Keyframe based video object deformation. In: Proc. International Conference on Cyberworlds 2008, pp. 142–149 (2008)

[7] Sand, P., Teller, S.: Particle video: Long-range motion estimation using point trajectories. In: Proc. of CVPR 2006, pp. 2195–2202 (2006)

[8] Agarwala, A., Hertzmann, A., Salesin, D.H., Seitz, S.M.: Keyframe-based tracking for rotoscoping and animation. ACM Trans. Graph. 23(3), 584–591 (2004)

[9] Zach, C., Pock, T., Bischof, H.: A Duality Based Approach for Realtime TV-L^1 Optical Flow. In: Hamprecht, F.A., Schnörr, C., Jähne, B. (eds.) DAGM 2007. LNCS, vol. 4713, pp. 214–223. Springer, Heidelberg (2007)

[10] Rao, C., Gritai, A., Shan, M., Syeda-mahmood, T.: View-invariant alignment and matching of video sequences. In: Proc. of ICCV 2003, pp. 939–945 (2003)

[11] Beier, T., Neely, S.: Feature-based image metamorphosis. In: Proc. of SIGGRAPH 1992, pp. 35–42 (1992)

[12] Igarashi, T., Moscovich, T., Hughes, J.F.: As-rigid-as-possible shape manipulation. In: ACM SIGGRAPH 2005 Papers, pp. 1134–1141 (2005)

[13] Schaefer, S., Mcphail, T., Warren, J.: Image deformation using moving least squares. ACM TOG 25(3), 533–540 (2006)

[14] Sun, J., Jia, J., Ang, C.-K., Shum, H.-Y.: Poisson matting. In: ACM SIGGRAPH 2004 Papers, pp. 315–321 (2004)

[15] Wei, X., Chai, J.: Videomocap: Modeling physically realistic human motion from monocular video sequences. ACM Transactions on Graphics 29(4) (2010)

[16] Vondrak, M., Sigal, L., Hodgins, J.K., Jenkins, O.: Video-based 3D Motion Capture through Biped Control. ACM Transactions on Graphics (Proc. SIGGRAPH 2012) (2012)

Large Scale Image Retrieval with Practical Spatial Weighting for Bag-of-Visual-Words

Fangyuan Wang[1,2], Hai Wang[1], Heping Li[1], and Shuwu Zhang[1]

[1] High-Tech Innovation Center, Institute of Automation, Chinese Academy of Sciences,
Beijing, China
[2] WaSu Media Group Co.Ltd, Hangzhou, China
{fangyuan.wang,hai.wang,heping.li,shuwu.zhang}@ia.ac.cn

Abstract. Most large scale image retrieval systems are based on Bag-of-Visual-Words (BoV). Typically, no spatial information about the visual words is used despite the ambiguity of visual words. To address this problem, we introduce a spatial weighting framework for BoV to encode spatial information inspired by Geometry-preserving Visual Phrases (GVP). We first interpret GVP method using this framework. We reveal that GVP gives too large spatial weighting when calculating L2-norm for images due to its implicit assumption of the independence of co-occurring GVPs. This makes GVP sensitive to images with small number of visual words. Then we propose an improved practial spatial weighting for BoV (PSW-BoV) to alleviate this effect while keep the efficiency. Experiments on Oxford 5K and MIR Flickr 1M show that PSW-BoV is robust to images with small number of visual words, and also improves the general retrieval accuracy.

Keywords: image retrieval, spatial weighting, bag-of-visual-words, geometry-preserving visual phrases.

1 Introduction

Large scale image retrieval is receiving more and more attentions owing to its great potential in application and importance of theory in research. The goal of an image retrieval system is to return the similar images in a ranked list for a query image.

In order to deal with large scale image dataset, most existing state-of-the-art image retrieval systems are based on bag-of-visual-words (BoV) model, which is firstly introduced as Video-Google in [3]. Numerous successful works have been proposed to improve the retrieval accuracy and efficiency based on this model. The vocabulary tree [4] and approximate nearest neighbor [5] increase the efficiency of building a large vocabulary, while soft matching [6] and hamming embedding [7] address the hard quantization problem of visual words. But, in most of these approaches, spatial information which is useful to alleviate the ambiguity of visual words is usually ignored. Several researches have been conducted to introduce spatial information into BoV model. The RANSAC [5] re-introduces spatial information in the post-processing step through geometry verification which is usually computationally expensive. Spatial

S. Li et al. (Eds.): MMM 2013, Part I, LNCS 7732, pp. 513–523, 2013.

Pyramid Matching [8] (SPM) encodes rigid spatial information by quantizing the image space and lacks the invariance to transformations. Spatial-bag-of-features [2] handle variances of SPM by changing the order of the histograms; the spatial histogram of each visual word is rearranged by starting from the position with the maximum frequency. But, the arrangement may not correspond to the true transformation. Geometry-Preserving Visual Phrases (GVP) [1] uses the co-occurring GVPs between images to encode both local and long-range spatial information. When calculating the number of co-occurring GVPs, it implicitly assumes all GVPs in an offset bin are independent, which makes this method sensitive to distracting images with small number of visual words. Some researchers also consider to encode spatial information through introducing spatial weighting for visual words, but their methods either need learning step[12] or are difficult to be facilitated by inverted files[13,14].

To address this problem, we introduce a spatial weighting framework for BoV inspired by GVP method. Using this framework, we reveal that GVP method is sensitive to images with small number of visual words. Further more to alleviate this effect, we propose a practical spatial weighting for BoV (PSW-BoV) which calculates the spatial weighting for visual words based on the following two principles:

(1) When the dependence of GVPs is not serious, which means the number of co-occurring visual words in an offset bin is not too big, we use similar spatial weighting for visual words as GVP method;

(2) When the dependence of GVPs is very serious, which means the number of co-occurring visual words in an offset bin is very big, we use a much smaller spatial weighting for visual words than GVP method;

Although PSW-BoV is quite simple, experiments on Oxford 5K[5] and MIR Flickr 1M datasets [11] demonstrate that it can alleviate the sensitive effect of GVP to a large extent and significantly improve the general retrieval accuracy.

The rest of the paper is organized as follows: section 2 introduces the spatial weighting framework for BoV; in section 3, we interpret and analyze GVP using the spatial weighting framework; section 4 introduces PSW-BoV; section 5 is the comparative experiments; finally we draw conclusions in section 6.

2   Spatial Weighting Framework for BoV

BoV typically represents an image I_i as a vector $V(I_i)$, with one component for each visual word in the vocabulary. The jth component $v_j(I_i)$ in the vector is the weight of the word j: the *tf-idf* weighting scheme [3] is usually used, which can be calculated using the following formular:

$$v_j(I_i) = \frac{n_{jI_i}}{n_{I_i}} \cdot \log(\frac{N}{n_j}) \tag{1}$$

where, n_{jI_i} is the number of word j in image I_i, n_{I_i} is the total number of words in image I_i, n_j is the number of images that contain word j and N is the total number of images in the whole dataset. The similarity of two images I_i and $I_{i'}$ is usually defined as the *cosine* similarity of the two vectors: $< V(I_i), V(I_{i'}) > / \| V(I_i) \| \cdot \| V(I_{i'}) \|$. With large vocabularies, BoV representation is very sparse and inverted files can be used to facilitate the searching.

Typical BoV model just ignores the spatial information of visual words. An instinctive method is to mimick the *tf-idf* weighting to consider the spatial weighting for visual words. Suppose we have already got the spatial weighting $\alpha_j(I_i)$ for word j in image I_i, then the weighting component $v_j(I_i)$ changes to:

$$v_j(I_i) = \alpha_j(I_i) \cdot \frac{n_{ji_i}}{n_{I_i}} \cdot \log(\frac{N}{n_j}) \tag{2}$$

This can be regarded as a framework because we can use different methods to calculate the spatial weighting for visual word.

3   Spatial Weighting Interpretation of GVP

3.1   Interpretation

According to [1], a geometry-preserving visual phrase (GVP) of length k is defined as k visual words in a certain spatial layout. To tolerate shape deformation, the image space is quantized into bins. Each image is represented as a vector of GVPs. Similar to BoV model, the vector representation $V^k(I)$ is defined as the histogram of GVP of length k (k-GVP), with the ith component representing the *tf-idf* weighting of phrases p_i. But, this kind of vector can be extremely long even when $k=2$ while a large vocabulary is used. However, if ignores the *idf* weights, the dot product of such vectors of two images equals the total number of co-occurring GVPs in these images, the L2-norm of a vector can be calculated by counting the co-occurring GVPs with itself, since $\| V^k(I) \| = \sqrt{V^k(I) \cdot V^k(I)}$. Then the *cosine* similarity can be calculated to measure the similarity between images as follows:

$$sim(I, I') = \frac{< V^k(I), V^k(I') >}{\| V^k(I) \| \cdot \| V^k(I') \|} = \frac{\sum\limits_{m_{I,I'} \geq k} \left\{ \sum\limits_{i=1}^{m_{I,I'}} \binom{m_{I,I'}}{k} \right\}}{\left(\sum\limits_{m_{I,I} \geq k} \left\{ \sum\limits_{i=1}^{m_{I,I}} \binom{m_{I,I}}{k} \right\} \right)^{1/2} \cdot \left(\sum\limits_{m_{I',I'} \geq k} \left\{ \sum\limits_{i=1}^{m_{I',I'}} \binom{m_{I',I'}}{k} \right\} \right)^{1/2}} \tag{3}$$

where, I, I' represents two images, $m_{I,I'}$, $m_{I,I}$ $m_{I',I'}$ is the co-occurring visual word number in an offset space bin between images I and I', I and itself, I' and itself respectively, $\binom{m_{I,I'}}{k}$, $\binom{m_{I,I}}{k}$, $\binom{m_{I',I'}}{k}$ is the number of co-occurring k-GVP in corresponding offset bin respectively.

If considering the *idf* weights of GVPs, the final similarity can be calculated as follows:

$$sim(I,I') = \frac{\sum\limits_{m_{I,I'} \geq k} \left\{ \sum\limits_{i=1}^{m_{I,I'}} \binom{m_{I,I'}-1}{k-1} \cdot idf^2(w_i) \right\}}{\left(\sum\limits_{m_{I,I} \geq k} \left\{ \sum\limits_{i=1}^{m_{I,I}} \binom{m_{I,I}-1}{k-1} \cdot idf^2(w_i) \right\} \right)^{1/2} \cdot \left(\sum\limits_{m_{I',I'} \geq k} \left\{ \sum\limits_{i=1}^{m_{I',I'}} \binom{m_{I',I'}-1}{k-1} \cdot idf^2(w_i) \right\} \right)^{1/2}} \qquad (4)$$

where, w_i is a visual word, $idf(w_i)$ is the idf weight of the visual word.

$\binom{m_{I,I'}-1}{k-1}$, $\binom{m_{I,I}-1}{k-1}$ and $\binom{m_{I',I'}-1}{k-1}$ in equation (4) can be regarded as the spatial weighting for the visual words as formular (2), so GVP method is essentially equivalent to a spatial weighting method for BoV.

3.2   Analysis

As shown in formular (3), GVP method directly uses the combination number $\binom{m}{k}$ of all visual words in an offset bin as the co-occurring GVPs number. Obviously, they assume that all co-occurring GVPs in the same offset bin are totally independent. However, this assumption is not true as illustrated below.

Suppose both image I and I' contain visual words A, B and C, the calculation of co-occurring visual words can be shown in Fig. 1.

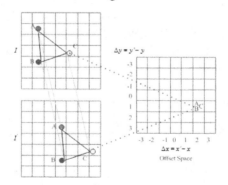

Fig. 1. Illustrative example for co-occurring GVPs. Different alphabets (A, B, C) represents different visual word.

As shown in Fig.1, the co-occurring GVPs are (AB), (BC), (AC). However, (AB) and (BC) share the same visual word B; (AB) and (AC) share the same visual word A; (BC) and (AC) share the same visual word C. This means different GVP may share the same visual word. Because all the visual words in (AC) are contained in (AB) and (BC), the spatial information encoded in (AC) is partially encoded in (AB) and (BC), vice versa. So they are also not spatially independent.

Therefore, the independence assumption of GVP is incorrect. This means the real number of GVPs should be less than the combination number of visual words.

Based on these analysis, we can infer that the spatial weighting of GVP (as shown in formular (4)) is prone to be bigger than ideal weighting (considered the dependence of GVPs), and the bias will increase as the number of co-occurring words increases.

In most cases of calculating co-occurring GVPs, due to the quantization effect of visual words (especially for a large vocabulary), the number of co-occurring visual words in an offset bin is not very big even for similar images. This means the dependence of GVPs is not very serious, so usually GVP method is still very effective.

However, GVP method is sensitive to distacting images with small number of visual words due to the independence assumption. Fig.2. is a real case example of the sensitivity effect of GVP.

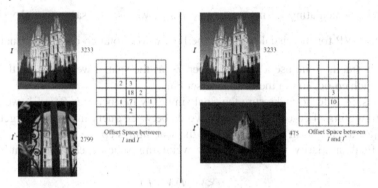

Fig. 2. Illustrative example of sensitivity of GVP to images with small number of visual words: image I, I',I^* has 3233,2799,475 visual words respectively, the numbers in bins are the number of co-occurring visual words. GVP method ranks I' and I^* incorrectly.

The reason can be explained as follows. As we known, the *consine* similarity needs to be normalized. When calculating the L2-norm for I', because the co-occurring visual words are generated with itself, the co-occurring number in the central offset bin is usually bigger than its total visual words number 2799 due to multiple times occurrence in I' of some words. If $k=2$, the number of GVP is more than 3915810 which is too large that the independece assumption is not reasonable any more. Therefore the spatial weighting ($m_{I',I'} - 1 = 2799 - 1 = 2798$) is too large. The case for I^* is similar, its spatial weigthing is 474. However, the bias of the spatial weighting of I' is much bigger than that of I^* according to the above analysis (if I^* has similar number of visual words with I', the bias effect can be roughly cancelled out). So after normalized using the biased L2-norms, the biased similarity of I' and I becomes smaller than that of I^* and I.

4   Practical Spatial Weighting for BoV

In order to alleviate the sensitive effect of GVP, the best method is to apply independence analysis for the co-occurring GVPs. But, the co-occurring GVPs are generated in the searching step, which means too much analysis will dramatically affect the efficiency of retrieval. Thus we propose a practical spatial weighting for BoV (PSW-BoV) to handle this issue which do not need extra analysis.

4.1   Practical Spatial Weigthing Scheme

As discussed in section 3, in most of cases the number of co-occurring GVPs is not very big, while the big number of co-occurring GVPs only occurs in the central offset bin when calculating the L2-norm for an image, in which the independence assumption is not acceptable. Based on these analyses, the practial spatial weighting scheme can be described as follows:

(1) When calculating the inner product of the co-occurring visual words between images, we use the same spatial weighting $\binom{m-1}{k-1}$ as GVP method;

(2) When calculating L2-norm of each image, we use the same spatial weighting $\binom{m-1}{k-1}$ as GVP for the visual words in the bins whose total co-occurring visual words number is small, while use a small number α as the spatial weighting for the visual words which co-ocurrs in the central offset bin;

Besides, in order to make sure the final similarity lies between 0 and 1, we use the spatial weighting of co-occurring visual words between query image and target image to re-adjust the L2-norm of each image in the searching step.

The final similarity with the spatial weighting scheme can be formulated as follows:

$$sim(I,I') = \frac{<V'^k(I),V'^k(I')>}{\|V'^k(I)\| \cdot \|V'^k(I')\|} \tag{5}$$

where,

$$<V'^k(I),V'^k(I')> = \sum_{m_{I,I'} \geq k} \left\{ \sum_{i=1}^{m_{I,I'}} \binom{m_{I,I'}-1}{k-1} \cdot idf^2(w_i) \right\} \tag{6}$$

$$\|V'^k(I)\| = \left(\|V^{*k}(I)\|^2 + <V'^k(I),V'^k(I')> - \sum_{m_{I,I'} \geq k} \left\{ \sum_{i=1}^{m_{I,I'}} \alpha \cdot idf^2(w_i) \right\} \right)^{\frac{1}{2}} \tag{7}$$

where,

$$\|V^{*k}(I)\| = \left(\sum_{k \leq m_{I,I'} < n_I} \left\{ \sum_{i=1}^{m_{I,I'}} \binom{m_{I,I'}-1}{k-1} idf^2(w_i) \right\} + \sum_{m_{I,I'} \geq n_I} \left\{ \sum_{i=1}^{m_{I,I'}} \alpha \cdot idf^2(w_i) \right\} \right)^{\frac{1}{2}} \tag{8}$$

$$\|V'^k(I')\| = \left(\|V^{*k}(I')\|^2 + <V'^k(I),V'^k(I')> - \sum_{m_{I,I'} \geq k} \left\{ \sum_{i=1}^{m_{I,I'}} \alpha \cdot idf^2(w_i) \right\} \right)^{\frac{1}{2}} \tag{9}$$

where,

$$\|V^{*k}(I')\| = \left(\sum_{k \leq m_{I',I'} < n_{I'}} \left\{ \sum_{i=1}^{m_{I',I'}} \binom{m_{I,I'}-1}{k-1} idf^2(w_i) \right\} + \sum_{m_{I',I'} \geq n_{I'}} \left\{ \sum_{i=1}^{m_{I',I'}} \alpha \cdot idf^2(w_i) \right\} \right)^{\frac{1}{2}} \tag{10}$$

Formular (5) is the final similarity, its numerator is calculated by formular (6) which corresponds to principle (1), its denominator is calculated by formular (7) and (9), which are the final L2-norms of I and I' re-adjusted based on the preliminary

L2-norms calculated by formular (8) and formular (10) respectively. k is moved from GVP, here we also only consider the visual words in the bins whose total co-occurring word number is $\geq k \cdot n_I$ and $n_{I'}$ are the total visual word numbers in image I and I'.

4.2   Searching with Practical Spatial Weighting

The practical spatial weighting scheme can be integrated into the searching step with inverted file which keeps one entry for each word occurrence with the image ID and word location [1]. For each image in database, we keep M bins to calculate the co-occurring words with query image, and another M bins to accumulate the summation of the *idf* weights, where M is the number of possible offsets.

(1) Initialize the two M bins for each image in the database to 0.

(2) For each word w in query image I, retrieve the image IDs and locations of the occurrences of w through the inverted files. For each retrieved word occurrence w' in image I_j, calculate the offset w and w', increment the corresponding offset bin of image I_j and accumlate the *idf* weighting in the offset bin.

$$N_{I_j, \Delta(x_w, x_{w'}), \Delta(y_w, y_{w'})} += 1 \tag{11}$$

$$D_{I_j, \Delta(x_w, x_{w'}), \Delta(y_w, y_{w'})} += idf^2(w) \tag{12}$$

where, N_{I_j} and D_{I_j} are the co-occurring word number matrix and *idf* summation matrix for image I_j.

(3) For each image I_j, traverse each bin m, calculate the scores as follows

$$S_{I_j} = \sum_{N_{I_j,m} \geq k} \binom{m-1}{k-1} \cdot D_{I_j,m} \tag{13}$$

$$S'_{I_j} = \sum_{N_{I_j,m} \geq k} \alpha \cdot D_{I_j,m} \tag{14}$$

where, S_{I_j} is corresponding to forumlar (6), S'_{I_j} is corresponding to the last part in formular (7) and (9).

(4) Suppose we pre-calculated the preliminary L2-norm $\| V^{*k}(I) \|$ and $\| V^{*k}(I_j) \|$, obtain the final score $\hat{S}_{I_j}$ by normalizing S_{I_j} as follows.

$$\hat{S}_{I_j} = \frac{S_{I_j}}{(\| V^{*k}(I) \|^2 + S_{I_j} - S'_{I_j})^{1/2} (\| V^{*k}(I_j) \|^2 + S_{I_j} - S'_{I_j})^{1/2}} \tag{15}$$

For $\| V^{*k}(I_j) \|$, it can be calculated usig similar steps. The difference is that in step (3), formular (13) changes to:

$$S_{I_j} = \sum_{k \leq N_{I_j,m} < n_{I_j}} \binom{m-1}{k-1} \cdot D_{I_j,m} + \sum_{n_{I_j} \leq N_{I_j,m}} \alpha \cdot D_{I_j,m} \tag{16}$$

Then,

$$\| V^{*k}(I_j) \| = \sqrt{S_{I_j}} \tag{17}$$

5   Experiments

5.1   Datasets and Evaluation Measure

Oxford 5K dataset is first introduced in [5] and has become a widely used evaluation benchmark. It contains 5062 images with more than 16M features. It also provides 55 test queries of 11 different Oxford landmarks with their ground thruth retrieval results.

MIR Flickr 1M dataset is provided by the ACM MIR Committee [11]. It contains roughly 1000,000 images retrieved from Flickr. Similar to [1,2,5], we add this dataset as distractors to the Oxford 5K dataset to test the scalability of our approach (this dataset is similar to that used in [5], the resolution of images for both datasets is roughly 500×333).

As in [1,2,5], we use the mean average precision (mAP) to evaluate the performance of all experiments.

5.2   Experimental Setting and Baseline

In order to be comparable with other methods, we use source descriptors (SIFT[9] on hessian affine regions[10]) and 1M vocabulary provided in [5], and the same AKM[5] method to train the other size vocabularies (50K, 100K, 250K, 500K). For MIR Flickr 1M dataset, we draw the same type descriptors using the tool available in [15] and the fastANN[5] to assign the visual word IDs. We use the same inverted file structure introduced in [1] to facilitate the searching and the same parameter setting as in [1].

We mainly consider BoV and GVP as our baseline. We also compare favorably with BOV+RANSAC and SBoF cited their reported results. We implemented BoV, GVP according to [1,5] respectively, we don't directly cite their reported results, because the trained vocabularies are different even they are got using the same tool.

5.3   Experimental Results

Firstly, we examine the effect of using different α as spatial weighting in formular (7), (8), (9),(10),(14) and (16). We verify on all the size of vocabularies. Table 1 shows the mAP scores using different α. When $\alpha = 0.8$, the mAP score on 500K and 1M vocabulary are larger than other coresponding values. When $\alpha > 0.8$, the mAP scores on all size vocabularies decrease as α increases. Therefore we set $\alpha = 0.8$ for the following experiments.

The value of best α is pretty small (much smaller than the visual word number in an image), this is quite reasonable because it means for most of visual words in an image it's no necessary to consider the spatial weigthing for them, and the reason it's smaller than 1 may due to the spatial weightings for other visual words have already been prone to be bigger than the ideal values.

Table 1. The effect of parameter changes under all size vocabularies on Oxford 5K dataset

α	50K	100K	250K	500K	1M
0.5	0.571	0.595	0.634	0.643	0.662
0.6	0.571	0.595	0.634	0.643	0.663
0.7	0.571	0.596	0.635	0.644	0.663
0.8	0.571	0.596	0.635	**0.645**	**0.663**
0.9	0.571	0.597	0.635	0.644	0.662
1	**0.571**	**0.597**	**0.636**	0.644	0.662
2	0.570	0.597	0.635	0.644	0.661
3	0.570	0.597	0.635	0.644	0.660
4	0.569	0.596	0.634	0.643	0.659
5	0.569	0.595	0.633	0.642	0.659

Secondly, we compare our approach with other methods under different size vocabularies on Oxford 5K dataset. Table 2 is the performance of different methods (The performance of BoV+RANSAC and SboF are cited from [2,5] respectively).

Table 2. Comparison of the performance of PSW-BoV with other methods using different size vocabularies on Oxford 5K

Vocab.	BoV	BoV+RANSAC	SboF	GVP	PSW-BoV
50K	0.486	0.569	0.523	0.551	**0.571**
100K	0.529	0.595	0.571	0.585	**0.596**
250K	0.574	0.633	^	0.627	**0.635**
500K	0.604	0.643	0.644	0.636	**0.645**
1M	0.617	0.645	0.651	0.654	**0.663**

Table 2 shows that our approach can outperform other methods on all size of vocabularies. Our approach significantly outperforms BoV method, more significant improvement is made on smaller vocabulary, because the visual words are more ambiguous. Compared with GVP, our approach further improves the mAP due to alleviate the sensitivity to distracting images with small number of visual words.

Thirdly, we verify our anlysis that GVP method is sensitive to distracting images with small number of visual words. We construct two distracting image datasets 65K_small and 65K_large from MIR Flickr 1M dataset. Where, 65K_small is composed by 65090 images where each image has less than 400 visual words; 65K_large is composed by 65090 images which are randomly selected from the images that have more than 1500 visual words. All experiments here are conducted on 1M vocabulary.The results are shown in Fig. 3.

The result in Fig. 3 shows that the difference of BoV on the two datasets is roughly 5.8%, the difference of GVP is roughly 6.3%, and the difference of PSW-BoV is roughly 0.6%. This promises the analysis that GVP is sensitive to the images with small number of visual words, while PSW-BOV alleviates this effect quite well.

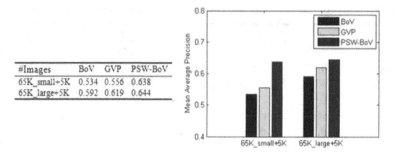

#Images	BoV	GVP	PSW-BoV
65K_small+5K	0.534	0.556	0.638
65K_large+5K	0.592	0.619	0.644

Fig. 3. The mAP scores on Oxford 5K+65K _small and Oxford 5K+65K_large datasets for BoV, GVP and PSW-BoV

Fourthly, we examine the performance for PSW-BoV on different size large scale datasets (100K, 200K, 500K, 1M, where the small datasets are randomly constructed from 1M dataset). The results are shown in Fig. 4.

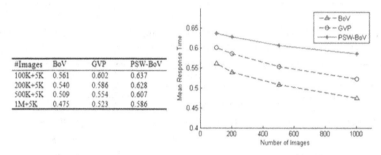

#Images	BoV	GVP	PSW-BoV
100K+5K	0.561	0.602	0.637
200K+5K	0.540	0.586	0.628
500K+5K	0.509	0.554	0.607
1M+5K	0.475	0.523	0.586

Fig. 4. The mAP scores on Oxford 5K+100K, Oxford 5K+ 200K, Oxford 5K+500K, Oxford 5K+1M datasets for BoV, GVP and PSW-BoV

Fig. 4 shows that PSW-BoV has a good scalability that can consistently improve the accuracy on different number of distracting images. The PSW-BoV method outperforms the BoV method roughly by 11.1%, GVP roughly by 6.3% on 1M dataset.

Finally, we report the efficiency of our approach. As PSW-BoV is improved from GVP, and uses different spatial weighting scheme, so the time efficency of PSW-BoV is smilar to GVP. In our experiments, a typical query on 200K+5k dataset consumes roughly 0.5s (CPU time in searching step, our CPU is 3.2GHz, main memory is 4G).

In our experiments, we do not directly compare with the existing spatial weighting methods introduced in [12,13,14]. According to their descriptions, they either need traning step or not suitable for large scale datasets. Our approach does not need training to get spatial weighting and can be facilated by inverted file for large datasets.

6   Conclusions

We first introduced a universal spatial weighting framework for BoV model. Then through analyzing GVP method using this framework, we reveal that GVP is sensitive to images with small number of visual words due to its implicit assumption of the independence of co-occurring GVPs. Finally to alleviate the sensitive effect of GVP, we proposed a practical spatial weighting for BoV (PSW-BoV) to encode more appropriate spatial information by considering the dependence influence while keep the efficiency.Experiments on Oxford 5K and MIR Flickr 1M datasets show that PSW-BoV can allevaite the sensitive effect of GVP to a large extent and further improve the general retrieval accuracy.

Acknowledgements. This wok is supported by National Key Technology R&D Program of China under Grant 2011BAH16B01, 2011BAH16B02 and the Cloud Computing Demonstration Project of National Development and Reform Commission.

References

1. Zhang, Z., Jia, Z., Chen, T.: Image Retrieval with Geometry-Preserving Visual Phrases. In: CVPR 2011, pp. 809–816. IEEE Computer Society, Colorado Spings (2011)
2. Cao, Y., Wang, C., Li, Z., Zhang, L., Zhang, L.: Spatial Bag-of-Features. In: CVPR 2010, pp. 3352–3359. IEEE Computer Society, San Francisco (2010)
3. Sivic, J., Zisserman, A.: Video google: A Text Retrieval Approach to Object Matching in Videos. In: ICCV 2003, pp. 1470–1477. IEEE Computer Society, Nice (2003)
4. Nister, D., Stewenius, H.: Scalable Recognition with A Vocabulary Tree. In: CVPR 2006, pp. 2161–2168. IEEE Computer Society, New York (2006)
5. Philbin, J., Chum, O., Isard, M., Sivic, J., Zisserman, A.: Object Retrieval with Large Vocabularies and Fast Spatial Matching. In: CVPR 2007, pp. 1–8. IEEE Computer Society, Minneapolis (2007)
6. Philbin, J., Chum, O., Isard, M., Sivic, J., Zisserman, A.: Lost in Quantization: Improving Particular Object Retrieval in Large Scale Image Databases. In: CVPR 2008, pp. 1–8. IEEE Computer Society, Anchorage (2008)
7. Jegou, H., Douze, M., Schmid, C.: Hamming Embedding and Weak Geometric Consistency for Large Scale Image Search. In: Forsyth, D., Torr, P., Zisserman, A. (eds.) ECCV 2008, Part I. LNCS, vol. 5302, pp. 304–317. Springer, Heidelberg (2008)
8. Lazebnik, S., Schmid, C., Ponce, J.: Beyond Bags of Features: Spatial Pyramid Matching for Recognizing Natural Scene Categories. In: CVPR 2006, pp. 2169–2178. IEEE Computer Society, New York (2006)
9. Mikolajczyk, K., Schmid, C.: Scale & Affine Invariant Interest Point Detectors. International Journal of Computer Vision 60, 63–86 (2004)
10. Lowe, D.: Distinctive Image Features From Scale-Invariant Interest Point Detectors. International Journal of Computer Vision 60, 91–110 (2004)
11. Huiskes, M., Thomee, B., Lew, M.: New Trends and Ideas in Visual Concept Detection. In: ACM MIR 2010, pp. 527–536. ACM, Pennsylvania (2010)
12. Marszalek, M., Schmid, C.: Spatial Weighting for Bag-of-Features. In: CVPR 2006, pp. 2118–2125. IEEE Computer Society, New York (2006)
13. Chen, X., Hu, X., Shen, X.: Spatial Weighting for Bag-of-Visual-Words and Its Application in Content-Based Image Retrieval. In: Theeramunkong, T., Kijsirikul, B., Cercone, N., Ho, T.-B. (eds.) PAKDD 2009. LNCS, vol. 5476, pp. 867–874. Springer, Heidelberg (2009)
14. Martinet, J., Urruty, T., Djeraba, C.: A new spatial weighting scheme for bag-of-visual-words. In: CBMI 2010, Grenoble, France, pp. 1–6 (2010)
15. http://www.robots.ox.ac.uk/~vgg/software/

Music Retrieval in Joint Emotion Space Using Audio Features and Emotional Tags

James J. Deng and C.H.C. Leung

Department of Computer Science
Hong Kong Baptist University, Hong Kong
{jdeng,clement}@comp.hkbu.edu.hk

Abstract. Emotion-based music retrieval provides a natural and humanized way to help people experience music. In this paper, we utilize the three-dimensional Resonance-Arousal-Valence emotion model to represent the emotions invoked by music, and the relationship between acoustic features and their emotional impact based on this model is established. In addition, we also consider the emotional tag features for music, and then represent acoustic features and emotional tag features jointly in a low dimensional embedding space for music emotion, while the joint emotion space is optimized by minimizing the joint loss of acoustic features and emotional tag features through dimension reduction. Finally we construct a unified framework for music retrieval in joint emotion space by the means of query-by-music or query-by-tag or together, and then we utilize our proposed ranking algorithm to return an optimized ranked list that has the highest emotional similarity. The experimental results show that the joint emotion space and unified framework can produce satisfying results for emotion-based music retrieval.

Keywords: Music retrieval, music emotion, dimensionality reduction, audio features, emotional tag, ranking.

1 Introduction

Music is a complex acoustic and physical product which is expressed by an art form and language. The mind, feeling, emotion, culture and other aspects of human beings are all encompassed by all kinds of music. In this sense, music plays a vital role in people's daily life, not only relieving stress, but also cultivating sentiment. With the astounding growth of music on the Internet, music retrieval is important for discovering music that fit listeners' tastes. Commonly used music services are based on music retrieval, and most of these applications use metadata information such as music title, genre, album, lyrics and biography, e.g. Yahoo Music, AllMusic, and MySpace. Furthermore, improved content-based approaches relying on music melody, rhythm, timbre, harmony greatly obtain better music retrieval results, e.g. Pandora.com, Themefinder, Echo Nest, and Musipedia. However, according to the study of musicology and psychology, music is regarded as the heart of our soul, while emotion is the core component of music

S. Li et al. (Eds.): MMM 2013, Part I, LNCS 7732, pp. 524–534, 2013.
© Springer-Verlag Berlin Heidelberg 2013

that expresses the inherent and high level spiritual quality and state of mind [15]. Recent years have seen a significant expansion in research on emotions induced by music. Therefore, emotion-based music retrieval provides a more natural and humanized way to better experience music.

However, We find that most of current music services ignore the emotion and sentiment influence, which are actually the core factors in human beings. Therefore, it is a valuable way in an attempt to provide music retrieval based on feelings or emotions when people search music. From the related works, we find that most works in this area focus on the relation between the physical properties of an acoustic signal and its invoked emotions. Hence, we construct a dimensional emotion model named Resonance-Arousal-Valence (RAV) to express music emotion. The relationship between acoustic features and their emotional impact based on this model has also been well investigated. In reality, the listeners are more comfortable with describing the emotions invoked by music through emotional tags such as happy, sad, calm, etc. Therefore, we consider both acoustic features and emotional tags for music emotion representation. In order to build the joint emotion space $\mathbb{R}^d$, we learn optimized projection functions for acoustic feature vectors and emotional tag vectors to a low dimensional embedding space, which is optimized by minimizing the joint loss of acoustic features and emotional tag features through joint subspace selection. Finally we construct an unified framework to rank music in the joint emotion space for music retrieval by the means of query-by-music or query-by-tag or together.

The structure of this paper is as follows: in Section 2 we give a review of related works on different music emotion models, and present our used Resonance-Arousal-Valence emotion model. In Section 3 we describe the joint emotion space learning for audio feature space and emotional tag space, and describe the joint emotion space selection. Section 4 presents our proposed unified framework for music retrieval in joint emotion space, and then explains our optimized ranking algorithm for emotion-based music retrieval. To this end, we describe the experiment and results in Section 5, and finally give the conclusion in Section 6.

2   Related Work

As emotion represents psychological and physiological human subjective experience, many researchers have explored emotion models rooted in two emotion theories: discrete emotion theory and dimensional emotion theory. The discrete emotion theory utilizes a number of emotional descriptors or adjectives to express basic emotions in human beings such as joy, sadness, and calm. Ortony et al. in [13] propose an OCC model to hierarchically describe 22 emotion type specifications. Specifically, 18 widely utilized mood categories are investigated through large ground truth set of music in [8]. The research community of Music Information Retrieval Evaluation eXchange also has classified music emotion into five categories by clustering different emotion labels [7]. However, the disadvantage of discrete emotion theory is that emotion terms or descriptors are ambiguous and not able to accurately describe emotions and measure their intensities. Conversely, the dimensional emotion theory believes that emotion should be regarded as continuous in a dimensional space. Thayer [18] suggested that the two underlying dimensions: energetic arousal and stress arousal, which is named arousal-valence emotion model widely utilized in music emotion recognition and proved effective in [11,22]. The disadvantage of this model is not able to distinguish certain emotions with subtle differences. A three-dimensional emotion model decomposes arousal into energy arousal and tension arousal in [16], which obtains the good performance in classification and regression. A layered model [5] suggests map emotions represented by the OCC model into a three-dimensional PAD space, whose dimensions are pleasure, arousal and dominance, where the dominance represents the controlling and dominant nature of the emotion. However, the dominance is not suitable for representing one aspect of emotions invoked by music. Bigand et al.[2] find that the third dimension seems to have an emotional character which is measured by musicological features, like continuity-discontinuity or melodic-harmonic contrast.

Since the two-dimensional Thayer's mood model has already been proved effective in other researchers' work, in this paper we inherit the merits of this model. However, in the context of music psychology and philosophy, we find that emotional resonance is also an important measurement for deep emotion expression, and emotion induced by music seems to have an association with musicology, like continuity-discontinuity or melody-harmony [2], which evoke resonant or dissonant emotional response. Therefore, we give a new terminology "Resonance" for representing the third dimension. Thus the three-dimensional emotion model Resonance-Arousal-Valence (**RAV**) is utilized to represent music emotion, where arousal refers to whether the music activate your emotion or not, valence represents whether the music evoke your pleasure or displeasure emotions, and resonance stands for emotional resonance to emotional dissonance.

There have been many studies in the area of emotion-based music retrieval. For example, Yang et al. [22] constructed a ranking-based emotion recognition system for music retrieval, which regresses the acoustic features to the two dimensional valence and arousal cartesian space to represent music emotion. Another approach is to reduce the acoustic features from high dimensional space to low dimensional space to represent emotional expression [14]. However, tags created by users are also play a significant role in music retrieval, and [19,12] combines audio features and social tags to compare music similarity for retrieval. Thus, emotional tags are also combined with acoustic features to represent music emotion in this paper, which is inspired by the work of Weston et al. [20] for measuring semantic similarities by model audio, artist name and tags. The contribution of this paper is to utilize acoustic features and emotional tags to construct a joint emotion space and a unified framework for music retrieval.

3   Joint Emotion Space Learning

We jointly utilize audio acoustic features and emotional tag features to construct a joint music emotion space $\mathbb{R}^d (d \geq 3)$, for compatibility with RAV model.

3.1 Audio Feature Space

The audio feature space consists of resonance-based features, arousal-based features and valence-based features, denoted by vectors γ, α, ν, respectively. Supposing there are N pieces of music, we formulate a music-to-acoustic feature matrix $X_{L \times N} = [\mathbf{R} \mid \mathbf{A} \mid \mathbf{V}]^T$. We aim to find the optimal matrix $A_{L \times d_L}$ to project the audio features to $\mathbb{R}^{d_L}$ by $X^T A$, where d_L denotes the dimension of the projected audio features. Graph Embedding attempts to find the projective map that optimal preserve the neighborhood structure of the original data set [21]. Given a graph $G = \{X, W\}$ with N vertices, each vertex represents a piece of music. Let W be a symmetric adjacency matrix with $W_{i,j}$ representing the weight, thus this graph embedding best preserves the relationship between different music, we aim to find the optimal low dimensional representation for this graph. Minimizing $\sum_{i,j} \|y_i - y_j\|^2 W_{i,j}$ can be transformed by the following optimization problem.

$$\hat{A} = argmin\{Tr(A^T X L X^T A)\} \tag{1}$$

$$s.t. \quad A^T X D X^T A = I \tag{2}$$

where D is a diagonal matrix with $D_{i,i} = \Sigma_j W_{i,j}$, and $L = D - W$ is the graph Laplacian matrix, and I is the identity matrix. Therefore, through computation of eigenvalues by $X L X^T A = \Lambda X D X^T A$, the optimal $\hat{A}$ are the eigenvectors corresponding to the maximum eigenvalue. As the similarity matrix W can be formulated by different similarity criteria such as Euclidean distance or cosine similarity etc. Currently there exist some popular dimension reduction methods such as Laplacian Eigenmap (LE), Locally Preserving Projections (LPP). We utilize the formula of $X^T X$ to measure all pairwise emotional similarity in audio feature space, and set nearest neighbors threshold to obtain W.

3.2 Emotional Tag Space

Apart from the audio features, we also consider the emotional tag features for music. Each piece of music may have a corresponding set of emotional tags. Given an emotional tag collection $T = \{T_1, T_2, \cdots, T_C\}$ with C elements, we assume music M implies an emotional tag T_i, thus the conditional probability of music M containing this emotional tag E_i is represented by

$$P(T_i|M) = \frac{P(M \cap T_i)}{P(M)} \tag{3}$$

Therefore, we represent emotional tag features for a piece of music by a probability vector $p = (P_1, P_2, \cdots, P_C)$. In order to obtain the probability vector for each piece of music, we compute the confidence for the above conditional probability by $\frac{f(M,T_i)}{f(M)}$. where $f(M)$ is the number of hit of Music M, and $f(M, T_i)$ is the number of hit containing both M and T_i.

Suppose there are N pieces of music, we formulate a music emotional tag-music matrix $Y_{N \times C}$ whose columns correspond to emotional tags and whose rows correspond to music. With music listening experience, the emotional tag features of a piece of music only contains a small set of corresponding emotional tags, thus Y is a spare matrix. The similarities based on emotional tags between pairwise of music are computed by inner product YY^T. For reducing emotional tag dimension to the joint emotion space, we learn a map function $\Phi(T) \to \mathbb{R}^d$.

we utilize Latent Semantic Analysis (LSA) approach to map the emotional tag features to the joint space $\mathbb{R}^d$ by using Singular Value Decomposition (SVD).

$$Y = U\Lambda S^T \tag{4}$$

where U is the $N \times r$ matrix, Λ is the $r \times r$ diagonal matrix consisting of singular values of Y in descending order, and S is the $C \times r$ matrix. We truncate the first d eigenvalues of Λ corresponding to the joint emotion dimensionality, and then assume $\hat{Y}_d$ is the closest approximation to Y, while $\hat{Y}_d$ minimizes the Frobenius norm over all rank-d matrices $||Y - Y_d||_F$. Therefore, the optimal emotional tag projection to the joint space is computed by $Y^T B$, where we denote $B = S_d$.

3.3   Joint Emotion Space Selection

At a given d for the joint emotion space dimensionality, we expect that music feature projection $X^T A$ and $Y^T B$ best approximate a finite N pieces of music. We denote the joint loss for mapping from the audio feature space and emotional tag space to the joint emotion space $\mathbb{R}^d$ by

$$J(W) = \sum_{k=1}^{\mathbb{K}} \sum_{i=1}^{N} J^k(w^k x_i^k, y_i^k) \tag{5}$$

where $\mathbb{K} = 2$. In order to minimize the joint loss, we utilize ℓ_1/ℓ_2 norm regularization for the above optimization problem [1].

$$\min \sum_{k=1}^{\mathbb{K}} \sum_{i=1}^{N} J^k(w^k x_i^k, y_i^k) + \lambda \|H\|_{l_1/l_2}^2 \tag{6}$$

$$s.t. \quad w^k = Qa^k, \ H = [a^1, a^2], \ Q^T Q = I \tag{7}$$

where $Q \in \mathbb{R}^{d \times d}$ is a common basis for the joint emotion space, and matrix H is penalized by the ℓ_1/ℓ_2 norm, and $a^k \in \mathbb{R}^d$, $k = 1, 2$. Then the above regularization scheme can be represented by using trace norm $\|W\|_{tr} = tr(W^T W)^{\frac{1}{2}}$.

$$\min \sum_{k=1}^{\mathbb{K}} \sum_{i=1}^{N} J^k(w^k x_i^k, y_i^k) + \lambda \|W\|_{tr}^2 \tag{8}$$

where $W \in \mathbb{R}^{\mathbb{K} \times d}$ and $W = [w^1, \cdots, w^{\mathbb{K}}]$. There are $\mathbb{K} = 2$ two parameter vectors, and we use Alternating Minimization (AM) algorithm [3] for optimization.

4   Emotion-Based Music Retrieval

We construct three ways such as query-by-music, query-by-tag and hybrid for music retrieval in the joint emotion space to return a ranked list of music.

Suppose given a piece of music m_i with corresponding audio features denoted by x_i, thus the audio features is represented in joint emotion space by $x_i^T A$, we define the rank function for query-by-music as follows.

$$\mathcal{F}_{\mathbb{A}}(m_i) = (x_i^T A)(x_j^T A)^T = x_i^T A A^T x_j \tag{9}$$

Similarly, suppose given an emotional tag vector y_i, and each element of y_i belongs to tag collection T, thus the emotion tag features is represented in joint emotion space by $y_i^T B$, we define the rank function for query-by-tag as follows.

$$\mathcal{F}_{\mathbb{T}}(y_i) = y_i^T B (y_j^T B)^T = y_i^T B B^T y_j \tag{10}$$

As the emotional tag collection T is finite with a small number of elements, some emotional tags not contained in our tag collection may also have influence for ranking music. To resolve this problem, we use Normalized Google Distance (NGD) [4] to measure the distance between the outer emotional tags in T^c and inner emotional tags in T for a piece of music M.

$$NGD(t_i, t_j | M) = \frac{\max\{\log f(t_i), \log f(t_j)\} - \log f(t_i, t_j)}{\log W - \min\{\log f(t_i), \log f(t_j)\}} \tag{11}$$

where $f(t_i)$ and $f(t_j)$ are the number of hits of emotional tags t_i and t_j with music M, respectively, and W is the total number of music M indexed. Thus the obtained $NGD(t_i, t_j | M)$ measures how close emotional tags t_i to t_j on a zero to infinity scale. Therefore, we choose the emotional tag with the smallest distance as the candidate emotional tag in query vector y_i.

Therefore, combining audio features and emotional tag features, we formulate a unified framework for music retrieval in our proposed joint emotion space $\mathbb{R}^d$ by the means of query-by-music or query-by-tag or together.

$$\mathcal{F}(\theta_i) = \mu \mathcal{F}_{\mathbb{A}}(m_i) + (1 - \mu) \mathcal{F}_{\mathbb{T}}(y_i) \tag{12}$$

where $\theta_i = (m_i, y_i)$, and μ is weight with $0 \le \mu \le 1$. Note that setting $\mu = 0$ recovers the standard rank function for query-by-tag, and setting $\mu = 1$ recovers the standard rank function for query-by-music. Setting μ to an appropriate value will optimize both feature sets.

4.1   Rank Music in Joint Emotion Space

The ultimate goal in emotion-based music retrieval is to return an optimized ranked list that has the highest emotional similarities. Therefore, we convert this rank problem to the optimization problem by minimizing emotional similarity sum of $\mathcal{F}(\theta_i)$ by $\mathbb{N}$ candidate pairs as follows.

$$\min \sum_{j=1}^{\mathbb{N}} \{\mu \mathcal{F}_{\mathbb{A}}(m_i) + (1 - \mu) \mathcal{F}_{\mathbb{T}}(y_i)\} + \frac{\eta}{2} \|\Theta\|_2^2 \tag{13}$$

$$s.t. \quad \Theta = [\theta_1, \theta_2, \cdots, \theta_{\mathbb{N}}]^T, \ 0 \le \mu \le 1 \tag{14}$$

We utilize minimization iteration methods to solve the above rank problem. The Algorithm 1 describes ranking procedure in detail.

Algorithm 1. Emotion-based Music Ranking

Input: Weight parameter w, regularization parameter η
Output: A ranked music list Θ_M with highest emotional similarities
1: initialize $\Theta_M \leftarrow$ RandomCandidateSet($\mathbb{N}$)
2: $\Theta_M \leftarrow sort(\Theta_M)$
3: $\Phi(\Theta_M) \leftarrow \sum_{j=1}^{\mathbb{N}} \mathcal{F}(\theta_i) + \frac{\eta}{2} \|\Theta_M\|_2^2$
4: **for** $i = 1 \rightarrow N$ **do**
5:    **if** $\Phi(\Theta_{M_i}) < \Phi(\Theta_M)$ **then**
6:       replace $\theta_{M_i} \leftarrow \theta_i$
7:       update $\Theta_M \leftarrow sort(\Theta_{M_i})$
8:    **end if**
9: **end for**
10: return Θ_M

---

5 Experiments

5.1 Data Sets and Evaluation Criteria

In our experiments, we have carefully collected 275 western classical music clips from Amazon website free preview music and StockMusic website. The styles of these music clips contain concerto, sonata, symphony, and string quartet, and the total of 10 composers of these classical music are Bach, Beethoven, Brahms, Chopin, Haydn, Mozart, Schubert, Schumann, Tchaikovsky, and Vivaldi. Then we choose representative music excerpts which are able to invoke emotions as experimental data. All these digital files are converted to a uniform format, with sampling rate 22050 Hz, 16 bits, 705 bit rate, and stereo channel. Each experimental music excerpt is truncated to keep 30 seconds length.

We use Mean Average Precision (MAP) and Normalized Discounted Cumulative Gain (NDCG) to measure the performance of emotion-based music ranking in joint emotion space. The mean average precision is based on the precision at position at k and average precision (AP). If there are r_k music emotional relevance in the top k music list, then $P@k = \frac{r_k}{k}$, thus the AP is given by

$$AP(q) = \frac{1}{r_q} \sum_{i=1}^{k} \{P@k \times rel(k)\} \tag{15}$$

where $rel(k)$ is an indicator function equaling 1 if music at rank k is emotional relevant, otherwise zero. MAP is obtained by meaning the AP. Normalized Discounted Cumulative Gain (NDCG) measures ranking quality and gain by

$$NDCG@k = Z_k \sum_{i=1}^{k} \frac{2^{rel_i} - 1}{\log(1 + i)} \tag{16}$$

where rel_i represents the graded relevance value at index i, k denotes the position, and Z_k represents a normalization factor.

5.2   Feature Extraction

According to the related work [11,9,17], we find that different music attributes such as timbre, rhythm, harmony and different acoustic properties such as energy, spectrum and tempo reflect different emotional expressions. The following parts give details about the arousal-based, valence-based and resonance-based features. We use MIR toolbox [10], Marsyas and CATBox library to extract the following mentioned audio features. Each 30 long seconds music excerpt is performed segmentation and frame decomposition with half overlapping. We also compute the statistical values of these selected features such as mean, standard variance and the difference. There are total 64 arousal-based features, 88 valence-based features, and 49 resonance-based features to extract in the experiment.

Arousal-based features: In musicology, intensity or dynamics represents loudness, which is correlated to arousal such as high intensity arousing excited or joyful feelings or emotions, while low dynamics arousing neutral or depress emotions. The acoustic feature that is often utilized to measure them is energy. The average energy of the given music is computed by root-mean-square (rms) method. Low energy and high energy are commonly used to express percentage of frames contrasted to average energy. High frequency energy measures the amount of energy above the cut-off certain frequency, which reflects the extent of brightness. In addition, since entropy is a useful tool to measure information, relative entropy of spectrum are also used to measure the degree of emotion arouse. Pitch represents fundamental frequency of a sound, thus high or low pitch represents different emotional expression such as active or inactive. Chroma features often describe energy distribution by twelve simitone (from A to G#).

Valence-based features: Timbre is a key and comprehensive factor to express different emotional expressions. A special timbre may inspire valence response or unpleasant feelings from the listener. The acoustic features often utilized to represent timbre are Mel-frequency cepstral coefficients (MFCCs), and statistical spectrum descriptors (spectral shape and spectral contrast). Spectral shape features are usually obtained by short-overlapping frames through Hanning window and discrete Fourier transform (DFT). Spectral shape features are consist of spectral centroid, flux, flatness and rolloff which represents the frequency less than a specific proportion of spectral distribution. Spectral contrast features describe the comparison or correlation of spectrum, such as spectral kurtosis, valley, skewness, regularity, spread and zero-crossing rate. We represent timbre features by using above MFCCs and spectrum characteristics. Rhythm also has important effect on invoking pleasure or displeasure feelings. As rhythm reflects different duration over a steady background of the beat, which is related with characteristics such as beat spectrum, beat onsets, onset rate, silence rate, fluctuation, event density and tempo having contribution to express rhythm.

Resonance-based features: In the paper, resonance means whether the music melodic or harmonic properties invokes your emotional resonance or dissonance. Melodic intervals usually refers to separately played music notes, thus melodic intervals are able to described by ascending or descending musical intervals. Harmony refers to simultaneously performed tones that represent mixture

Table 1. Music Emotional Tags

Anger	Angst	Aggression	Brooding	Calm		Confident	Desire	Distress	Dreamy
Earnest	Exciting	Fear	Grief	Happy		Hope	Joy	Pessimism	Pleased
Pride	Relief	Romantic	Sad	Satisfaction	Shame	Upbeat			

Table 2. Comparison ratio of emotional similarity of two emotion models

Emotion Model	Mean Υ	Std Υ
Arousal-Valence	0.4860	0.3286
Resonance-Arousal-Valence	0.3907	0.2538

sounds such as muddy, sharp, and smooth. A harmony chord often reflects consonance, while an unharmony chord often reflects dissonance, thus they have effect on invoking resonant or dissonant emotional feelings. Consonance features are defined by the peaks of spectrum and their space of spectral peaks, thus we use spectral peaks and roughness to describe music consonance characteristic. Tonality describes the hierarchical pitch relationship between center key; thus key, key clarity and tonal fusion represented by frequency ration of the component tone, are often utilized to represent tonality.

Emotional Tag features: Emotional tag features are represented by the vector space model. We utilize discrete emotion theory to express the emotional tag features, thus we choose some common emotional tags expressed by music in [13,15,8]. Therefore, we collect 25 emotional tags in our experiment as shown on Table 1. After that we utilize Youtube music search results to compute the probability of each piece of music associated with emotional tags, and finally convert them to music-to-tag frequency matrix.

5.3 Results

We first compare the utilized RAV emotion model with the two dimensional emotion model utilized in [6,22]. We define the ratio $\Upsilon = \frac{||e_i - e_j||}{||e_i||}$ to evaluate the emotional similarity, where e_i, e_j represents the emotion of music i and j respectively. Table 2 shows the comparison results of emotional similarity in a small dataset, where $||e_i||$ is the largest, thus the ratio is confined by $0 \leq \Upsilon < 1$.

Then we investigate different dimensions for joint emotion space. Figure 3 shows the results of joint emotion space selection, where the vertical axis represents the accuracy of music emotion recognition in joint emotion space, and the horizontal axis represents the dimensionality of the joint emotion space. There are two training datasets in the experiment. One training set contains 75 pieces of music, and the other contains 150 pieces of music. Suppose the joint emotion space dimensionality $d = 3$, we use Support Vector Regression (SVR) to map the arousal-based features, valence-based features and resonance-based features to rav values. It is shown that when the range of the joint emotion space dimensionality is from 12 to 15, we can obtain the optimized projection of audio features and emotional tag features to a low dimensional embedding space.

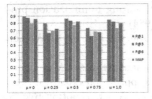 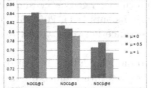 

Fig. 1. Average ranking accuracy by P@k

Fig. 2. Average ranking accuracy by NDCG@k

Fig. 3. Accuracy of the selected joint emotion space

The effect of the weight parameter u is quite evident for ranking accuracy. u values greater than 0.5 means that audio features play a dominant role on music rank, while u values less than 0.5 means that emotional tag features play a leading role. We set $\eta = 0.01$ to iteratively obtain the minimization of $\Phi(\Theta_M)$. The Figure 1 shows the music rank results in the joint emotion space at $d = 12$ by precision at position 1, 3, and 6, and MAP. It shows that emotional tag features have better results than audio features in the given dataset. Furthermore, we also use three graded emotional relevance values $r = 3, 2, 1$, which represents definitely emotional relevant, slightly emotional relevant, and no emotional relevant respectively, to compute NDCG in Figure 2. From the observations, it shows that our proposed unified framework for music retrieval in optimized joint emotion space is effective, with over 80% mean precision.

6   Conclusions

In this paper, we utilize the three-dimensional Resonance-Arousal-Valence emotion model to represent the emotions invoked by music, and the relationship between acoustic features and their emotional impact based on this model has also been established. Apart from audio features, we also consider the emotional tag features for music emotion representation. Then we represent acoustic features and emotional tag features jointly in a low dimensional embedding space for music emotion, while the joint emotion space is optimized by minimizing the joint loss of acoustic features and emotional tag features through dimension reduction. Finally we construct a unified framework for music retrieval in joint emotion space by the means of query-by-music or query-by-tag or together, and then we utilize our proposed ranking algorithm to return an optimized ranked list that has the highest emotional similarities. The results show that the selected joint emotion space is highly effective in emotion-based music retrieval.

References

1. Argyriou, A., Evgeniou, T., Pontil, M.: Convex multi-task feature learning. Machine Learning 73(3), 243–272 (2006)
2. Bigand, E., Vieillard, S., Madurell, F., Marozeau, J., Dacquet, A.: Multidimensional scaling of emotional responses to music: The effect of musical expertise and of the duration of the excerpts. Cognition & Emotion 19(8), 1113–1139 (2005)

3. Csiszár, I., Tusnády, G.: Information geometry and alternating minimization procedures. Statistics and Decisions suppl (1), 205–237 (1984)
4. Evangelista, A.J.: Google distance between words. Frontiers A Journal of Women Studies, 1–3 (2006)
5. Gebhard, P.: Alma: a layered model of affect. In: Autonomous Agents & Multiagent Systems/Agent Theories, Architectures, and Languages, pp. 29–36 (2005)
6. Han, B.-J., Rho, S., Dannenberg, R.B., Hwang, E.: SMERS: Music Emotion Recognition Using Support Vector Regression. In: International Society for Music Information Retrieval, Number Ismir, pp. 651–656 (2009)
7. Hu, X., Downie, J.S.: Exploring mood metadata: Relationships with genre, artist and usage metadata. In: International Symposium on Music Information Retrieval (2007)
8. Hu, X., Downie, J.S., Ehmann, A.F.: Lyric text mining in music mood classification. Information Retrieval 183(Ismir), 411–416 (2009)
9. Kim, Y.E., Schmidt, E.M., Migneco, R., Morton, B.G., Richardson, P., Scott, J., Speck, J.A., Turnbull, D.: Music emotion recognition: a state of the art review. Information Retrieval (Ismir), 255–266 (2010)
10. Lartillot, O., Toiviainen, P.: Mir in matlab (ii): A toolbox for musical feature extraction from audio. Spectrum (Ii), 127–130 (2007)
11. Lu, L.L.L., Liu, D., Zhang, H.-J.Z.H.-J.: Automatic mood detection and tracking of music audio signals (2006)
12. Nanopoulos, A., Karydis, I.: Know thy neighbor: Combining audio features and social tags for effective music similarity. In: International Conference on Acoustics, Speech, and Signal Processing, pp. 165–168 (2011)
13. Ortony, A., Clore, G.L., Collins, A.: The Cognitive Structure of Emotions, vol. 18. Cambridge University Press (1988)
14. Ruxanda, M.M., Chua, B.Y., Nanopoulos, A., Jensen, C.S.: Emotion-based music retrieval on a well-reduced audio feature space. In: International Conference on Acoustics, Speech, and Signal Processing, pp. 181–184 (2009)
15. Scherer, K.: Which emotions can be induced by music? what are the underlying mechanisms? and how can we measure them? Journal of New Music Research 33(3), 239–251 (2004)
16. Schimmack, U., Reisenzein, R.: Experiencing activation: energetic arousal and tense arousal are not mixtures of valence and activation. Emotion 2(4) (2002)
17. Schmidt, E.M., Turnbull, D., Kim, Y.E.: Feature selection for content-based, time-varying musical emotion regression categories and subject descriptors. Spectrum, 267–273 (2010)
18. Thayer, R.: The biopsychology of mood and arousal. Oxford University Press (1989)
19. Turnbull, D.R., Barrington, L., Lanckriet, G.R.G., Yazdani, M.: Combining audio content and social context for semantic music discovery. In: Research and Development in Information Retrieval, pp. 387–394 (2009)
20. Weston, J., Bengio, S., Hamel, P.: Large-scale music annotation and retrieval: Learning to rank in joint semantic spaces. CoRR, abs/1105.5196 (2011)
21. Yan, S., Xu, D., Zhang, B., Zhang, H.-J., Yang, Q., Lin, S.: Graph embedding and extensions: A general framework for dimensionality reduction. IEEE Transactions on Pattern Analysis and Machine Intelligence 29(1), 40–51 (2007)
22. Yang, Y., Chen, H.: Ranking-based emotion recognition for music organization and retrieval. IEEE Transactions on Audio Speech and Language (2010)

Analyzing Favorite Behavior in Flickr

Marek Lipczak[1,*], Michele Trevisiol[2], and Alejandro Jaimes[2]

[1] Dalhousie University, Halifax, Canada, B3H 1W5
`lipczak@cs.dal.ca`
[2] Yahoo! Research, Barcelona, Spain
`{trevi,ajaimes}@yahoo-inc.com`

Abstract. Liking or marking an object, event, or resource as a favorite is one of the most pervasive actions in social media. This particular action plays an important role in platforms in which a lot of content is shared. In this paper we take a large sample of users in Flickr and analyze logs of their favorite actions considering factors such as time period, type of connection with the owner of the photo, and other aspects. The objective of our work is, on one hand to gain insights into the "liking" behavior in social media, and on the other hand, to inform strategies for recommending items users may like. We place particular focus on analyzing the relationship between recent photos uploaded by user's connections and the favorite action, noting that a direct application of our work would lead to algorithms for recommending users a subset of these "recently uploaded" photos that they might favorite. We compare several features derived from our analysis, in terms of how effective they might be in retrieving favorite photographs.

1 Introduction

Sharing and marking objects as favorites are fairly new phenomena in social media and many questions remain open on the behavior of users in relation to liking or marking an object as a favorite. Questions include *what* they favorite, *when* they favorite, and *how* they are related to the owner[1] of such object (e.g., in Flickr users can be "contacts", "friends", or "family").

The problem of understanding the dynamics of such actions is of extreme importance given the pervasiveness of sharing and like/favorite actions in many social media platforms. It is important because when users express such preferences explicitly, they are implicitly contributing to the building of more accurate user models of themselves. Such models have applications in a wide range of areas: they can be used to recommend content, to improve user experience in terms of interaction design, for advertising purposes, and for recommending other users. In addition, one of the basic functionalities of social media platforms is providing easy access to content added by friends and other types of connections. It is common in some social media platforms (e.g., Facebook, Twitter) to

---

* Research conducted during an internship at Yahoo! Research Barcelona.
[1] We use the term "owner" to refer to the user who uploads the photo.

S. Li et al. (Eds.): MMM 2013, Part I, LNCS 7732, pp. 535–545, 2013.

rank photos and updates based on their recency and other features, but due to the increasing amount of shared content, and the size of personal networks, a simple recency based ranking is insufficient. Gaining insights into the favorite actions can contribute to designing novel ranking and recommendation algorithms, and to developing new functionalities around surfacing content users may like or favorite.

In this paper we present an analysis of favorite behavior on a large Flickr dataset. We analyze over 110 million favorite actions, focusing most of our study on a set of 24,000 users[2]. In particular, we examine temporal factors, user profiles derived from tags, and photo and photo-owner features, as well as the relationship between favorite actions and different link types between the users performing the actions and the owners of the photos. Finally, we perform experiments using several features to gain insights into their suitability for building algorithms for recommending photos to "favorite."

We examine the following: (1) whether users tend to favorite photos of people connected to them more than of people who are not connected, (2) whether users tend to favorite recent photos more than non-recent photos, and (3) whether the favorite activity happens in bursts. In addition, we evaluate several features for predicting favorite actions.

2   Related Work

Valafar et al.[13] performed a study of favorites in Flickr and found that 10% of users are responsible for 80 − 90% of all favorites, and that the favorite action exhibits 50% overlap and 15% reciprocity between users. These statistics are confirmed by many other studies (e.g. [2,9,11]). Cha et al.[2] investigated how an image spreads through the social network and highlighted how propagation varies considerably with the duration of exposure to new photos. In some cases, it takes a long time for photos to propagate from one user to another (i.e. [1,3]) as there is an initial phase of exponential growth in the number of users that favorite a photo, followed by a phase of slow and linear growth over the years.

Lee et al.[8] studied reciprocity in Twitter, and also in Flickr around favorites by dividing users into three groups: those that only browse, those that also upload photos, but do not participate in social activities, and those that participate in social activities. van Zwol et al.[15] presented a multi-modal, machine learned approach that combines social, visual and textual signals to predict favorite photos, while Lu and Li [10] exploited the photos previously marked as favorites by friends, in order to build a personalized search model to assist users in getting access to photos of interest.

Wonyong et al.[6] recommend tags for newly uploaded images, taking advantage of the tags assigned to favorite images of the user who uploaded the image, and combining tags with visual similarity. A similar work presented by Chen et al.[16] used favorite photos in order to extract representative tags, under the assumption that favorite images are better annotated.

---

[2] All analysis was aggregate, anonymous, and only on public photos.

Gursel and Sen [7] proposed an online photo recommendation system based on metadata and comments, assuming these two sources are highly related to the user's interests. De Choudhury et al.[5] developed a recommendation framework to connect image content with communities in online social media. They used visual features, user generated tags, and social interaction (i.e. comment actions) in order to recommend the most suitable group for a given image. Finally, Trevisiol et al.[12] presented an image ranking technique based on aggregating browsing patterns of about one million Flickr users.

A significant number of papers have been published using Flickr data, so we focused only on citing those that specifically deal with favorites and that are more relevant to our work. We are not aware, however, of any large-scale favorite action analysis such as the one we present in this paper.

3   The Dataset

Our dataset consists of a snapshot of Flickr until May 2008, which includes the explicit social network at the time and all interactions on public Flickr photos: over 110 million favorite actions made by over 1 million users.

Most users favorited photos only a few times (expected long tail of the distribution), more than $140K$ users favorited at least 100 photos, and the most active users favorited almost 100 thousand photos (heavy tail of the distribution). Sine long tail and heavy tail users are not representative for most of the favorite actions, we discarded them from most of the experiments. As we show later, the origin and recency of photos are very important factors. Therefore, with some exceptions, for the rest of the paper we consider only favorite actions made on photos uploaded to the system by the user's *connections* within 10 hours of the favorite action recorded. In order to limit the impact of the long and heavy tail we constrain the set of users for which we run the experiments to users with more than 100 and less than 2,500 favorites. We refer to this set as the "sample." We end up with 24,000 users that chose 8.6 million favorites among 1.194 billion photos.

4   Data Analysis

In Flickr, each photo has an *owner*, and users can be linked by more than one relationship type (*contacts, friends, family*, or any combination of those three). In the rest of the paper we will use the word *connection* to refer to any of the relationship types, but when we use the word *contact* we refer only to the contact relationship.

4.1   Photo Origin

We calculated the number of favorites with respect to link types (Table 1). The largest number of favorite photos come from *contacts*, both in terms of absolute number and average number of favorites per link, while *family* links have the fewest number of favorites. At the same time, nearly half of all favorites come from linked users: users tend to favorite photos of users that they are linked to, especially of their *contacts*.

Table 1. Social links statistics. *All favorites* includes favorites from users that are not linked by any of the relationship types to the user performing the favorite action.

links type	nr of favorites	avg favorites per link
contacts	29,642,943	0.90
friends	28,125,595	0.79
family	4,577,669	0.63
any link type	59,206,180	0.83
all favorites	112,177,317	10^{-5} (estimated)

4.2 Recency

Recency of a photo has been found to be important in image retrieval (van Zwol [14]). An analysis of our entire dataset of 110 million favorite actions shows that 20% of favorites happen within 10 hours, 30% within 24 hours, and 50% within a week from upload time.

4.3 Time Span of Favorite Actions

With *time span* we denote the number of days between the first and last time a user favorites photos. We examined the following sets of users:

  i. **all users:** the entire set of users in the initial dataset (over 1 million users).
 ii. **users with over 200 favorites:** 80% of all favorite actions are performed by the users in this set, which is obtained by filtering the 1 million users by selecting only those that have more than 200 favorites.
iii. **sample:** the set of $24K$ users described in Section 3, where we considered only favorite actions performed within 10 hours of uploading of a favorited photo.

Fig. 1(a) shows the cumulative distribution of users who performed favorite actions in a time span of k days. The distribution for *all users* is strongly biased by the long tail of users with a small number of favorites. In group (ii), the distribution resembles a normal distribution with high variance, where 80% of users have a time span of over 200 days. In group (iii) the ratio is even higher, around 90%.

Time Span of Owner-User Interactions. We created a histogram of favorite actions for all user-owner pairs in our data set. Note that with our notation, a user selects a photo as a favorite and an owner is the one who uploads that photo.

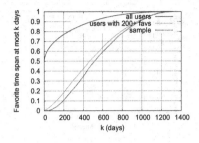 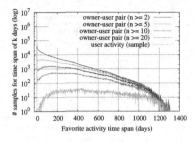

(a) Cumulative distribution of the ratio of time span of favorite action

(b) Time span of interaction between the owner of the photo and the user performing favorite actions

Fig. 1. Time span of user's activity

We analyzed only user-owner pairs with at least n favorites in total. As Fig. 1(b) shows, there are high peaks in very short time periods (less than a day). Note that we are considering only users that are connected by any of the relationship types, therefore the analysis shows that the favorite action happens in bursts and in many cases users do not return to favorite more photos of those owners: users tend to favorite photos of connections in short bursts.

Temporal Locality of User's Interests. We analyzed favorite actions that were close to each other in time and observed how many of the favorites shared a particular feature (*e.g.*, were uploaded by the same owner). In Fig. 2(a) we see that a large number of photos favorited in less than an hour are likely to be from the same owner, or group. Unexpectedly, in Fig. 2(a) we can observe subsequent daily peaks for owners and tags. In Fig. 2(b) similar peaks are observed for weeks. One interpretation of this is that when a user is interested in a picture with a certain feature, pictures that share this feature are more likely to be favorited.

4.4   Favorite Sessions

Another interesting aspect of favorite actions is their *burstiness*, in other words, measuring whether favorite actions occur uniformly over time or in bursts. We analyzed favorite actions within sessions, assuming that favorite actions are performed in the same session if the time difference between each pair of consecutive actions was lower than 30 minutes. Given this constraint, we measured the size of each session, comparing the two last groups described in Section 4.3: users with more than 200 favorites in total, and the 24k sampled users. The size of the sessions is much smaller for the sampled users, for which we considered only favorites performed within 10 hours of adding a photo.

We found that approximately 70% of the sessions have favorited photos of no more than 3 different owners. In almost 21% of the sessions the photos selected

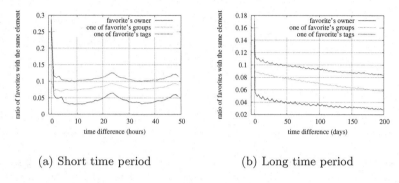

(a) Short time period   (b) Long time period

Fig. 2. Likelihood of favoriting a photo with the same user, group or tags

as favorites are from a single owner, and in about 35% of the sessions from two owners.

5   Computational Features

Below we describe a number of features derived from the outcomes of the data analysis and perform a simple evaluation of their suitability in terms of how well they are able to recall favorited photos. We also build a baseline favorite recommender as a proof of concept.

Since we have information on which photos have been favorited, we perform the evaluation by assuming that we want to predict a particular favorite action. In other words, let's say that at time t a user favorites a photo p. When the user favorites that photo, he choses it from a set of photos S. In our analysis we simply consider all photos in set S and examine which features might be more useful in predicting photo p, the one that was selected as a favorite.

We will use the following notation:

- **Recipient** – a user who is receiving photo recommendations.
- **Owner** – the owner of the photo that is recommended to the recipient.
- **Recommendation event** – the moment in time (t) in which the system presents the recommendation to the recipient. We assume that the favorite action takes place when a photo is recommended (i.e., not much later, for instance, not a week later).
- **Search space** – the set of photos S that are considered for recommendation in a single recommendation event. In this analysis we focus on photos that were uploaded by *connections* of the recipient, at most 10 hours before the recommendation event.

5.1   Photo Based Features

Photo based features are extracted from each photo that is considered for recommendation (i.e., each photo in the search space S defined above).

- **Photo recency** – time stamp of the upload of the photo (users are more likely to favorite recently uploaded photos (see Section 4)).
- **Number of favorites** – the number of times a photo was favorited by other users prior to the recommendation event.
- **Number of comments** – could be indicative of interest in the photo.

5.2 Photo Owner Features

Owner based features are related to the owner of the photo that is considered for recommendation, so all photos uploaded by a single user (owner) have the same owner features.

- **Likelihood of favoriting owner's photo** – the number of times a photo from the user was favorited divided by the total number of user's uploads.
- **Inverted batch size** – the inverted number of photos by the same user in the search space.
- **Recency of connection link** – time stamp of establishing a connection between two users: users might be more curious about photos of recent connections.

5.3 Feature Evaluation

In the following experiments we used the favorite actions of 100 users randomly chosen from the sample set of 24k users described in Section 4.3. Therefore, we considered only favorites done on photos from *connections* uploaded at most 10 hours before the favorite action.

The objective is to rank all the photos that were uploaded within that time frame by the user's connections, so the favorited photo is in the top of the ranking. Each favorite action was considered a separate recommendation event. In this setting, the dataset contained $38,211$ recommendation events in which the total number of photos was $4,632,013$ (on average 121 photos per recommendation event). We split the data into training and test sets, where the first 80 favorites of each user correspond to the training set (8000 recommendation events).

Since photos uploaded within 10 hours are considered, it is possible that some of them may have already been marked as favorites by the user. Given that photos cannot be favorited more than once by the same user, these were omitted (on average 1.8 photo per recommendation event). All feature values in training and test sets were calculated on the information that was available prior to the recommendation event (following the timestamps of user actions).

We used the average recall@k metric to determine the accuracy of features. Recall@k is the number of true positive instances among the first k results of the ranking, divided by the total number of positive instances. In each test case there is always one true positive instance (the favorited photo), and averaged recall@k represents the ratio of recommendation events for which the ranking was able to place the favorited photo among the first k photos from the search space.

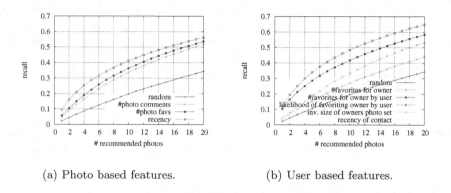

(a) Photo based features.   (b) User based features.

Fig. 3. Accuracy of features in photo recommendation task

Accuracy of Photo Based Features. Among the three photo based features (i.e., *photo recency, number of favorites, number of comments*), the recency of the photo turns out to be the most accurate (Fig. 3(a)). This feature is also the third most accurate among all tested features. Good performance of this feature could be predicted observing the relation between the recency and the ratio of favorited photos. However, it is also possible that the performance of the recency feature is biased by the fact that the most recent photos of a user are shown first in the Flickr interface.

The number of favorites and number of comments prior to the recommendation event represent actions by other users (i.e., those that do not own the photo and those that are not receiving the recommendation). Such actions are commonly used in standard recommendation techniques based on collaborative filtering. Both features have lower accuracy than the recency of a photo. 90% of the photos in the search space that are already marked as favorites were favorited less than 10 times. This is expected given that we consider only photos that were uploaded at most ten hours before the recommendation event.

Accuracy of User Based Features. The main feature describing an owner of a photo is the number of favorites of his/her photos prior to the recommendation event. We tested three features: total number of favorites for the owner's photos (#favorites for owner), total number of favorites by the recipient for the owner (#favorites for owner by recipient), and likelihood of owner's photo being favorited by a specific user given the photos in the search spaces of all recommendation events prior to the current one (likelihood of favoriting owner by user).

The first metric has the highest coverage, the last is most likely to have the best precision. Surprisingly, we can observe a very large difference between the total count of favorites and the personal count of favorites. The former has very low accuracy, which is unexpected.

The third feature is clearly superior. It measures the likelihood of an owner's photo being favorited by a recipient. The feature is personalized, which means that a separate likelihood value is calculated for each recipient. The high accuracy of this feature suggests that users tend to have a set of owners whose photos they frequently favorite. Indeed, in the sample, 40% of favorites are from owners who were favorited five times already (the total ratio of photos in the search space from these users is 13%). On the other hand, owners with no prior favorites contribute with 41% photos in the search space, but only 20% of favorites come from them.

The inverted size of owners' photos has reasonably good accuracy, suggesting that users who submit large sets of photos are in general, less likely to submit interesting photos. The last user based feature – the recency of user connections has very low accuracy.

Accuracy of Similarity Based Features. In addition to photo and user based features we tested a range of features calculated based on the similarity of users and photos. We found that comparing to other features similarity between recipients and owners/photos has low favorite photo prediction accuracy. It appears that that tags and groups are not as important in choosing favorited photos as who the photo owners are. Two additional reasons for low accuracy of similarity based metrics is the sparsity of tags and groups and the fact that users often assign the same set of tags to a large group of photos.

6   Discussion

The analysis of the favorite action can be a rather complex task in any social media platform. On one hand, the meaning of a favorite action may vary from user to user, and on the other hand, it may change over time. In Flickr, for example, some users may favorite photos to show their connections that they appreciate "something" about the photos uploaded by them. This could be qualities of the photo itself (e.g., they "like" the photo). They could also simply "like" the event depicted in the photo (e.g., they want to show a special connection to an event), or even use it as a way to show appreciation for a particular user (e.g., by marking your photos as a favorite, I indicate that I am interested in your work).

Other motivations may include gaining visibility for their own photos, or of their favorite collection– in Flickr, users are shown favorite "sets" a current photo appears in, so if I favorite a popular photo, my set of favorites could be shown next to the photo I favorite, to users who view that photo.

In addition, the use of favorites can also vary widely. Users may favorite photos to, in some sense, collect them, without necessarily using the collection for anything (i.e., some users may not even view those photos again, so the favorite action is an ephemeral one when it is performed), and such favorite collections may change over time in terms of size and scope.

We did not study the motivations for the favorite actions and that would most likely require a combination of quantitative and qualitative methods. In that respect, our analysis deals with the tip of the iceberg of possibilities in analyzing favorite actions, and any assumptions about why photos are marked as favorites should be thought out carefully and properly validated.

7   Conclusions and Future Work

We presented an analysis of the favorite action in Flickr. The results show that users tend to favorite recent pictures of their connections, and in particular of their contacts; favorite actions tend to happen in bursts, particularly when considering individual user-owner pairs (i.e., it is common for a user who favorites a photo of a connection to favorite several photos of that connection in a very short period of time). We also examined different features (for owners and photos) to determine how useful they might be in a recommendation task. In particular, we used the results of the analysis to build a number of computational features and tested their suitability in determining which photographs may be marked as favorites. Our work contributes to gaining insights into the "liking" behavior in social media (at least in the specific case of Flickr), and to inform strategies for recommending items users may like.

Future work includes performing user studies and surveys to gain deeper insights into the reasons people favorite photos (e.g., as a form of appreciation with their connections, in order to collect photos, etc.), as well as building and evaluating a recommender system based on the findings of this study. The analysis itself could also be expanded and, for example, we could maybe find that different types of Flickr users exhibit different types of behavior.

Acknowledgements. This research is partially supported by the European Community's Seventh Framework Programme FP7/2007-2013 under the AR-COMEM and SOCIALSENSOR projects, and by the Spanish Centre for the Development of Industrial Technology under the CENIT program, project CEN-20101037 (www.cenitsocialmedia.es), "Social Media."

References

1. Cha, M., Benevenuto, F., Ahn, Y.-Y., Gummadi, K.P.: Delayed information cascades in flickr: Measurement, analysis, and modeling. Computer Networks 56(3), 1066–1076 (2012)
2. Cha, M., Mislove, A., Adams, B., Gummadi, K.P.: Characterizing social cascades in flickr. In: Proceedings of the First Workshop on Online Social Networks, WOSP 2008, p. 13 (2008)
3. Cha, M., Mislove, A., Gummadi, K.P.: A measurement-driven analysis of information propagation in the flickr social network. In: Proceedings of the 18th International Conference on World Wide Web, WWW 2009, p. 721 (2009)
4. Chen, J., Nairn, R., Nelson, L., Bernstein, M., Chi, E.: Short and tweet: experiments on recommending content from information streams. In: Proceedings of the 28th International Conference on Human Factors in Computing Systems, CHI 2010, pp. 1185–1194. ACM, New York (2010)

5. De Choudhury, M., Sundaram, H., Lin, Y.-R., John, A., Seligmann, D.D.: Connecting content to community in social media via image content, user tags and user communication. In: IEEE International Conference on Multimedia and Expo (ICME), pp. 1238–1241 (June 2009)
6. Eom, W., Lee, S., De Neve, W., Ro, Y.M.: Improving image tag recommendation using favorite image context. In: 2011 18th IEEE International Conference on Image Processing (ICIP), pp. 2445–2448 (September 2011)
7. Gürsel, A., Sen, S.: Producing timely recommendations from social networks through targeted search. In: Proceedings of the 8th International Conference on Autonomous Agents and Multiagent Systems, vol. 2, pp. 805–812. International Foundation for Autonomous Agents and Multiagent Systems (2009)
8. Lee, J.G., Antoniadis, P., Salamatian, K.: Faving Reciprocity in Content Sharing Communities: A Comparative Analysis of Flickr and Twitter, pp. 136–143. IEEE (2010)
9. Lerman, K., Jones, L.: Social browsing on flickr. In: International Conference on Weblogs and Social Media, pp. 1–4 (2007)
10. Lu, D., Li, Q.: Personalized search on Flickr based on searcher's preference prediction. In: WWW, pp. 81–82 (2011)
11. Prieur, C., Cardon, D., Beuscart, J.-S., Pissard, N., Pons, P.: The Stength of Weak cooperation: A Case Study on Flickr. arxiv.org 65(8), 610–613 (2008)
12. Trevisiol, M., Chiarandini, L., Aiello, L.M., Jaimes, A.: Image ranking based on user browsing behavior. In: SIGIR, pp. 445–454 (2012)
13. Valafar, M., Rejaie, R., Willinger, W.: Beyond friendship graphs: a study of user interactions in Flickr. In: Proceedings of the 2nd ACM Workshop on Online Social Networks, pp. 25–30. ACM (2009)
14. van Zwol, R.: Flickr: Who is looking? In: Proceedings of the IEEE/WIC/ACM International Conference on Web Intelligence, WI 2007, pp. 184–190. IEEE Computer Society, Washington, DC (2007)
15. van Zwol, R., Rae, A., Pueyo, L.G.: Prediction of favourite photos using social, visual, and textual signals. In: ACM Multimedia 2010, pp. 1015–1018 (2010)
16. Xian, C., Hyoseop, S.: Extracting Representative Tags for Flickr Users. In: IEEE International Conference on Data Mining, pp. 312–317 (2010)

Unequally Weighted Video Hashing for Copy Detection

Jiande Sun[1,2,3], Jing Wang[2], Hui Yuan[2], Xiaocui Liu[2], and Ju Liu[2,3]

[1] School of Electronics Engineering and Computer Science, Peking University, Beijing, China
[2] School of Information Science and Engineering, Shandong University, Jinan, China
[3] The Hisense State Key Laboratory of Digital-Media Technology, Qingdao, China
`wangjing1987122520@163.com,`
`{jd_sun,huiyuan,juliu}@sdu.edu.cn, liuxiaocui777@126.com`

Abstract. In this paper, an unequally weighted video hashing algorithm is presented, in which visual saliency is used to generate the video hash and weight different hash bits. The proposed video hash is fused by two hashes, which are the spatio-temporal hash (ST-Hash) generated according to the spatio-temporal video information and the visual hash (V-Hash) generated according to the visual saliency distribution. In order to emphasize the contribution of visual salient regions to video content, Weighted Error Rate (WER) is defined as an unequally weighted hash matching method to take the place of BER. The WER, unlike BER, gives hash bits unequal weights according to their corresponding visual saliency in hash matching. Experiments verify the robustness and discrimination of the proposed video hashing algorithm and show that the WER-based hash matching is helpful to achieve better precision rate and recall rate.

Keywords: Video Copy Detection, Video Hashing, Ordinal Features, Visual Saliency, Unequally Weighted Hash.

1   Introduction

Recently video copy detection has received increasing attention as various digital videos can be distributed and transmitted easily through the Internet. It has increasing applications in the fields of video authentication, retrieval, copyright protection, etc.

Video hashing is the main method of video copy detection, in which the video hashing based on ordinal feature is proved to be suitable for copy detection by Law-To in [1]. Mohan derived the video hash from an ordinal feature based on the difference in intensity and color between the blocks in each frame [2]. Hampapur obtained the video hash from an ordinal feature based on the spatio-temporal distribution of motion, color and luminance in videos [3]. Job Oostveen proposed to divide each frame of a video clip into blocks, used the difference of between spatially adjacent blocks and the difference between temporally consecutive frames as the ordinal feature [4]. Esmaeili proposed a TIRI-DCT-based video hash algorithm, which extracted the content-based ordinal feature from the temporally informative representation image (TIRI) [5]. Though all the mentioned algorithms have good performance in video copy detection, they ignored the importance of visually salient

S. Li et al. (Eds.): MMM 2013, Part I, LNCS 7732, pp. 546–557, 2013.

regions in video content representation. During hash matching of these algorithms, Bit Error Rate (BER), or similarly Hamming Distance, is used as the benchmark to measure the difference between video hashes, in which each hash bit has equal weight. That is to say, the hash matching can not reflect the importance of the hash bits corresponding to the visually salient regions. In addition, the attacks to video copy detection are usually performed the nonsalient regions in order to preserve the salient content. A robust video fingerprinting based on visual attention regions is presented in [6], in which visual information is used for generating fingerprints. Sun and Wang proposed to weight normal video hash according to visual attention, which can improve the recall rate greatly [7, 8]. Though they proved that the visual attention can improve the performance of video hashing algorithms, the visual attention is used only in hash generation or hash matching. Therefore if we take the visual attention into account in both video hash generation and hash matching, the video hashing algorithms based on it can be expected to achieve better performance.

In this paper, a video hashing algorithm based on visual saliency is proposed. In the proposed algorithm, the video hash is obtained by combining spatio-temporal hash (ST-Hash) and visual hash (V-Hash). The ST-Hash is obtained according to the ordinal features which derived from the intensity difference between adjacent temporally informative representation image (TIRI) blocks in the order of Hilbert curve. V-Hash is obtained according to the ordinal features of representative saliency map (RSM) in the same way as TIRI. During hash matching each bit of the obtained hash is weighted according to the RSM, and weighted error rate (WER) is used to measure the similarity of different hashes. The experimental results verify that the proposed video hashing algorithm can achieve better performance. Fig. 1 shows the framework of the proposed video hashing algorithm.

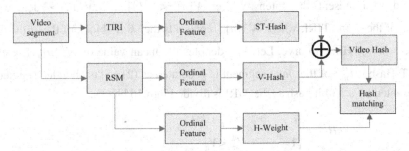

Fig. 1. The framework of the proposed video hashing algorithm

2   Unequally Weighted Video Hashing

2.1   Spatio-Temporal Hashing

The ST-Hash is obtained from the temporally informative representation image (TIRI) of video, which is proposed in [5]. TIRI summarizes the frames in a video segment temporally into an image to represent both spatial and temporal video information. This image is called temporally informative representation image (TIRI) and it is:

$$F(m,n) = \sum_{k=1}^{K} w_k F(m,n,k) \qquad (1)$$

where $F(m,n)$ is the luminance value of pixels of TIRI. $F(m,n,k)$ is the luminance value of the (m,n)th pixel of the kth frame in a video segment with K frames. w_k is the weight coefficient and equal to γ^k. γ is set to 0.6 empirically.

Fig.2 (a) shows three example TIRIs of the experimental video. Since TIRI can describe the spatial and temporal content of the video segment at the same time, the TIRIs are robust to the temporal video processing, e.g., frame exchanging, frame dropping, fps changing. The TIRIs are divided into blocks, and the ordinal feature of a TIRI is formed according to the difference in intensity between the blocks. The blocks are arranged in the order of Hilbert Curve [9] and the red line shows the Hilbert curve used in the TIRI in Fig. 2(b).

(a)   (b)

Fig. 2. (a) Three example TIRIs and (b) TIRI blocks in Hilbert curve order

Here the TIRI is divided into 16 blocks. The ST-Hash is generated according to the intensity difference between adjacent TIRI blocks. The mean value of each block is adopted to represent the intensity. Let $V[n] = <V^1[n],..........V^k[n]>$ denote the blocks of the nth TIRI, where $V^k[n]$ represents the kth block of the nth TIRI in the order of Hilbert curve. Let V_n^k denotes the mean value of $V^k[n]$. Therefore the ST-Hash $H_n'^k$ of the nth segment is generated as (2) and each video segment is represented by a 16-bit hash as the TIRI is divided into 16 blocks.

$$H_n'^k = \begin{cases} 0 & V_n^k \geq V_n^{k+1} \\ 1 & V_n^k < V_n^{k+1} \end{cases} \qquad (2)$$

2.2   Visual Hashing

V-Hash is obtained according to the ordinal features of representative saliency map (RSM) in the same way as TIRI. The visual attention model proposed in [10] is improved to generate the saliency map for each frame of a video segment. In [10], the saliency map **SM** is obtained with $x = 0.5$ as following:

$$\omega = xN(S\_Local) + (1-x)N(S\_Global) \qquad (3)$$

$$SM = \omega \cdot S\_Global \qquad (4)$$

where ω is a weighting matrix with the range of value in the matrix is [0, 1], called weight model, and the range of **S_Global** is also normalized to [0, 1]. x denotes the weight coefficient. $N(\cdot)$ denotes normalization.

In this paper, after obtaining the local and global saliency as in [10], the simulated annealing (SA) is adopted to find the optimal combination coefficients of the local and global layer in constructing the final VAM. The simulated annealing (SA) is a probabilistic method proposed by Kirkpatrick et al. for finding the global minimum of a cost function [12, 13]. Here, the cost function is defined as (5):

$$\omega = xN(S\_Local) + (1-x)N(S\_Global)$$

$$x^* = \underset{x \in [0,1]}{\arg\min}(-roc\_area(\omega)) \tag{5}$$

where x^* denotes the optimized weight coefficient. The local saliency and global saliency are normalized to the same range using the maximum normalization operator $N(\cdot)$ in [14]. $roc\_area(\cdot)$ is a function of calculating the area under the ROC curves, i.e. AUC, which is also used for evaluating the performance of VAM [15]. The larger the AUC value, the better the performance of VAM, and vice versa.

The result of optimization process on a test image is shown in Fig. 3. The x-axis indicates the number of iteration and the y-axis is the AUC value. It can be seen that after 25 iterations, the AUC reaches maximum and remains stable and the corresponding x is the optimal weight coefficient.

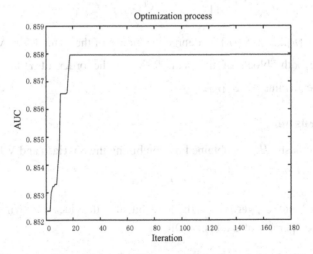

Fig. 3. The curve for weighting coefficients optimization of visual saliency map

After the optimal coefficient x is determined, the final saliency map **SM** can be obtained by (3) and (4). Then the representative saliency map (RSM) of the video segment is generated as follows:

$$RSM(m,n) = \sum_{k=1}^{K} w_k SM(m,n,k) \tag{6}$$

Where $SM(m,n,k)$ is the luminance value of the (m,n)th pixel of the kth saliency map of the video segment, which has K frames. w_k is the same weight coefficient as the one in (1). $RSM(m,n)$ is the luminance value of pixels of RSM.

$$(a) \qquad\qquad\qquad (b)$$

Fig. 4. (a) Three example RSMs corresponding to Fig. 2(a) and (b) RSM blocks and the corresponding hash weights in Hilbert curve order

Fig. 4(a) shows three example RSMs of the experimental video. In the experimental video, the moving man is the most salient object. From Fig. 4 (a), it can be seen that the regions of the moving man is detected out exactly.

The V-Hash is generated the same as ST-Hash. Therefore the V-Hash S'^k_n of the nth segment is generated as follows:

$$S'^k_n = \begin{cases} 0 & S^k_n \geq S^{k+1}_n \\ 1 & S^k_n < S^{k+1}_n \end{cases} \tag{7}$$

Let $S[n] = <S^1[n],\ldots\ldots S^k[n]>$ denote the blocks of the nth RSM, where $S^k[n]$ represents the kth block of the nth RSM in the order of Hilbert curve. S^k_n denotes the mean value of $S^k[n]$.

2.3   Hash Fusion

The final video hash H^k_n is obtained by combining the ST-Hash and V-Hash:

$$H^k_n = H'^k_n \oplus S'^k_n \tag{8}$$

where H^k_n, H'^k_n and S'^k_n are the kth hash bit of nth video segment respectively. $\oplus$ denotes the operator of Exclusive-OR.

3   Unequally Weighted Hash Matching

3.1   Hash Weights Generation

As visually salient regions play the principal role in the representation of video content, it is reasonable that the hash bits corresponding to salient regions should be assigned high weights during hash matching.

The hash weights are calculated according to the RSM for the corresponding video hash bits. As shown in Fig. 4(b), the hash weights for different RSM blocks of the n th video segment are generated as follows:

$$W_n(k) = \frac{\displaystyle\sum_{i=1}^{h_0}\sum_{j=1}^{w_0} RSM(i,j,k)}{\displaystyle\sum_{k=1}^{16}\sum_{i=1}^{h_0}\sum_{j=1}^{w_0} RSM(i,j,k)} \tag{9}$$

Where $RSM(i,j,k)$ denotes the luminance value of the (i,j) th pixel of the k th block, h_0 is the height of the block, and w_0 is the width of the block. $W_n(k)$ is the weight for the k th hash bit, where $\displaystyle\sum_{k=1}^{16} W_n(k) = 1$.

3.2   Weighted Error Rate (WER)

In order to reflect the importance of the salient regions of the video during hash matching, the hash bits corresponding to salient regions should be assigned high weights during hash matching. In order to reflect the contribution of visual salient regions to video content, the weighted error rate (WER) is defined as an unequally weighted hash matching method.

$$WER = \frac{1}{N}\sum_{n=1}^{N}\sum_{k=1}^{16} W_n(k)(H_n^k \oplus Q_n^k) \tag{10}$$

where H_n^k and Q_n^k are the k th hash bit of n th segment in the reference and query video respectively. $\oplus$ denotes the operator of Exclusive-OR. $W_n(k)$ is the hash weight corresponding to H_n^k. N is the number of segments in the video. If the hash weights are same, WER is equal to BER.

Here a threshold T is set to determine whether the test video is a video copy or not. If WER is higher than T, the test video is not a video copy and vice versa.

4   Experimental Analysis

In order to verify the robustness and discrimination of the proposed algorithm, **517** videos with temporal and spatial attacks are used in the experiments and the experimental videos are downloaded partly from CC_Web_Video database [16], www.open-video.org, www.reefvid.org and www.youku.com. Total **7238** videos are involved in the experiments including the videos with 14 attacks. The videos are classified into ten kinds and shown in Fig. 5, which are (a) animation movie, (b) music video, (c) cartoon video, (d) standard test video, (e) sports video, (f) Micro-film, (g) public service video, (h) news video, (i) live video, (j) documentary movie. The temporal attacks include random frame dropping, frame rate conversion, and frame exchanging. However these temporal attacks can cause few errors the hash as the analysis in Section 2.1. The spatial attacked frames are shown in Fig. 6 (b)-(l), which are (b) Gaussian noising, (c) 3 × 3median filtering, (d) histogram equalization with 64 levels, (e) contrast decreased by 25%, (f) contrast increased by 25%, (g) insertion of a logo, (h) Gaussian noising and insertion of a small logo, (i) Gaussian noising and decreased by 25%, (j) letter-box, (k) resizing from CIF to QCIF, (l) retargeting. Fig. 6 (a) shows the original frame. In the following experiments, every video is processed by all these attacks. Recall rate and precision rate defined in [17] are used as the metrics in video copy detection.

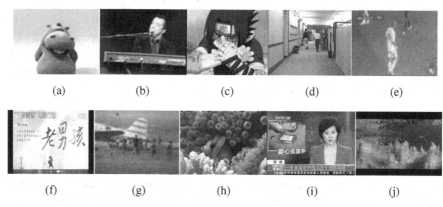

(a)   (b)   (c)   (d)   (e)

(f)   (g)   (h)   (i)   (j)

Fig. 5. Example experimental videos

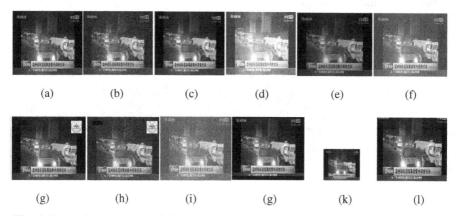

(a)   (b)   (c)   (d)   (e)   (f)

(g)   (h)   (i)   (g)   (k)   (l)

Fig. 6. Comparison between original video frame and the frames with various video attacks

In the following experiments, the video hash proposed in this paper is called fused hash. The algorithm, in which the fused hash matched under the metric of WER, is denoted as the fused hash with WER here, i.e. the proposed algorithm. The algorithm, in which the fused hash matched under the metric of BER, is denoted as the fused hash with BER.

Fig. 7 shows the comparison among the fused hash proposed in this paper, the hash in [5] and the hash in [11]. In Fig. 7, recall rate and precision rate are calculated based on BER. It can be seen from the curves that the recall and precision rate of the fused hash proposed in this paper are better than those of the hashes in [11] and [5] respectively. In addition, the mean of the recall rate of the fused hash proposed in this paper is approximate **10%** and **4%** higher than that of hash in [11] and [5] respectively. The mean of the precision rate of the fused hash is approximate **4%** and **2%** higher than that of hash in [11] and [5] respectively. These results show that the fused hash proposed in this paper performs better on both recall rate and precision rate under the metric of BER.

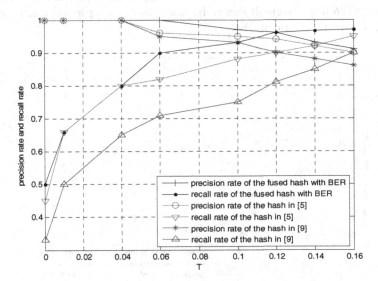

Fig. 7. The comparison on recall rate and precision rate among different hashes based on BER

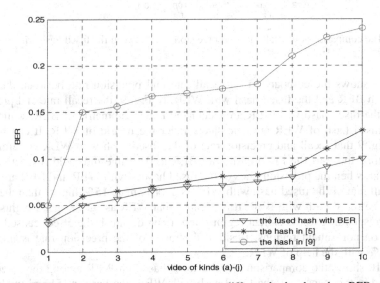

Fig. 8. The comparison on robustness among different hashes based on BER

Fig. 8 shows the comparison on robustness based on BER among the fused hash proposed in this paper, the hash in [5] and the hash in [11]. In Fig. 8, each point on the curves is corresponding to the experimental result of each kind of videos. No.1 to 10 in horizontal axis are corresponding to video kinds (a) to (j). In this experiment, the videos of the kinds from (a) to (j) are processed by the attacks (b)-(l). It can be seen that the proposed algorithm has the lowest BER, whose mean value is about **0.078**. However the corresponding mean values of the algorithms in [5] and [11] are about **0.092** and **0.176** respectively. It means the fused hash in this paper can achieve better robustness under the metric of BER.

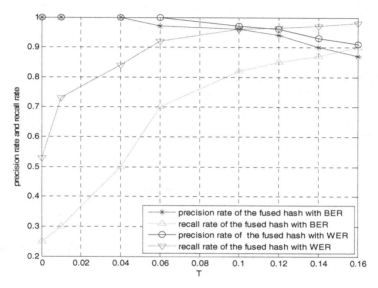

Fig. 9. The comparison on recall rate and precision rate between the fused hash with WER and BER

Fig. 9 shows the comparison on recall rate and precision rate between the fused hash with BER and the fused hash with WER. In Fig. 9, the recall rate and precision rate of the fused hash with BER are calculated under the metric of BER, while those of the fused hash of WER are calculated under the metric of WER. It can be seen from Fig. 9 that recall and precision rate of the fused hash with WER are almost all higher than those of the fused hash with BER at the same threshold respectively. In particular, when the precision rates of the fused hashes with WER and BER are equal, the recall rate of the fused hash with WER is approximate **15%** higher than the recall rate of the fused hash with BER. It is because that WER assigns larger weights to the bits corresponding to the salient regions and it can diminish the effect caused by the attacks on non-salient regions. Besides, the mean of the precision rate is improved about **2%** with the help of WER.

Fig. 10 shows the comparison on robustness based on WER among the fused hash proposed in this paper, i.e. the fused hash with WER, the hash in [5] and the hash in [11]. In Fig. 10, each point on the curves is corresponding to the experimental result of each kind of videos. No.1 to 10 in horizontal axis are corresponding to video kinds (a) to (j). In this experiment, the videos of the kinds from (a) to (j) are processed by the attacks (b)-(l). It can be seen that the proposed algorithm has the lowest WER, whose mean value is about **0.058**. However the corresponding mean values of the algorithms in [5] and in [11] are about **0.092** and **0.176** respectively, which are the same as the BERs shown in Fig. 8. The comparison between Fig. 8 and Fig. 10 demonstrates that WER can bring about **0.01** reduction of error rate. That is to say, WER is helpful for improving the robustness of video hash.

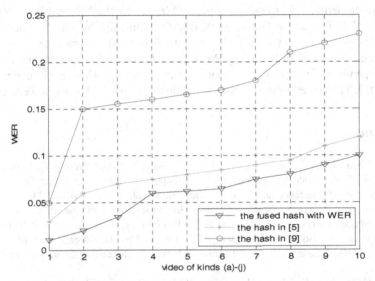

Fig. 10. The comparison on robustness among different hashes based on WER

In addition, the mean WER of the proposed algorithm is **0.501** and the variances are **0.0033**. It shows that the mean of WER of the proposed algorithm is close to 0.5 and has litter fluctuation. It shows that the proposed algorithm has better discrimination. Therefore if the video hash is generated according to the features with different saliency, the hash bits should be given different weights during hash matching as WER can improve overall performance of the same video hash. That can be seen from the abovementioned experiment results.

5   Conclusion

In this paper, a novel video hashing algorithm is proposed. The attention model is considered in the video hash generation, and the Weighted Error Rate (WER) is defined for video hash matching. The experiment results show that the proposed algorithm can achieve high precision rate and recall rate and it has good robustness and discrimination. The comparison experiments demonstrate that the proposed unequally weighted hash matching method, can improve both the recall rate and robustness, which is extremely important for video authentication and illegal video filtering, for instance. The comparison experiments also demonstrate that if the video hash is generated according to the features with different saliency, the hash bits should be given different weights during hash matching as WER can bring more improvements in the performance of video copy detection than BER. In the following research, how to optimize the weights for different hash bits spatio-temporally will be studied to make further improvement.

Acknowledgments. This work was supported in part by the National Basic Research Program of China under Grant 2009CB320905, in part by the National Nature Science Foundation of China under Grants 61001180, in part by Postdoctoral Innovation Foundation of Shandong Province under Grant 201103004, and in part by the Independent Innovation Foundation of Shandong University under Grant 2011JC027. The corresponding author is Jiande Sun (Email: jd_sun@sdu.edu.cn).

References

1. Law-To, J., Chen, L., Joly, A., Laptev, I., Buisson, O., Gouet-Brunet, V., Boujemaa, N., Stentiford, F.: Video Copy Detection: A Comparative Study. In: Proceedings of ACM International Conference on Image and Video Retrieval, pp. 371–378 (2007)
2. Mohan, R.: Video Sequence Matching. In: Proceedings of the IEEE International Conference on Acoustics, Speech and Signal Processing, vol. 6, pp. 3697–3700 (1998)
3. Hampapur, A., Bolle, R.M.: VideoGREP: Video Copy Detection Using Inverted File Indices, IBM Research Division Thomas, T.J. Watson Research Center, Technical Report (2001)
4. Job C.O., Ton, K., Jaap, H.: Visual Hashing of Digital Video: Applications and Techniques. In: Proceeding of SPIE, vol. 4472, p. 121 (2001)
5. Esmaeili, M.M., Fatourechi, M., Ward, R.K.: A Robust and Fast Video Copy Detection System Using Content-Based Fingerprinting. IEEE Transactions on Information Forensics and Security 6(1), 213–226 (2011)
6. Su, X., Huang, T.J., Gao, W.: Robust Video Fingerprinting Based on Visual Attention Regions. In: IEEE International Conference on Acoustics, Speech and Signal Processing, pp. 1525–1528 (2009)
7. Sun, J.D., Wang, J., Zhang, J., Nie, X.S., Liu, J.: Video Hashing Algorithm with Weighted Matching Based on Visual Saliency. IEEE Signal Processing Letters 19(6), 328–331 (2012)
8. Wang, J., Sun, J.D., Liu, J., Nie, X.S., Yan, H.: A Visual Saliency Based Video Hashing Algorithm. In: International Conference on Image Processing, pp. 645–648 (2012)
9. Butz, A.R.: Alternative Algorithm for Hilbert's Space-Filling Curve. IEEE Transactions on Computers 20(4), 424–426 (1971)
10. Zhang, J., Sun, J.D., Yan, H.: Visual Attention Model with Cross-Layer Saliency Optimization. In: IEEE International Conference on Intelligent Information Hiding and Multimedia Signal Processing, pp. 240–243 (2011)
11. Nie, X.S., Liu, J., Sun, J.D., Liu, W.: Robust Video Hashing Based on Double-Layer Embedding. IEEE Signal Processing Letters 18(5), 307–310 (2011)
12. Rutenbar, R.A.: Simulated Annealing Algorithms: An Overview. IEEE Circuits and Devices Magazine 5, 19–26 (1989)
13. Le Meur, O., Chevet, J.-C.: Relevance of A Feed-Forward Model of Visual Attention for Goal-Oriented and Free-Viewing Tasks. IEEE Transactions on Image Processing 19(11), 2801–2813 (2010)

14. Kirkpatrick, S., Gelatt, C.D., Vecchi Jr., M.P.: Optimization by Simulated Annealing. Science 220, 621–630 (1983)
15. Itti, L., Koch, C.: Feature Combination Strategies for Saliency-Based Visual Attention Systems. Journal of Electronic Imaging 10(1), 161–169 (2001)
16. Wu, X., Ngo, C.-W., Hauptmann, A.G., Tan, H.-K.: Real-Time Near-Duplicate Elimination for Web Video Search with Content and Context. IEEE Transactions on Multimedia 11(2), 196–207 (2009)
17. Law-To, J., Buisson, O., Gouet-Brunet, V.: ViCopT: A Robust System for Content-Based Video Copy Detection in Large Databases. Multimedia Systems 15, 337–353 (2009)

Author Index